Advances in Experimental Medicine and Biology

Volume 1046

More information about this series at http://www.springer.com/series/5584

Jun Aruga
Editor

Zic family

Evolution, Development and Disease

Springer

Editor
Jun Aruga
Graduate School of Biomedical Sciences
Nagasaki University
Nagasaki, Nagasaki, Japan

ISSN 0065-2598 ISSN 2214-8019 (electronic)
Advances in Experimental Medicine and Biology
ISBN 978-981-10-7310-6 ISBN 978-981-10-7311-3 (eBook)
https://doi.org/10.1007/978-981-10-7311-3

Library of Congress Control Number: 2017963871

Printed on acid-free paper

This Springer imprint is published by Springer Nature
The registered company is Springer Nature Singapore Pte Ltd.
The registered company address is: 152 Beach Road, #21-01/04 Gateway East, Singapore 189721, Singapore

Preface

Studies conducted in the last two decades indicate that *Zic* family genes are essential components of the animal genome. They code for zinc finger proteins that control cell differentiation in many developmental processes. These processes include maintenance of pluripotency; generation of the mesoderm, neuroectoderm, and neural crest; formation of the body pattern, the neural tube, and the skeletal patterns; and establishment of a platform for higher brain functions. Since the involvement of *Zic* family genes in early developmental processes has been established across major animal models, they have now become attractive research targets not only in the field of developmental biology but also in the field of evolutionary biology. Most multicellular animals, except sponges and comb jellies, possess *Zic* genes, and these genes may be involved in the developmental programs of these animals. A compilation of such information would allow us to infer the role of *Zic* genes in the evolution of animal body plans, such as in the emergence and diversification of the central nervous system. Thus, publications dealing with the role of *Zic* family genes in development and evolution have steadily contributed to the field of basic biology.

Zic family genes have often been the subject of medical research. They have been recognized as the causal genes for various genetic disorders such as holoprosencephaly, heterotaxy, Dandy-Walker malformation, and craniosynostosis. The importance of these genes in medicine is being increasingly appreciated after the recent discovery of their critical roles in stem cell regulation, both in normal and oncogenic processes. Therefore, *Zic* family-related medical research has been published in journals of human genetics, stem cell biology, cancer biology, and diseases of the nervous system and is expected to contribute further to the rapidly growing field of translational medicine, which includes stem cell applications, drug development, and cancer diagnosis and treatment.

Despite the enrichment of research on the *Zic* family genes, it is becoming difficult to visualize the overall picture of *Zic* family-related studies. This is probably because research has been conducted on diverse topics, and only a few researchers may be motivated to study the genes themselves. This book aims to provide essential and comprehensive information on existing research on *Zic* family genes for future researchers and graduate students in the fields of biology and medicine. The contributing authors were selected based on their critical publications in the field of *Zic* biology. Their articles have been written such that they are easily comprehensible by researchers working in other fields as well. In Part I (Chaps. 1, 2, 3, 4, 5, 6, 7, 8, 9, 10, and 11) of the book, research topics have been ordered according to experimental animal models; a wide range of animals, including cnidarians, ecdysozoans, lophotrochozoans, and chordates, have been covered. In Part II (Chaps. 12, 13, 14, 15, 16, 17, and 18), topics closely related to clinical research have been included. To provide warp threads to the story, summaries from the viewpoints of comparative genomics and biochemical properties have been included at the beginning and at the end, respectively.

This book has been prepared with an interdisciplinary approach, to be useful to both developmental biologists and medical researchers. For instance, the updated results of human genetics research would be of interest not only to medical researchers but also to those working in the fields of developmental biology and evolutionary biology. Similarly, the information on the roles *Zic* family genes play in certain animal models is expected to inspire researchers in the fields of evolutionary biology and medicine. While editing this book, I came across many important unresolved questions and hints for future studies. For instance, important findings regarding the molecular function of the *Zic* genes have been reported in the last couple of years. The notion of Zic proteins as major enhancer function mediators, in conjunction with chromatin remodeling factors, is becoming popular, and it sheds new light on *Zic* biology and being resolved questions concerning the position of *Zic* genes regulatory network controlling problems ensures the importance of the *Zic* gene family in the regenerative medicine.

I thank all the contributing authors for sharing their significant findings and insights. Some authors graciously complied with my repeated editorial requests, which I sincerely appreciate. Lastly, I thank Mr. Yasushi Okazaki, Member, Springer Nature, for encouraging me to publish this book on the *Zic* family.

Nagasaki, Japan Jun Aruga
August 2017

Contents

Part II Zic Family in Medicine

Contributors

Kota Abe Department of Biological Sciences, Graduate School of Science, University of Tokyo, Tokyo, Japan

Ruth M. Arkell Early Mammalian Development Laboratory, John Curtin School of Medical Research, The Australian National University, Canberra, ACT, Australia

Jun Aruga Department of Medical Pharmacology, Nagasaki University, Graduate School of Biomedical Sciences, Nagasaki, Japan

Kristen S. Barratt Early Mammalian Development Laboratory, John Curtin School of Medical Research, The Australian National University, Canberra, ACT, Australia

Helen M. Bellchambers Department of Pediatrics, Indiana University School of Medicine, Indianapolis, IN, USA

Vincent Bertrand Aix-Marseille Université, CNRS, Institut de Biologie du Développement de Marseille, Marseille, France

Guillaume Bordet Aix-Marseille Université, CNRS, Institut de Biologie du Développement de Marseille, Marseille, France

Koula E.M. Diamand Early Mammalian Development Laboratory, John Curtin School of Medical Research, The Australian National University, Canberra, ACT, Australia

Jennifer Forecki Department of Cell Biology and Neuroscience, Montana State University, Bozeman, MT, USA

Minoru Hatayama Department of Medical Pharmacology, Nagasaki University, Graduate School of Biomedical Sciences, Nagasaki, Japan

Eloísa Herrera Instituto de Neurociencias de Alicante, Consejo Superior de Investigaciones Científicas-Universidad Miguel Hernández, CSIC-UMH, Alicante, Spain

Rob Houtmeyers Department of Oncology, Katholieke Universiteit Leuven, Leuven, Belgium

Deborah A. Hursh Division of Cell and Gene Therapy, Center for Biologics Evaluation and Research, Food and Drug Administration, Silver Spring, MD, USA

Kaoru S. Imai Department of Biological Sciences, Graduate School of Science, Osaka University, Toyonaka, Japan

Toru Kawanishi Department of Systems Biology, Harvard Medical School, Boston, MA, USA

Hisato Kondoh Faculty of Life Sciences, Kyoto Sangyo University, Kyoto, Japan

Vladimir Korzh International Institute of Molecular and Cell Biology, Warsaw, Poland

Michael J. Layden Department of Biological Sciences, Lehigh University, Bethlehem, PA, USA

Christa Merzdorf Department of Cell Biology and Neuroscience, Montana State University, Bozeman, MT, USA

Kathleen J. Millen Department of Pediatrics, University of Washington School of Medicine, Seattle Children's Hospital Research Institute, Center for Integrative Brain Research, Seattle, WA, USA

Yutaka Satou Department of Zoology, Graduate School of Science, Kyoto University, Kyoto, Japan

Jabob Souopgui Department of Oncology, Katholieke Universiteit Leuven, Leuven, Belgium

Department of Embryology and Biotechnology, Université Libre de Bruxelles, Gosselies, Belgium

Brian G. Stultz Division of Cell and Gene Therapy, Center for Biologics Evaluation and Research, Food and Drug Administration, Silver Spring, MD, USA

Hiroyuki Takeda Department of Biological Sciences, Graduate School of Science, University of Tokyo, Tokyo, Japan

Sabine Tejpar Department of Oncology, Katholieke Universiteit Leuven, Leuven, Belgium

Stephanie M. Ware Department of Pediatrics, Indiana University School of Medicine, Indianapolis, IN, USA

Department of Medical and Molecular Genetics, Indiana University School of Medicine, Indianapolis, IN, USA

Cecilia Lanny Winata International Institute of Molecular and Cell Biology, Warsaw, Poland

Max-Planck Institute for Heart and Lung Research, Bad Nauheim, Germany

Part I
Zic Family in Animal Evolution and Development

Chapter 1
Comparative Genomics of the *Zic* Family Genes

Jun Aruga and Minoru Hatayama

Abstract *Zic* family genes encode five C2H2-type zinc finger domain-containing proteins that have many roles in animal development and maintenance. Recent phylogenetic analyses showed that *Zic* family genes are distributed in metazoans (multicellular animals), except Porifera (sponges) and Ctenophora (comb jellies). The sequence comparisons revealed that the zinc finger domains were absolutely conserved among the *Zic* family genes. Zic zinc finger domains are similar to, but distinct from those of the *Gli*, *Glis*, and *Nkl* gene family, and these zinc finger protein families are proposed to have been derived from a common ancestor gene. The Gli-Glis-Nkl-Zic superfamily and some other eukaryotic zinc finger proteins share a tandem CWCH2 (tCWCH2) motif, a hallmark for inter-zinc finger interaction between two adjacent C2H2 zinc fingers. In Zic family proteins, there exist additional evolutionally conserved domains known as ZOC and ZFNC, both of which may have appeared before cnidarian-bilaterian divergence. Comparison of the exon-intron boundaries in the *Zic* zinc finger domains revealed an intron (A-intron) that was absolutely conserved in bilaterians (metazoans with bilateral symmetry) and a placozoan (a simple nonparasitic metazoan). In vertebrates, there are five to seven *Zic* paralogs among which *Zic1*, *Zic2*, and *Zic3* are generated through a tandem gene duplication and carboxy-terminal truncation in a vertebrate common ancestor, sharing a conserved carboxy-terminal sequence. Several hypotheses have been proposed to explain the Zic family phylogeny, including their origin, unique features in the first and second zinc finger motif, evolution of the nuclear localization signal, significance of the animal taxa-selective degeneration, gene multiplication in the vertebrate lineage, and involvement in the evolutionary alteration of the animal body plan.

Keywords Zinc finger protein · Gene family · Molecular evolution · Eumetazoa · Evo-devo · Bauplan

---

J. Aruga (✉) · M. Hatayama
Department of Medical Pharmacology, Nagasaki University Graduate School of Biomedical Sciences, Nagasaki, Japan
e-mail: aruga@nagasaki-u.ac.jp

J. Aruga (ed.), *Zic family*, Advances in Experimental Medicine and Biology 1046,
https://doi.org/10.1007/978-981-10-7311-3_1

1.1 Distribution of the *Zic* Genes in the Animal Phylogenetic Tree

Both *Zic* family-targeted comparative genomics studies (Aruga et al. 2006; Layden et al. 2010) and whole-genome sequencing projects (Srivastava et al. 2008, 2010; Moroz et al. 2014) provided us a draft picture with which to consider the distribution of *Zic* homologues in animal phylogeny. Figure 1.1 indicates the distribution of *Zic* family in the current animal phylogeny. The *Zic* family can be found in some metazoans, but not in the other living organisms, including those phylogenetically close to metazoans, such as choanoflagellates and fungi (Hatayama and Aruga 2010). In the metazoans, the *Zic* family is distributed broadly, except in Ctenophora and Porifera. The metazoan animal phyla in which the presence of *Zic* has been verified include Placozoa (*Trichoplax adhaerens*) (Srivastava et al. 2008) and Dicyemida (*Dicyema acuticephalum*) (Aruga et al. 2007) that had been classified as "Mesozoans" (Brusca and Brusca 2003), which are characterized by very simple body plans.

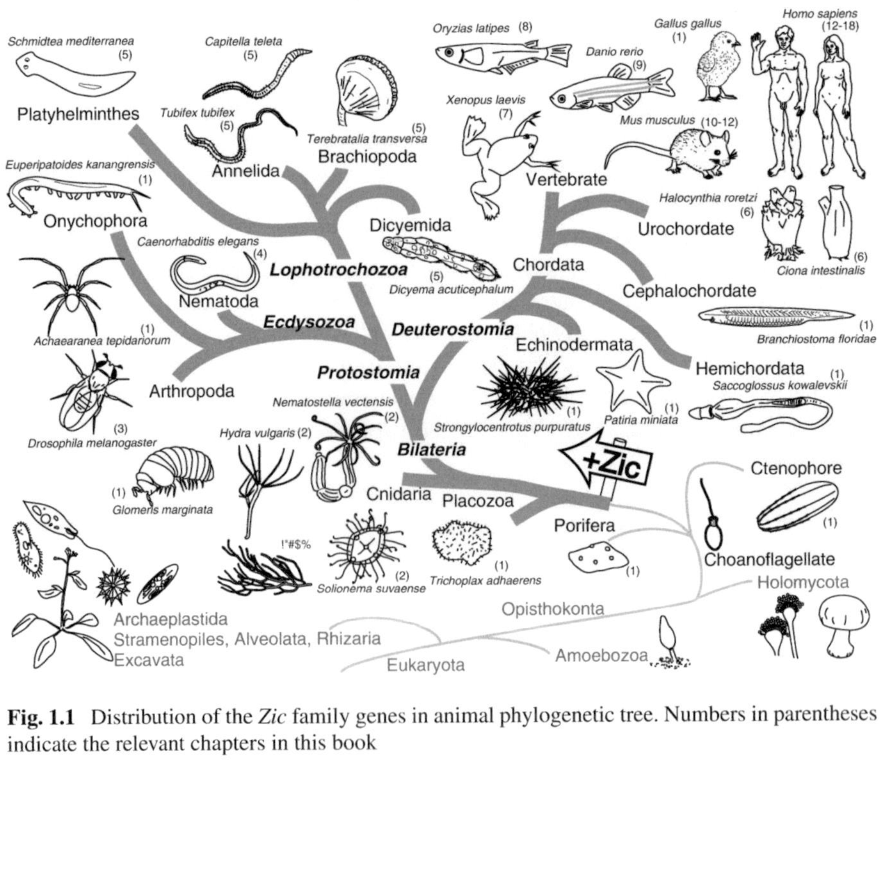

Fig. 1.1 Distribution of the *Zic* family genes in animal phylogenetic tree. Numbers in parentheses indicate the relevant chapters in this book

1.2 Cardinal Amino Acid (aa) Sequence Features of Zic Family Proteins

Zic family zinc finger (ZF) proteins share unique features in the aa sequences (Fig. 1.2) (Aruga et al. 2006, 2007; Layden et al. 2010; Hatayama and Aruga 2010). We first summarize the aa sequence feature in the evolutionary conserved domains of Zic family proteins.

1.2.1 Zinc Finger Domain (ZFD)

ZFDs are found in a large number of eukaryotic proteins (Wolfe et al. 2000; Krishna et al. 2003). Among them, Zic family proteins contain five tandemly repeated Cys2His2 (C2H2)-type ZFDs (Fig. 1.2). Here, the C2H2 ZF units are designated as ZF1 through ZF5 from the amino-terminal to carboxy-terminal. The sequence conservation was particularly higher in ZF2-ZF5 than in ZF1 (Aruga et al. 2006). ZF1 has a notable feature in which the number of aa sequences between the two cysteine residues is variably increased (ZF1, 6-38; ZF2-5, 4, or 2). ZF1 and ZF2 contain absolutely conserved tryptophan residues in the +2 positions from the first cysteine residue. This sequence feature has been designated tandem CWCH2 (tCWCH2) and is predicted to mediate the interaction between ZF1 and ZF2 (Hatayama and Aruga 2010). The number of aa sequences between the two histidine residues was consistent among the published sequences (ZF1, 4; ZF2-5, 3). The sequences following ZF2 and ZF3 C2H2 motifs are TGEKP, which conform to the consensus sequence found in the GLI-Kruppel family zinc finger proteins (Ruppert et al. 1988). ZF2-5 contain additional absolutely conserved aa sequences. Several functions have been assigned to Zic ZFD, including DNA binding, protein binding, transcriptional activation, and nuclear localization (see Chap. 18).

1.2.2 ZOC Domain

The ZOC domain is located at the amino-terminal end, being distant from ZFD. ZOC was first described as a conserved domain between mouse Zic and fly Odd-paired (Zic-*O*pa *c*onserved domain) (Aruga et al. 1996). The consensus sequence motif was either (S/T)RDFLxxxR (Aruga et al. 2006) or RDFL-(1-2 aa)-RR (Layden et al. 2010), and the latter motif is more prevalent throughout metazoans. As a known function, the ZOC domain is involved in the interaction with the I-mfa protein (Mizugishi et al. 2004) and Pax3 (Himeda et al. 2013) (see Chap. 18).

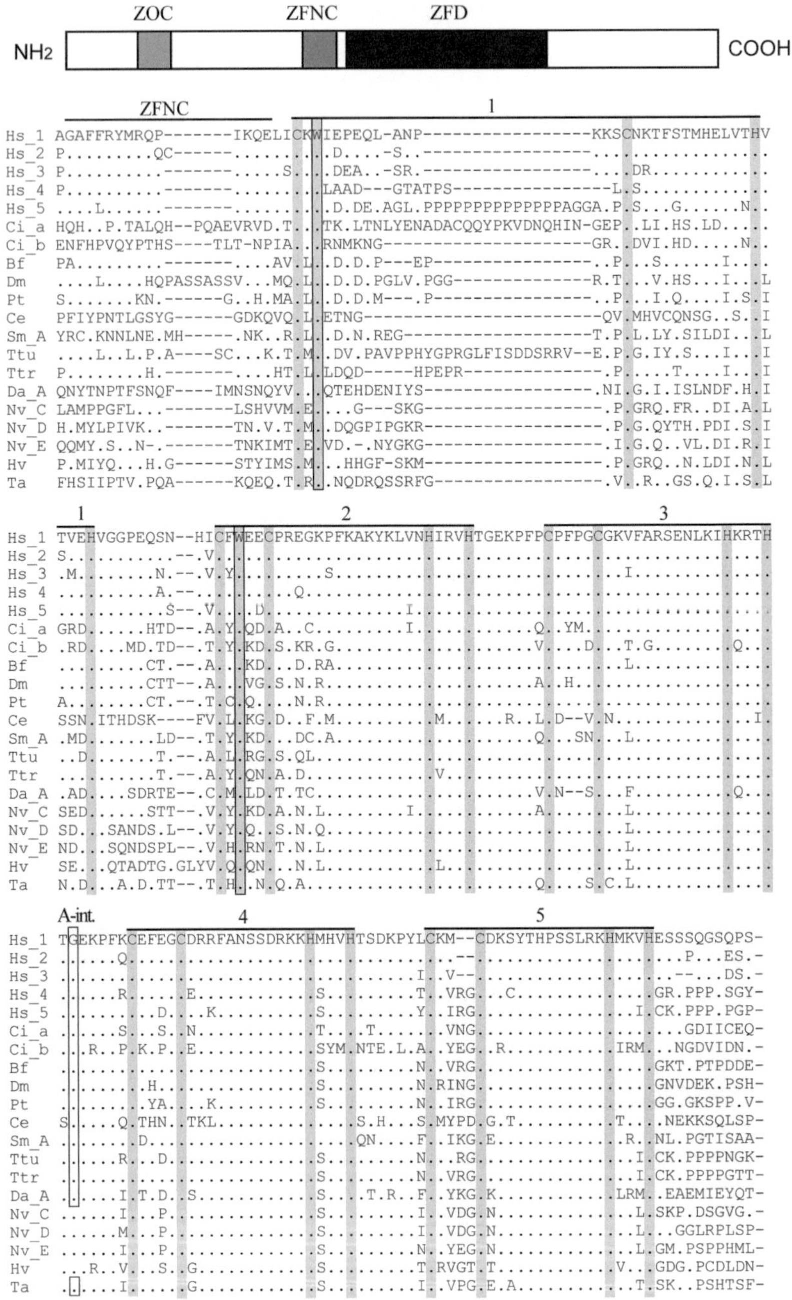

Fig. 1.2 Structure of Zic proteins. *Top*, domain structure; *bottom*, sequence alignment of ZFD and ZFNC. Hs *Homo sapiens*, Ci *Ciona intestinalis*, Bf *Branchiostoma floridae*, Dm *Drosophila melanogaster*, Pt *Parasteatoda tepidariorum*, Ce *Caenorhabditis elegans*, Sm *Schmidtea mediterranea*, Ttu *Tubifex tubifex*, Ttr *Terebratalia transversa*, Da *Dicyema acuticephalum*, Nv *Nematostella vectensis*, Hv *Hydra vulgaris*, Ta *Trichoplax adhaerens*. *1–5*, ZF1-ZF5; *ZFNC*, ZFNC region; dots, residues identical to top line (human ZIC1); *shade*, cysteine (C) and histidine (H) forming C2H2 motif; *outlined shade*, tryptophan (W) forming tCWCH2 motif; *open box*, location of A-intron (phase 2, G/GN)

1.2.3 ZFNC Domain

The ZFNC domain is the conserved sequence immediately amino-terminal to ZFD (*ZF*-*N*-terminal *c*onserved) (Aruga et al. 2006). The consensus sequence is GAF(F/L) RYMRQP-(0-7aa)-IKQE. Function has not been assigned to the ZFNC domain. Clearly, conserved ZFNCs have been identified in Zic aa sequences from Mollusca, Annelida, Acoelomorpha, Arthropoda, Echinodermata, and Chordata (Vertebrata and Cephalochordata), but not in Placozoa, Cnidaria, Platyhelminthes, Dicyemida, Nematoda, or Urochordata (Aruga et al. 2006, 2007; Layden et al. 2010).

1.2.4 Other Conserved Sequences

The carboxy-terminal domain in vertebrates Zic1-3 is conserved where the consensus sequence is NFNEWYV. Transcriptional activity repression function has been assigned to the domain (Kuo et al. 1998; Twigg et al. 2015).

1.3 Exon-Intron Organization of *Zic* Family Genes

The overall picture of exon-intron organization of *Zic* family was clarified by BAC and fosmid-based cloning and sequencing approach (Fig. 1.3) (Aruga et al. 2006). In the ZFDs, five locations of intron insertion (A to E) have been reported. Among them, an intron insertion site (A) between *ZF3* and *ZF4* is absolutely conserved in 27 bilaterian *Zic* genes, but not in the seven cnidarian *Zic* genes (Aruga et al. 2006). The other intron sites in *Zic* ZFD were each found in a small number of animals. B-intron was found in vertebrate *Zic1-3* and some arthropods (Hexapoda). Although both intron loss and intron gain dynamically occur in the course of evolution, the bilaterian-specific distribution of A-introns suggests the presence of unknown constraints in the conservation of A-introns (Aruga et al. 2006). Recent studies indicate that a bilaterian ancestor was rich in introns and that differences in intron number largely reflect intron loss (Roy 2006), raising a possibility that the A-intron possessed some protective feature against intron loss. Interestingly in a placozoan, *Trichoplax adhaerens*, Zic genes possessed A-introns (Figs. 1.3 and 1.4c). Therefore, the distribution of A-introns in the metazoans is bilaterian 27 species + placozoan 1 species, but not in cnidarian 6 species (*Hydra*, *Nematostella*, *Exaiptasia*, *Acropora*, *Orbicella*, and *Scolionema*). However, in Hydra *Zic3* (Hemmrich et al. 2012), a remnant of A-intron was observed where the splicing donor site seems to be replaced by a new one emerging in the intron (J.A. unpublished observation). The distribution of the A-intron can be explained in two ways. If the Placozoa is a sister taxon of (bilaterian + cnidarian) taxon, the A-intron appeared in an eumetazoan ancestor and was lost in the cnidarian ancestor (Fig. 1.4c top). If (placozoan + bilaterian)

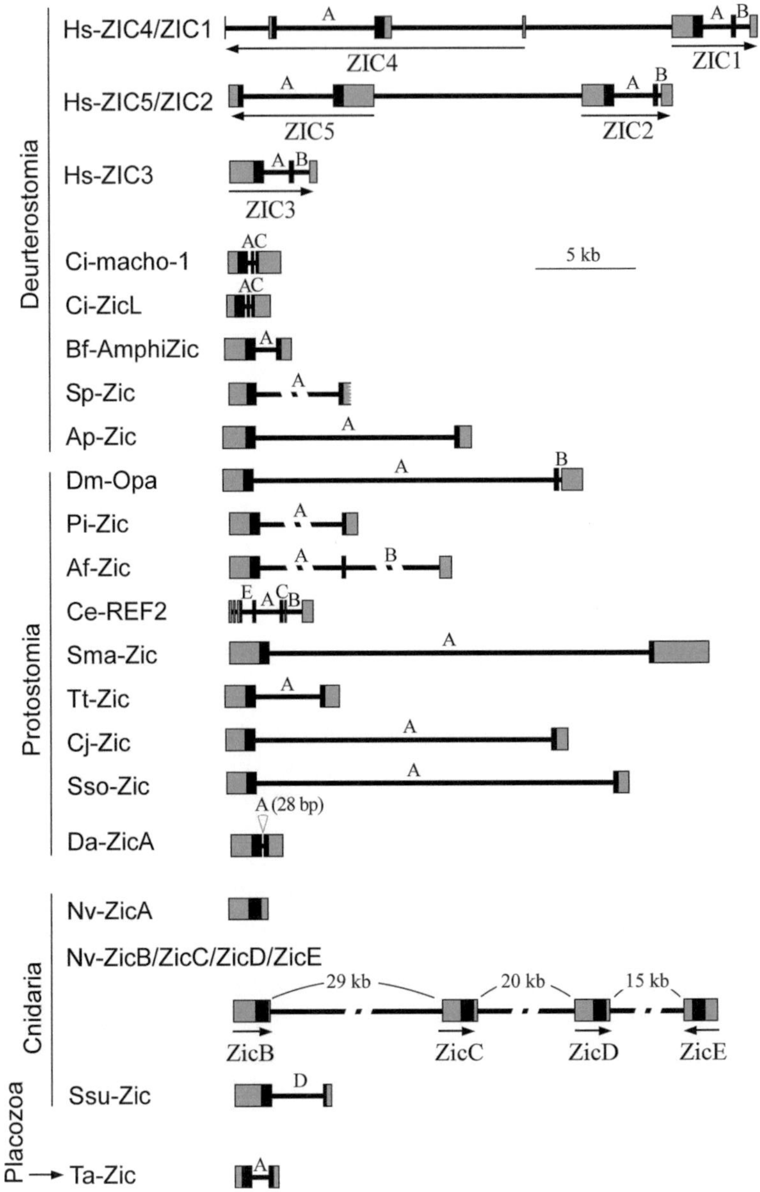

Fig. 1.3 Exon-intron organization of the Zic genes from various species. Abbreviation not in Fig. 1.2. Ssu *Scolionema suvaense*, Sma *Schistosoma mansoni*, Pi *Pandinus imperator*, Af *Artemia franciscana*, Sp *Strongylocentrotus purpuratus*, Ap *Asterina pectinifera*

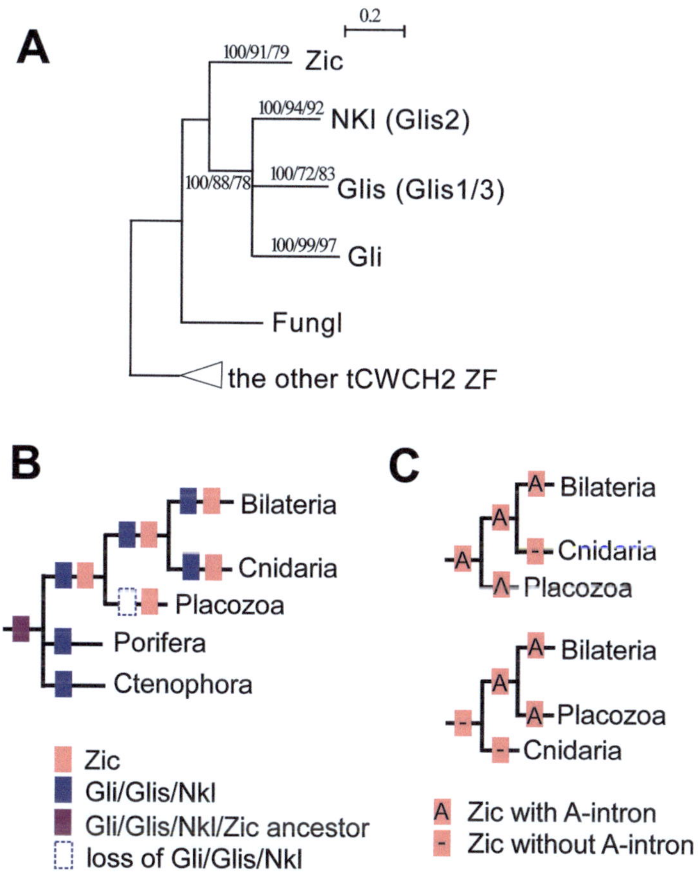

Fig. 1.4 Origin of Zic genes. (**a**) Phylogenetic tree showing the relationship among the Gli, Glis, Nkl, and Zic superfamily genes (Hatayama and Aruga 2010). The tree pattern is based on the Bayesian inference, maximal likelihood, and neighbor-joining methods (consensus). Scores for each branch indicate the statistical support values obtained in each phylogenetic tree construction method {BI (postprobability)/ML (bootstrap value)/NJ (bootstrap value)}. (**b**) Hypothesis concerning the appearance of Zic in the animal phylogeny. (**c**) Appearance of Zic genes containing the A-intron in animal phylogeny

taxon is the sister taxon of cnidaria, the A-intron appeared after diverging from a cnidarian ancestor (Fig. 1.4c bottom). A similar proposed phylogenetic tree is in Fig. 1.4c top (Srivastava et al. 2008), and the presence of remnant A-intron in a cnidarian Zic may favor the former hypothesis.

1.4 Conservation in the Nonprotein-Coding Region

Based on the sequence comparisons of the evolutionary conserved domains and the exon-intron structures, hypothesis elucidating the evolutionary processes has been proposed. Some conserved noncoding sequences are shown to act as the cis-regulatory elements for gene expression regulation of vertebrates *Zic1* (Aruga et al. 2000; Sassa et al. 2007) and *Zic3* (Garnett et al. 2012).

1.5 Hypotheses for the Evolutionary Processes of *Zic* Family Genes

1.5.1 Origin of Zic *Genes*

Recent molecular phylogenetic analysis using the conserved ZFDs (Hatayama and Aruga 2010) (Layden et al. 2010) suggests the common ancestor of the *Zic*, *Gli*, and *Glis* (Fig. 1.4a). Concerning *Glis2*, it was not confirmed to group with the other families of Glis (Glis1 and Glis3) (Hatayama and Aruga 2010; Layden et al. 2010); therefore, it would be appropriate to refer to *Glis2* genes as in the *Nkl* family (Layden et al. 2010). *Gli, Glis* (*Glis1* and *Glis3*), and *Nkl* (*Glis2*) families are likely to have had a common ancestor after *Zic* establishment (Fig. 1.4a).

The distribution of the *Zic* family in the animal phylogenetic tree is suggestive because of the controversial phylogenetic relationships among Porifera (sponges), Cnidaria (jellyfishes, corals, and related species), Ctenophora (comb jellies), Placozoa (*Trichoplax*), and Bilateria (all other animal phyla). In the animal phylogenetic tree, there are two opposing hypotheses: (1) Ctenophora branched off first (Ctenophora-sister hypothesis) and (2) Porifera branched off first (Porifera-sister hypothesis) (Telford et al. 2016). In either hypothesis, the presence of *Gli* and *Glis* and the absence of *Zic* family in Porifera and Ctenophora (Fig. 1.1) indicate that the *Zic* family originated in the presumptive Cnidaria-Placozoa-Bilateria common ancestor after branching off of Ctenophora and Porifera organisms (Fig. 1.4b). However, an alternative scenario that assumes the innovation of Zic genes in a metazoan ancestor and the loss in the Ctenophora and Porifera is possible, considering the limited whole-genome information in these phyla.

1.5.2 Acquisition of the Unique Features in the First and Second Zinc Finger Motif

Based on the sequence comparison of the *Zic* family genes (Aruga et al. 2006) and structural analysis of the human ZIC3 (Hatayama et al. 2008), we discovered "tandem CWCH2 (tCWCH2)," a protein structural motif that is conserved in many

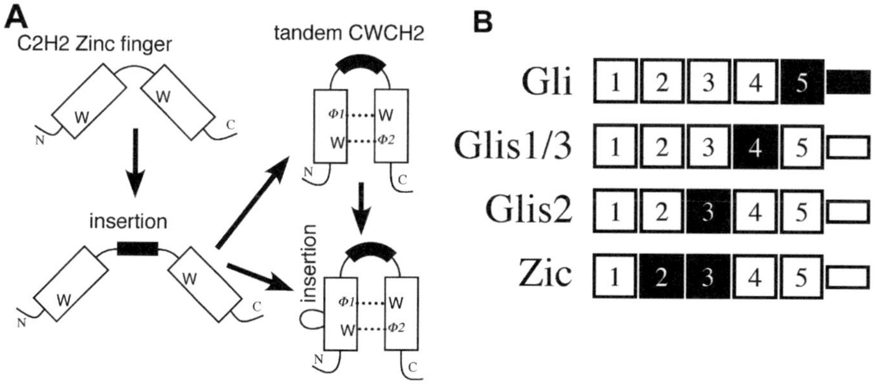

Fig. 1.5 Evolutionary processes of Zic family proteins. (**a**) Evolution of tCWCH2 domain (first and second ZF units of Gli/Glis/Nkl/Zic superfamily genes. The tCWCH2 motifs were generated from classical C2H2 sequences concurrently with the acquisition of the additional sequence features (see texts) (Hatayama and Aruga 2010). (**b**) Location of NLSs in Gli, Glis, and Zic ZF domains. Each ZF is depicted as a *white box* and the NLS-coding ZF is depicted as a black box (Hatayama and Aruga 2012)

eukaryotic C2H2-type zinc finger proteins (Figs. 1.2 and 1.5a; see Chap. 18 for 3D structure). In cases of ZIC3, GLI1, and Zap1 (a yeast zinc finger protein) where two amino-terminal C2H2 zinc finger motifs conform to the tCWCH2 motif, the two zinc finger motifs form a single structural unit by sharing a hydrophobic core. The tryptophan residues in the second aa residue from the first cysteine residue in each C2H2 zinc finger motif form the center of the hydrophobic core by their hydrophobic side chains.

The absolute conservation of the tCWCH2 motif in Zic family proteins suggests that the fused ZF1 and ZF2 are an indispensable feature of the Zic family zinc finger domains. Based on the analysis of 587 tCWCH2-containing genes from public databases, we deduced additional aa sequence features in the tCWCH2-type zinc finger pairs in comparison to general C2H2 zinc finger pairs (Fig. 1.5a) (Hatayama and Aruga 2010): (1) site-specific aa frequencies, (2) longer linker sequence between two C2H2 ZFs, and (3) frequent extra sequences within the C2H2 ZF motifs. Viewing the Zic ZFD from these points, the longer and variable sizes of the intervening sequences between the two cysteine residues in ZF1 represent the third feature. The ZF1-ZF2 linker sequence length were mostly 10, whereas those in ZF2-ZF3, ZF3-ZF4, and ZF4-ZF5 are absolutely 7. The carboxy-terminally adjacent positions of the first histidine residues (φ1, φ2) are V, L, and I (M). The addition of an extra sequence is thought to reflect a reduced structural constraint as a consequence of specialization in the inter-ZF interaction. This idea is supported by lower conservation of ZF1 than ZF2-ZF5 of the Zic family proteins. The longer linker sequences may be required for the opposed positioning of ZF units. The aa residue selection in φ1 and φ2 is likely because they are participating in the hydrophobic core formation together with the conserved tryptophan.

1.5.3 Evolution of the Nuclear Localization Signal (NLS)

Zic proteins can be localized in cell nuclei (Koyabu et al. 2001; Bedard et al. 2007; Hatayama et al. 2008). In eukaryotes, macromolecules like Zic proteins are transported into and out of the nucleus through nuclear pore complexes. Importin α-mediated pathways are known to play critical roles in the nuclear import of Zic proteins (Fig. 18.3) (Hatayama et al. 2008). Importin α binds to NLS of Zic proteins. Importin α-Zic is thought to bind another component, importin β, which is transported into the nucleus with the support of the RanGDP-RanGTP gradient. In the nucleus, the complex is disassembled, and the Zic protein is released. Protein nuclear localization is also affected by nuclear export signals, which mediate the binding of adaptor proteins (e.g., CRM1) involved in the export of proteins from the nucleus through nuclear pore complexes.

Previous studies showed conserved features of the NLS-containing aa sequences. Some NLSs contained one or two clusters of basic aa residue, called the monopartite or bipartite type of classical NLSs (Lange et al. 2007). NLS of human ZIC3 has been mapped to several lysine and arginine residues in ZF2 and ZF3 through deletion and mutation analyses. But in its 3D conformation, the basic residues are positioned similar to those in the bipartite classical NLS (Fig. 18.1) (Hatayama et al. 2008). Conversely, the GLI NLS has been mapped to ZF5 and the region flanking the C-terminus of the zinc finger domain that is not conserved in Zic. GLI1 ZF2 + ZF3 does not show NLS activity. In the superimposition of the 3D models between ZIC3 and GLI1, the residues that form ZIC3 NLS (K312, K349) do not overlap with that of ZIC3, and the residues adjacent to NLS-forming residues were not conserved in GLI1 protein (Hatayama et al. 2008). Therefore, the NLS is not conserved between ZIC3 and GLI1. In addition, Glis (Glis3) NLS and Nkl (Glis2) NLS have been assigned to ZF4 and ZF3, respectively. Therefore, the NLSs in the Gli, Glis, Nkl, and Zic families are not identical to each other (Fig. 1.4b). The evolutionary history of the NLSs in Gli-Glis-Nkl-Zic superfamily may reflect the variable structural constraint of ZFs because of their multifunctional properties (Hatayama and Aruga 2012).

Interestingly the ZIC3 nuclear export signal (NES) was also mapped in the ZF2-ZF3 region (Bedard et al. 2007). When treated with leptomycin B (nuclear export inhibitor that inactivates CRM1), ZIC3 nuclear export was inhibited in an NES-dependent manner (Bedard et al. 2007). Some of NES-responsible residues (L315, I319, V321) were mapped adjacent to the NLS-responsible residue R320, indicating the multifunctional property of the highly conserved carboxy half of ZF2.

1.5.4 Animal Taxa-Selective Sequence Degeneration

When the aa sequence divergence rate was compared among the phyla, the variances of the divergence rates from the deduced ancestral sequences were significantly high in comparison to the housekeeping genes such as ATPS and aldolase

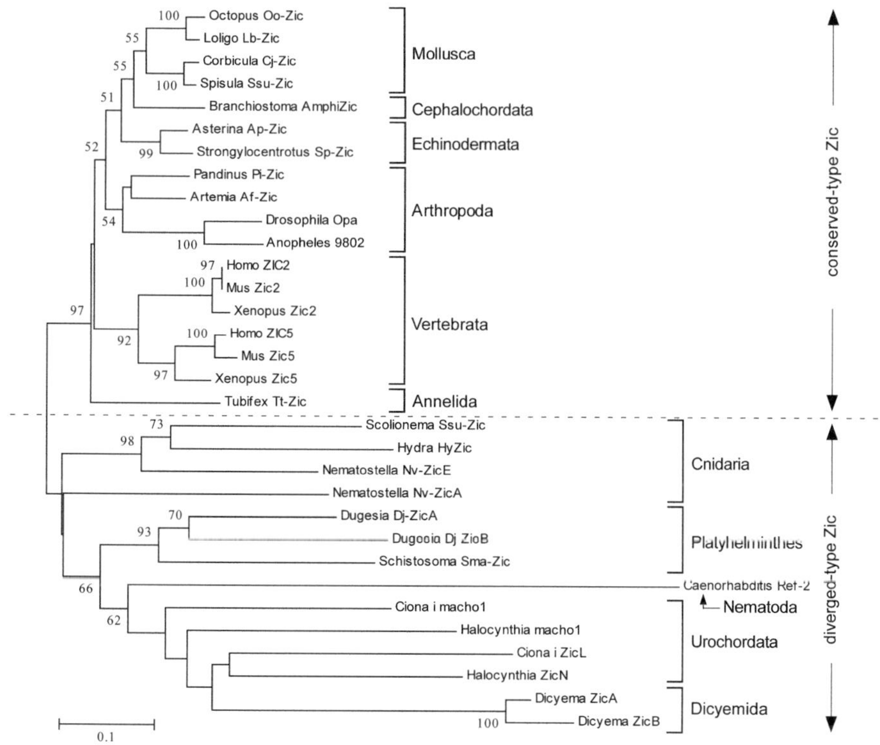

Fig. 1.6 The conservation and degeneration of the Zic family proteins during evolution. Neighbor-joining tree of Zic family ZFD aa sequences. Numbers in internal branches indicate the percentage value of bootstrap test. Scale bar represents evolutionary distance in substitutions/aa residue. Classification of Zic genes according to a previous study (Aruga et al. 2006) is indicated by broken line and arrows

(Aruga et al. 2007). The divergence rate was high in Cnidaria, Platyhelminthes, Nematoda, Urochordata, and Dicyemida, whereas it was low (evolutionary conserved) in Mollusca, Vertebrata, Arthropoda, Echinodermata, and Cephalochordata. This feature of bipolarized divergence rates results in the skewed molecular phylogenetic tree that does not reflect proper animal phylogeny when the tree is constructed using a distance-based method (e.g., neighbor-joining tree, Fig. 1.6). To describe the bipolarized feature, we classified Zic proteins into "conserved-type Zic" and "diverged-type Zic" (Fig. 1.6, Table 1.1) (Aruga et al. 2006, 2007). This classification is based on the neighbor-joining tree using the Zic zinc finger domain, including its flanking sequences, and the presence or absence of ZOC in strict criteria (Aruga et al. 2006). An intriguing point of the classification is that both conserved type and diverged-type exist in each superphylum of bilaterians (Deuterostomia, Ecdysozoa of Protostomia, and Lophotrochozoa of Protostomia).

The animal taxa-selective sequence degeneration was not limited to *Zic* genes, but also found in another toolkit gene, *Msx*, that is closely related to Zic protein in

Table 1.1 Classification of Zic proteins from the viewpoint of sequence conservation

	Conserved-type Zic	Diverged-type Zic
Deuterostomia	Vertebrata, Cephalochordata, Echinodermata	Urochordata
Ecdysozoa (Protostomia)	Arthropoda	Nematoda
Lophotrochozoa (Protostomia)	Mollusca, Annelida, Dicyemida	Platyhelminthes
Non-bilaterians		Cnidaria, Placozoa

Table 1.2 Classification of conserved or diverged Msx

	Conserved-type Msx	Diverged-type Msx
Chordata	Vertebrata, Cephalochordata,	Urochordata
Cnidaria	Anthozoa	Hydrozoa

some developmental contexts (see Chaps. 5 and 12). In the *Msx* phylogeny, the bipolarized sequence degeneration can be clearly seen in Chordata and Cnidarian lower taxa (Table 1.2) (Takahashi et al. 2008a).

Based on the studies on the *Zic* and *Msx* phylogeny, we speculated that selective loss of the conserved domains in certain toolkit genes reflected, at least partly, the evolutionary constraints that are associated with the animal body organization. Developmental toolkit genes are utilized in many biological contexts during animal development. It would be rational to consider that structural constraints depend on how frequently or how tightly the toolkit gene products are utilized in maintaining species. The validity of this hypothesis would be evaluated by examining whether other genes fundamental to the organization of the animal body show similar evolutionary tendencies.

1.5.5 Zic Evolutionary Processes

In a previous study, we summarized *Zic* evolutionary processes in terms of conserved domain and intron (Fig. 1.7a). A *Gli-Glis-Nkl-Zic* superfamily ancestor may be present in the metazoan ancestor. The earliest *Zic* genes may have appeared in the early metazoans (in common ancestor of bilaterian, placozoan, and cnidarian) after Porifera and Ctenophora diverged. A-intron may be inserted into the placozoan-cnidarian-bilaterian common ancestor, but may have been lost in many cnidarians. A-introns were kept thereafter in all bilaterians. ZFNC and ZOC domains appeared before the bilaterian-cnidarian divergence. Additional intron sequences were inserted into Zic genes in cnidarian, nematodes, arthropods, urochordates, and vertebrates; the positions of the introns diverged. The urochordate (ascidian) has a unique C-intron in both types of *Ci-Zic* genes, *macho-1* (*Zic-r.a*) and *ZicL* (*Zic-r.b*), suggesting that a gene duplication event occurred after insertion of the C-intron in the urochordate clade. In nematodes, five additional introns, three of which were located within the ZF domain, seemed to have been inserted.

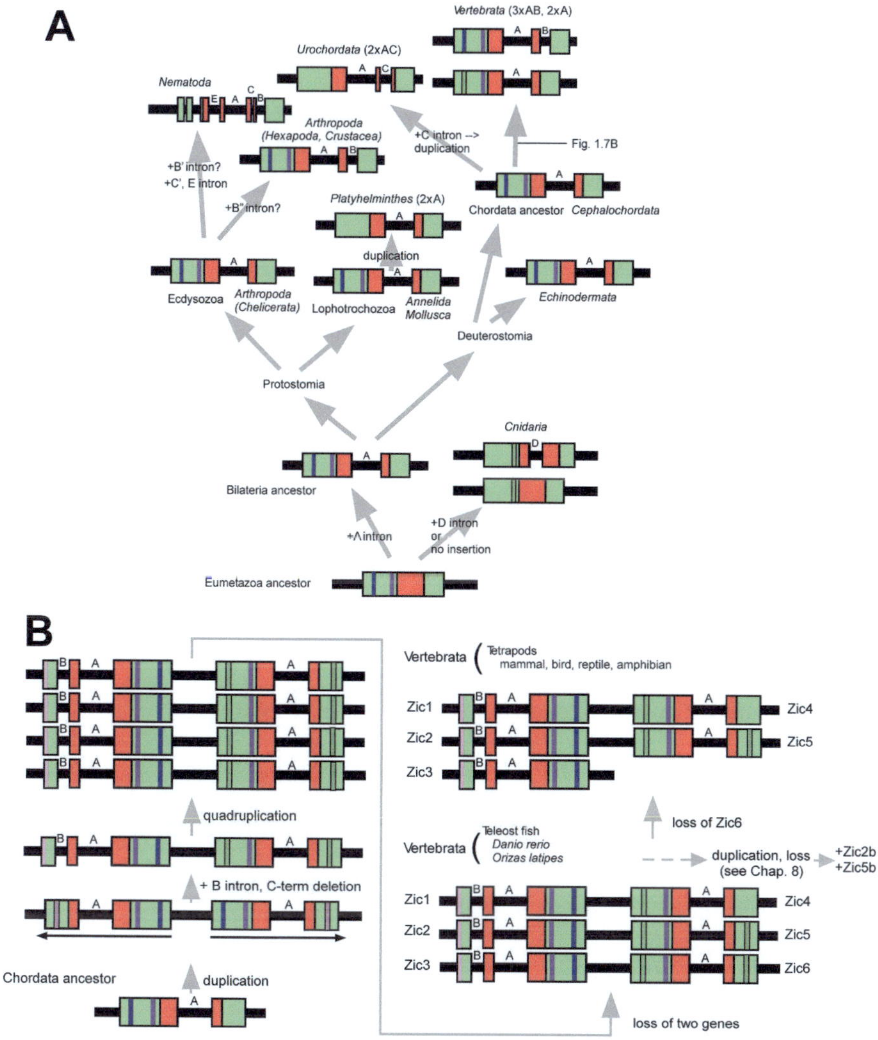

Fig. 1.7 Evolutionary process of Zic genes. (**a**) Whole animal wide. (**b**) Vertebrate-wide. Boxes indicate protein-coding regions. Thick horizontal bars indicate intron and flanking region. Green indicates protein-coding regions without clearly conserved domain structures; red indicates ZFD; dark blue indicates ZOC; light blue indicates ZFNC. Thin horizontal arrows in (**b**) represent the direction of transcription (A figure in Aruga et al. 2006 is modified)

Evolutionary processes in the vertebrate lineage have been proposed as follows (Fig. 1.7b) (Aruga et al. 2006). First, gene duplication occurred in a vertebrate ancestor, resulting in head-to-head array of *Zic* genes with the A intron. Then, the B-intron was inserted in one of the two *Zic* genes; this was followed by duplication twice (quadruplication) of the duplicated *Zic*, concomitant with the entire genome duplication events postulated in accordance with the DNA dose per haploid genome and the molecular phylogenetic analysis of many genes (Ohno 1970; Holland 1999). Of the resulting eight *Zic* genes, three (one AB-intron gene and two A-intron genes) may have disappeared in the course of evolution (duplication-loss model) (Aruga et al. 2006). In case of teleost fish, there exist a teleost-specific member, *zic6*, that is arrayed head-to-head with *zic3* (Keller and Chitnis 2007) and teleost subgroup-specific members, *zic2b* and *zic5b* (see Chap. 8 in detail).

The vertebrates Zic1, Zic2, and Zic3 have conserved sequences within this subfamily of *Zic* genes. In particular, the conservation of their carboxy-terminal sequence region, NFNEWYV, is clear. Similar sequences were found in the midst of the region between the ZF domain and the C-terminal ends of vertebrate Zic5 and conserved-type Zic proteins (Zic proteins from arthropods, annelids, molluscs, and brachiopod). The Zic1/2/3 conserved carboxy-terminus may have acquired a role to regulate transcriptional regulatory capacity of Zic proteins as indicated in vertebrate Zic proteins (Kuo et al. 1998; Twigg et al. 2015). Therefore, the Zic1/2/3 C-terminus may be generated through the truncation of prototype Zic genes after the first duplication and before the quadruplication (Fig. 1.7b) and endowed new functional properties to prototype Zic. Considering the major role of Zic1/2/3 in the neural and mesodermal development, the C-terminal truncation may have a considerable effect on the establishment of vertebrate body plan.

1.6 Evolutionary Conserved Roles in Zic Proteins

Zic family genes have been utilized variously in the animals as in the other chapters of this book. Here, we note the research subjects that have been addressed from the evolutionary viewpoint.

1.6.1 Generation of Neural Cells

The initial major finding in Zic biology was their involvement in the neural induction (neuroectodermal differentiation) (Fig. 1.8) (reviewed in Aruga 2004; Merzdorf 2007). Some leading results were obtained in studies using amphibian embryos (Nakata et al. 1997, 1998; Mizuseki et al. 1998) (Fig. 1.8a, Chap. 7). Thereafter the role of Zic family in neurogenesis has been investigated in many species (Chaps. 2, 3, 4, 5, 6, 8, 9, and 10). As an example, the dorsal restriction of the Zic protein in chick spinal cord is regulated by sonic hedgehog and BMP4/7, and the Zic1 enhance the expansion of neuronal precursors by inhibiting neuronal differentiation (Fig. 1.8b, d) (Aruga et al. 2002 #116).

In metazoan-wide consideration from evolutionary point of view, the roles of Zic family in neural development may be divided into lineage commitment and neuronal subtype specification (Fig. 1.8c). The lineage commitment here refers to the differentiation of neural cells from nonneural cells including differentiation of ectoderm

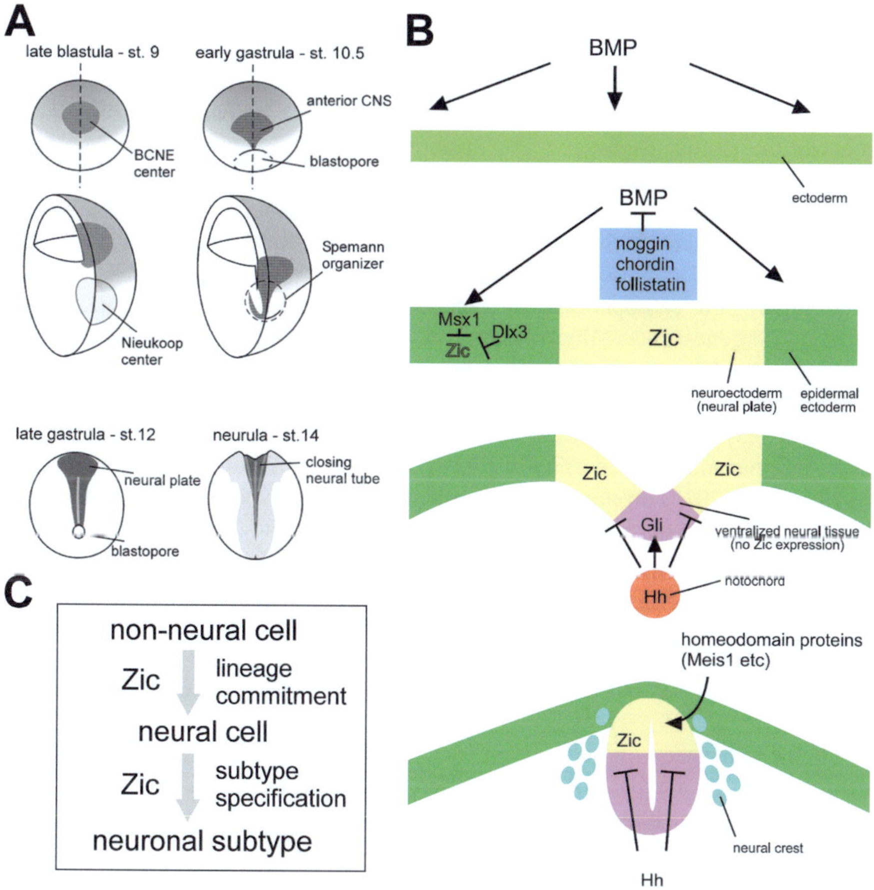

Fig. 1.8 Roles of Zic genes in neural development. (**a**) Neuroectodermal differentiation in *Xenopus* embryos. BCNE (blastula chordin- and noggin-expressing center) contains the prospective neuroectoderm and Spemann's organizer precursor cells. Nieuwkoop center secretes nodal-related factors (potent mesoderm inducers) and cerberus (a wnt/nodal/BMP antagonist) and forms the anterior endomesodermal cells. Spemann's organizer itself differentiates into dorsal mesoderm-derived tissues or organs and emanates neural inducers (noggin, chordin, follistatin) (Reprinted from Aruga and Mikoshiba 2011). (**b**) Yellowish green indicates naïve ectoderm; dark green, epidermal ectoderm; yellow, (presumptive) neuroectoderm where Zic is expressed; pink represents neuroectoderm where Zic expression has disappeared; orange, notochord; blue, migrating neural crest cells (Reprinted from Aruga 2004 with permission). (**c**) Generalized use of Zic genes in neural development

into neuroectoderm (neural plate). For the role in neuronal subtype specification, the requirement may have been clearly shown at the lateral CNS in vertebrates and nematodes (see 6.3) and neuronal migration and wiring deficits (Chap. 11) and cerebellar granule neuronal differentiation (Chap. 13) in rodents. Their involvement in lineage commitment has been shown by gain-of-function experiments in amphibian embryos and cultured mammalian cells (reviewed in Aruga 2004). Absolute necessity for all metazoan neural lineage commitment is not validated because the nematode Zic homologue REF-2 loss of function shows a globally normal organization of the nervous system (Chap. 4) (Bertrand and Hobert 2009). It is still an open question as to how extensively Zic genes are used for the neural lineage commitment in the animal phylogeny. For example, the expression of Zic genes during neural lineage commitment is commonly observed in Deuterostomia Zic genes (Fig. 1.9) (Chaps. 6, 7, 9, and 10). However, there have been no complete loss-of-function experiments including the paralogs that show functional redundancy (e.g., Fig. 12.2) in the Deuterostomia animals.

In this regard, ctenophorans are exceptional animals that have nervous system without *Zic* genes. Interestingly, there is a hypothesis that the ctenophore nervous system evolved independently from the nervous system in bilaterians and cnidarians (Moroz et al. 2014; Moroz and Kohn 2016), and ctenophore Gli and Glis homologues are expressed in a presumptive sensory organ (apical organ, containing balance sensing and photoreceptive cells) (Layden et al. 2010) (see also Chap. 2). Layden et al. (2010) proposed that the role of *Zic* genes in neural development may have been shared with a *Gli/Glis/Nkl/Zic* ancestral gene, together with the fact that *Gli*, *Glis*, and *Nkl* genes play roles in metazoan neurogenesis (Aruga 2004; Merzdorf 2007). This hypothesis would be verified by further evidence on functional significance of the Gli/Glis/Nkl/Zic genes in basal metazoans. Based on the currently available information, it would be appropriate to speculate that *Zic* family genes play significant roles in the establishment of the bilaterian-cnidarian-type nervous system.

On the other hand, in vertebrate-wide consideration, we should take the role in neuronal precursor expansion into consideration additionally. This is because vertebrate CNS contains huge number of cells, and the loss of function of vertebrate Zic genes generally causes hypoplastic change of the CNS (Chaps. 7, 9, 10, 11, 12, 13, 14, and 15). Furthermore, in the evolutionary process of early vertebrate, animals acquired three new *Zic* paralogs with a novel structural and, presumably, functional feature (5.5). We hypothesize here that the *Zic* family may have been involved in the enlargement of vertebrate nervous system.

Fig. 1.9 (continued) plate. (**h**) Hemichordate *Saccoglossus* zic expression. From left to right column, pairs of illustration and picture indicate progressing stages (late blastula, midgastrula, early postgastrula, enterocoely, neurula, and juvenile). In *top diagrams*, *red*, mesoderm; *blue*, ectoderm; *yellow*, endoderm; *mixed yellow and red*, prospective endomesoderm; *A* anterior, *An* animal pole, *D* dorsal, *P* posterior, *V* ventral, *Veg* vegetal pole. (**i–m**) Echinodermata zic expression. Viewed laterally; animal pole is up and oral side is right. *Colors* and *vertical bars* indicate molecularly distinct territories. (**i–k**) Starfish. (**l, m**) Sea urchin. (**i, l**) blastulae. (**j, m**) Gastrulae. (**k**) Early larva. (**n**) Expression of the corresponding markers' in vertebrate CNS. Echinoderm gene names (quotations) indicate their vertebrate homologues. (**a–g** Reprinted from Gostling and Shimeld 2003 with permission. **h** Reprinted from Green et al. 2013 with permission. **i–n** Reprinted from Yankura et al. 2010 with permission)

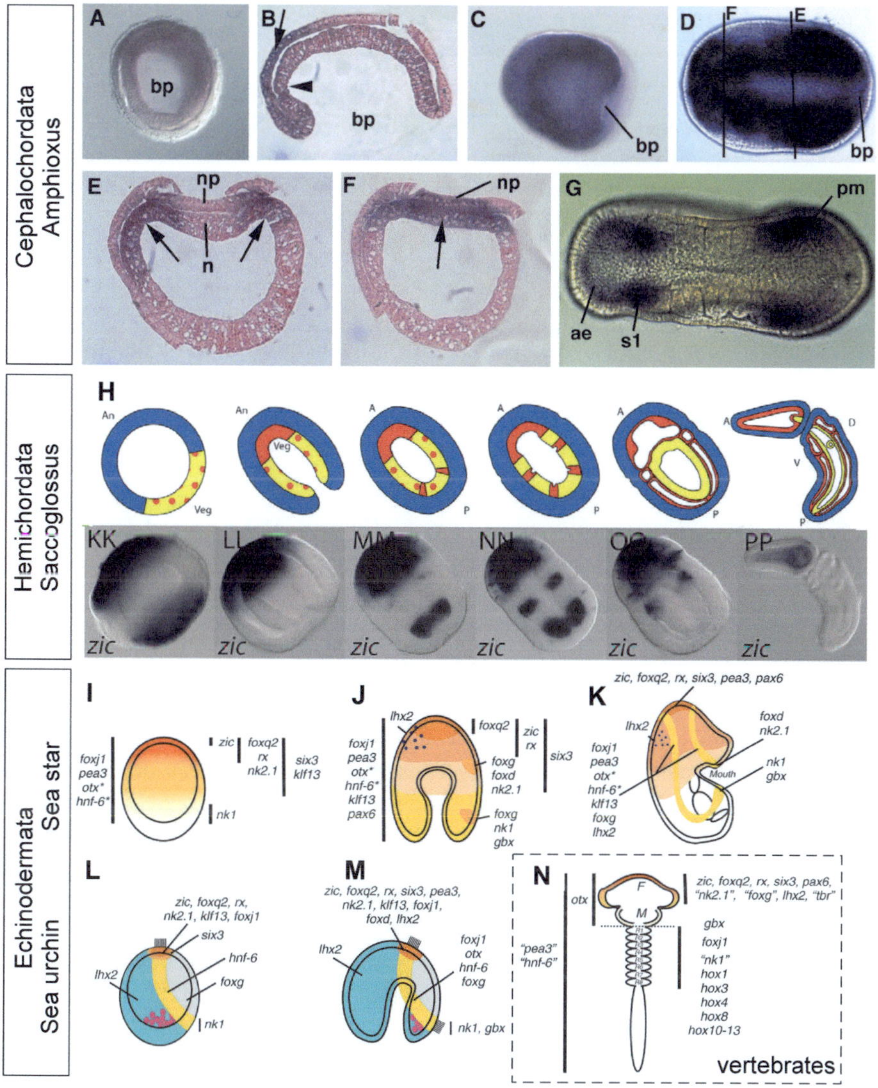

Fig. 1.9 Zic expression in Deuterostomia embryos. (**a**–**g**) Cephalochordate amphioxus AmphiZic expression. (**a**) Posterior view of a gastrula stained for AmphiZic. *bp* blastopore. Dorsal is to the top. (**b**) Section of a gastrula. Dorsal to the left. Dorsal ectoderm (*arrow*) and dorsal mesendoderm (*arrowhead*) express AmphiZic. (**c**) Dorsal view of late gastrula. (**d**) Dorsal view of an early neurula. AmphiZic expression is restricted to lateral neural and mesodermal cells and excluded from the midline except of the anterior most. The approximate levels of the sections shown in (**c**, **f**) are indicated. (**e**) AmphiZic is expressed in the future somites (*arrows*) and lateral neural plate (*np*) but is excluded from the medial neural plate and future notochord (*n*). Note that the surface ectoderm on one side of this embryo has become detached. (**f**) At an anterior level, AmphiZic is expressed in the neural plate and dorsal mesendoderm, including midline cells. (**g**) Dorsal view of mid neurulae with five somites. Expression is confined to the anterior endoderm (*ae*), the anterior part of the first somite (*s1*), and posterior unsegmented mesoderm (*pm*) and is completely absent from the neural

1.6.2 Generation of Mesoderm

Mesodermal expression of *Zic* genes is described in many bilaterian model animals. Functional significance of *Zic* family has been shown in the development of naïve mesoderm, visceral mesoderm, paraxial mesoderm, somites, notochord, and their derivatives. In addition, the studies using annelids and brachiopod animals suggest that mesodermal expression of *Zic* is included in the molecular basis for lophotrochozoan hallmarks, such as chaetae (Chap. 5). These studies support the bilaterian-wide importance of the Zic family in mesodermal development. Also in Cnidaria, there exists a bifunctional endomesodermal tissue (gastrodermis) where *Zic* genes are expressed (Chap. 2). Layden et al. (2010) provided a hypothesis that bilaterian mesodermal *Zic* expression is derived, at least in part, from the endomesodermal expression that existed in the common ancestor of the Bilateria and Cnidaria.

1.6.3 Generation of Neural Crest Cells or CNS Lateral Border Cells

The neural crest has been considered unique to vertebrates (Bronner and LeDouarin 2012). However, recent studies indicate the presence of rudimentary neural crest cells in a common ancestor of vertebrates and urochordates (Chap. 12). The involvement of Zic family genes in the neural crest cell specification has been investigated in several vertebrates, and evolutionary conserved roles have been described (see Chap. 12 in detail). A recent report proposed that a gene regulatory network including Zic family genes is conserved in lateral neural border specification in bilaterians, including nematodes (Li et al. 2017) (Fig. 1.10). They showed that nematode *Zic* ortholog *ref-2* is expressed in the lateral neuroblast of the worm together with the other members of vertebrate neural plate border specification gene set (*Msx/vab-15*, *Pax3/7/pax-3*, and *Zic/ref-2*). Furthermore, conditional knockout of either gene impaired the migration and differentiation of P-neuroblasts that are lateral neuroblasts located between the CNS and skin. These results raise the possibility that the *Zic* genes had been included in the lateral neural blast specifying gene set (module) in the bilaterian common ancestor. This idea is also supported by the *Zic*, *Msx*, and *Pax3/7* expression in the lateral neuroblasts in polychaete annelids (Layden et al. 2010; Denes et al. 2007). These studies were insightful and considered the role of Zic genes in CNS prototype establishment in the bilaterian ancestor.

1.6.4 Establishment of Metamerism

The *Zic* family may be utilized for segmentation processes in many bilaterians. The expression and/or function of *Zic* family genes during body axis segmentation has been described in vertebrates (Chap. 12), annelids (Chap. 5), *Drosophila* (Chap. 3),

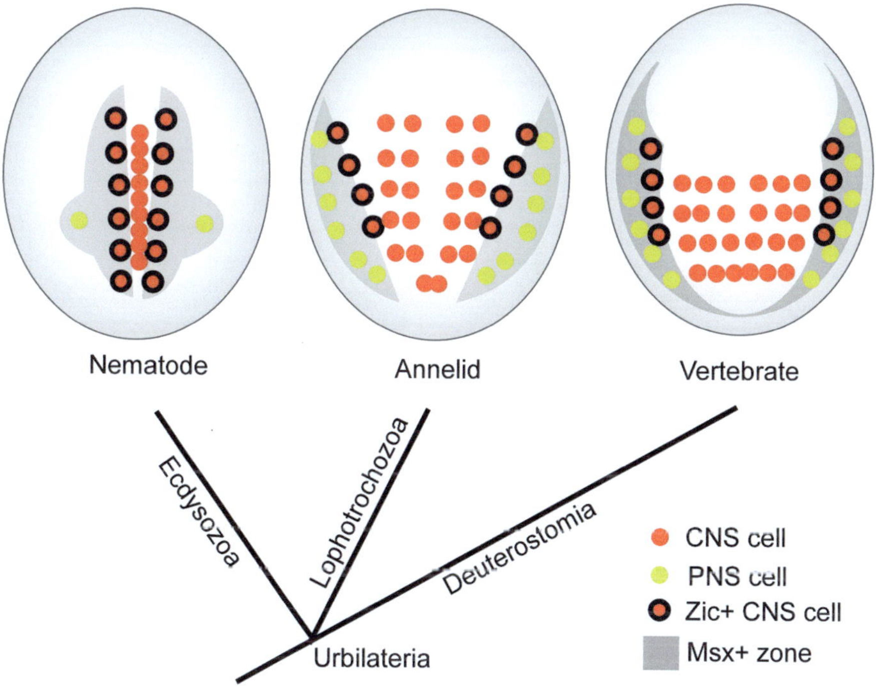

Fig. 1.10 Conserved role or expression of Zic genes at CNS lateral border region. Figure is based on a figure in Li et al. (2017). The expression of Zic is based on (Nematoda Ref2 Li et al. 2017; Capitella Zic Layden et al. 2010; chick Zic1 Simoes-Costa et al. 2012). Msx and Zic are expressed in the lateral border of CNS neural cells in a nematode, annelid, and vertebrate species

spiders (Damen et al. 2005; Kanayama et al. 2011; Akiyama-Oda and Oda 2016), myriapods (Janssen et al. 2011), and onychophorans (Janssen and Budd 2013) (Fig. 1.11).

A summary of the embryonic segmentation modes (Fig. 1.9) based on Kanayama et al. (2011) indicates that the mechanisms are largely different among the species. In case of fruit fly segmentation gene hierarchy, *Opa* is the only broadly expressed pair-rule gene that controls the even-numbered parasegment boundaries by altering other segmentation gene expression (Clark and Akam 2016). In spider head, Opa is required to maintain the head regenerative zone and to promote stripe splitting in this zone (Kanayama et al. 2011). In segmentation of clitellate annelids (e.g., earthworm and leech), the teloblasts are located at the posterior end of the embryo and generate daughter cells one by one, each of which corresponds to the embryonic segment. A clitellate *Zic* is expressed in the first daughter cells of the mesodermal teloblasts (Takahashi et al. 2008b). In mammalian somitogenesis, *Zic2* and *Zic3* are expressed and required for the fidelity of the somite boundaries (Inoue et al. 2007). Despite the apparent dissimilarity of the modes, a conserved feature of the Zic family genes would be the temporally controlled gene expression and the interaction with the partially overlapping gene regulatory network (e.g., components of Notch, hedgehog, Wnt signaling).

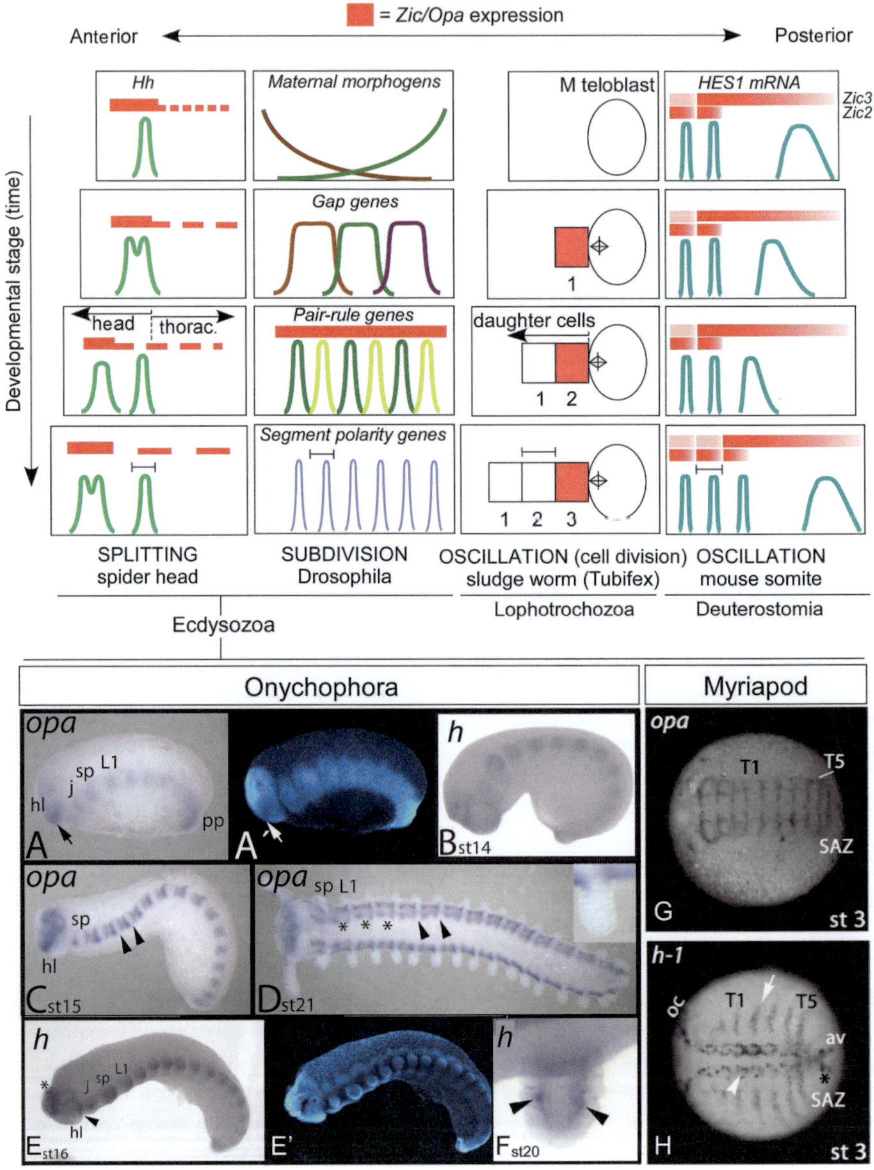

Fig. 1.11 Zic expression in segmentation. *Top*, Zic expression during embryonic segment formation. In each figure the red color indicates Zic/Opa expressing cell or zone. A figure in Kanayama et al. (2011) is modified by adding Zic expression profiles based on (Kanayama et al. 2011; Benedyk et al. 1994; Takahashi et al. 2008b; Inoue et al. 2007). *Bottom*, expression of Opa (*opa*) and hairy (*h-1* or *h*, Notch downstream transcription factor, HES ortholog) in Onychophora (*Euperipatoides*) and myriapod (*Glomeris*). In both species, Opa expression occurs in tissue at the base of the limbs and in stripes between the limbs, but not in the limbs or ventral tissue aligned with the limbs (The pictures are reprinted from Janssen and Budd 2013; Janssen et al. 2011 with copyright permission)

The segmental expression and/or functional analysis suggests that the *Zic* family may also be involved in the development of taxon-specific segmental structures, such as limbs of onychophoran (velvet worm) and myriapod (pill millipede) (Fig. 1.9), the chaetes in annelids (Chap. 5), and vertebrae, costae, and sternebrae in mammalians (Chap. 12).

1.6.5 Dorsoventral Patterning of Somites

Zic1/4 are involved in the determination of trunk dorsoventral patterning in teleost fishes (Moriyama et al. 2012; Kawanishi et al. 2013). The role has been investigated with an evolutionally point of view (see Chap. 7 in detail).

1.6.6 Establishment of Binocular Vision

The binocular vision capability and the presence of ipsilaterally projecting neurons were investigated in various vertebrates using Zic2 as a key molecule (Herrera et al. 2003). The involvement of Zic2 in the laterality selection in neural circuit development has now been extended widely into the central nervous system (see Chap. 11 in detail).

References

Akiyama-Oda Y, Oda H (2016) Multi-color FISH facilitates analysis of cell-type diversification and developmental gene regulation in the Parasteatoda spider embryo. Develop Growth Differ 58(2):215–224. https://doi.org/10.1111/dgd.12263

Aruga J (2004) The role of Zic genes in neural development. Mol Cell Neurosci 26(2):205–221. https://doi.org/10.1016/j.mcn.2004.01.004

Aruga J, Mikoshiba K (2011) Role of BMP, FGF, calcium signaling, and Zic proteins in vertebrate neuroectodermal differentiation. Neurochem Res 36(7):1286–1292. https://doi:10.1007/s11064-011-0422-5

Aruga J, Nagai T, Tokuyama T, Hayashizaki Y, Okazaki Y, Chapman VM, Mikoshiba K (1996) The mouse zic gene family. Homologues of the Drosophila pair-rule gene odd-paired. J Biol Chem 271(2):1043–1047

Aruga J, Shimoda K, Mikoshiba K (2000) A 5′ segment of the mouse Zic1 gene contains a region specific enhancer for dorsal hindbrain and spinal cord. Brain Res Mol Brain Res 78(1–2):15–25

Aruga J, Tohmonda T, Homma S, Mikoshiba K (2002) Zic1 promotes the expansion of dorsal neural progenitors in spinal cord by inhibiting neuronal differentiation. Dev Biol 244(2):329–341. https://doi.org/10.1006/dbio.2002.0598

Aruga J, Kamiya A, Takahashi H, Fujimi TJ, Shimizu Y, Ohkawa K, Yazawa S, Umesono Y, Noguchi H, Shimizu T, Saitou N, Mikoshiba K, Sakaki Y, Agata K, Toyoda A (2006) A wide-range phylogenetic analysis of Zic proteins: implications for correlations between protein structure conservation and body plan complexity. Genomics 87(6):783–792. https://doi.org/10.1016/j.ygeno.2006.02.011

Aruga J, Odaka YS, Kamiya A, Furuya H (2007) Dicyema Pax6 and Zic: tool-kit genes in a highly simplified bilaterian. BMC Evol Biol 7:201. https://doi.org/10.1186/1471-2148-7-201

Bedard JE, Purnell JD, Ware SM (2007) Nuclear import and export signals are essential for proper cellular trafficking and function of ZIC3. Hum Mol Genet 16(2):187–198. https://doi.org/10.1093/hmg/ddl461

Benedyk MJ, Mullen JR, DiNardo S (1994) Odd-paired: a zinc finger pair-rule protein required for the timely activation of engrailed and wingless in Drosophila embryos. Genes Dev 8(1):105–117

Bertrand V, Hobert O (2009) Linking asymmetric cell division to the terminal differentiation program of postmitotic neurons in C. elegans. Dev Cell 16(4):563–575. https://doi.org/10.1016/j.devcel.2009.02.011

Bronner ME, LeDouarin NM (2012) Development and evolution of the neural crest: an overview. Dev Biol 366(1):2–9. https://doi.org/10.1016/j.ydbio.2011.12.042

Brusca RC, Brusca GJ (2003) Mesozoan bauplans. In: Invertebrates, 2nd edn. Sinauer Associates, Sunderland, pp 210–215

Clark E, Akam M (2016) Odd-paired controls frequency doubling in Drosophila segmentation by altering the pair-rule gene regulatory network. Elife 5. https://doi.org/10.7554/eLife.18215

Damen WG, Janssen R, Prpic NM (2005) Pair rule gene orthologs in spider segmentation. Evol Dev 7(6):618–628. https://doi.org/10.1111/j.1525-142X.2005.05065.x

Denes AS, Jekely G, Steinmetz PR, Raible F, Snyman H, Prud'homme B, Ferrier DE, Balavoine G, Arendt D (2007) Molecular architecture of annelid nerve cord supports common origin of nervous system centralization in bilateria. Cell 129(2):277–288. https://doi.org/10.1016/j.cell.2007.02.040

Garnett AT, Square TA, Medeiros DM (2012) BMP, Wnt and FGF signals are integrated through evolutionarily conserved enhancers to achieve robust expression of Pax3 and Zic genes at the zebrafish neural plate border. Development 139(22):4220–4231. https://doi.org/10.1242/dev.081497

Gostling NJ, Shimeld SM (2003) Protochordate Zic genes define primitive somite compartments and highlight molecular changes underlying neural crest evolution. Evol Dev 5(2):136–144

Green SA, Norris RP, Terasaki M, Lowe CJ (2013) FGF signaling induces mesoderm in the hemichordate Saccoglossus kowalevskii. Development 140(5):1024–1033. https://doi.org/10.1242/dev.083790

Hatayama M, Aruga J (2010) Characterization of the tandem CWCH2 sequence motif: a hallmark of interzinc finger interactions. BMC Evol Biol 10:53. https://doi.org/10.1186/1471-2148-10-53

Hatayama M, Aruga J (2012) Gli protein nuclear localization signal. Vitam Horm 88:73–89. https://doi.org/10.1016/B978-0-12-394622-5.00004-3

Hatayama M, Tomizawa T, Sakai-Kato K, Bouvagnet P, Kose S, Imamoto N, Yokoyama S, Utsunomiya-Tate N, Mikoshiba K, Kigawa T, Aruga J (2008) Functional and structural basis of the nuclear localization signal in the ZIC3 zinc finger domain. Hum Mol Genet 17(22):3459–3473. https://doi.org/10.1093/hmg/ddn239

Hemmrich G, Khalturin K, Boehm AM, Puchert M, Anton-Erxleben F, Wittlieb J, Klostermeier UC, Rosenstiel P, Oberg HH, Domazet-Loso T, Sugimoto T, Niwa H, Bosch TC (2012) Molecular signatures of the three stem cell lineages in hydra and the emergence of stem cell function at the base of multicellularity. Mol Biol Evol 29(11):3267–3280. https://doi.org/10.1093/molbev/mss134

Herrera E, Brown L, Aruga J, Rachel RA, Dolen G, Mikoshiba K, Brown S, Mason CA (2003) Zic2 patterns binocular vision by specifying the uncrossed retinal projection. Cell 114(5):545–557

Himeda CL, Barro MV, Emerson CP Jr (2013) Pax3 synergizes with Gli2 and Zic1 in transactivating the Myf5 epaxial somite enhancer. Dev Biol 383(1):7–14. https://doi.org/10.1016/j.ydbio.2013.09.006

Holland PW (1999) Gene duplication: past, present and future. Semin Cell Dev Biol 10(5):541–547. https://doi.org/10.1006/scdb.1999.0335

Inoue T, Ota M, Mikoshiba K, Aruga J (2007) Zic2 and Zic3 synergistically control neurulation and segmentation of paraxial mesoderm in mouse embryo. Dev Biol 306(2):669–684. https://doi.org/10.1016/j.ydbio.2007.04.003

Janssen R, Budd GE (2013) Deciphering the onychophoran 'segmentation gene cascade': gene expression reveals limited involvement of pair rule gene orthologs in segmentation, but a highly conserved segment polarity gene network. Dev Biol 382(1):224–234. https://doi.org/10.1016/j.ydbio.2013.07.010

Janssen R, Budd GE, Prpic NM, Damen WG (2011) Expression of myriapod pair rule gene orthologs. EvoDevo 2(1):5. https://doi.org/10.1186/2041-9139-2-5

Kanayama M, Akiyama-Oda Y, Nishimura O, Tarui H, Agata K, Oda H (2011) Travelling and splitting of a wave of hedgehog expression involved in spider-head segmentation. Nat Commun 2:500. https://doi.org/10.1038/ncomms1510

Kawanishi T, Kaneko T, Moriyama Y, Kinoshita M, Yokoi H, Suzuki T, Shimada A, Takeda H (2013) Modular development of the teleost trunk along the dorsoventral axis and zic1/zic4 as selector genes in the dorsal module. Development 140(7):1486–1496. https://doi.org/10.1242/dev.088567

Keller MJ, Chitnis AB (2007) Insights into the evolutionary history of the vertebrate zic3 locus from a teleost-specific zic6 gene in the zebrafish, Danio rerio. Dev Genes Evol 217(7):541–547. https://doi.org/10.1007/s00427-007-0161-4

Koyabu Y, Nakata K, Mizugishi K, Aruga J, Mikoshiba K (2001) Physical and functional interactions between Zic and Gli proteins. J Biol Chem 276(10):6889–6892. https://doi.org/10.1074/jbc.C000773200

Krishna SS, Majumdar I, Grishin NV (2003) Structural classification of zinc fingers: survey and summary. Nucleic Acids Res 31(2):532–550. https://doi.org/10.1093/nar/gkg161

Kuo JS, Patel M, Gamse J, Merzdorf C, Liu X, Apekin V, Sive H (1998) Opl: a zinc finger protein that regulates neural determination and patterning in Xenopus. Development 125(15):2867–2882

Lange A, Mills RE, Lange CJ, Stewart M, Devine SE, Corbett AH (2007) Classical nuclear localization signals: definition, function, and interaction with importin alpha. J Biol Chem 282(8):5101–5105. https://doi.org/10.1074/jbc.R600026200

Layden MJ, Meyer NP, Pang K, Seaver EC, Martindale MQ (2010) Expression and phylogenetic analysis of the zic gene family in the evolution and development of metazoans. EvoDevo 1(1):12. https://doi.org/10.1186/2041-9139-1-12

Li Y, Zhao D, Horie T, Chen G, Bao H, Chen S, Liu W, Horie R, Liang T, Dong B, Feng Q, Tao Q, Liu X (2017) Conserved gene regulatory module specifies lateral neural borders across bilaterians. Proc Natl Acad Sci U S A 114(31):E6352–E6360. https://doi.org/10.1073/pnas.1704194114

Merzdorf CS (2007) Emerging roles for zic genes in early development. Dev Dyn 236(4):922–940. https://doi.org/10.1002/dvdy.21098

Mizugishi K, Hatayama M, Tohmonda T, Ogawa M, Inoue T, Mikoshiba K, Aruga J (2004) Myogenic repressor I-mfa interferes with the function of Zic family proteins. Biochem Biophys Res Commun 320(1):233–240. https://doi.org/10.1016/j.bbrc.2004.05.158

Mizuseki K, Kishi M, Matsui M, Nakanishi S, Sasai Y (1998) Xenopus Zic-related-1 and Sox-2, two factors induced by chordin, have distinct activities in the initiation of neural induction. Development 125(4):579–587

Moriyama Y, Kawanishi T, Nakamura R, Tsukahara T, Sumiyama K, Suster ML, Kawakami K, Toyoda A, Fujiyama A, Yasuoka Y, Nagao Y, Sawatari E, Shimizu A, Wakamatsu Y, Hibi M, Taira M, Okabe M, Naruse K, Hashimoto H, Shimada A, Takeda H (2012) The medaka zic1/zic4 mutant provides molecular insights into teleost caudal fin evolution. Curr Biol 22(7):601–607. https://doi.org/10.1016/j.cub.2012.01.063

Moroz LL, Kohn AB (2016) Independent origins of neurons and synapses: insights from ctenophores. Philos Trans R Soc Lond Ser B Biol Sci 371(1685):20150041. https://doi.org/10.1098/rstb.2015.0041

Moroz LL, Kocot KM, Citarella MR, Dosung S, Norekian TP, Povolotskaya IS, Grigorenko AP, Dailey C, Berezikov E, Buckley KM, Ptitsyn A, Reshetov D, Mukherjee K, Moroz TP, Bobkova Y, Yu F, Kapitonov VV, Jurka J, Bobkov YV, Swore JJ, Girardo DO, Fodor A, Gusev F, Sanford R, Bruders R, Kittler E, Mills CE, Rast JP, Derelle R, Solovyev VV, Kondrashov FA, Swalla BJ, Sweedler JV, Rogaev EI, Halanych KM, Kohn AB (2014) The ctenophore genome and the evolutionary origins of neural systems. Nature 510(7503):109–114. https://doi.org/10.1038/nature13400

Nakata K, Nagai T, Aruga J, Mikoshiba K (1997) Xenopus Zic3, a primary regulator both in neural and neural crest development. Proc Natl Acad Sci USA 94(22):11980–11985

Nakata K, Nagai T, Aruga J, Mikoshiba K (1998) Xenopus Zic family and its role in neural and neural crest development. Mech Dev 75(1-2):43–51

Ohno S (1970) Evolution by gene duplication. Springer-Verlag, Berlin

Roy SW (2006) Intron-rich ancestors. Trends Genet 22(9):468–471. https://doi.org/10.1016/j.tig.2006.07.002

Ruppert JM, Kinzler KW, Wong AJ, Bigner SH, Kao FT, Law ML, Seuanez HN, O'Brien SJ, Vogelstein B (1988) The GLI-Kruppel family of human genes. Mol Cell Biol 8(8):3104–3113

Sassa T, Aizawa H, Okamoto H (2007) Visualization of two distinct classes of neurons by gad2 and zic1 promoter/enhancer elements in the dorsal hindbrain of developing zebrafish reveals neuronal connectivity related to the auditory and lateral line systems. Dev Dyn 236(3):706–718. https://doi.org/10.1002/dvdy.21084

Simoes-Costa MS, McKeown SJ, Tan-Cabugao J, Sauka-Spengler T, Bronner ME (2012) Dynamic and differential regulation of stem cell factor FoxD3 in the neural crest is encrypted in the genome. PLoS Genet 8(12):e1003142. https://doi.org/10.1371/journal.pgen.1003142

Srivastava M, Begovic E, Chapman J, Putnam NH, Hellsten U, Kawashima T, Kuo A, Mitros T, Salamov A, Carpenter ML, Signorovitch AY, Moreno MA, Kamm K, Grimwood J, Schmutz J, Shapiro H, Grigoriev IV, Buss LW, Schierwater B, Dellaporta SL, Rokhsar DS (2008) The Trichoplax genome and the nature of placozoans. Nature 454(7207):955–960. https://doi.org/10.1038/nature07191

Srivastava M, Simakov O, Chapman J, Fahey B, Gauthier ME, Mitros T, Richards GS, Conaco C, Dacre M, Hellsten U, Larroux C, Putnam NH, Stanke M, Adamska M, Darling A, Degnan SM, Oakley TH, Plachetzki DC, Zhai Y, Adamski M, Calcino A, Cummins SF, Goodstein DM, Harris C, Jackson DJ, Leys SP, Shu S, Woodcroft BJ, Vervoort M, Kosik KS, Manning G, Degnan BM, Rokhsar DS (2010) The Amphimedon queenslandica genome and the evolution of animal complexity. Nature 466(7307):720–726. https://doi.org/10.1038/nature09201

Takahashi H, Kamiya A, Ishiguro A, Suzuki AC, Saitou N, Toyoda A, Aruga J (2008a) Conservation and diversification of Msx protein in metazoan evolution. Mol Biol Evol 25(1):69–82. https://doi.org/10.1093/molbev/msm228

Takahashi H, Shimizu T, Aruga J (2008b) Expression pattern of annelid Zic in embryonic development of the oligochaete Tubifex tubifex. Dev Genes Evol 218(10):553–560. https://doi.org/10.1007/s00427-008-0252-x

Telford MJ, Moroz LL, Halanych KM (2016) Evolution: a sisterly dispute. Nature 529(7586):286–287. https://doi.org/10.1038/529286a

Twigg SR, Forecki J, Goos JA, Richardson IC, Hoogeboom AJ, van den Ouweland AM, Swagemakers SM, Lequin MH, Van Antwerp D, McGowan SJ, Westbury I, Miller KA, Wall SA, Consortium WGS, van der Spek PJ, Mathijssen IM, Pauws E, Merzdorf CS, Wilkie AO (2015) Gain-of-function mutations in ZIC1 are associated with coronal craniosynostosis and learning disability. Am J Hum Genet 97(3):378–388. https://doi.org/10.1016/j.ajhg.2015.07.007

Wolfe SA, Nekludova L, Pabo CO (2000) DNA recognition by Cys2His2 zinc finger proteins. Annu Rev Biophys Biomol Struct 29:183–212. https://doi.org/10.1146/annurev.biophys.29.1.183

Yankura KA, Martik ML, Jennings CK, Hinman VF (2010) Uncoupling of complex regulatory patterning during evolution of larval development in echinoderms. BMC Biol 8:143. https://doi.org/10.1186/1741-7007-8-143

Chapter 2
Cnidarian *Zic* Genes

Michael J. Layden

Abstract To understand the ancestral and evolved roles of *zic* homologs, it is important to reconstruct the putative roles of ancient *zic* homologs in the animal phylogeny. Most studies of *zic* genes have been conducted in model systems that are members of the bilaterian phylum. However, two additional phyla have *zic* homologs encoded in their genomes. The three animal phyla that contain *zic* homologs all share a common ancestor and collectively are termed the parahoxozoans (cnidarians (corals, sea anemones, and jellyfish), placozoans (*Trichoplax adhaerens*), and bilaterians (chordates, insects, nematodes, annelids, echinoderms, etc.). In this chapter we briefly discuss our understanding of *zic* genes in the parahoxozoans with a particular focus on how expression of cnidarian *zic* homologs in the medusozoan *Hydra vulgaris* and the anthozoan *Nematostella vectensis* informs our understanding of the putative ancestral roles zic homologs played in the cnidarian-bilaterian common ancestor.

Keywords Cnidarian · *Nematostella* · *Hydra* · *Zic* · *Hyzic* · *Nvzic* · Neurogenesis · Neuronal

2.1 Generalized Bilaterian *Zic* Expression and Function

Later chapters provide detailed descriptions of zic functions in bilaterians. However, it is necessary to briefly generalize *zic* functions in bilaterians to serve as a point of comparison with cnidarian expression patterns. Bilaterian *zic* genes have widespread roles in neural development (Aruga 2004). Vertebrate *zic* homologs are broadly expressed in the forming neural ectoderm indicating early roles nearer the top of a neurogenic cascade (Mizuseki et al. 1998; Merzdorf 2007; Nakata et al. 2000). *Zic* expression later resolves to regulate formation of neural crest cells and to pattern neuronal fates in the dorsal neural tube (Elms et al. 2003; Aruga 2004). Within *C. elegans*, *Drosophila*, and the lophotrochozoans *C. telata* and *T. tubifex*,

M. J. Layden (✉)
Department of Biological Sciences, Lehigh University, Bethlehem, PA, USA
e-mail: layden@lehigh.edu

J. Aruga (ed.), *Zic family*, Advances in Experimental Medicine and Biology 1046,
https://doi.org/10.1007/978-981-10-7311-3_2

zic homologs are expressed in subsets of the neural progenitors and developing neurons where they regulate proliferation and differentiation (Alper and Kenyon 2002; Benedyk et al. 1994; Takahashi et al. 2008; Layden et al. 2010). The diversity of neurogenic functions in bilaterians implies that *zic* homologs are not necessarily a core component of neurogenic cascades but rather suggests that early in evolution *zic* genes were co-opted to pattern already neuralized tissue. Over time its role has evolved in different lineages to take on key upstream function as in the vertebrates or control subtype patterning and neural progenitor biology.

Bilaterian *zic* genes also function during development of mesodermal tissues. The *Drosophila zic* gene *odd-paired* is known function in visceral mesoderm to regulate midgut development (Cimbora and Sakonju 1995). In *vertebrates* zic genes are expressed in a segmentally reiterated pattern within somites (Nagai et al. 1997; Inoue et al. 2007). In ascidians, maternally derived *macho-1* is segregated into and required for muscle development (Gyoja 2006). In annelids *zic* homologs are expressed broadly in the mesoderm and in a segmentally reiterated pattern in the mesoderm surrounding the chaetal sacs (Layden et al. 2010; Takahashi et al. 2008). Although, many of the bilaterian expression patterns are associated with segmented structures, there is little evidence to suggest that they indicate a conserved role regulating segmentation in animal patterning. Vertebrate and ascidian *zic* homologs are also important for specifying parts of the mesodermally derived notochord (Wada and Saiga 2002). Similar to the variety of functions for *zic* genes in neurogenesis, it seems likely that *zic* homologs were co-opted into mesodermal patterning early on, but that *zic* genes are not central to specify mesodermal identity.

2.2 Cnidarian and Bilaterian *Zic* Homologs Derived from Single Ancestral Gene

Parahoxozoans (Placozoa, Cnidaria, and Bilateria) all have clear *zic* homologs leaving the Ctenophora (comb jellies) and Porifera (sponges) as the only metazoan phyla lacking definitive *zic* genes (Fig. 2.1) (Aruga et al. 2006; Layden et al. 2010). The current understanding of the animal tree of life places either ctenophores or poriferans at the base of the metazoan tree, arguing *zic* emergence occurred in the common ancestor of the Parahoxozoa. Within the parahoxozoans, placozoans represent the most ancient group to diverge. The cnidarians and bilaterians are sister taxa that diverged between 600 and 750 mya (Dunn et al. 2008; Hejnol et al. 2009). The current available draft genome of *Trichoplax adhaerens* (the only known placozoan species) identified only a single zic gene, but the expression or function of *Tad-zic* is not known (Srivastava et al. 2008; Layden et al. 2010).

Evidence suggests that a single ancestral *zic* homolog in the cnidarian-bilaterian common ancestor independently radiated giving rise to larger *zic* families in extant taxa of both lineages. In addition to the characteristic C2H2 zinc finger domains, cnidarian and bilaterian *zic* homologs possess ZOC and ZF-NC domains that were

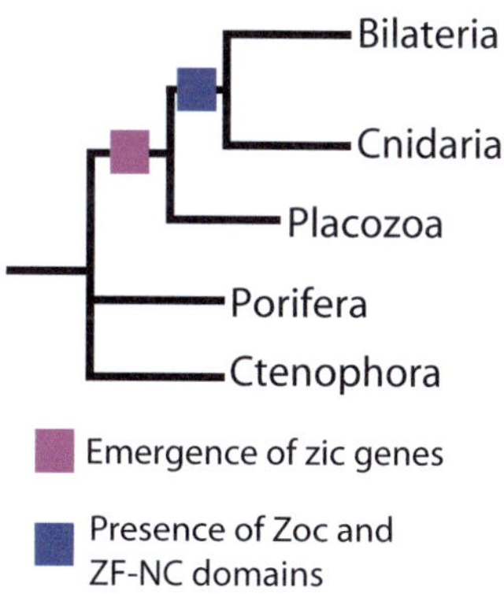

Fig. 2.1 Map of zic genes on animal phylogeny. The emergence of zic genes (purple square) and presence of ZOC and ZF-NC domains (blue square) are mapped onto metazoan phylogeny

not present in the *Tad-zic* but do exist in cnidarian *zic* homologs (Fig. 2.1). The parsimonious explanation is that the elaborated domain structure of *zic* homologs arose once in a common ancestor of the cnidarians and bilaterians (Layden et al. 2010). Shared intron-exon structures across bilaterian *zics* suggest they arose from a single ancestral *zic*, which argues for a single *zic* homolog present at the base of the bilaterian lineage (Aruga et al. 2006). The Cnidaria is comprised of two main subphyla, the Anthozoa (corals, sea anemones, sea pens) and the Medusozoa (hydras, jellyfish, siphonophores) (Darling et al. 2005). Within the cnidarians, a single *zic* gene has been reported in *Hydra vulgaris* and the jellyfish *Scolionema suvaense* (Lindgens 2004; Aruga et al. 2006). Five *zic* homologs are present in the genome of the sea anemone *Nematostella vectensis*, and six *zic-like* genes exist in the genome of the coral *A. millepora* (Layden et al. 2010; Shinzato et al. 2011). However, the tandem arrangement of four of the five *zic* genes in the *Nematostella* genome and the fact that the two of six coral genes are pseudogenes suggest that there was likely an independent expansion of *zics* in the anthozoans (Shinzato et al. 2011; Aruga et al. 2006). These observations coupled with the fact that only a single *zic* homolog is found in the draft *Trichoplax* genome supports a single ancestral *zic* gene possessing ZOC and ZF-NC domains existed in the cnidarian-bilaterian common ancestor. Thus, comparisons between cnidarian and bilaterian *zic* gene expression and function can be used to infer the ancestral functions of early *zic* homologs.

2.3 The Cnidarian Body Plan

To better compare *zic* expression and potential functions, it is necessary to understand the relationship of cnidarian and bilaterian body plans. Bilaterian animals are characterized by the fact that they are bilaterally symmetric. The long anterior-posterior (A-P) axis is intersected by an orthogonally situated dorsal-ventral axis that bisects the A-P axis along the midline to create two mirror image left and right

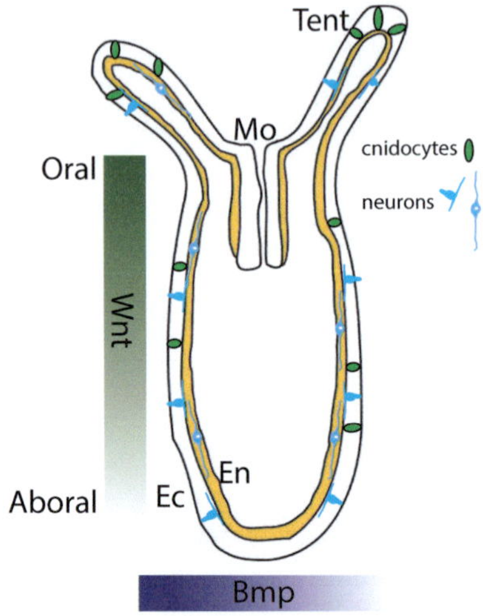

Fig. 2.2 Schematic of basic cnidarian body plan. Outline of basic cnidarian polyp body plan. Wnt gradients pattern the oral-aboral axis and BMP gradients pattern the directive axis. Differentiated cells such as neurons and cnidocytes are intermixed with other cell types. *Mo* Mouth, *Ec* ectoderm, *En* endoderm, *Tent.* tentacles

halves. The bilaterians are also triploblastic animals, which means that they possess three germ layers. The innermost layer is the endoderm, which gives rise to the gut and visceral organs. The middle mesodermal layer gives rise to the musculature, bone, blood, and connective tissues. The outermost layer, the ectoderm, gives rise to the skin, nervous system, sensory organs and in vertebrates the neural crest cells. The three germ layers segregate during gastrulation in the early embryo. Lastly, within the germ layers, discrete organs form, which effectively partition biological functions into specialized tissues.

Cnidarians have sessile polyp and free-swimming medusa body plans that share the same basic architecture. Here we focus on the polyp stage because it is believed to represent the ancestral cnidarian body plan, and no *zic* expression is reported in any medusa stage animals to date (Fig. 2.2). Polyps have a single oral opening to a saclike gut. The mouth is surrounded by tentacles, which are densely packed with cnidocyte stinging cells used to capture prey. Cnidocytes are a cnidarian-specific cell type with mechanosensory neuronal properties (Hausmann and Holstein 2004; Brinkmann et al. 1996). Opposite the oral end is the foot often referred to as the aboral pole. The oral-aboral axis is the long axis of the animal, and like anterior-posterior axis of bilaterian animals, it is patterned by a Wnt gradient (Leclère et al. 2016; Sinigaglia et al. 2013). Updated morphological descriptions now suggest that although some cnidarians are radially symmetric, many species are either bilaterally or biradially symmetric. The second body axis that breaks the perfect radial symmetry is termed the directive axis. Functional data suggests the conserved bilaterian dorsal ventral patterning genes chordin and BMP2/4 also pattern the directive axis in cnidarians (Saina et al. 2009; Leclère and Rentzsch 2014).

Cnidarians are diploblastic animals meaning that they only possess two germ layers. They generate an ectoderm (outer) and endoderm (inner) germ layer, which also segregate during gastrulation. The endoderm is a bifunctional gastrodermis sometimes referred to as the endomesoderm or mesendoderm due to the mixture of cell types that are typically restricted to either endoderm (absorptive digestive cells) or mesoderm (muscle-like) in bilaterians. However, cnidarians lack a true mesodermal germ layer. Differentiated cell types such as neural, secretory, myoepithelial, and absorptive cells are scattered among epithelial cells in both layers and for the most part are not partitioned into tissues with singular dedicated function.

2.4 *Hyzic* Expression in Cnidocyte Development in Adult *Hydra vulgaris* Polyps

Expression analysis of *Hyzic* in the *Hydra vulgaris* genome identified a role for *zic* homologs regulating cnidocyte development (Fig. 2.3). Work by Lindgens et al. determined that *Hyzic* is expressed in a scattered pattern throughout the adult body column in single and paired interstitial cells (i-cells) as well as in nematoblast nests (Lindgens et al. 2004). I-cells are a hydrozoan-specific stem cell that gives rise to multiple cell types in adult animals, but the majority of cells derived from i-cells are neurons and cnidocytes (Bode and David 1978). Nematoblasts are cnidocyte precursors that generate cnidocytes within the body column, which then migrate to their final positions in the body. Nematoblast nests are derived from a single nematoblast that proliferates into nests of 4, 8, 16, and 32 cells. Differentiation of the entire nest into mature cnidocytes can occur at any time (David and Challoner 1974). Greater than 90% of *Hyzic* expression in nematoblast nests is detected in 1-, 2-, 4-, and 8-cell nests indicating a role in a more upstream position of cnidocyte differentiation. This upstream role is further supported by the observation that *Hyzic* is co-expressed with the proliferative cell marker BrdU but is excluded from post-mitotic cnidocytes expressing either *Cnash* or *Nowa* (Lindgens et al. 2004). BrdU-positive cells could be co-labeled with *Nowa* and *Cnash* only after long exposures suggesting BrdU in *Nowa*- or *Cnash*-positive cells was inherited during division of a *Hyzic* expressing nematoblast precursor cell. Together these data all argue in favor of *Hyzic* acting to promote the formation of cnidocyte mechanosensory cells by acting in precursor cells upstream of *Cnash* and *Nowa*. Even though functional data does not accompany *Hyzic* expression in *Hydra*, the well-established stereotypy of cnidocyte development provides high confidence that *zic* plays a positive role in promoting proper development of the cnidocytes.

The expression *Hyzic* upstream *Cnash* in cnidocyte development hints at a conserved role for zic genes upstream of bHLH proneural genes during neurogenesis in the cnidarian-bilaterian ancestor. In bilaterians *zic* homologs have been shown to act upstream of bHLH proneural genes. *Cnash* is a bHLH proneural gene homolog of the *achaete-scute* gene family (Hayakawa et al. 2004). Together *achaete-scute* and *atonal* family bHLH proneural genes promote neurogenesis in bilaterian and cnidarian species (Bertrand et al. 2002; Layden et al. 2012; Richards and Rentzsch 2015). Regardless of whether *Hyzic* regulates *Cnash*, its expression in precursor cells is significant. The expression of vertebrate *zic* genes in the neural plate ectoderm, *Ref-*

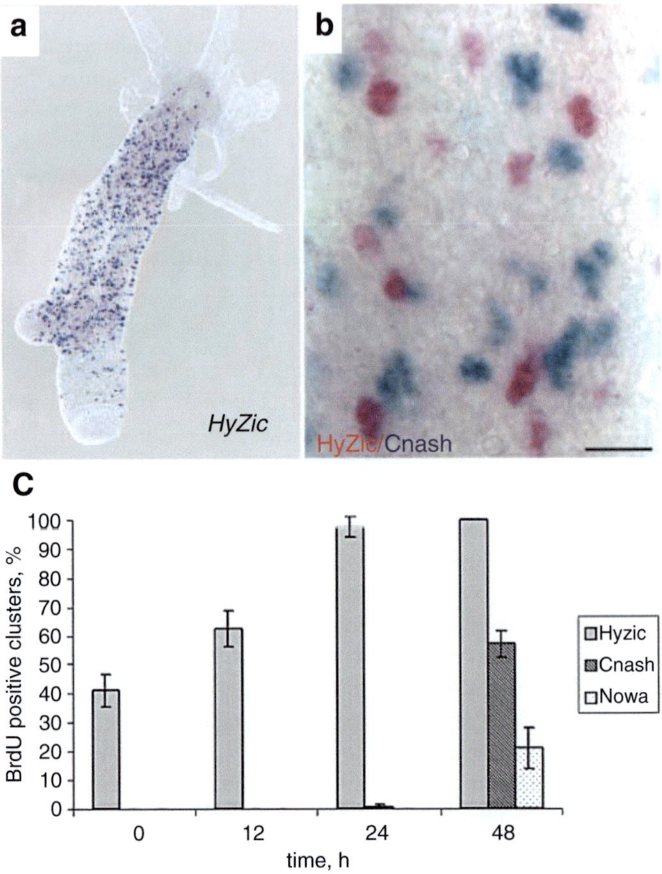

Fig. 2.3 *Hyzic* expression in *Hydra vulgaris*. *Hyzic* expression is expressed in a scattered pattern in adult polyps (**a**). *Hyzic* (red) and *Cnash* (blue) do not co-express in adult polyps (**b**). *Hyzic* co-expresses with BrdU in short and long pulses, but *Cnash* and *Nowa* require extended pulses of BrdU to be labeled (**c**) (This Figure is adapted from Figures in Lindgens et al. (2004) with permission)

2 in *C. elegans* Pn.p. derivatives, *opa* in *Drosophila* neural ectoderm, *AmphiZic* in the presumptive neural plate, *Ct-zic* in *C. teleta* anterior neural ectoderm, and *Hr-ZicN* in all *H. roretzi* neural tube precursors, suggests expression and function of *zic* homologs in neural precursors are widespread throughout the bilaterians (Layden et al. 2010; Wada and Saiga 2002; Alper and Kenyon 2002; Gostling and Shimeld 2003; Cimbora and Sakonju 1995). Together with *Hyzic* expression, it argues that the ancestral *zic* function in neurogenesis is linked to neural precursor biology. It is still unclear if *Hyzic* expression in i-cells regulates development of other neuronal cells in *Hydra*. Minimally, the work by Lindgens et al. demonstrates a role for a cnidarian *zic* homolog in nematoblast neuronal precursor cells and in specification of the cnidocyte neural subtype. Functional studies linking neurogenic *zic* expres-

sion to downstream targets are the next necessary step to better determine degree to which the neurogenic role for cnidarian and bilaterian *zic* homologs is conserved.

2.5 Cnidarian Development

Expression of *zic* homologs during cnidarian development has been best addressed in the anthozoan sea anemone *Nematostella vectensis* (Layden et al. 2010). Here we describe development of the sea anemone *Nematostella vectensis* (Fig. 2.4), because it is the only species whose developmental expression of *zic* genes is known, and *Nematostella* development is similar to development of other cnidarian species.

Nematostella undergoes irregular cleavage prior to forming a blastula. Gastrulation occurs at the oral pole, and the site of gastrulation forms the future mouth of the polyp. The nervous system of *Nematostella* initially forms in the embryonic ectoderm at approximately mid-blastula stage (Richards and Rentzsch 2015). Neural precursors are scattered throughout the body column and intermixed with other cell types. At the beginning of the planula stage, the apical sensory organ

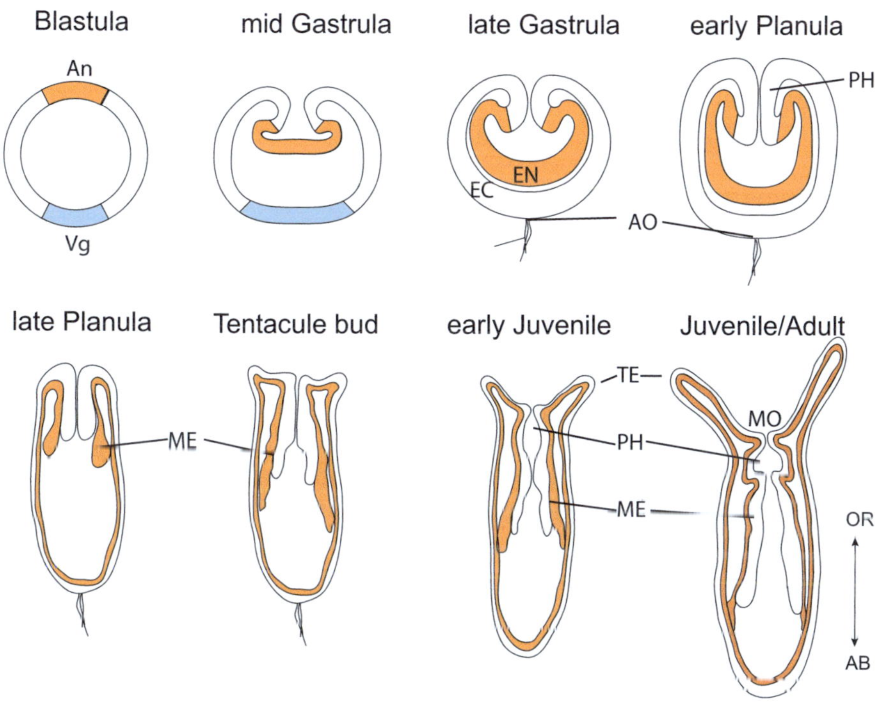

Fig. 2.4 Schematic of *Nematostella* developmental stages. Developmental stages of *Nematostella vectensis*. *An* Animal pole, *Vg* vegetal pole, *Ec* ectoderm, *En* endoderm, *PH* pharynx, *AO* apical organ, *ME* mesentery, *TE* tentacle, *Or* oral, and *Ab* aboral

situated at the aboral pole is present. *Nematostella* planula maturation occurs over roughly 3 days. During this time, overt morphological complexity does not dramatically increase, but numerous structures are forming. The pharynx forms from the ectoderm and folds in to form the mouth, and two primary mesenteries form off of the aboral end of the pharynx, followed by six more mesenteries, which are more or less distributed in a radially symmetric pattern. The musculature begins to form, endodermal neurogenesis initiates, and the nervous system continues to mature (Layden et al. 2016b). The planula larval stages end at the tentacle bud stage. Species-dependent cues thought to be mediated by the apical organ then drive settlement and metamorphosis into a juvenile polyp. During metamorphosis the tentacle buds grow into mature tentacles and the body column elongates. The juvenile polyp continues to add tentacles (16 in *Nematostella)* and grows until sexual maturity.

2.6 *Nvzic* Expression During Development of *Nematostella vectensis*

Developmental *NvzicC-NvzicE* expression at larval stages is consistent with roles in neurogenesis and patterning the forming tentacles (Fig. 2.5) (Layden et al. 2010). *NvzicE* expression is the first *zic* gene detected by mRNA in situ hybridization in *Nematostella. NvzicE* is initially expressed throughout the ectoderm in late gastrula stages. The failure to detect any *Nvzic* homologs prior to late gastrula implies that they are not involved in the earliest steps of embryonic neurogenesis.

NvzicE expression is maintained throughout the ectoderm in early planula stages, but becomes enriched in the presumptive tentacular domain and in the apical sensory organ ectoderm by late planula. Aboral *NvzicE* expression is lost as the apical organ dissolves, but tentacular expression is maintained in the tentacle bud stage. *NvzicC* and *NvzicD* expression is first detected in overlapping domains with *NvzicE* in the presumptive tentacular forming regions of early planulae. *NvzicD* expression is limited to only the tentacular ectoderm, whereas *NvzicC* is expressed in the tentacular endoderm and ectoderm, and it is expressed in the presumptive tentacles as well as the presumptive intertentacular space. *NvzicC* and *NvzicD* expression is maintained in the tentacular region throughout larval and tentacular bud stages. *NvzicD* expression is also detected in individual cells in the larval ectoderm and endoderm throughout planula stages. The exact role of *Nvzic* genes in tentacle development is not clear, but many genes regulated by *zic* in other species are highly enriched in the forming tentacles. For example, *NvashA* a *Nematostella* homolog of the *achaete-scute* family that includes *Cnash* is expressed in the forming tentacle ectoderm at the bud stage (Layden et al. 2012). *NvashA* has been linked directly to neurogenesis, but not cnidocyte development. Additionally, neurogenic expression of *Nvelav1*, *Nvmushashi*, and the cnidocyte markers *Nvminicoll–4* all show strong enrichment in presumptive tentacle buds during late planula stages and tentacle bud stage (Marlow et al. 2009; Zenkert et al. 2011). The early *NvzicC-E* expression in

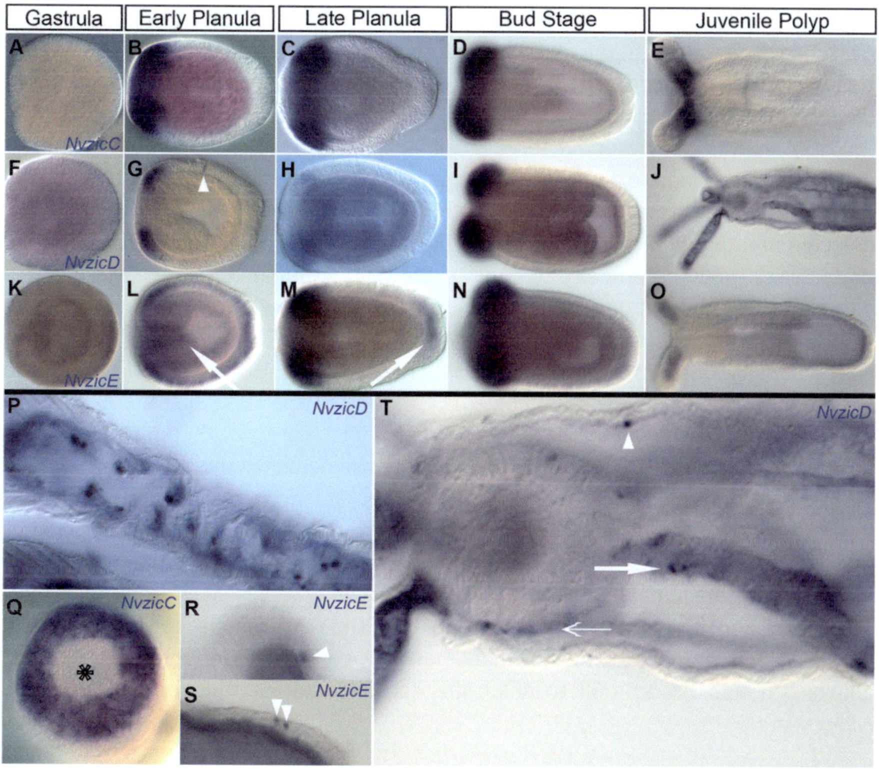

Fig. 2.5 *Nvzic* expression in *Nematostella vectensis.* This figure was originally published in Layden et al. 2010. Expression of *NvzicC* (A-E, Q), *NvzicD* (F-J, P, S, T), and *NvzicE* (K-O, R). Late gastrula (A,F,K), early (B,G,L) and late (C,H,M,Q) planula, bud (D,I,N), and polyp (E, J, O, R, S, T) stages are shown. All images are lateral views with oral to the left except Q, which is an oral view with mouth indicated by an asterisk. The endoderm is indicated by en, ectoderm by ec, pharyngeal endoderm by phen, and mesentery endoderm by msen. *NvzicC* is expressed in presumptive tentacle and tentacle bud ectoderm and endoderm (B, C, and D) but not in oral ectoderm (Q) in planula stages. The polyp *NvzicC* expression is in the tentacle proximal to the polyp body, but not in distal regions (E). *NvzicD* is expressed in the presumptive tentacle ectoderm (G and H) and individual ectodermal cells (G, arrowhead) of the planula. *NvzicD* is also expressed in tentacle buds (I), tentacular endoderm (J and P), pharyngeal endoderm (T, open arrow), and in distinct cells in the endodermal component of the directive mesenteries (T, closed arrow) in the polyps. The polyp *NvzicD* expression is occasionally observed in the endodermal cells (T, arrowhead) and the ectodermal cells (S, arrowheads) of the polyp. *NvzicE* expression is expressed in the aboral pole of the gastrula embryo (K, arrow). *NvzicE* is expressed in the oral ectoderm (L, arrow), presumptive tentacle (L and M) and the apical tuft (M, arrow), the planula and tentacle buds (N), tentacle endoderm (O), and occasional individual ectodermal cells in the tentacles (R, arrowhead)

the presumptive tentacle region precedes the neuronal gene expression suggesting that *NvashA*, cnidocyte specification, and other neural genes are good candidates to be regulated by one or more *Nvzic* genes during tentacle formation. *NvzicE* expression in the forming apical tuft sensory organ also hints at an upstream role for *Nvzic* genes during neurogenesis within this structure. However, it should be noted that in the sensory organ and in the tentacles, neurons are intermixed with nonneural cells. It is likely that any neurogenic role for *Nvzics* is linked to larger role in formation of the tentacles and the apical organ and not solely neuronal in *Nematostella*.

Nvzic expression observed throughout development is downregulated in their developmental spatial expression domains, with the exception of *NvZicC*, which is maintained in a stripe at the base of the tentacles. *NvzicD* expression becomes detected in individual cells in the tentacular endoderm, within stripes of cells in the mesenteries, and sporadically in cells of the body column ectoderm. Mesenteries contain multiple cell types including neurons, cnidocytes, and gametes, and they also have cnidoglandular tracts that give rise to motile collections of cnidocytes called nematosomes (Williams 1975, 1979; Tucker et al. 2011). Based on the role for *Hyzic* in cnidocyte development, it is tempting to speculate that the expression of *NvzicD* in tracts down the mesenteries represents expression in the cnidoglandular tracts that are giving rise to the cnidocyte dense nematosomes.

The potential role for *NvzicC* in the tentacular endoderm is unclear. Unlike bilaterian animals, *Nematostella* forms neurons from endodermal tissue, suggesting that *NvzicC* might control neural development within tentacular endoderm. Alternatively, the tentacular expression of *NvzicC* may represent early mesodermal-like expression. The tentacles are capable of multiple movements and consequently contain a fair number of myoepithelial muscle cells. Without further functional analysis, it will be difficult to determine the endodermal function of *NvzicC* in *Nematostella*.

2.7 Comparing *Hyzic* and *NvzicA-E* Expression Suggests an Ancestral Role for *Zic* Homologs Was Conserved Neural Role for *Zics* in Cnidarians

Cnidarian *zic* genes have been studied in the medusozoan *Hydra vulgaris* and the anthozoan *Nematostella vectensis* providing at least one example of *zic* homolog expression in the two main cnidarian clades. Based on expression patterns in both species, it is likely that *zic* genes play a role in cnidarian neurogenesis. Considering that bilaterian *zic* are widely reported to regulate neural development, the cnidarian expression implies neurogenic functions represent a conserved ancestral role for *zic* family genes. The neurogenic programs are conserved between cnidarians and bilaterians supporting the idea that *zic* homologs were incorporated into basic neural programs present in the cnidarian-bilaterian common ancestor (Layden and Martindale 2014; Layden et al. 2016a, 2012; Richards and Rentzsch 2014, 2015; Rentzsch et al. 2016; Watanabe et al. 2014; Hayakawa et al. 2004; Nakanishi et al. 2012). What is unclear is whether the neurogenic role for *zic* predated or evolved in the cnidarian-bilaterian ancestor. *Trichoplax adhaerens* does not possess any neurons, but the parahoxozoan common ancestor may have had a nervous system that was lost in the placozoan lineage. Ctenophores possess neurons, and sponges have sensory-like cells with neuronal properties (Ludeman et al. 2014) arguing that a primitive neural-like cell may have been present in the parahoxozoan. Determining if roles in neurogenesis or neuronal-like cells represent the earliest functions for *zic* genes will require characterization of *Tad-zic*.

Additional studies of developmental stages in cnidarians are necessary to better determine if the tentacular endodermal expression of *NvzicC* represents a conserved or *Nematostella* specific expression. There has been a recent expansion in the number of viable cnidarian models. Two species of particular interest are the hydrozoans *Clytia hemisphaerica* and *Hydractinia echinata* (Houliston et al. 2015; Plickert et al. 2012). Obtaining embryos to analyze *zic* expression during development of these two species will provide much needed insight about the cnidarian *zic* endodermal expression.

2.8 Concluding Remarks

The expression studies initiated in the two cnidarians represent an excellent start to begin to understand the role of ancestral *zic* genes in animals. However, functional studies are necessary to establish definitive roles for *Hyzic* and *Nvzic* genes. Additionally, with functional studies and modern molecular approaches now possible in cnidarians, it will be possible to determine the molecular basis of cnidarian *zic* phenotypes. Comparisons of the downstream *zic* targets and identification of molecules that interact with cnidarian *zic* genes will provide key evidence to determine the degree to which *zic* genes function through conserved pathways in cnidarians and bilaterians. Lastly, efforts to better investigate the function and expression of *Tad-zic* will provide a glimpse into the roles the original ancestral *zic* gene may have played in animal evolution.

References

Alper S, Kenyon C (2002) The zinc finger protein REF-2 functions with the Hox genes to inhibit cell fusion in the ventral epidermis of C. elegans. Development 129:335–3348

Aruga J (2004) The role of Zic genes in neural development. Mol Cell Neurosci 26(2):205–221

Aruga J et al (2006) A wide-range phylogenetic analysis of Zic proteins: implications for correlations between protein structure conservation and body plan complexity. Genomics 87(6):783–792

Benedyk MJ, Mullen JR, DiNardo S (1994) Odd-paired: a zinc finger pair-rule protein required for the timely activation of engrailed and wingless in Drosophila embryos. Genes Dev 8(1):105–117

Bertrand N, Castro DS, Guillemot F (2002) Proneural genes and the specification of neural cell types. Nat Rev Neurosci 3(7):517–530

Bode HR, David CN (1978) Regulation of a multipotent stem cell, the interstitial cell of Hydra. Prog Biophys Mol Biol 33:189–206

Brinkmann M, Oliver D, Thurm U (1996) Mechanoelectric transduction in nematocytes of a hydropolyp (Corynidae). J Comp Phys 178:125–138
Cimbora DM, Sakonju S (1995) *Drosophila* midgut morphogenesis requires the function of the segmentation gene *odd-paired*. Dev Biol 169:580–595
Darling JA et al (2005) Rising starlet: the starlet sea anemone, *Nematostella vectensis*. BioEssays 27(2):211–221
David CN, Challoner D (1974) Distribution of interstitial cells and differentiating Nematocytes in nests in Hydra attenuata. Am Zool 14:537–542
Dunn CW et al (2008) Broad phylogenomic sampling improves resolution of the animal tree of life. Nature 452(7188):745–749
Elms P et al (2003) Zic2 is required for neural crest formation and hindbrain patterning during mouse development. Dev Biol 264(2):391–406
Gostling NJ, Shimeld SM (2003) Protochordate Zic genes define primitive somite compartments and highlight molecular changes underlying neural crest evolution. Evol Dev 5(2):136–144
Gyoja F (2006) Expression of a muscle determinant gene, macho-1, in the anural ascidian Molgula tectiformis. Dev Genes Evol 216(5):285–289
Hausmann K, Holstein TW (2004) Sensory receptor with bilateral symmetrical polarity. Naturwissenschaften 72:145–146
Hayakawa E, Fujisawa C, Fujisawa T (2004) Involvement of Hydra achaete-scute gene CnASH in the differentiation pathway of sensory neurons in the tentacles. Dev Genes Evol 214(10):486–492
Hejnol A et al (2009) Assessing the root of bilaterian animals with scalable phylogenomic methods. Proc R Soc B Biol Sci 276(1677):4261–4270
Houliston E, Momose T, Manuel M (2015) Clytia hemisphaerica: a jellyfish cousin joins the laboratory. Trends Genet 26(4):159–167
Inoue T et al (2007) Zic2 and Zic3 synergistically control neurulation and segmentation of paraxial mesoderm in mouse embryo. Dev Biol 306(2):669–684
Layden MJ, Martindale MQ (2014) Non-canonical Notch signaling represents an ancestral mechanism to regulate neural differentiation. EvoDevo 5(1):1–14
Layden MJ et al (2010) Expression and phylogenetic analysis of the zic gene family in the evolution and development of metazoans. EvoDevo 1(1):12. Available at:
Layden MJ, Boekhout M, Martindale MQ (2012) *Nematostella vectensis* achaete-scute homolog NvashA regulates embryonic ectodermal neurogenesis and represents an ancient component of the metazoan neural specification pathway. Development 139(5):1013–1022. Available at:
Layden MJ, Johnston H, et al (2016a) MAPK signaling is necessary for neurogenesis in Nematostella vectensis. BMC Biol 1–19
Layden MJ, Rentzsch F, Röttinger E (2016b) The rise of the starlet sea anemone Nematostella vectensis as a model system to investigate development and regeneration. Wiley Interdiscip Rev: Dev Biol 1–21
Leclère L Rentzsch F (2014) RGM regulates BMP-mediated secondary axis formation in the sea anemone *Nematostella vectensis*. *CellReports*, pp 1–11
Leclère L et al (2016) Development of the aboral domain in Nematostella requires β-catenin and the opposing activities of Six3/6 and Frizzled5/8. Development 143(10):1766–1777
Lindgens D, Holstein TW, Technau U (2004) Hyzic, the Hydra homolog of the zic/odd-paired gene, is involved in the early specification of the sensory nematocytes. Development 131(1):191–201
Ludeman DA et al (2014) Evolutionary origins of sensation in metazoans: functional evidence for a new sensory organ in sponges. BMC Evol Biol 14:1–11
Marlow HQ et al (2009) Anatomy and development of the nervous system of Nematostella vectensis, an anthozoan cnidarian. Dev Neurobiol 69(4):235–254
Merzdorf CS (2007) Emerging roles forzic genes in early development. Dev Dyn 236(4):922–940
Mizuseki K et al (1998) Xenopus Zic-related-1 and Sox-2, two factors induced by chordin, have distinct activities in the initiation of neural induction. Development 125:579–587

Nagai T et al (1997) The expression of the mouse Zic1, Zic2, and Zic3 gene suggests an essential role for Zic Genesin body pattern formation. Dev Biol 182:299–313
Nakanishi N et al (2012) Nervous systems of the sea anemone Nematostella vectensis are generated by ectoderm and endoderm and shaped by distinct mechanisms. Development 139(2):347–357
Nakata K et al (2000) A novel member of the Xenopus Zic family, Zic5, mediates neural crest development. Mech Dev 99:83–91
Plickert G, Frank U, Müller WA (2012) Hydractinia, a pioneering model for stem cell biology and reprogramming somatic cells to pluripotency. Int J Dev Biol 56(6-7-8):519–534
Rentzsch F, Layden M, Manuel M (2016) The cellular and molecular basis of cnidarian neurogenesis. Wiley Interdiscip Rev: Dev Biol 1–19
Richards GS, Rentzsch F (2014) Transgenic analysis of a SoxB gene reveals neural progenitor cells in the cnidarian Nematostella vectensis. Development 141(24):4681–4689
Richards GS, Rentzsch F (2015) Regulation of Nematostella neural progenitors by SoxB, Notch and bHLH genes. Development 142(19):3332–3342
Saina M et al (2009) BMPs and chordin regulate patterning of the directive axis in a sea anemone. Proc Natl Acad Sci U S A 106(44):18592–18597
Shinzato C et al (2011) Using the Acropora digitifera genome to understand coral responses to environmental change. Nature 476(7360):320–323
Sinigaglia C et al (2013) The bilaterian head patterning gene six3/6 controls aboral domain development in a cnidarian. PLoS Biol 11(2):e1001488
Srivastava M et al (2008) The Trichoplax genome and the nature of placozoans. Nature 454(7207):955–960
Takahashi H, Shimizu T, Aruga J (2008) Expression pattern of annelid Zic in embryonic development of the oligochaete Tubifex tubifex. Dev Genes Evol 218(10):553–560
Tucker RP, Shibata B, Blankenship TN (2011) Ultrastructure of the mesoglea of the sea anemone Nematostella vectensis (Edwardsiidae). Invertebr Biol 130(1):11–24
Wada S, Saiga H (2002) HrzicN, a new Zic family gene of ascidians, plays essential roles in the neural tube and notochord development. Development 129(24):5597–5608
Watanabe H et al (2014) Sequential actions of β-catenin and Bmp pattern the oral nerve net in Nematostella vectensis. Nat Commun 5:5536–5514
Williams RB (1975) A redescription of the brackish-water sea anemone *Nematostella vectensis* Stephenson, with an appraisal of congeneric species. J Nat Hist 9(1):51–64
Williams RB (1979) Studies on the nematosomes of *Nematostella vectensis* Stephenson. J Nat Hist 13(1):69–80
Zenkert C et al (2011) Morphological and molecular analysis of the *Nematostella vectensis* Cnidom. Dearden PK (ed). PLoS ONE 6(7):e22725

Chapter 3
Odd-Paired: The *Drosophila* Zic *Gene*

Deborah A. Hursh and Brian G. Stultz

Abstract Zinc finger in the cerebellum (Zic) proteins are a family of transcription factors with multiple roles during development, particularly in neural tissues. The founding member of the Zic family is the *Drosophila odd-paired* (*opa*) gene. The Opa protein has a DNA binding domain containing five Cys2His2-type zinc fingers and has been shown to act as a sequence-specific DNA binding protein. Opa has significant homology to mammalian Zic1, Zic2, and Zic3 within the zinc finger domain and in two other conserved regions outside that domain. *opa* was initially identified as a pair-rule gene, part of the hierarchy of genes that establish the segmental body plan of the early *Drosophila* embryo. However, its wide expression pattern during embryogenesis indicates it plays additional roles. Embryos deficient in *opa* die before hatching with aberrant segmentation but also with defects in larval midgut formation. Post-embryonically, *opa* plays important roles in adult head development and circadian rhythm. Based on extensive neural expression, *opa* is predicted to be involved in many aspects of neural development and behavior, like other proteins of the Zic family. Consensus DNA binding sites have been identified for Opa and have been shown to activate transcription in vivo. However, there is evidence Opa may serve as a transcriptional regulator in the absence of direct DNA binding, as has been seen for other Zic proteins.

Keywords *odd-paired* · *opa* · Zic · *Drosophila* · Segmentation · Visceral mesoderm · Head development · Zinc finger transcription factor

3.1 Introduction

The *Drosophila* pair-rule gene *odd-paired* (*opa*) is the founding member of the Zic family. *opa* was identified as an essential gene required for correct anterior posterior patterning in the groundbreaking genetic screen for mutations affecting the *Drosophila* body plan carried out by Nusslein-Vollhard, Weischaus, and colleagues

D. A. Hursh (✉) · B. G. Stultz
Division of Cell and Gene Therapy, Center for Biologics Evaluation and Research, Food and Drug Administration, Silver Spring, MD, USA
e-mail: deborah.hursh@fda.hhs.gov

J. Aruga (ed.), *Zic family*, Advances in Experimental Medicine and Biology 1046,
https://doi.org/10.1007/978-981-10-7311-3_3

(Jurgens et al. 1984). It was designated a "pair-rule" gene because embryos lacking *opa* die during embryogenesis with cuticular pattern defects in every other segment. *opa* forms part of a larger class of pair-rule genes that includes the helix-loop-helix transcription factor *hairy*, the homeodomain transcription factors *even-skipped* (*eve*) and *fushi tarazu* (*ftz*), and the founding member of the Runx family of transcription factors, *runt* (*run*). Pair-rule genes are part of a hierarchy of zygotic genes that create the basic metameric body plan of the *Drosophila* embryo.

3.2 Opa Protein Structure

opa's predicted protein sequence contains five zinc fingers and, at the time the gene was cloned, its closest homology was to other proteins containing similar zinc finger domains such as members of the Krüppel and Gli families (Benedyk et al. 1994; Cimbora and Sakonju 1995). A more closely related set of vertebrate homologs was later identified whose founding members have expression in the neural tube and cerebellum and were thus given the name zinc finger protein of the cerebellum (Zic). Both the protein structure and conservation of genomic organization indicate that *opa* and the Zic proteins derive from a common ancestor (Aruga et al. 1996). There is only one Zic homolog in *Drosophila*, so *opa* presumably serves all Zic functions in flies.

Proteins in the Zic family all contain a conserved zinc finger domain made up of five Cys2His2-type zinc fingers. These zinc fingers are highly conserved between Opa and the five vertebrate Zic family members, but the closest conservation is with the vertebrate Zic1, Zic2, and Zic3 proteins (Aruga et al. 2006). The zinc finger region is also highly similar to that of the Gli family, transcription factors of the Hedgehog signaling pathway. The zinc finger region serves as a sequence-specific DNA binding domain (Pavletich and Pabo 1993; Mizugishi et al. 2001; Sen et al. 2010). Outside the zinc finger domain, two other regions of conservation exist. The first is an N-terminal flanking region, ZF-NC, and the second is the Zic-Opa conserved or ZOC domain (Fig. 3.1a). The function of the ZF-NC is unknown, but the ZOC domain has been proposed to be involved in transcriptional activation (Mizugishi et al. 2004).

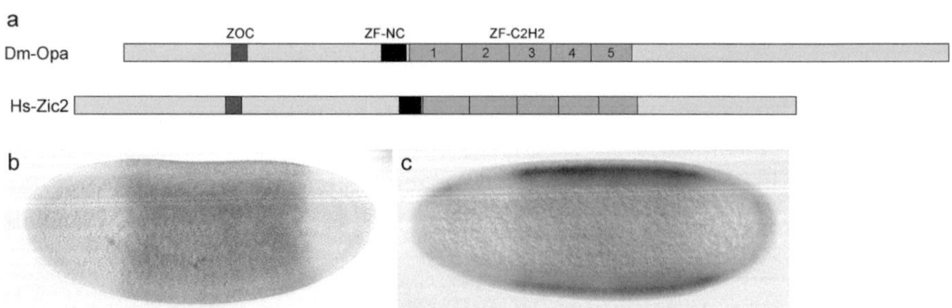

Fig. 3.1 Protein domains are conserved between *Drosophila* Opa and Human Zic. (**a**) A diagram of the two proteins shows the relative locations of the conserved ZOC, ZF-NC, and five zinc finger (ZF-C2H2) domains. Stage 5 *Drosophila* embryos show the early expression pattern of *opa* RNA by hybridization in situ (**b**) or protein by immunohistochemistry (**c**). Embryos are oriented with anterior to the left

3.3 *opa* Gene Structure and Expression Pattern

The *opa* transcription unit spans 17.2 kb of genomic sequence, including three introns, one that is 14 kb and two that are very small (*D. melanogaster* release 6.13; Fig. 3.2a) (Benedyk et al. 1994). There is conservation of several intron-exon boundaries between *opa* and vertebrate Zic1, Zic2, and Zic3 genes (Aruga et al. 1996), supporting their evolutionary relationship. *Drosophila* genome sequence data indicates that the *opa* gene product is a single protein made from a single RNA transcript. *opa* RNA levels are at their highest in the early embryo, consistent with *opa*'s role in segmentation (Benedyk et al. 1994; Cimbora and Sakonju 1995), but RNA-seq analysis by the modENCODE project reports expression from the *opa* gene in all embryonic, larval, and adult stages tested (Graveley et al. 2011). By in situ analysis, RNA and protein expression are seen uniformly in ectoderm and presumptive mesoderm throughout the entire segmented region of the early stage 5 embryo, when cellularization of the early embryonic syncytium occurs (Fig. 3.1b, c). During gastrulation, this expression becomes weakly periodic. *opa's* expression is distinct from other members of the pair-rule family, all of whom have strong repetitive striped expression patterns that align with segmental boundaries. *opa*

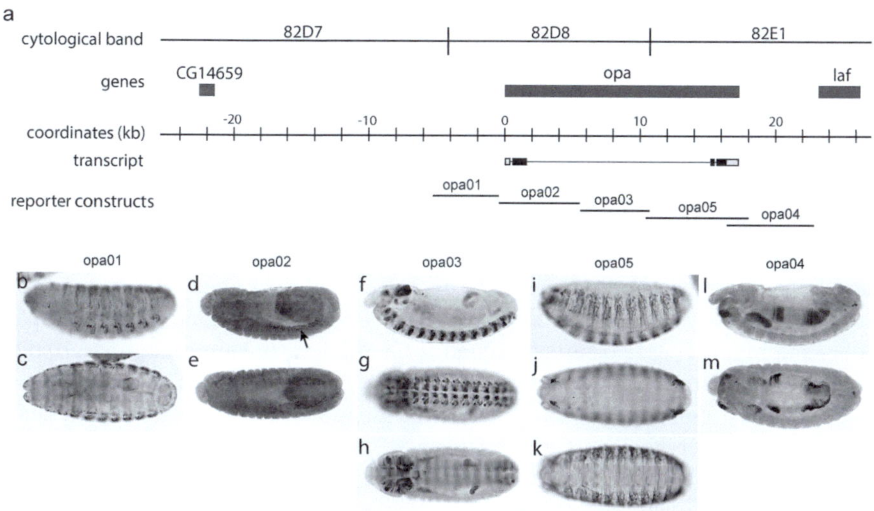

Fig. 3.2 Diagram of the *opa* genomic region. (**a**) A 50 kb region of the genome, with coordinates relative to the transcription start site of *opa*, shows the two closest gene neighbors and the location of regions used to make five *β-galactosidase* reporter constructs (opa01-opa05). Solid areas in the schematic of the transcript indicate the coding region. (**b–m**) The five reporter constructs have tissue-specific embryonic *lac-Z* expression at various stages of development. (**b, c**) opa01 at stage 16: muscle attachment sites and midgut. (**d–e**) opa02 at stage 12: faint mesoderm expression, as indicated by arrow. (**f–h**) opa03 at stage 14; CNS and Malpighian tubules. (**i–k**) opa05 at stage 16: muscle attachment sites. (**l–m**) opa04 at stage 14: visceral mesoderm and salivary glands. Panels **b**, **d**, **I**, **f**, and **l** are lateral views, dorsal up. Panels **c**, **e**, **g**, **j**, and **m** are ventral views. Panels **h** and **k** are dorsal views. All embryos are oriented with anterior to the left

expression disappears from the presumptive mesoderm and ectoderm by stage 9 and initiates in a spatially specific manner in the developing visceral mesoderm surrounding the developing gut at stage 11 (Cimbora and Sakonju 1995). Data from the large-scale RNA in situ embryonic expression database maintained by the Berkeley *Drosophila* Genome Project indicate that *opa* mRNA can be found later in development in the embryonic head, particularly in sensory structures and the visual primordium. Expression is also reported in the embryonic central nervous system (CNS) (Tomancak et al. 2002, 2007). Tissue-specific RNA expression levels have been measured using array-based methods: FlyAtlas reports tissue-specific *opa* mRNA expression in adult brain, head, thoracicoabdominal ganglion, and midgut tissues and during the third instar larval stage in the CNS and midgut (Chintapalli et al. 2007). Whole transcriptome RNA-seq analysis of isolated neuroblasts found enrichment of *opa* transcripts in larval cerebral mushroom bodies, which are insect brain structures involved in learning and memory (Yang et al. 2016). We examined *opa* RNA expression in situ in the larval eye-antennal disc and found expression in an area that forms the first antennal segment and also in the squamous peripodial epithelium (Lee et al. 2007). Limited expression was also seen in other imaginal discs at third instar, in particular in the primordia of the mesopleura and both the dorsal and ventral hinge of the wing disc (Hursh, unpublished observations) (Butler et al. 2003).

Based on the positions of the nearest genes, the *opa* gene could occupy a region as large as 44 kb. We created five large abutting *β-galactosidase* (*lac-Z*) reporter constructs starting at approximately 5 kb upstream and extending 5 kb downstream of the *opa* transcript region (Fig. 3.2a). Three of these have significant expression in the eye-antennal disc, which will form the adult eye, and most of the external head structures (Lee et al. 2007). All the constructs have embryonic expression (Fig. 3.2b–m) that largely agrees with RNA in situ data for *opa*. A collection of *Gal4* activating transgenic constructs containing DNA from the flanking upstream, downstream, and large intronic region of *opa* drive expression in embryonic, larval, and adult nervous system, in particular the subesophageal neurons, mushroom bodies, optic lobe, and brain. These constructs extend from 20 kb upstream of the transcription start site of *opa* to 5 kb downstream (FlyLight project) (Pfeiffer et al. 2008). Binding sites for many transcription factors have been identified over this region using chromatin immunoprecipitation (ChIP) by the modENCODE project (Negre et al. 2011). In aggregate, these data suggest that *opa* is a large gene with a complex cis-regulatory structure.

3.4 Role of *opa* During Development

Genetic screens have identified *opa* as having various potential roles in *Drosophila* development. *opa* was recovered in a screen for genes controlling germ cell migration and gonad formation (Moore et al. 1998). However, it is not clear that this is a direct role or secondary to *opa's* function as a pair-rule gene. *opa* was also recovered in a screen for *Notch* pathway modifiers through a genetic interaction with *Presenilin* (*Psn*) mutations (Mahoney et al. 2006), but the mechanism by which *opa* interacts with *Psn* is unknown. Only a few biological roles of *opa* have been subject to detailed analysis by genetic or molecular methods. These are summarized below.

3.4.1 Embryonic Segmentation

As predicted by its recovery in a screen for essential genes specifying anterior-posterior structure of the embryo, *opa* plays a role in the process of segmentation. Unlike in vertebrates, where segmentation is a sequential operation, with somites being added one by one, the segmental repeats of *Drosophila* embryos are formed simultaneously. This metameric pattern is set up early in development by the action of a hierarchy of zygotic gene activities that divides the anterior-posterior axis of the embryo into progressively smaller domains. Gap genes interpret patterns set down maternally and divide the embryo into large domains. These domains are read by the second group of genes, the pair-rule genes, which specify the periodic repeats of the thoracic and abdominal portions of the body plan. This pattern is further refined by segment polarity genes, which define the boundaries and polarity of given segmental elements (Fig. 3.3).

opa is a member of the intermediate group of the hierarchy, the pair-rule genes. Most of the pair-rule genes are transcription factors. Combinatorial expression pat-

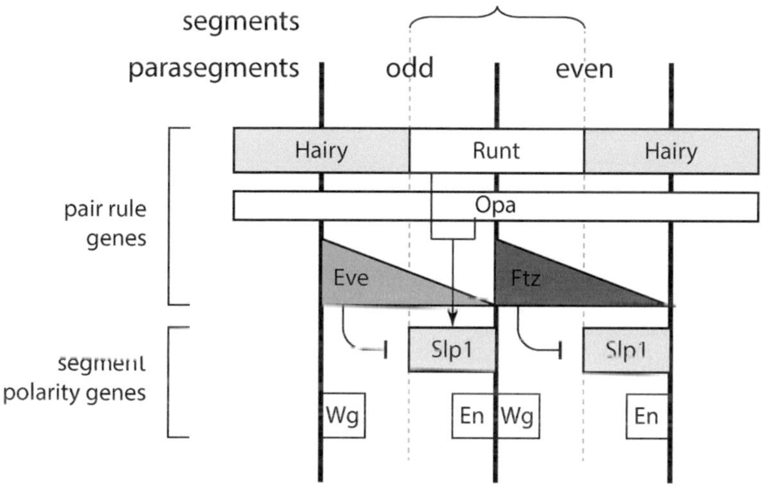

Fig. 3.3 *opa* in conjunction with other pair-rule genes regulates segment polarity genes. *Opa* and *run* induce the expression of *slp1* in odd segments, while *eve* and *ftz* negatively regulate *slp1*. *opa* is also required for *wg* and *en* expression; however whether this is a direct interaction is unclear (Modified from Swantek and Gergen 2004). Bracket indicates segmental register, solid lines identify parasegments

terns of these transcriptional activators and repressors establish the repeating segmental pattern of the next level in the hierarchy, the segment polarity genes (Swantek and Gergen 2004; Schroeder et al. 2011; Jaynes and Fujioka 2004). Some of this expression follows a repeating organization that is offset from the metameric body segments of the embryo; these units are referred to as parasegments (PS). *opa* is genetically required for the expression of *wingless* (*wg*), the major *Drosophila* Wnt homolog, and the homeodomain transcription factor, *engrailed* (*en*); both these genes act as critical segment polarity genes (Benedyk et al. 1994). It is not clear whether this is a direct effect of *opa* or part of the complex gene regulatory network that mediates the segmentation gene network. However, in concert with the pair-rule gene *run*, *opa* activates yet another pair-rule gene, the forkhead domain transcription factor *sloppy paired 1* (*slp1*) (Fig. 3.3), indicating its direct transcriptional role in the segmentation process (Swantek and Gergen 2004; Sen et al. 2010).

Interestingly, all the identified pair-rule genes, with the exception of *opa*, are initially expressed in seven evenly spaced but phase-shifted stripes. The overlap between the stripes sets up the unique transcriptional code that specifies the 14 segments of the larva (Schroeder et al. 2011). However, the expression pattern of pair-rule genes is both dynamic and transitory. Recent work has finely dissected the phases of pair-rule gene expression and suggests that as the only pair-rule gene not expressed in stripes, *opa* plays a unique role in regulating a shift in the frequency of stripe patterns of pair-rule genes. This shift causes a concomitant shift in regulatory interactions that set up the activation of segment polarity genes (Clark and Akam 2016). This study notes that other pair-rule genes have small transcription units (<3.5 kb), while *opa's* transcription unit is much larger, due to the presence of the 14 kb intron. This could result in the *opa* protein being fully translated later than other pair-rule genes, allowing it to serve as a timer for this shift. It is clear from these analyses that *opa's* role in embryonic segmentation is critical, and has yet to be fully elucidated.

3.4.2 Embryonic Midgut Formation

The internal mesodermal and endodermal structures of the *Drosophila* embryo are formed after gastrulation and the elongation and subsequent retraction of the germ band. The embryonic midgut is formed when a continuous sheet of visceral musculature encircles migrating endoderm. This visceral musculature is responsible for the formation of three constrictions around the endoderm at invariant locations, resulting in the stereotypical four-lobed midgut structure. Four specialized outgrowths called gastric caeca evaginate from the anterior-most portion of this gut structure.

opa is expressed in founder cells of the midgut visceral mesoderm starting at stage 11 (Cimbora and Sakonju 1995; Bilder and Scott 1998; Reim et al. 2017). Genes expressed in the midgut largely follow a parasegmental pattern of expression. *opa* expression is seen in PS 3–5 (weakly in PS 3) and PS 9–12 (Fig. 3.4a) and per-

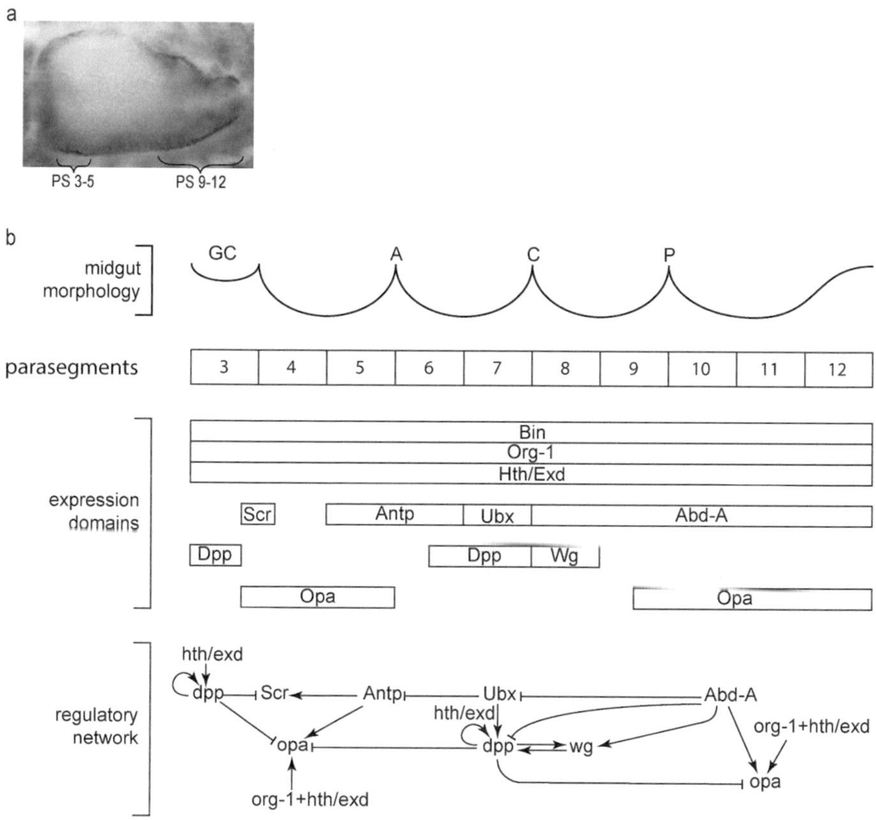

Fig. 3.4 Larval midgut gene regulation. (**a**) A stage 15 embryo shows the expression domains of *opa* in the developing midgut by immunohistochemistry. Anterior is oriented to the left. (**b**) A schematic diagram of midgut morphology depicts the gastric caeca (GC) and three constrictions: anterior (A), central (C), and posterior (P). Directly below the midgut diagram are the corresponding parasegments (PS) and expression domains of various genes involved in midget formation including *opa*. These genes are regulated by a complex network of interactions indicated by arrows (←) for a positive interaction and bars (⊢) to indicate negative regulation

sists through gut development (Cimbora and Sakonju 1995; Schaub and Frasch 2013). Loss of function (LOF) *opa* mutations fail to make the three characteristic midgut constrictions (Cimbora and Sakonju 1995). The positions of these midgut constrictions and gastric caeca are determined by a regulatory network of transcription factors and signaling proteins expressed in tissue- and parasegment-specific domains of the visceral mesoderm (Fig. 3.4b). Visceral mesoderm-specific transcription factors, such as the Fox-F protein *binou*, the *Drosophila* T-box factor *optomotor-blind-related-gene 1* (*org-1*), and the TALE homeobox factors *extradenticle* (*exd*) and *homothorax* (*hth*), provide tissue-specific gene activation, while the homeotic transcription factors Sex *combs reduced* (*Scr*), *Antennapedia* (*Antp*), *Ultrabithorax* (*Ubx*), and *abdominal-A* (*abd-A*) are organized in specific domains along the anterior-posterior axis of the visceral mesoderm and provide parasegment-specific transcriptional regulation. In addition, *wg* and the *Drosophila* bone morphogenetic protein (BMP) ortholog, *decapentaplegic* (*dpp*), also impart parasegment-specific regulatory inputs. *opa* transcription in the visceral mesoderm requires *org-1*, *exd*, and *hth* (Schaub and Frasch 2013). In the anterior parasegments, *opa* also requires *Antp* for activation, while in the posterior parasegments, *abd-A* plays that role (Cimbora and Sakonju 1995). *opa* is confined to its parasegmental domains by repression by *dpp*, which is itself regulated by *hth*, *exd*, *bin*, *Ubx*, and *wg* (Cimbora and Sakonju 1995; Graba et al. 1997; Bilder et al. 1998; Zaffran et al. 2001; Stultz et al. 2006a). The position of *opa* in PS 3–5 and PS 9–12 suggests that *opa* has a direct role in the formation of the first and third midgut constrictions formed in those anterior and posterior regions. However, the loss of the second midgut constriction in *opa* mutants implies that *opa* may affect gene expression indirectly in PS 6–8. The NK class homeobox transcription factor *bagpipe* (*bap*) is misexpressed in *opa* mutations, as is the expression of PS 7-specific *Ubx* (Cimbora and Sakonju 1995; Azpiazu et al. 1996), itself required for *dpp* expression. These results suggest that *opa* may provide some feedback regulation to other transcription factors in the visceral mesoderm.

3.4.3 Adult Ventral Head Morphogenesis

Mutations in *opa* were recovered in a screen for genes involved in *Drosophila* adult head formation, revealing a postembryonic role for the gene (Lee et al. 2007). All external adult structures in the fly are formed from saclike primordia called imaginal discs. The adult head is constructed from three pairs of these discs, the labial, clypeo-labral, and eye-antennal discs (Fig. 3.5a). Of these, the eye-antennal discs form the majority of head tissue. The adult *Drosophila* head is formed at metamorphosis with the fusion of the separate imaginal discs into the complete head structure. The paired labial, clypeo-labral, and eye-antennal discs fuse along the left-right midline to form the anterior portion, or "face," of the fly head.

The *Drosophila* BMP, *dpp*, plays a role in the morphogenesis of the adult ventral head via its expression in the eye-antennal discs (Stultz et al. 2006b). This *dpp* expression persists into the pupal period and is found aligning along the edges of the anterior domain of fusion (Fig. 3.5a). Failure to express *dpp* in the ventral head leads to apoptotic cell death and loss of head tissue, including sensory structures (Fig. 3.5b, c).

dpp head mutations can have incomplete inheritance and variable expressivity (Stultz et al. 2005) and are highly sensitive to the dose of BMP signaling. For instance, the effects of a single copy of a *dpp* head defect mutation can be enhanced by single copy LOF mutations in other members of the BMP pathway, such as the BMP type 1 receptor *thickveins*. This two-gene contribution to a phenotype is known as digenic inheritance. A genome-wide screen for additional genes that displayed this same type of dominant genetic interactions with *dpp* head mutations

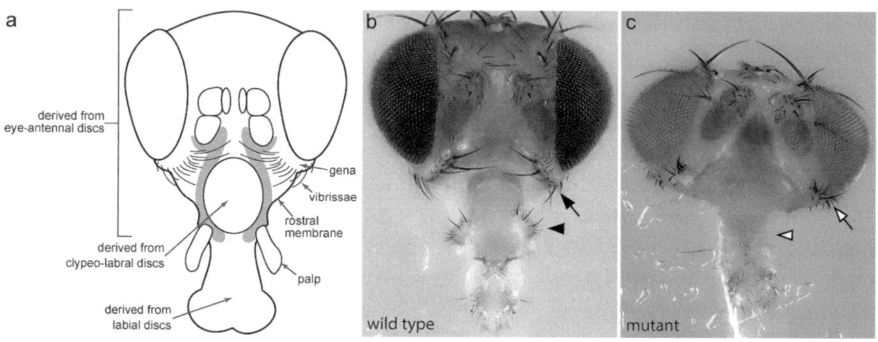

Fig. 3.5 A *Drosophila* model for midline defects. (**a**) A diagram of the *Drosophila* adult head shows the expression domain of *dpp* along the regions where the six imaginal discs that form the adult head fuse together (shaded area). (**b**) A wild-type adult head is compared to a (**c**) mutant adult head resulting from *dpp* head mutation. $opa^{LOF}/dpp^{head\ mutations}$ have identical mutant phenotypes, as do *opa* temperature-sensitive alleles moved to nonpermissive temperature after second larval instar. Note the disruption of ventral head structures including a smaller eye, loss of cuticle tissue, disordered sensory vibrissae, and missing maxillary palps. Solid arrow indicates wild-type vibrissae and solid arrowhead indicates wild-type palp. Open arrow indicates mutant vibrissae, and open arrowhead indicates missing palp

recovered LOF mutations in *opa*. These genetic interactions of *dpp* and *opa* are identical to those seen with *dpp* mutations and BMP receptors, i.e., one copy each of recessive *opa* and *dpp* mutations produces flies with adult head defects (Lee et al. 2007). A temperature-sensitive *opa* mutation, when shifted to nonpermissive temperature after embryogenesis, also displays the same ventral head defect observed in crosses of *opa* LOF mutations to *dpp* mutations. These data demonstrate that *opa* plays a role with *dpp* in ventral adult head formation.

The *dpp* expression responsible for head formation is controlled by a specific transcriptional enhancer in the 5′ end of the *dpp* gene; *dpp* head mutations delete this enhancer without disrupting the *dpp* coding region (Stultz et al. 2006b). Analysis of the behavior of this enhancer using reporter constructs in transgenic flies indicates that *opa* is required cell autonomously for its expression (Lee et al. 2007) and suggests that *opa* is acting upstream of *dpp* as a transcription factor to activate the head-specific enhancer. Sequences specifically bound by the *opa* zinc finger protein domain were determined by systematic evolution of ligands by exponential enrichment (SELEX), and a preferred binding site identified. However, similar sequences are not found in the *dpp* head enhancer nor does an expressed *opa* zinc finger protein domain interact with enhancer DNA in electrophoretic mobility shift assays (EMSA) (Sen et al. 2010). Thus while *opa* seems to be necessary to activate transcription from this enhancer, it appears to do this indirectly, without requiring direct DNA binding.

BMP pathway and Zic family members also interact genetically in vertebrates suggesting that this genetic network might be conserved evolutionarily. Mutations in BMP pathway members and Zic genes are associated with vertebrate head abnormalities, in particular holoprosencephaly (Brown et al. 1998; Fernandes et al. 2007).

Holoprosencephaly is a craniofacial abnormality of vertebrates characterized by defects in forebrain and midline facial structures (Muenke and Beachy 2000; Dubourg et al. 2007; Petryk et al. 2015). Holoprosencephaly is also characterized by digenic inheritance, incomplete penetrance, and variable expressivity (Ming and Muenke 2002; Petryk et al. 2015). Modifier genes and multiple interacting loci can also influence the penetrance and severity of these defects (Fernandes and Hebert 2008). While arthropod and vertebrate heads are constructed by different embryological mechanisms, it is intriguing to speculate that aspects of the genetic regulatory network that specifies midline morphogenesis may have been conserved through metazoan evolution.

3.4.4 **opa's** ***Role in the*** **Drosophila** ***Nervous System***

Zic proteins play numerous roles in vertebrate neural development (Aruga 2004), and this role appears to be ancestral for the Zic family (Layden et al. 2010). *opa* is differentially expressed during the formation of neural progenitors (Eroglu et al. 2014), and it seems likely that it plays a role in neuronal development. In *Drosophila*, data from in situ analysis, tissue-specific arrays, and whole genome RNA-seq of isolated cells, as outlined above, indicates that *opa* mRNA is robustly expressed in both the embryonic and adult nervous system. For example, expression is found in mushroom bodies, which control memory and learning, and the subesophageal ganglia, which receive gustatory and neurosecretory inputs. Interestingly, a recent genome-wide association study identified *opa* as one of a set of genes with polymorphisms related to variations in olfactory perception (Arya et al. 2015), suggesting that *opa* plays a role in the synaptic connectivity related to odor perception. However, no developmental genetic studies have as yet demonstrated *opa's* function in neural development, despite its expression in many parts of the developing fly nervous system. Additional work is necessary to elucidate the role of *opa* in neurogenesis and determine if those roles correspond to observed functions of vertebrate Zic genes.

3.4.5 **opa** ***Is Part of the Circadian Clock***

Fruit flies have been an important model for the study of circadian rhythms, since the identification of the first circadian rhythm gene, the *Drosophila period* (*per*) (Konopka and Benzer 1971). The genes involved in circadian rhythms in *Drosophila* have been studied extensively (Hardin 2011). The circadian pacemaker is initiated when two E-box transcription factors, *clock* (*clk*) and *cycle* (*cyc*), form a heterodimer to induce transcription of the genes *per* and *timeless* (*tim*). Per and Tim in turn inhibit Clk/Cyc activity via phosphorylation. Per and Tim degrade over the day, which allows resetting of the pacemaker through resumption of activity of Clk and

Cyc. This regular oscillation in gene expression allows the expression of both behavioral and metabolic genes to be linked to the circadian clock (Tataroglu and Emery 2014). *opa* may be involved in providing tissue specificity to the clock. In a genome-wide ChIP-seq analysis for targets of Clk and Cyc, Stark and colleagues found that *opa* is additionally required for expression of circadian transcriptional targets in the head (Meireles-Filho et al. 2014). Another GATA family transcription factor, *serpent*, serves a similar purpose in the body of the fly. Further analysis of targets that require Opa/Clk/Cyc for transcription in the fly head will likely reveal additional complexity to the role of *opa* in the circadian clock of *Drosophila*.

3.5 Opa as a Transcription Factor

3.5.1 Opa Is a Sequence-Specific DNA Binding Protein

Zic proteins, including Opa, belong to the Krüppel-like zinc finger protein family. These proteins contain five Cys2His2 zinc finger motifs that bind DNA in a sequence-specific manner (Merzdorf 2007; Lichti-Kaiser et al. 2012). An optimal binding site was identified by SELEX for Zic1, Zic2, and Zic3 (Mizugishi et al. 2001). This site was nearly identical to the binding site for the related Cys2His2 zinc finger protein, Gli (Fig. 3.6d, e). The zinc finger domains involved in DNA binding for both Zic and Gli families are highly conserved (Pavletich and Pabo 1993; Aruga et al. 1996). Sequence-specific DNA binding has been demonstrated for Zic1, Zic2, and Zic3 by genome-wide ChIP-Chip and ChIP-seq (Lim et al. 2010; Winata et al. 2013; Luo et al. 2015; Sankar et al. 2016). Specific binding sites for Opa have been determined by us and others using yeast-1 hybrid, SELEX, and high-throughput SELEX (Fig. 3.6a–c) (Noyes et al. 2008; Sen et al. 2010; Nitta et al. 2015). These Opa sites are similar to the defined Zic2 site and are also nearly identical to the Gli consensus site, which is recognized by *Cubitus interruptus* (*Ci*), the *Drosophila* Gli homolog (Alexandre et al. 1996; Hepker et al. 1999). These data indicate that Zic proteins, including Opa, have sequence-specific DNA binding capability.

3.5.2 Opa May Not Require DNA Binding to Activate Transcription in All Cases

In vitro, purified Zic zinc finger protein constructs display weak binding affinity for their optimal sequences, much lower than the affinity of Gli zinc finger protein constructs for the identical site (Mizugishi et al. 2001). The preferred sites for Opa differ slightly from the identified Gli/Ci site (Fig. 3.6), and purified Opa zinc finger protein constructs bind more strongly to the consensus Opa site than to the Gli/Ci

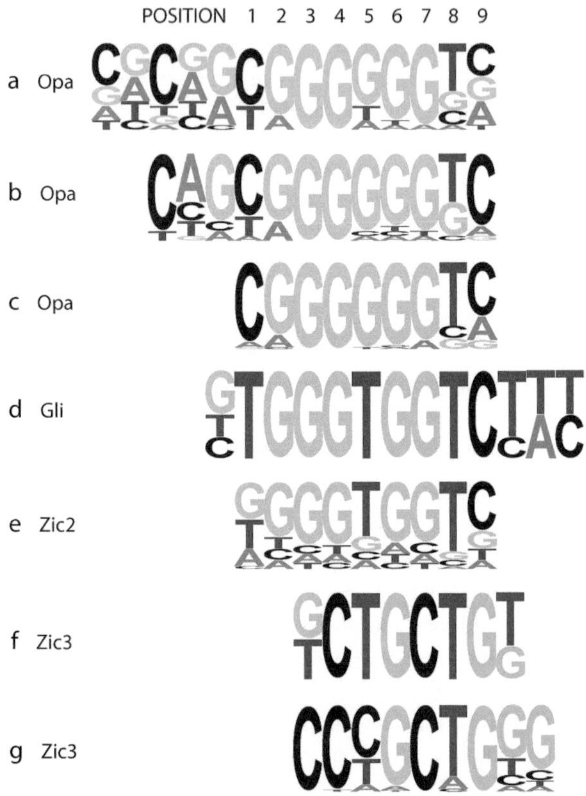

Fig. 3.6 Experimentally determined binding motifs for Opa, Gli, and Zic family members are similar. The Opa DNA binding motif as determined by high-throughput SELEX (**a**) (Nitta et al. 2015), a yeast 1-hybrid screen (**b**) (Noyes et al. 2008), and SELEX (**c**) (Sen et al. 2010) all contain a core CGGGGGGTC sequence. Closely related transcription factor binding sites, including Gli, as determined by DNA footprinting (**d**) (Kinzler and Vogelstein 1990), Zic2 by SELEX (**e**) (Mizugishi et al. 2001), Zic3 by ChIP-seq (**f**) (Winata et al. 2013), and Zic3 by ChIP-chip (**g**) (Lim et al. 2010) are all similar

site. However, purified Ci zinc finger protein constructs bind to both the optimal Gli/Ci site and the optimal Opa site with substantially higher affinity than Opa (Sen et al. 2010). Zic proteins thus appear to be weak interactors with DNA and may not always depend on direct DNA binding for activity in the way observed for most DNA sequence-specific transcription factors.

The behavior of our SELEX-determined Opa consensus site in vivo illustrates this point. We multimerized this binding site, placed it upstream of *lac-Z* in a reporter construct, and introduced it into flies transgenically (Sen et al. 2010). Multiple chromosomal insertions of this construct had minimal endogenous expression and responded robustly in vivo to ubiquitous full-length Opa expression. They only weakly responded to ubiquitous expression of Ci. Ubiquitous expression of Opa also resulted in spatially specific expression, indicating the requirement for additional factors. An equivalent transgenic construct in which nucleotides critical for in vitro Opa binding had been mutated was also examined: it was not activated by either Opa or Ci in vivo. Therefore, while Opa zinc finger domains do not compete with Ci zinc finger domains for binding to optimal Opa sites in vitro, in vivo, full-length Opa more robustly activates gene expression from its optimal binding sites than full-length Ci, and for this in vivo interaction, DNA binding is essential (Sen et al. 2010).

When the *Drosophila* genome was interrogated for sites with homology to our SELEX determined Opa site (Fig. 3.6c) (Sen et al. 2010), a site matching the consensus was found in *slp1*, a segmentation gene that genetically requires *opa* for expression (Swantek and Gergen 2004). This site bound an Opa zinc finger protein in EMSA. Two short (518 bp, 202 bp) *lac-Z* reporter constructs bearing this site were introduced transgenically into flies and examined in embryos. Similarly to the multimerized site above, these constructs responded to ectopic Opa expression but not to Ci. However, when the same nucleotide substitutions were introduced into the *slp1* site as were used to eliminate Opa binding in the multimerized SELEX site, they did not abrogate the ability of Opa to activate the reporter construct. Thus a single Opa binding site in its native context in vivo apparently has different requirements for DNA binding than an artificial multimerized version (Sen et al. 2010).

Zic proteins can activate gene expression from constructs bearing no obvious Zic sites (Mizugishi et al. 2001), and some Zic3 binding sites are unable to induce reporter expression (Winata et al. 2015). Even in the absence of DNA binding, Opa is able to produce tissue-specific expression cell autonomously in vivo from reporter constructs containing the *dpp* head enhancer element described above (Sen et al. 2010). This observation, in combination with the *slp1* data, suggests that Opa, like Zic proteins, may not absolutely require DNA binding for transcriptional activity. In both cases, additional known DNA binding factors are required to activate transcription; for the head capsule enhancer, the Hox gene *labial* is required (Stultz et al. 2012), while *slp1* requires *run* transcriptional activity (Swantek and Gergen 2004). Opa may use protein-protein interactions in some cases to activate gene expression, or other transcription factors could introduce a chromatin configuration more conducive to Opa binding. The range of mechanisms by which Opa influences transcription remains to be elucidated.

Opa may also function by collaborating with developmental signaling pathways. Zic proteins bind Tcf, the transcriptional effector of the Wnt pathway (Fujimi et al. 2012; Murgan et al. 2015). They also bind Gli proteins, the terminal effectors of the Hedgehog pathway (Koyabu et al. 2001). Opa activates gene expression from consensus Ci binding sites in vivo, and ectopic co-expression of Opa and Ci in the fly head primordia produces morphogenetic defects that are intermediate to either one alone (Sen et al. 2010), suggesting that the two proteins are capable of interaction in vivo. In such instances, Zic proteins may influence transcription by altering the output of signal transduction.

3.5.3 Opa Displays Concentration-Dependent Effects

In vivo, Opa can potentiate Run-dependent regulation of *slp1* in a concentration-dependent manner (Swantek and Gergen 2004). Clark and Akam present evidence that during segmentation, enhancers of different segmentation genes show sensitivity in their response to the concentration of Opa in the nucleus (Clark and Akam 2016). In our analysis of ventral head development, Opa displayed dominant genetic interactions; it also produced dramatic head defects when overexpressed in third instar eye-antennal discs (Lee et al. 2007). This genetic behavior also suggests that Opa concentration influences its function. Understanding the role of concentration in Opa's transcriptional behavior may clarify how it functions as a regulator of gene expression.

3.6 Summary

opa plays documented important roles in embryonic development and adult morphogenesis. Its appearance in many classical genetic screens and recent genome-wide molecular screens suggests that it plays more critical roles that are only beginning to be identified. Zic proteins are associated with human craniofacial defects and developmental abnormalities (Grinberg and Millen 2005). Further analysis of *opa* in *Drosophila*, with its abundance of genetic resources, may shed further light on the roles of Zic proteins in human health and disease.

Acknowledgments The authors would like to thank Dr. Heuijung Lee, who initiated the *opa* project in our lab, for her contributions to this review. We thank Drs. Annette Parks, Brent McCright, Malcolm Moos, and Mark Mortin for comments that improved the manuscript.

References

Alexandre C, Jacinto A, Ingham PW (1996) Transcriptional activation of hedgehog target genes in Drosophila is mediated directly by the cubitus interruptus protein, a member of the GLI family of zinc finger DNA-binding proteins. Genes Dev 10(16):2003–2013

Aruga J (2004) The role of Zic genes in neural development. Mol Cell Neurosci 26(2):205–221

Aruga J, Nagai T, Tokuyama T, Hayashizaki Y, Okazaki Y, Chapman VM, Mikoshiba K (1996) The mouse zic gene family. Homologues of the Drosophila pair-rule gene odd-paired. J Biol Chem 271(2):1043–1047

Aruga J, Kamiya A, Takahashi H, Fujimi TJ, Shimizu Y, Ohkawa K, Yazawa S, Umesono Y, Noguchi H, Shimizu T, Saitou N, Mikoshiba K, Sakaki Y, Agata K, Toyoda A (2006) A wide-range phylogenetic analysis of Zic proteins: implications for correlations between protein structure conservation and body plan complexity. Genomics 87(6):783–792. https://doi.org/10.1016/j.ygeno.2006.02.011

Arya GH, Magwire MM, Huang W, Serrano-Negron YL, Mackay TF, Anholt RR (2015) The genetic basis for variation in olfactory behavior in Drosophila melanogaster. Chem Senses 40(4):233–243. https://doi.org/10.1093/chemse/bjv001

Azpiazu N, Lawrence PA, Vincent JP, Frasch M (1996) Segmentation and specification of the Drosophila mesoderm. Genes Dev 10(24):3183–3194

Benedyk MJ, Mullen JR, Dinardo S (1994) Odd-paired – a zinc-finger pair rule protein required for the timely activation of engrailed and wingless in drosophila embryos. Genes Dev 8(1):105–117

Bilder D, Scott MP (1998) Hedgehog and wingless induce metameric pattern in the Drosophila visceral mesoderm. Dev Biol 201(1):43–56. https://doi.org/10.1006/dbio.1998.8953
Bilder D, Graba Y, Scott MP (1998) Wnt and TGFbeta signals subdivide the AbdA Hox domain during Drosophila mesoderm patterning. Development 125(9):1781–1790
Brown SA, Warburton D, Brown LY, Yu CY, Roeder ER, Stengel-Rutkowski S, Hennekam RCM, Muenke M (1998) Holoprosencephaly due to mutations in ZIC2, a homologue of Drosophila odd-paired. Nat Genet 20(2):180–183
Butler MJ, Jacobsen TL, Cain DM, Jarman MG, Hubank M, Whittle JR, Phillips R, Simcox A (2003) Discovery of genes with highly restricted expression patterns in the Drosophila wing disc using DNA oligonucleotide microarrays. Development 130(4):659–670
Chintapalli VR, Wang J, Dow JA (2007) Using FlyAtlas to identify better Drosophila melanogaster models of human disease. Nat Genet 39(6):715–720. https://doi.org/10.1038/ng2049
Cimbora DM, Sakonju S (1995) Drosophila midgut morphogenesis requires the function of the segmentation gene odd-paired. Dev Biol 169(2):580–595
Clark E, Akam M (2016) Odd-paired controls frequency doubling in Drosophila segmentation by altering the pair-rule gene regulatory network. Elife 5. https://doi.org/10.7554/eLife.18215
Dubourg C, Bendavid C, Pasquier L, Henry C, Odent S, David V (2007) Holoprosencephaly. Orphanet J Rare Dis 2:8
Eroglu E, Burkard TR, Jiang Y, Saini N, Homem CC, Reichert H, Knoblich JA (2014) SWI/SNF complex prevents lineage reversion and induces temporal patterning in neural stem cells. Cell 156(6):1259–1273. https://doi.org/10.1016/j.cell.2014.01.053
Fernandes M, Hebert JM (2008) The ups and downs of holoprosencephaly: dorsal versus ventral patterning forces. Clin Genet 73(5):413–423. https://doi.org/10.1111/j.1399-0004.2008.00994.x
Fernandes M, Gutin G, Alcorn H, McConnell SK, Hebert JM (2007) Mutations in the BMP pathway in mice support the existence of two molecular classes of holoprosencephaly. Development 134(21):3789–3794. https://doi.org/10.1242/dev.004325
Fujimi TJ, Hatayama M, Aruga J (2012) Xenopus Zic3 controls notochord and organizer development through suppression of the Wnt/beta-catenin signaling pathway. Dev Biol 361(2):220–231. https://doi.org/10.1016/j.ydbio.2011.10.026
Graba Y, Aragnol D, Pradel J (1997) Drosophila Hox complex downstream targets and the function of homeotic genes. BioEssays 19(5):379–388
Graveley BR, Brooks AN, Carlson JW, Duff MO, Landolin JM, Yang L, Artieri CG, van Baren MJ, Boley N, Booth BW, Brown JB, Cherbas L, Davis CA, Dobin A, Li R, Lin W, Malone JH, Mattiuzzo NR, Miller D, Sturgill D, Tuch BB, Zaleski C, Zhang D, Blanchette M, Dudoit S, Eads B, Green RE, Hammonds A, Jiang L, Kapranov P, Langton L, Perrimon N, Sandler JE, Wan KH, Willingham A, Zhang Y, Zou Y, Andrews J, Bickel PJ, Brenner SE, Brent MR, Cherbas P, Gingeras TR, Hoskins RA, Kaufman TC, Oliver B, Celniker SE (2011) The developmental transcriptome of Drosophila melanogaster. Nature 471(7339):473–479. https://doi.org/10.1038/nature09715
Grinberg I, Millen KJ (2005) The ZIC gene family in development and disease. Clin Genet 67(4):290–296
Hardin PE (2011) Molecular genetic analysis of circadian timekeeping in Drosophila. Adv Genet 74:141–173. https://doi.org/10.1016/B978-0-12-387690-4.00005-2
Hepker J, Blackman RK, Holmgren R (1999) Cubitus interruptus is necessary but not sufficient for direct activation of a wing-specific decapentaplegic enhancer. Development 126(16):3669–3677
Jaynes JB, Fujioka M (2004) Drawing lines in the sand: even skipped et al. and parasegment boundaries. Dev Biol 269(2):609–622. https://doi.org/10.1016/j.ydbio.2004.03.001
Jurgens G, Wieschaus E, Nussleinvolhard C, Kluding H (1984) Mutations affecting the pattern of the larval cuticle in Drosophila-melanogaster. 2. Zygotic loci on the 3rd chromosome. Wilhelm Rouxs Arch DevBiol 193(5):283–295

Kinzler KW, Vogelstein B (1990) The GLI gene encodes a nuclear protein which binds specific sequences in the human genome. Mol Cell Biol 10(2):634–642

Konopka RJ, Benzer S (1971) Clock mutants of Drosophila melanogaster. Proc Natl Acad Sci U S A 68(9):2112–2116

Koyabu Y, Nakata K, Mizugishi K, Aruga J, Mikoshiba K (2001) Physical and functional interactions between Zic and Gli proteins. J Biol Chem 276(10):6889–6892

Layden MJ, Meyer NP, Pang K, Seaver EC, Martindale MQ (2010) Expression and phylogenetic analysis of the zic gene family in the evolution and development of metazoans. EvoDevo 1(1):12. https://doi.org/10.1186/2041-9139-1-12

Lee H, Stultz BG, Hursh DA (2007) The Zic family member, odd-paired, regulates the Drosophila BMP, decapentaplegic, during adult head development. Development 134(7):1301–1310

Lichti-Kaiser K, ZeRuth G, Kang HS, Vasanth S, Jetten AM (2012) Gli-similar proteins: their mechanisms of action, physiological functions, and roles in disease. Vitam Horm 88:141–171. https://doi.org/10.1016/B978-0-12-394622-5.00007-9

Lim LS, Hong FH, Kunarso G, Stanton LW (2010) The pluripotency regulator Zic3 is a direct activator of the Nanog promoter in ESCs. Stem Cells 28(11):1961–1969. https://doi.org/10.1002/stem.527

Luo Z, Gao X, Lin C, Smith ER, Marshall SA, Swanson SK, Florens L, Washburn MP, Shilatifard A (2015) Zic2 is an enhancer-binding factor required for embryonic stem cell specification. Mol Cell 57(4):685–694. https://doi.org/10.1016/j.molcel.2015.01.007

Mahoney MB, Parks AL, Ruddy DA, Tiong SY, Esengil H, Phan AC, Philandrinos P, Winter CG, Chatterjee R, Huppert K, Fisher WW, L'Archeveque L, Mapa FA, Woo W, Ellis MC, Curtis D (2006) Presenilin-based genetic screens in Drosophila melanogaster identify novel notch pathway modifiers. Genetics 172(4):2309–2324

Meireles-Filho AC, Bardet AF, Yanez-Cuna JO, Stampfel G, Stark A (2014) cis-regulatory requirements for tissue-specific programs of the circadian clock. Curr Biol 24(1):1–10. https://doi.org/10.1016/j.cub.2013.11.017

Merzdorf CS (2007) Emerging roles for zic genes in early development. Dev Dyn 236(4):922–940. https://doi.org/10.1002/dvdy.21098

Ming JE, Muenke M (2002) Multiple hits during early embryonic development: digenic diseases and holoprosencephaly. Am J Hum Genet 71(5):1017–1032

Mizugishi K, Aruga J, Nakata K, Mikoshiba K (2001) Molecular properties of Zic proteins as transcriptional regulators and their relationship to GLI proteins. J Biol Chem 276(3):2180–2188

Mizugishi K, Hatayama M, Tohmonda T, Ogawa M, Inoue T, Mikoshiba K, Aruga J (2004) Myogenic repressor I-mfa interferes with the function of Zic family proteins. Biochem Biophys Res Commun 320(1):233–240. https://doi.org/10.1016/j.bbrc.2004.05.158

Moore LA, Broihier HT, Van Doren M, Lunsford LB, Lehmann R (1998) Identification of genes controlling germ cell migration and embryonic gonad formation in Drosophila. Development 125(4):667–678

Muenke M, Beachy PA (2000) Genetics of ventral forebrain development and holoprosencephaly. Curr Opin Genet Dev 10(3):262–269

Murgan S, Kari W, Rothbacher U, Iche-Torres M, Melenec P, Hobert O, Bertrand V (2015) Atypical transcriptional activation by TCF via a Zic transcription factor in C. elegans neuronal precursors. Dev Cell 33(6):737–745. https://doi.org/10.1016/j.devcel.2015.04.018

Negre N, Brown CD, Ma L, Bristow CA, Miller SW, Wagner U, Kheradpour P, Eaton ML, Loriaux P, Sealfon R, Li Z, Ishii H, Spokony RF, Chen J, Hwang L, Cheng C, Auburn RP, Davis MB, Domanus M, Shah PK, Morrison CA, Zieba J, Suchy S, Senderowicz L, Victorsen A, Bild NA, Grundstad AJ, Hanley D, MacAlpine DM, Mannervik M, Venken K, Bellen H, White R, Gerstein M, Russell S, Grossman RL, Ren B, Posakony JW, Kellis M, White KP (2011) A cis-regulatory map of the Drosophila genome. Nature 471(7339):527–531. https://doi.org/10.1038/nature09990

Nitta KR, Jolma A, Yin Y, Morgunova E, Kivioja T, Akhtar J, Hens K, Toivonen J, Deplancke B, Furlong EE, Taipale J (2015) Conservation of transcription factor binding specificities across 600 million years of bilateria evolution. Elife 4. https://doi.org/10.7554/eLife.04837

Noyes MB, Meng X, Wakabayashi A, Sinha S, Brodsky MH, Wolfe SA (2008) A systematic characterization of factors that regulate Drosophila segmentation via a bacterial one-hybrid system. Nucleic Acids Res 36(8):2547–2560. https://doi.org/10.1093/nar/gkn048

Pavletich NP, Pabo CO (1993) Crystal structure of a five-finger GLI-DNA complex: new perspectives on zinc fingers. Science 261(5129):1701–1707

Petryk A, Graf D, Marcucio R (2015) Holoprosencephaly: signaling interactions between the brain and the face, the environment and the genes, and the phenotypic variability in animal models and humans. Wiley Interdiscip Rev Dev Biol 4(1):17–32. https://doi.org/10.1002/wdev.161

Pfeiffer BD, Jenett A, Hammonds AS, Ngo TT, Misra S, Murphy C, Scully A, Carlson JW, Wan KH, Laverty TR, Mungall C, Svirskas R, Kadonaga JT, Doe CQ, Eisen MB, Celniker SE, Rubin GM (2008) Tools for neuroanatomy and neurogenetics in Drosophila. Proc Natl Acad Sci U S A 105(28):9715–9720. https://doi.org/10.1073/pnas.0803697105

Reim I, Frasch M, Schaub C (2017) Chapter six – T-box genes in Drosophila mesoderm development. In: Manfred F (ed) Current topics in developmental biology, vol 122. Academic, pp 161–193

Sankar S, Yellajoshyula D, Zhang B, Teets B, Rockweiler N, Kroll KL (2016) Gene regulatory networks in neural cell fate acquisition from genome-wide chromatin association of Geminin and Zic1. Sci Rep 6:37412. https://doi.org/10.1038/srep37412

Schaub C, Frasch M (2013) Org-1 is required for the diversification of circular visceral muscle founder cells and normal midgut morphogenesis. Dev Biol 376(2):245–259. https://doi.org/10.1016/j.ydbio.2013.01.022

Schroeder MD, Greer C, Gaul U (2011) How to make stripes: deciphering the transition from non-periodic to periodic patterns in Drosophila segmentation. Development 138(14):3067–3078. https://doi.org/10.1242/dev.062141

Sen A, Stultz BG, Lee H, Hursh DA (2010) Odd paired transcriptional activation of decapentaplegic in the Drosophila eye/antennal disc is cell autonomous but indirect. Dev Biol 343(1–2):167–177

Stultz BG, Ray RP, Hursh DA (2005) Analysis of the shortvein cis-regulatory region of the decapentaplegic gene of Drosophila melanogaster. Genesis 42(3):181–192

Stultz BG, Jackson DG, Mortin MA, Yang X, Beachy PA, Hursh DA (2006a) Transcriptional activation by extradenticle in the Drosophila visceral mesoderm. Dev Biol 290(2):482–494

Stultz BG, Lee HJ, Ramon K, Hursh DA (2006b) Decapentaplegic head capsule mutations disrupt novel peripodial expression controlling the morphogenesis of the Drosophila ventral head. Dev Biol 296(2):329–339

Stultz BG, Park SY, Mortin MA, Kennison JA, Hursh DA (2012) Hox proteins coordinate peripodial decapentaplegic expression to direct adult head morphogenesis in Drosophila. Dev Biol 369(2):362–376

Swantek D, Gergen JP (2004) Ftz modulates runt-dependent activation and repression of segment-polarity gene transcription. Development 131(10):2281–2290

Tataroglu O, Emery P (2014) Studying circadian rhythms in Drosophila melanogaster. Methods 68(1):140–150. https://doi.org/10.1016/j.ymeth.2014.01.001

Tomancak P, Beaton A, Weiszmann R, Kwan E, Shu S, Lewis SE, Richards S, Ashburner M, Hartenstein V, Celniker SE, Rubin GM (2002) Systematic determination of patterns of gene expression during Drosophila embryogenesis. Genome Biol 3(12):RESEARCH0088

Tomancak P, Berman BP, Beaton A, Weiszmann R, Kwan E, Hartenstein V, Celniker SE, Rubin GM (2007) Global analysis of patterns of gene expression during Drosophila embryogenesis. Genome Biol 8(7):R145. https://doi.org/10.1186/gb-2007-8-7-r145

Winata CL, Kondrychyn I, Kumar V, Srinivasan KG, Orlov Y, Ravishankar A, Prabhakar S, Stanton LW, Korzh V, Mathavan S (2013) Genome wide analysis reveals Zic3 interaction with

distal regulatory elements of stage specific developmental genes in zebrafish. PLoS Genet 9(10):e1003852. https://doi.org/10.1371/journal.pgen.1003852
Winata CL, Kondrychyn I, Korzh V (2015) Changing faces of transcriptional regulation reflected by Zic3. Curr Genomics 16(2):117–127. https://doi.org/10.2174/1389202916666150205124519
Yang CP, Fu CC, Sugino K, Liu Z, Ren Q, Liu LY, Yao X, Lee LP, Lee T (2016) Transcriptomes of lineage-specific Drosophila neuroblasts profiled by genetic targeting and robotic sorting. Development 143(3):411–421. https://doi.org/10.1242/dev.129163
Zaffran S, Kuchler A, Lee HH, Frasch M (2001) biniou (FoxF), a central component in a regulatory network controlling visceral mesoderm development and midgut morphogenesis in Drosophila. Genes Dev 15(21):2900–2915

Chapter 4
Zic Genes in Nematodes: A Role in Nervous System Development and Wnt Signaling

Guillaume Bordet and Vincent Bertrand

Abstract Transcription factors of the Zic family play important roles during animal development, and their misregulation has been implicated in several human diseases. Zic proteins are present in nematodes, and their function has been mostly studied in the model organism *C. elegans*. *C. elegans* possesses only one Zic family member, REF-2. Functional studies have shown that this factor plays a key role during the development of the nervous system, epidermis, and excretory system. In addition, they have revealed that the *C. elegans* Zic protein acts as an atypical mediator of the Wnt/β-catenin pathway. In other animals including vertebrates, Zic factors are also regulators of nervous system development and modulators of Wnt signaling, suggesting that these are evolutionary ancient functions of Zic proteins.

Keywords Zic · Nematodes · *C. elegans* · Transcription · Development · Nervous system · Wnt signaling

4.1 Introduction

The phylum Nematoda is a very large and diverse group of animals that belongs to the clade Ecdysozoa. It contains both free-living and parasitic animals including human pathogens. The most studied nematode is *Caenorhabditis elegans*. Since the late 1970s, it has become a major model organism for research in various fields including genetics, molecular biology, development, and neurobiology. Because of its short life cycle, it is a powerful model to conduct genetic screens (Brenner 1974). In addition, with its transparency, small number of somatic cells (959 in the adult hermaphrodite) and fixed lineage, it is an excellent system to study development in vivo at single cell resolution (Sulston and Horvitz 1977; Sulston et al. 1983).

G. Bordet · V. Bertrand (✉)
Aix-Marseille Université, CNRS, Institut de Biologie du Développement de Marseille, Marseille, France
e-mail: vincent.bertrand@univ-amu.fr

J. Aruga (ed.), *Zic family*, Advances in Experimental Medicine and Biology 1046,
https://doi.org/10.1007/978-981-10-7311-3_4

The sequencing of the *C. elegans* and human genomes has revealed that 60–80% of human genes have a homolog in *C. elegans*, and *C. elegans* has been extensively used to characterize the function of many of these conserved gene families (Kaletta and Hengartner 2006). A single member of the Zic transcription factor family is present in *C. elegans*, making it easy to study. This family plays key roles during the development of various animals including vertebrates. In addition, misregulation of Zic factors is involved in several human diseases. Here, we will review the studies that have been conducted on the *C. elegans* Zic factor. We will first present its structure and then describe its function during development of the nervous system, the epidermis, and the excretory system. Finally, we will present its role as mediator of Wnt signaling.

4.2 Structure of the *Zic* Gene of *C. elegans*, *ref-2*

C. elegans has only one *Zic* family gene, *ref-2* (Alper and Kenyon 2002; Aruga et al. 2006). The name *ref-2* (*regulator of fusion-2*) corresponds to the phenotype of the first mutant identified for this gene, which presents a defect in the fusion pattern of ventral epidermal cells (Alper and Kenyon 2002). An analysis of several other nematode species for which the genome is available (*Brugia malayi*, *Onchocerca volvulus*, *Pristionchus pacificus*, *Panagrellus redivivus*, *Strongyloides ratti*) suggests that they also possess only one *Zic* family gene. In *C. elegans*, the Zic protein REF-2 presents two isoforms: a short isoform REF-2a (303 amino acids) and a long isoform REF-2b (315 amino acids) (Fig. 4.1a). They result from the use of alternative transcription start sites, REF-2b having 12 additional amino acids at the N-terminal end. The most abundant isoform is REF-2a. The *C. elegans* REF-2 proteins possess the five C2H2-type zinc fingers characteristic of the Zic protein family (Fig. 4.1b). The degree of conservation of the zinc fingers 2–5 is very high, while zinc finger 1 is less conserved. Some Zic family proteins in various animal species also contain two additional conserved domains: ZOC (Zic/odd-paired conserved) and ZF-NC (zinc finger N-terminal flanking conserved). The ZOC domain is only present in the long isoform of the *C. elegans* protein (REF2-b), while the ZF-NC domain is absent in both isoforms (Aruga et al. 2006; Layden et al. 2010).

Several mutations in the *ref-2* gene have been isolated. The first mutation isolated, *mu218*, is a gain of function mutation in a cis-regulatory element located just 3′ to the coding region (Fig. 4.1a) (Alper and Kenyon 2002). This mutation leads to an ectopic expression of *ref-2* in some ventral epidermal cells. Two loss-of-function mutations have subsequently been identified: *ot327* is a point mutation in zinc finger 4 that converts a key cysteine into a tyrosine and is predicted to result in a loss of DNA-binding activity for zinc finger 4 (Fig. 4.1a, b), and *gk178* is a deletion resulting in a frame shift that removes the five zinc fingers and is likely a null allele (Fig. 4.1a) (Bertrand and Hobert 2009). Both loss-of-function alleles give identical phenotypes (Bertrand and Hobert 2009).

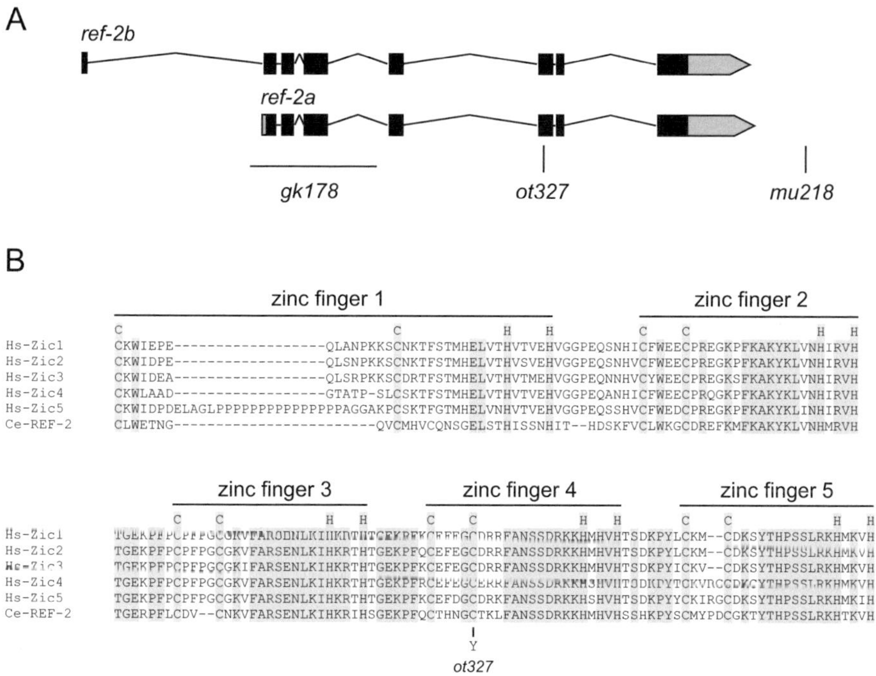

Fig. 4.1 Structure of the *C. elegans Zic* gene *ref-2*. (**a**) Organization of the *ref-2* locus. (**b**) Alignment of the zinc finger region of the REF-2 protein with the corresponding region of the five human Zic proteins

4.3 Role of REF-2/Zic During *C. elegans* Development

4.3.1 Nervous System Development

Functional studies have demonstrated that the *C. elegans Zic* gene *ref-2* plays an important role during the development of the nervous system. In *C. elegans*, the nervous system is composed of 302 neurons in the hermaphrodite (Hobert 2016). Most of them are generated during embryogenesis by a series of invariant asymmetric divisions (Sulston et al. 1983). In the embryo, REF-2 is expressed in many (but not all) neuronal precursors before their terminal divisions (Bertrand and Hobert 2009). The REF-2 protein subsequently disappears from most neuronal lineages, and its expression is restricted to only a few postmitotic neurons at larval and adult stages.

In *ref-2* loss-of-function mutants, the global organization of the nervous system is normal. While many neuronal subtypes are correctly specified, the generation of a few subtypes is defective (Bertrand and Hobert 2009). For example, there is a defect in the specification of the interneurons AIY and AVK. The mode of action of REF-2 has been best characterized in the case of the AIY cholinergic interneuron

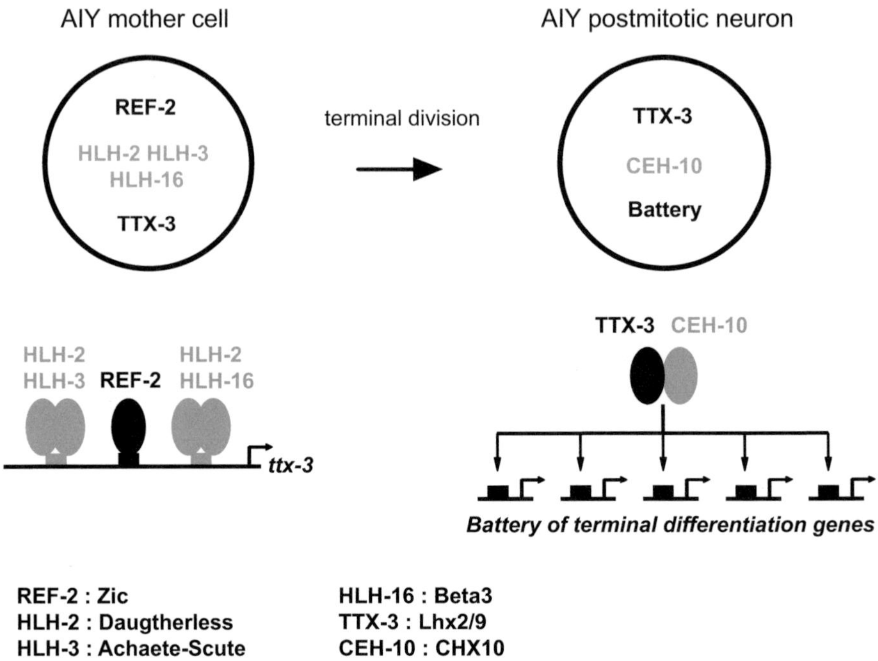

Fig. 4.2 Role of the *C. elegans* Zic protein REF-2 in the initiation of the terminal differentiation program of the AIY cholinergic interneuron. In the AIY mother cell, REF-2 cooperates with proneural bHLH factors (HLH-2, HLH-3, and HLH-16) to initiate the expression of the terminal transcription factor TTX-3. After terminal division, TTX-3 activates the expression of the terminal transcription factor CEH-10 in the postmitotic AIY neuron. TTX-3 and CEH-10 then activate and maintain the expression of a battery of terminal differentiation genes throughout the life of the AIY neuron

(Fig. 4.2) (Bertrand and Hobert 2009; Bertrand et al. 2011; Murgan et al. 2015). The identity of the AIY neuron is specified by two terminal selector transcription factors of the homeodomain family: TTX-3 (a LHX2/9 ortholog) and CEH-10 (a CHX10 ortholog) (Hobert et al. 1997; Altun-Gultekin et al. 2001; Wenick and Hobert 2004). TTX-3 and CEH-10 are expressed in the postmitotic AIY neuron throughout the life of the animal, where they directly activate and maintain the expression of a large battery of AIY subtype-specific terminal differentiation genes such as specific neurotransmitter receptors or ion channels (Wenick and Hobert 2004). In *ref-2* mutants, the AIY neuron still expresses pan-neuronal markers but loses its specific subtype identity (Bertrand and Hobert 2009). The expression of TTX-3, of CEH-10 and of the battery of AIY-specific terminal differentiation genes is lost. During embryonic development REF-2 is transiently expressed in the mother cell of the AIY neuron before its terminal division but then disappears from the AIY lineage after the terminal division. REF-2 directly activates the expression of *ttx-3* in the mother cell of AIY by binding to a Zic site present in the *ttx-3* cis-regulatory region. To activate *ttx-3* expression, REF-2 cooperates with several proneural bHLH transcription factors (HLH-2, HLH-3, and HLH-16) that bind bHLH sites close to the Zic site in the *ttx-3* cis-regulatory region (Bertrand and Hobert 2009; Bertrand et al. 2011; Murgan et al. 2015). *ttx-3* then activates *ceh-10* expression in the postmitotic AIY neuron, and *ttx-3* and *ceh-10* subsequently automaintain their expression throughout the life of the neuron. Therefore, the Zic factor REF-2 acts transiently in a neuronal precursor to initiate a neuronal subtype-specific differentiation program.

Interestingly, Zic factors also play a role in the development of the nervous system in many animal species. *Zic* genes are expressed in the developing nervous system in deuterostomes (vertebrates, amphioxus, ascidians) (Aruga 2004), protostomes (*Drosophila* and annelids) (Cimbora and Sakonju 1995; Takahashi et al. 2008; Layden et al. 2010), and cnidarians (Lindgens et al. 2004; Layden et al. 2010). Thus, the role of Zic factors in neuronal specification observed in nematodes may reflect an ancient metazoan feature.

4.3.2 Epidermis Development

The *Zic* gene *ref-2* was initially isolated in a genetic screen for mutants affecting the fusion of the Pn.p ventral epidermal cells (Alper and Kenyon 2002). In the *C. elegans* larva, some of the ventral epidermal cells fuse with a large epidermal syncytium (hyp7), while some others don't fuse. REF-2 is expressed in the ventral epidermal cells, and its expression lasts longer in cells that do not fuse. When *ref-2* function is reduced using RNA interference, all ventral epidermal cells fuse. Conversely, an overexpression of REF-2 using a ubiquitous driver or a gain-of-function mutation in the *ref-2* cis-regulatory regions (*mu218*) prevents ventral epidermal cells from fusing. Therefore, the REF-2 protein keeps ventral epidermal cells from fusing. A stronger reduction of *ref-2* function using higher concentrations of double-stranded RNA leads to defects in the generation of the ventral epidermal cells. The *ref-2* gene genetically interacts with two *Hox* genes, *lin-39* and *mab-5*, to modulate ventral epidermal cell fusion (Alper and Kenyon 2002). However, the detailed molecular mechanism through which the REF-2 protein regulates cell fusion remains to be determined.

4.3.3 Excretory System Development

The *C. elegans* Zic factor REF-2 also plays a role during the development of the excretory system (Bertrand and Hobert 2009). The excretory system of *C. elegans* is composed of three cell types: the excretory cell, the duct cell, and the pore cell (Sundaram and Buechner 2016). This system ensures the osmotic balance as well as the elimination of waste from the body. The excretory cell collects fluids that are then sent outside by the duct cell and the pore cell. During development three different cells successively fulfill the function of the pore cell (G1, G2, and then its daughter G2p). REF-2 is expressed during development in G1 and in the G2 lineage (Bertrand and Hobert 2009). In addition, *ref-2* loss-of-function mutants die at early larval stages with a phenotype characteristic of defects in the function of the excretory system (the pseudocoelome fills with fluid) (Bertrand and Hobert 2009; Nelson and Riddle 1984). Two gland cells, the excretory gland cells, are associated with the excretory system (Sundaram and Buechner 2016). These cells are not required for the viability of the animal, and their function remains obscure. REF-2 is expressed in the excretory gland cells and is needed for the expression of the excretory gland marker *lim-6* (Bertrand and Hobert 2009). Therefore, REF-2 seems required for the correct development and function of the excretory system, but the detailed molecular mechanism awaits further characterization.

4.4 Role of REF-2/Zic in Wnt Signaling

The Wnt/β-catenin signaling pathway plays key roles during animal development (MacDonald et al. 2009; Clevers and Nusse 2012). This pathway is present in all animals from sponges to human. When this pathway is active, the transcriptional coactivator β-catenin is stabilized and interacts with a transcription factor of the T-cell factor (TCF) family. This β-catenin:TCF complex activates the transcription of "classic" target genes of the pathway via TCF-binding sites present in their cis-regulatory regions (Fig. 4.3). When the pathway is inactive, β-catenin is degraded

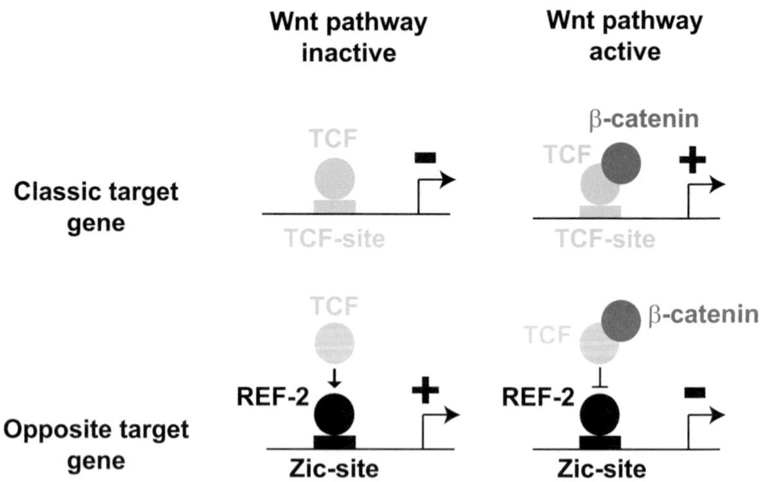

Fig. 4.3 Role of the *C. elegans* Zic protein REF-2 in response to Wnt signaling. Classic target genes are activated by the TCF:β-catenin complex and repressed by TCF in the absence of β-catenin. Opposite target genes are activated by a REF-2:TCF complex, and this activation is blocked by β-catenin

and TCF represses the expression of the "classic" target genes. While most direct target genes follow this "classic" type of regulation, a few other direct target genes have been observed to follow an "opposite" type of regulation. These "opposite" target genes are repressed by the β-catenin:TCF complex and activated by TCF in the absence of β-catenin (Fig. 4.3) (Hoverter and Waterman 2008; Cadigan 2012; Murgan and Bertrand 2015). Interestingly, it has been recently observed in *C. elegans* that REF-2 is involved in this atypical regulation by TCF during the asymmetric divisions of neuronal precursors in the embryo (Murgan et al. 2015).

In the *C. elegans* embryo, several asymmetric divisions oriented along the anteroposterior axis are regulated by a variant of the Wnt/β-catenin cascade named the Wnt/β-catenin asymmetry pathway (Kaletta et al. 1997; Lin et al. 1998; Mizumoto and Sawa 2007; Phillips and Kimble 2009; Sawa and Korswagen 2013; Bertrand 2016). This pathway regulates transcription via the formation of a complex between a TCF transcription factor named POP-1 and a β-catenin named SYS-1. Following asymmetric division, this pathway is active in the posterior daughter cell but not in the anterior daughter cell. The expression of the *ttx-3* transcription factor gene is activated in the AIY mother cell (anterior daughter cell), where the Wnt/β--catenin pathway is inactive, but not in the posterior daughter cell, where the pathway is active. This "opposite" regulation by the Wnt/β-catenin pathway of the *ttx-3* target gene is mediated by REF-2 (Fig. 4.3) (Murgan et al. 2015). In the anterior daughter cell, the TCF protein forms a complex with the REF-2 protein, and this TCF:REF-2 complex activates *ttx-3* expression via a Zic-binding site present in its cis-regulatory regions. In the posterior daughter cell, β-catenin binding to TCF blocks the activation by the TCF:REF-2 complex. This is an atypical mode of action for a TCF transcription factor, and whether other "opposite" target genes are regulated in a similar manner remains to be determined.

Interestingly, Zic factors have been shown to interact with the Wnt/β-catenin pathway in other organisms as well. In vertebrates, Zic proteins bind TCF proteins (Zic2 binds TCF4 and Zic3 binds TCF1 and TCF3) (Pourebrahim et al. 2011; Fujimi et al. 2012). In addition, Zic factors block the expression of classic target genes of the Wnt/β-catenin pathway in mammalian cell cultures and in *Xenopus* embryos (Pourebrahim et al. 2011; Fujimi et al. 2012). The mechanism of repression has been characterized for Zic2 (Pourebrahim et al. 2011). Zic2 binds the TCF-4:β-catenin complex at the level of cis-regulatory elements containing TCF sites, and this Zic2 binding blocks the ability of the TCF-4:β-catenin complex to activate transcription. Although this mechanism occurs on TCF sites, it is reminiscent of the situation observed in *C. elegans* on Zic sites: in both cases the Zic:TCF:β-catenin association is unable to activate transcription. Interestingly, it has been recently observed that in ascidian also a Zic protein physically interacts with a TCF factor (Oda-Ishii et al. 2016). This suggests that modulating the response to the Wnt/β--catenin pathway via a direct interaction with TCF is a general function of Zic factors in animals.

4.5 Conclusion

For many years *C. elegans* has been used as a model organism to characterize the function of several conserved gene families (Kaletta and Hengartner 2006). *C. elegans* has one Zic family member, and so far only few functional studies have been conducted to determine its role during development. However, these studies have revealed several interesting properties that seem to be shared with other animals. The *C. elegans* Zic factor plays an important role during nervous system development, a property conserved in many animals from cnidarians to vertebrates, suggesting that it represents an ancient function of Zic factors. In *C. elegans*, Zic is not required for the generation of neurons in general, but it plays a role in the specification of some neuronal subtypes. Its mode of action has been characterized in one specific cholinergic lineage where it acts transiently in the neuronal precursor to initiate the terminal differentiation program. It will be interesting to determine whether Zic acts at a similar step in the specification cascade of the other neuronal subtypes that it regulates.

C. elegans studies have also revealed that Zic directly interacts with TCF and plays a role in the transcriptional regulation by Wnt/β-catenin signaling. Zic factors bind TCF and modulate the activity of the Wnt/β-catenin pathway in other animals as well, pointing toward an ancestral property of Zic proteins. In *C. elegans*, Zic interacts with TCF to regulate the expression of "opposite" target genes of the Wnt pathway. This atypical mode of action for TCF has so far only been observed for one target gene. It will be important to determine whether a similar mechanism regulates the expression of other "opposite" target genes in *C. elegans* and if this mode of action is conserved in other animals.

Acknowledgments We thank Harold Cremer, Antoine Barrière, and Sabrina Murgan for comments on the manuscript. Work in the V.B. lab is funded by an ATIP/Avenir start-up grant from CNRS/INSERM, grants from the Fédération pour la Recherche sur le Cerveau, and from the Agence Nationale de la Recherche (ANR-17-ERC2-0018, ANR-14-CE11-0001 and ANR-11-LABX-0054). G.B. is supported by a PhD fellowship from the Fondation pour la Recherche Médicale (FDT20160435125).

References

Alper S, Kenyon C (2002) The zinc finger protein REF-2 functions with the Hox genes to inhibit cell fusion in the ventral epidermis of C. elegans. Development 129(14):3335–3348

Altun-Gultekin Z, Andachi Y, Tsalik EL, Pilgrim D, Kohara Y, Hobert O (2001) A regulatory cascade of three homeobox genes, ceh-10, ttx-3 and ceh-23, controls cell fate specification of a defined interneuron class in C. elegans. Development 128(11):1951–1969

Aruga J (2004) The role of Zic genes in neural development. Mol Cell Neurosci 26(2):205–221. https://doi.org/10.1016/j.mcn.2004.01.004

Aruga J, Kamiya A, Takahashi H, Fujimi TJ, Shimizu Y, Ohkawa K, Yazawa S, Umesono Y, Noguchi H, Shimizu T, Saitou N, Mikoshiba K, Sakaki Y, Agata K, Toyoda A (2006) A wide-range phylogenetic analysis of Zic proteins: implications for correlations between protein structure conservation and body plan complexity. Genomics 87(6):783–792. https://doi.org/10.1016/j.ygeno.2006.02.011

Bertrand V (2016) Beta-catenin-driven binary cell fate decisions in animal development. Wiley Interdiscip Rev Dev Biol 5(3):377–388. https://doi.org/10.1002/wdev.228

Bertrand V, Hobert O (2009) Linking asymmetric cell division to the terminal differentiation program of postmitotic neurons in C. elegans. Dev Cell 16(4):563–575. https://doi.org/10.1016/j.devcel.2009.02.011

Bertrand V, Bisso P, Poole RJ, Hobert O (2011) Notch-dependent induction of left/right asymmetry in C. elegans interneurons and motoneurons. Curr Biol 21(14):1225–1231. https://doi.org/10.1016/j.cub.2011.06.016

Brenner S (1974) The genetics of Caenorhabditis elegans. Genetics 77(1):71–94

Cadigan KM (2012) TCFs and Wnt/beta-catenin signaling: more than one way to throw the switch. Curr Top Dev Biol 98:1–34. https://doi.org/10.1016/B978-0-12-386499-4.00001-X

Cimbora DM, Sakonju S (1995) Drosophila midgut morphogenesis requires the function of the segmentation gene odd-paired. Dev Biol 169(2):580–595. https://doi.org/10.1006/dbio.1995.1171

Clevers H, Nusse R (2012) Wnt/beta-catenin signaling and disease. Cell 149(6):1192–1205. https://doi.org/10.1016/j.cell.2012.05.012

Fujimi TJ, Hatayama M, Aruga J (2012) Xenopus Zic3 controls notochord and organizer development through suppression of the Wnt/beta-catenin signaling pathway. Dev Biol 361(2):220–231. https://doi.org/10.1016/j.ydbio.2011.10.026

Hobert O (2016) A map of terminal regulators of neuronal identity in Caenorhabditis elegans. Wiley Interdiscip Rev Dev Biol 5(4):474–498. https://doi.org/10.1002/wdev.233

Hobert O, Mori I, Yamashita Y, Honda H, Ohshima Y, Liu Y, Ruvkun G (1997) Regulation of interneuron function in the C. elegans thermoregulatory pathway by the ttx-3 LIM homeobox gene. Neuron 19(2):345–357

Hoverter NP, Waterman ML (2008) A Wnt-fall for gene regulation: repression. Sci Signal 1(39):pe43. https://doi.org/10.1126/scisignal.139pe43

Kaletta T, Hengartner MO (2006) Finding function in novel targets: C. elegans as a model organism. Nat Rev Drug Discov 5(5):387–398. https://doi.org/10.1038/nrd2031

Kaletta T, Schnabel H, Schnabel R (1997) Binary specification of the embryonic lineage in Caenorhabditis elegans. Nature 390(6657):294–298. https://doi.org/10.1038/36869

Layden MJ, Meyer NP, Pang K, Seaver EC, Martindale MQ (2010) Expression and phylogenetic analysis of the zic gene family in the evolution and development of metazoans. EvoDevo 1(1):12. https://doi.org/10.1186/2041-9139-1-12

Lin R, Hill RJ, Priess JR (1998) POP-1 and anterior-posterior fate decisions in C. elegans embryos. Cell 92(2):229–239

Lindgens D, Holstein TW, Technau U (2004) Hyzic, the hydra homolog of the zic/odd-paired gene, is involved in the early specification of the sensory nematocytes. Development 131(1):191–201. https://doi.org/10.1242/dev.00903

MacDonald BT, Tamai K, He X (2009) Wnt/beta-catenin signaling: components, mechanisms, and diseases. Dev Cell 17(1):9–26. https://doi.org/10.1016/j.devcel.2009.06.016

Mizumoto K, Sawa H (2007) Two betas or not two betas: regulation of asymmetric division by beta-catenin. Trends Cell Biol 17(10):465–473. https://doi.org/10.1016/j.tcb.2007.08.004

Murgan S, Bertrand V (2015) How targets select activation or repression in response to Wnt. WormBook 4(4):e1086869

Murgan S, Kari W, Rothbacher U, Iche-Torres M, Melenec P, Hobert O, Bertrand V (2015) Atypical transcriptional activation by TCF via a zic transcription factor in C. elegans neuronal precursors. Dev Cell 33(6):737–745. https://doi.org/10.1016/j.devcel.2015.04.018

Nelson FK, Riddle DL (1984) Functional study of the Caenorhabditis elegans secretory-excretory system using laser microsurgery. J Exp Zool 231(1):45–56. https://doi.org/10.1002/jez.1402310107

Oda-Ishii I, Kubo A, Kari W, Suzuki N, Rothbacher U, Satou Y (2016) A maternal system initiating the zygotic developmental program through combinatorial repression in the ascidian embryo. PLoS Genet 12(5):e1006045. https://doi.org/10.1371/journal.pgen.1006045

Phillips BT, Kimble J (2009) A new look at TCF and beta-catenin through the lens of a divergent C. elegans Wnt pathway. Dev Cell 17(1):27–34. https://doi.org/10.1016/j.devcel.2009.07.002

Pourebrahim R, Houtmeyers R, Ghogomu S, Janssens S, Thelie A, Tran HT, Langenberg T, Vleminckx K, Bellefroid E, Cassiman JJ, Tejpar S (2011) Transcription factor Zic2 inhibits Wnt/beta-catenin protein signaling. J Biol Chem 286(43):37732–37740. https://doi.org/10.1074/jbc.M111.242826

Sawa H, Korswagen HC (2013) Wnt signaling in C. elegans. WormBook:1–30. https://doi.org/10.1895/wormbook.1.7.2

Sulston JE, Horvitz HR (1977) Post-embryonic cell lineages of the nematode, Caenorhabditis elegans. Dev Biol 56(1):110–156

Sulston JE, Schierenberg E, White JG, Thomson JN (1983) The embryonic cell lineage of the nematode Caenorhabditis elegans. Dev Biol 100(1):64–119. doi:0012-1606(83)90201-4 [pii]

Sundaram MV, Buechner M (2016) The Caenorhabditis elegans excretory system: a model for tubulogenesis, cell fate specification, and plasticity. Genetics 203(1):35–63. https://doi.org/10.1534/genetics.116.189357

Takahashi H, Shimizu T, Aruga J (2008) Expression pattern of annelid Zic in embryonic development of the oligochaete Tubifex tubifex. Dev Genes Evol 218(10):553–560. https://doi.org/10.1007/s00427-008-0252-x

Wenick AS, Hobert O (2004) Genomic cis-regulatory architecture and trans-acting regulators of a single interneuron-specific gene battery in C. elegans. Dev Cell 6(6):757–770. https://doi.org/10.1016/j.devcel.2004.05.004

Chapter 5
Lophotrochozoan *Zic* Genes

Jun Aruga

Abstract Lophotrochozoa is a sister taxon of Ecdysozoa in the Protostomia that includes mollusks, annelids, brachiopods, and platyhelminths. Recent studies have clarified the structure, expression, and roles of lophotrochozoan *Zic* family genes. *Zic* genes in oligochaete annelid *Tubifex tubifex* (freshwater sludge worm) and polychaete annelid *Capitella teleta* (bristle worm) are commonly expressed in a subset of developing brain and mesoderm derivatives. The latter includes the naïve mesoderm and the associated chaetal sacs in each body segment, although the segmentation processes differ between the two species. Furthermore, in brachiopod *Terebratalia transversa* (lamp shell), Zic is expressed in the anterior ectodermal domains and mesodermal derivatives, including those associated with the chaetal sacs. This result suggests the common involvement of *Zic* genes in the development of chaetae, a lophotrochozoan novelty acquired in the course of evolution. In addition, the highly simplified lophotrochozoan *Dicyema acuticephalum* (dicyemid mesozoan, a cephalopod endoparasite), which lost its gut, nervous system, and muscles during evolution, expresses its *Zic* genes in hermaphroditic gonads, highlighting the role of *Zic* genes in germ cell development. The role of Zic in head regeneration was revealed in studies on platyhelminth *Schmidtea mediterranea* (freshwater planarian). Planarian Zic expression was induced in a subpopulation of neoblasts that includes adult pluripotent stem cells. It is needed for head regeneration and production of an anterior signaling center. Suppression of Wnt-β-catenin signaling underlies Zic-mediated head regeneration, reminiscent of Wnt-β-catenin suppression by vertebrate *Zic* genes. Taken together, studies on the lophotrochozoan *Zic* genes are essential to understanding not only the roles of these genes in body plan evolution but also the molecular mechanism underlying adult stem cell regulation.

Keywords Zic · Lophotrochozoa · Annelida · Brachiopoda · Dicyemida · Platyhelminthes · Development · Regeneration · Gene expression

---

J. Aruga (✉)
Department of Medical Pharmacology, Nagasaki University Graduate School of Biomedical Sciences, Nagasaki, Japan
e-mail: aruga@nagasaki-u.ac.jp

J. Aruga (ed.), *Zic family*, Advances in Experimental Medicine and Biology 1046,
https://doi.org/10.1007/978-981-10-7311-3_5

5.1 Lophotrochozoans

The lophotrochozoans are a monophyletic group of animals that includes the annelids, mollusks, brachiopods, platyhelminths, and other animals. Lophotrochozoa is one of three major clades of Bilateria (animals with bilateral symmetry) and is a sister taxon of Ecdysozoa (arthropods and nematodes and their relatives). The first part of the name Lophotrochozoa derives from members of the group (brachiopods and phoronids) that have a ciliated feeding organ called a lophophore. The second part of the name derives from the remaining groups (mollusks, annelids, and platyhelminths), which have a larval stage known as a trochophore (Telford 2006). None of the lophotrochozoan phyla have both defining characteristics, meaning that there is no single morphological characteristic uniting the Lophotrochozoa (Telford 2006). However, the monophyly of lophotrochozoans is well supported by several molecular phylogenetic studies, including those of 18S and/or 28S ribosomal subunit genes (Halanych et al. 1995; Mallatt and Winchell 2002) and expressed sequence tags (Dunn et al. 2008).

Recent studies suggest the presence of two major groups of Lophotrochozoa. One is the Trochozoa, which includes the mollusks, annelids, brachiopods, phoronids, and nemerteans. The other group, Platyzoa, has been proposed to include the platyhelminths and others (Halanych 2016). However, whether or not Platyzoa can be considered a monophyletic group is still under debate (Struck et al. 2014).

One difficulty in lophotrochozoan phylogeny is the reduced morphology and modified life cycles that result from their parasitic lifestyle (Halanych 2016). In this regard, dicyemid mesozoans might be a model of body plan simplification in lophotrochozoan evolution. Dicyemids are endoparasites or endosymbionts of cephalopods, and their bodies usually consist of 10–40 cells with neither body cavities nor differentiated organs (Furuya and Tsuneki 2003). Based on the presence of a lophotrochozoan-type peptide motif in a dicyemid *Hox* gene (*DoxC*), dicyemids are proposed to derive from a lophotrochozoan ancestor (Kobayashi et al. 1999).

Although the relationships between some lophotrochozoan phyla are still controversial (Telford 2006; Halanych 2016), molecular phylogenetic studies of the developmentally critical toolkit genes have strongly contributed to a better understanding of lophotrochozoan animals from an evolutionary point of view. These include components of the TGF-β signaling pathway (Kenny et al. 2014), basic helix-loop-helix-type transcription factors (Bao et al. 2017), and *Hox* genes (Lee et al. 2003; Steinmetz et al. 2011; Biscotti et al. 2014; Schiemann et al. 2017).

5.2 Comparisons of the Expression Patterns of *Zic* Genes in Polychaete and Oligochaete Annelids

5.2.1 Annelid Zic *Genes*

We will first look at *Zic* genes in the annelids, also known as ringed worms or segmented worms. Phylum Annelida includes two main classes: Polychaeta and Clitellata (Ruppert et al. 2004). The latter includes Oligochaeta and Hirudinea as subclasses. Polychaeta, Oligochaeta, and Hirudinea are represented by the rag worm, earthworm, and leech, respectively. Their body is composed of three regions: the prostomium, trunk, and pygidium (Fig. 5.1). The elongate trunk consists of a longitudinal series of similar body units: the segments. At the anterior end of the trunk is the prostomium, which contains the brain and sense organs. The pygidium includes the anus at the posterior end of the body. In each segment, there are chaetae (chitinous bristles) that project outward from the epidermis to provide traction and perform other tasks. Typically, they appear in bilaterally paired bundles, with one dorsolateral and one ventrolateral pair per segment (Fig. 5.1). Each chaeta arises from a pit-like epidermal follicle composed of follicle cells and a single chaetoblast at the base of the follicle. The nervous system consists of the anterior dorsal brain in the prostomium and a ventral pair of longitudinal nerve cords. The nerve consists of the suprapharyngeal ganglia and subpharyngeal ganglia, which are connected circumferentially around the pharynx (Fig. 5.1). In each segment, the ventral nerve cords have a pair of segmental ganglions from which segmental nerves enter the body wall and innervate the body wall musculature (Fig. 5.1). The overall structure of the annelid nervous system is known as a segmental nervous system.

Zic homologues have been isolated from polychaete *Capitella teleta* and oligochaete *Tubifex tubifex,* and their expression profiles during development have been described (Layden et al. 2010; Takahashi et al. 2008). Although their embryonic development, in particular the segmentation process, largely differs between these two species, similarities in their developmental program have been pointed out in recent studies. This includes the colinearity of *Hox* gene expression (Endo et al. 2016) and axis organizers (Seaver 2014). Furthermore, the development of both animals starts with spiral egg cleavage in a similar fashion. However, the segmentation processes are largely different. In polychaete *Platynereis*, the segments are acquired after the trochophore (swimming plankton) stage (Fig. 5.1). The segmentation process is regulated by the hedgehog signaling pathway (Dray et al. 2010), similar to that in *Drosophila*. The hedgehog signaling components (hh, ptc, and Gli) are expressed in segment polarity-like patterns, and the chemicals that antagonize the hedgehog signaling disrupt segment formation (Dray et al. 2010). On the other hand, segmentation in oligochaete *Tubifex* derives from cell bandlet-produced prog-

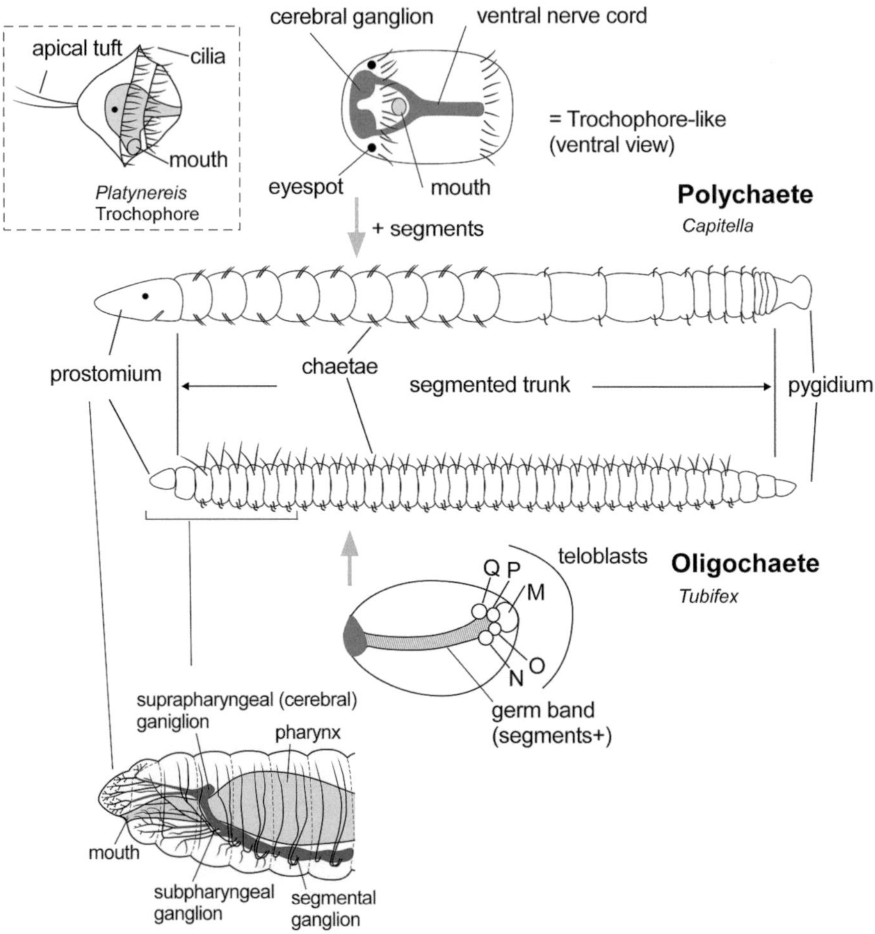

Fig. 5.1 Annelid body plan. Polychaete annelids acquire segments after a trochophore-like stage, whereas oligochaete annelids directly develop an adult-like segmented body without metamorphosis. Gross CNS organization is similar. All cartoons except Capitella embryo (top middle, ventral view) are lateral view. Left side is the anterior in all cartoons

enies of the large posterior teloblast (Figs. 5.1 and 5.2). There are five pairs of teloblasts arranged bilaterally (M, N, O, P, and Q). The teloblasts generate daughter cells one by one through asymmetrical cell division, and the older daughter cells are located on the anterior side of each bandlet. Among the teloblasts, the M blast lineage contributes solely to the mesodermal structures (Goto et al. 1999).

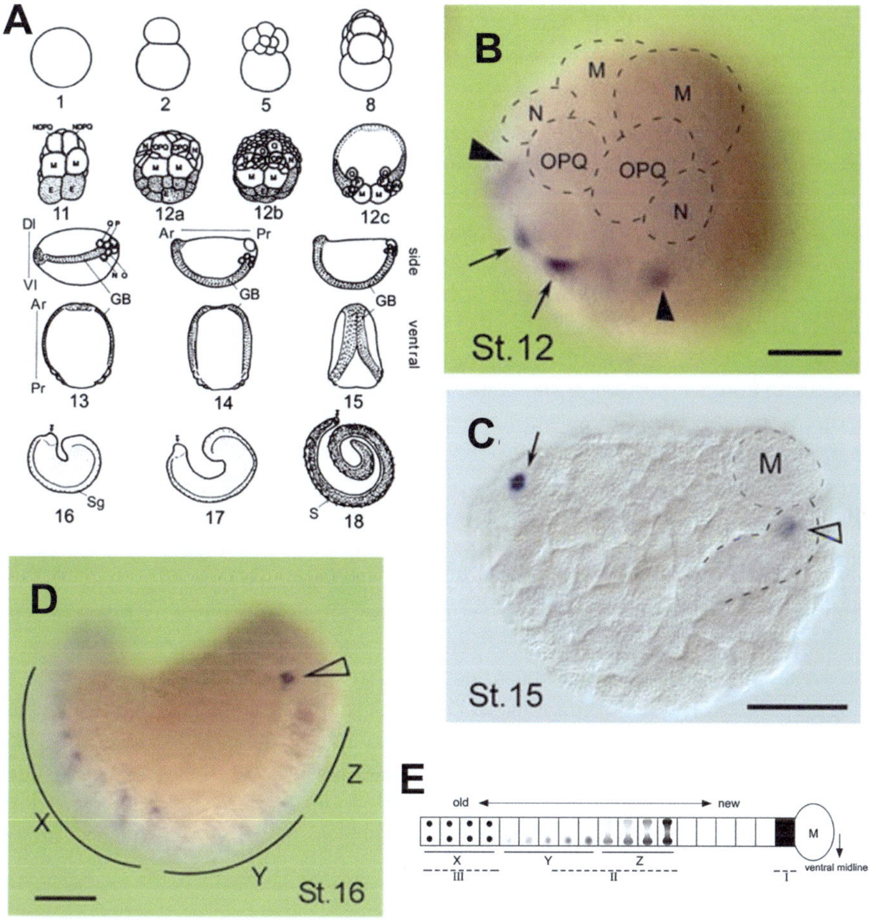

Fig. 5.2 *Tubifex* Zic expression. (**a**) Summary of *Tubifex* development (for details, see Shimizu 1982). *Ar*, anterior; *Dl*, dorsal; *E*, endodermal cells; *GB*, germ band; *Pr*, posterior; *S*, setae; *Sg*, segment; *Vl*, ventral; *double arrowheads*, prostomium. M, N, O, P, and Q indicate teloblasts, and NOPQ and OPQ indicate their precursors. (**b**) Antero-dorsal view of a stage 12 embryo shortly after division of NOPQ into N and OPQ. *Arrows* indicate a pair of anteriorly located surface cells. *Solid arrowheads* indicate stained cells located internally. (**c**) Stage 15 embryos viewed from the left side (anterior to left, cleared specimen). Mesodermal germ bands are demarcated by broken lines. Circles of broken lines indicate M teloblasts. *Arrow* indicates CNS progenitor-like signal; *open arrowhead* indicates a primary m-blast cell. (**d**) Stage 16 embryo side view (anterior to left). *X*, *Y*, and *Z* indicate the three domains; *open arrowhead* indicates a primary m-blast cell. (**e**) Schematic illustration of Zic expression in the mesodermal germ bands derived from M teloblasts. Each square represents a mesodermal segment that coincides with the body segment. Anterior is to the left; dorsal is at the top. The three domains *X*, *Y*, and *Z* correspond to those indicated in (**d**). *Roman numerals* represent the tentative triphasic expression pattern of Zic (Reprinted from Takahashi et al. 2008)

5.2.2 Comparison of Mesodermal Expression in Oligochaetes and Polychaetes

In oligochaete mesodermal development, *Zic* expression was first detected in the newest M daughter cells (called m-blasts) and, later, in differentiating mesodermal cells (Takahashi et al. 2008). The earliest expression is transient and ceases upon the first cell division (Fig. 5.2c–e). The later phase of expression was observed initially as a dumbbell-shaped structure and subsequently as intensely stained paired dots (Fig. 5.2d, e). The dots are thought to be mesodermal tissues surrounding chaetal sacs (Takahashi et al. 2008).

In polychaetes, mesodermal *Zic* expression has been detected in one or a few cells in an anterior portion of the bilateral mesodermal bands, in the visceral mesoderm surrounding the foregut, in the head mesoderm at the anterior tip of the animal, and in mesoderm associated with forming chaetae. The earliest mesodermal expression was found in the anterior portion of the bilateral mesodermal domain (Fig. 5.3a, b) (Layden et al. 2010). However, its expression in the bilateral mesodermal band seems to cease by the following stage, and segmentally reiterated *Zic*-expressing cell cluster expression can be seen in the lateral bands (Fig. 5.3c, d). The number of segmental clusters increases along the anterior-posterior axis as new segments form. Later, the expression of bilateral mesoderm derivatives can be seen in cells wrapping the chaetal sacs (Layden et al. 2010).

Thus, the expression of polychaete *Zic* in the segmented mesoderm is somewhat similar to that of oligochaete *Zic*. It is remarkable that the expression profiles are conserved despite the largely different segmentation processes in polychaetes and oligochaetes. The expression of oligochaete *Zic* in visceral mesoderm surrounding the foregut and in the head mesoderm has not been described but awaits further expression analysis.

5.2.3 Comparisons of Zic Expression in Oligochaete and Polychaete Neural Development

In oligochaete *Tubifex*, the brain (cerebral ganglions) is known to derive from micromeres. In developing oligochaete embryos, one of the earliest Zic expressions is in a pair of anterior surface micromere clusters (Fig. 5.2b). It is possible that these cells include primitive cerebral ganglions. However, no lineage-tracing data exists for brain-forming cells in oligochaetes today, and thus their cell identity remains uncertain. Zic-expressing anterior end cell clusters can be detected during the following stages and appear to be internalized (Fig. 5.2c). The *Zic*-expressing cell clusters in the anterior tip (prostomium) partly overlap with *Emx* expression (Takahashi et al. 2008). Considering that *Emx* homologues are expressed in early embryonic brains in both *Drosophila* and mouse (Lichtneckert and Reichert 2005), oligochaete *Zic* is likely to be expressed in the cerebral ganglions, at least in part.

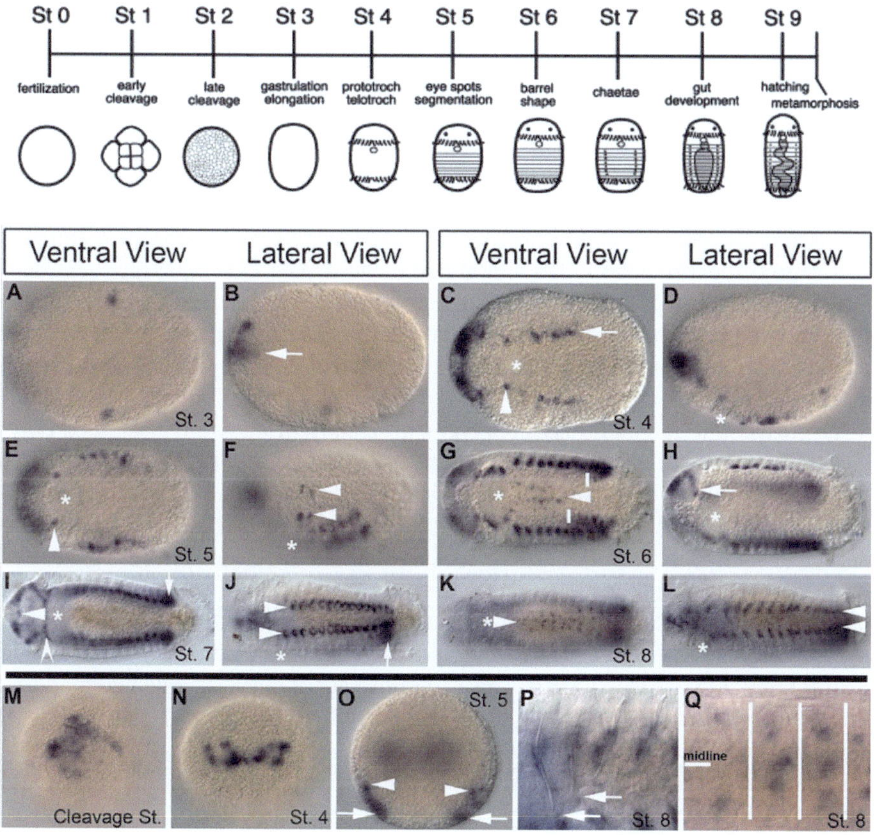

Fig. 5.3 *Capitella* Zic expression. Expression of Zic at cleavage (M), stage 3 (A and B), stage 4 (C, D, and N), stage 5 (E, F, and O), stage 6 (G and H), stage 7 (I and, J), and stage 8 (K, L, P and Q). Views are as indicated except that H is a ventral-lateral view; M is an animal view; N and O are anterior views; P is a lateral view; and Q is a ventral view. In all lateral and ventral views, the anterior is to the *left*, and ventral is down in all lateral views. An *asterisk* (*) indicates the relative mouth position. Zic is detectable in micromeres on an animal pole (M). Zic mesodermal expression is detected in the anterior mesodermal band on the ventral side of stage 3 animals (A and B) and in two bilateral mesoderm domains adjacent to the foregut at stages 4 and 5 (C and E, *arrowheads*). The mesodermal expression expands around the foregut during stages 6 (G) and 7 (I, notched *arrowhead*). At stage 5, four lateral mesodermal domains (F, *arrowheads*) are detected. At stage 7, lateral mesoderm domains form two longitudinal rows along the anterior-posterior (A-P) axis (J, arrowheads). At stage 8, the longitudinal domains are clearly mesoderm associated with chaetae (L, *arrowheads*, and P), but not the chaetal sac (P, *arrows*). Zic is expressed in the presumptive brain ectoderm at all stages (B, *arrow*, C-L, and N). Ventrolateral expression at the lateral edge of ventral neural ectoderm is detected at stage 4 (C, *arrow*) and stage 5 (F and O, *arrows*). The lateral neural ectoderm domain is downregulated in an A-P wave apparent at stage 6 (G, *white line* demarcates anterior border of expression) but remains expressed in the growth zone (I and I, *arrows*). Zic is also detectable in neurons in the brain (H, *arrow*) and, beginning at stage 6, in the ventral nerve cord (VNC; G, *white arrowhead*). VNC expression persists through stage 8 (K, *arrowhead*, and Q). Medial and lateral ganglions are detected in each segment (Q, which is a higher magnification of L). In Q, the *white lines* demarcate segmental boundaries (Reprinted from Layden et al. 2010)

In polychaete *Capitella*, neuroectodermal expression of *Zic* is detected at the bilaterally symmetrical anterior neuroectoderm and in a subregion of the developing brain. Expression in the anterior neuroectoderm region is maintained throughout development. Initially, *Zic*-expressing cells are thought to form brain neurons, based on previous fate-mapping studies (Meyer and Seaver 2009; Meyer et al. 2010). The expression of the *Capitella* bHLH-type neurogenic gene *ash* seems to partly overlap with the *Zic*-expressing region (Layden et al. 2010; Meyer and Seaver 2009). In addition to the anterior neuroectoderm, *Zic*-expressing cells exist in the ventrolateral neuroectoderm to form the ventral nerve chord. The *Zic*-expressing ventrolateral neuroectoderm partly overlaps with that of *ash* expression and is proposed to be the lateral edge of the presumptive ventral neuroectoderm (Layden et al. 2010). At a later stage, *Zic* expression is observed at the medial and lateral ganglions in each segment of the ventral nerve cord.

Expression in the brain may be a common feature of the oligochaete and polychaete *Zic* genes. Although properties of the early neuroectoderm in the oligochaete have not been described, expression profiles prior to brain establishment look similar in oligochaete and polychaete animals. In the polychaete *Capitella*, neuroectodermal development has been described. The neuronal precursor cells are generated by ingression as single cells from the anterior surface cell layer (ectoderm), and prior to this process, *ash* is expressed (Meyer and Seaver 2009). *Capitella Zic* is likely to be co-expressed in the neural ectoderm at the gastrulation stage (Meyer and Seaver 2009; Layden et al. 2010). This is reminiscent of the vertebrate neural induction process (Nakata et al. 1997; Kuo et al. 1998). Together with the sequential expression of *Zic* and *ash* in cnidarians (Chap. 2), studies on annelid *Zic* implicate the eumetazoan-wide usage of *Zic* family genes in the initial stage of neurogenesis.

5.3 Zic Expression in Brachiopods

Despite the large number (more than 140,000) of animal species included in this major animal taxon (Ruppert et al. 2004), lophotrochozoan *Zic* expression profiles during embryonic development are limited. However, a recent study that includes the *Zic* expression profile of the brachiopod *T. transversa* sheds light on the conserved expression features of different lophotrochozoan phyla (Schiemann et al. 2017). Brachiopods are commonly known as lamp shells and resemble bivalve mollusks. In the brachiopod's late gastrula stage, Zic is expressed in the mesoderm of the chaetal sacs, apical lobe mesoderm, and anterior ectoderm, and in early larvae, expression is detected in the chaetal sacs, pedicle lobe, anterior mesoderm, and anterior ectoderm (Fig. 5.4c, d). Schiemann et al. (2017) found brachiopod *Zic* expression in the mesodermal cells surrounding the chaetal sacs, similar to annelid *Zic* expression surrounding the chaetal sacs. They also showed that the homeodomain-containing transcription factor *Arx* (Fig. 5.4a, b) and Notch signaling components (*Notch*, *Delta*, *Hes1*, and *Hes2*) are co-expressed in the chaetal sacs or associated

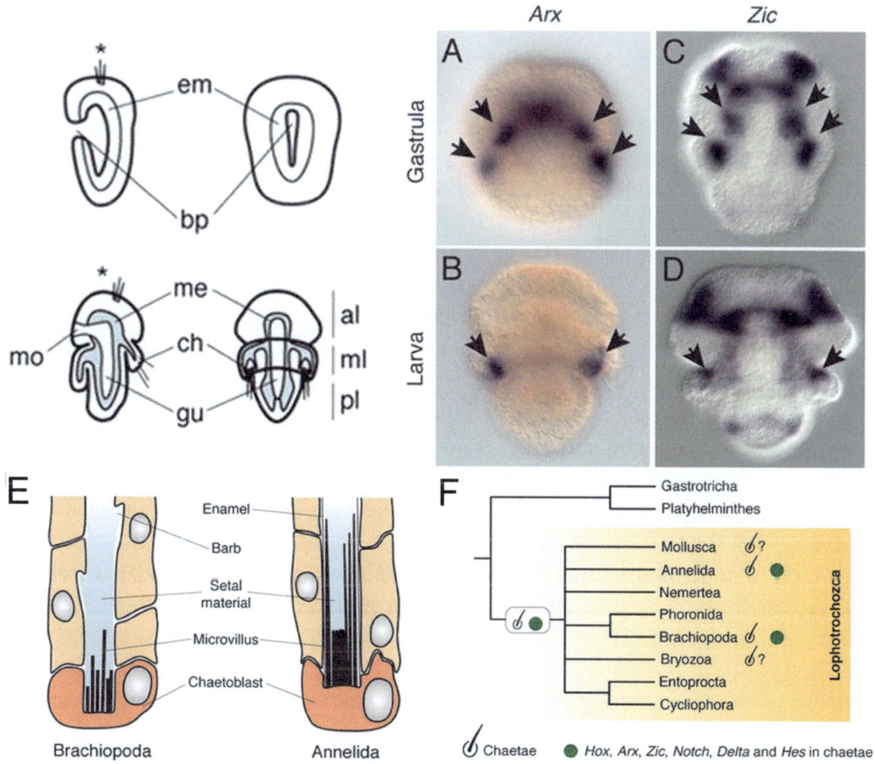

Fig. 5.4 Similarity of chaetae between brachiopods and annelids. Expression of Arx (A, B) and Zic (C, D) genes in gastrula (A, C) embryos and larvae (B, D) of *Terebratalia*. (**a**) In mid-gastrulae, Arx is expressed in the ectoderm of the prospective chaetal sac territories (*black arrows*) and in a ventral domain. (**b**) In early larvae, Arx is expressed in the chaetal sacs (*black arrows*). (**c**) In late gastrulae, Zic is expressed in the mesoderm of the chaetal sacs (*black arrows*), apical lobe mesoderm, and anterior ectoderm. (**d**) In early larvae, Zic is detected in the chaetal sacs (*black arrows*), in a domain in the pedicle lobe, and in the anterior mesoderm and anterior ectoderm. *Asterisk*, anterior/animal pole; *al*, apical lobe; *bp*, blastopore; *ch*, chaetae; *em*, endomesoderm; *gu*, gut; *me*, mesoderm; *ml*, mantle lobe; *mo*, mouth; *pl*, pedicle lobe. (**e**) Morphological similarities between brachiopod and annelid chaetae. (**f**) The shared morphological and molecular characters of chaetae in Brachiopoda and Annelida, together with the presence of chaetae-like structures (chaetae sign with a question mark) in the Mollusca and Bryozoa, support the homology of this lophotrochozoan novelty (**a–d**, **f** Reprinted from Schiemann et al. 2017 with permission. **e** Adapted from Gustus and Cloney 1972 with permission. The modification was done in Schiemann et al. 2017)

tissue. Based on the conserved fashion of toolkit gene usage and the structural similarity between annelid chaetae and brachiopod chaetae (Fig. 5.4e, f), they argued that brachiopod and annelid chaetae are homologous. Because chaeta-like structures are observed in other lophotrochozoan phyla, such as mollusks, further studies on Zic expression profiles in mollusks are expected.

5.4 Dicyemid *Zic* Genes

Dicyemid mesozoans are considered to be highly simplified bilaterians. Some molecular phylogenetic studies show an affinity to the Lophotrochozoa (Kobayashi et al. 1999; Suzuki et al. 2010). Dicyemid mesozoans (phylum Dicyemida) are parasites found in the kidneys of cephalopod mollusks (Furuya and Tsuneki 2003). Their body consists of 9–41 cells, having neither a body cavity nor differentiated organs such as nervous or gastrointestinal systems (Fig. 5.5a, b). Their life history consists of two phases (Fig. 5.5b, d). One is the vermiform stage, in which vermiform embryos are formed asexually in the renal sac of the host. The other is the infusoriform larva, which develops from a fertilized egg produced around hermaphroditic gonads called infusorigens and can escape from the host into seawater. Because both fertilization and embryonic development can be seen in the worm body, dicyemids are good subjects for developmental analysis.

We identified two *Zic* homologues, *ZicA* and *ZicB*, in the dicyemid genome (Aruga et al. 2007). The genes possessed a bilaterian-specific intron (see Chap. 1) that was very short (27 or 28 nt). Uniform short intron length is not limited to *Zic* genes, but is a general feature of the dicyemid genes (Ogino et al. 2010). Among 210 introns from 39 dicyemid genes, the mean ± SD length was 26.1 ± 19.2 nt, and 99.6% of introns ranged from 21 to 32 bp. Concerning the biological significance of the short/uniform sized introns, natural selection appears to favor transcripts with sort-intron sequences to minimize the cost and time of gene expression in metazoans.

Dicyemid *Zic* genes are expressed in the hermaphroditic gonads (Fig. 5.5c, d), and therefore their possible roles are proposed to be gametogenesis or differentiation of gametes (Aruga et al. 2007). Zic involvement in gametogenesis has not been described for other animals. The results may indicate the versatility of Zic by extending the developmental contexts where Zic can be involved. More importantly, they raise a new question about the involvement of the Zic family in gametogenesis of other animals. Interestingly, mouse *Zic3* is highly expressed in the primordial germ cells of both male and female gonads (NCBI GDS/10599812).

Fig. 5.5 (continued) nematogen, rhombogen, and vermiform embryo. The development of infusorigens, gametogenesis around the infusorigen, and development of the two types of embryo all proceed within the axial cell. *AG* agamete, *AX* axial cell, *CL* calotte, *D* developing infusoriform embryo, *DV* developing vermiform embryo, *F* fertilized egg, *H* hermaphroditic gonad (infusorigen), *IN* infusoriform embryo, *O* oogonium, *P* peripheral cell, *PO* primary oocyte, *S* spermatogonium, *SP* sperm, *V* vermiform embryo, *2C* 2-cell-stage embryo. (**c**) ZicA expression in infusorigen. (**d**) ZicB expression in the infusorigen and 2-cell-stage embryo. (**e**) Summary of *Dicyema* ZicA, ZicB, Brachyury homologue (Do-bra), Otx homologue (Do-otx), Hox homologue (DoxC) expression in the dicyemid life cycle (**a**–**d** Reprinted from Aruga et al. 2007. **e** Modified from Kobayashi et al. 2009)

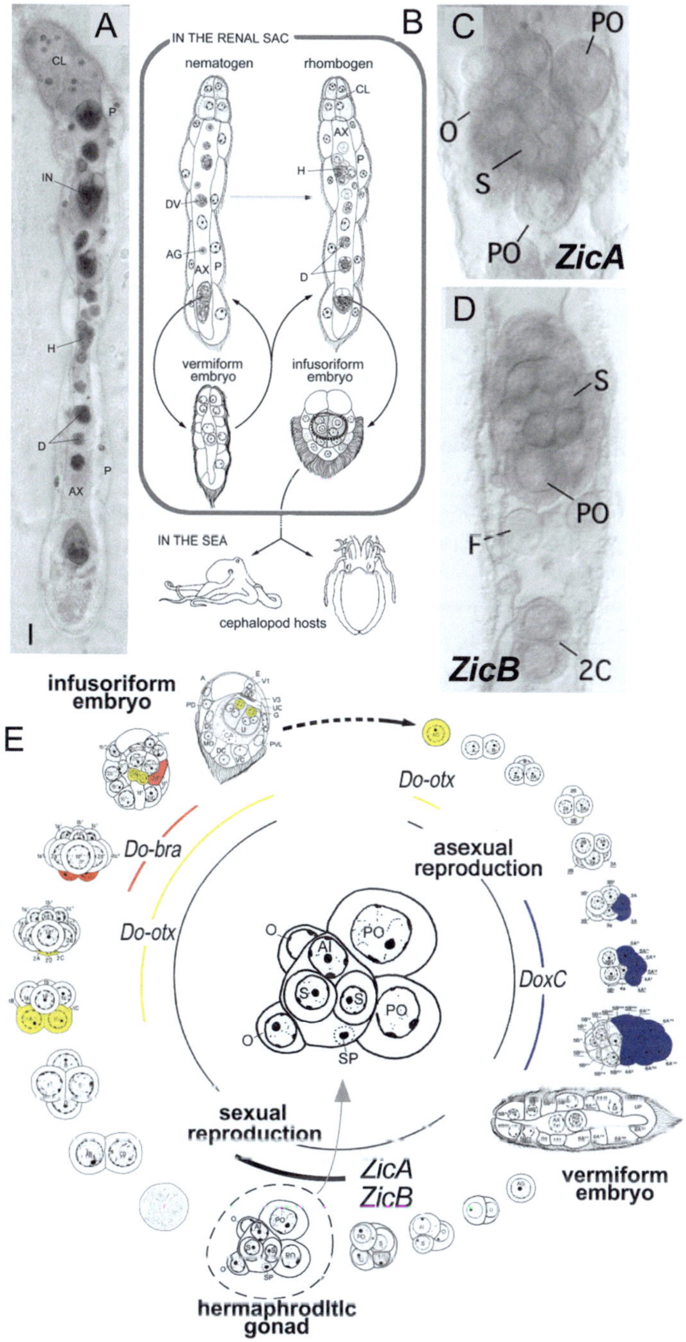

Fig. 5.5 Dicyemid zic expression. (**a**) Light micrograph of a rhombogen. Note that embryos develop in the axial cell. Scale bar, 10 μm. (**b**) Life cycle of the dicyemid (After Furuya and Tsuneki 2003). The dotted line indicates an unknown process. The vermiform stage includes the

5.5 Roles of *Zic* Genes in Planarian Head Regeneration

Planarians are flatworms belonging to the Turbellaria class of phylum Platyhelminthes. Biologists have been fascinated by planarians' enormous regenerative abilities. Following pioneering studies to establish freshwater planarian *Dugesia* species (e.g., *S. mediterranea* [also called *Dugesia mediterranea*], *Dugesia tigrina*, and *Dugesia japonica*) as an experimental model system for current molecular genetic analyses (Garcia-Fernandez et al. 1993; Umesono et al. 1997), important findings have been accumulating to explain the molecular mechanisms underlying planarian regeneration. Planarian adult pluripotent stem cells, called neoblasts, produce all differentiated cell types needed for whole-body regeneration (Wagner et al. 2011). Neoblasts are the only known proliferating cells in planarians (Newmark and Sanchez Alvarado 2000) and are instructed to proliferate, migrate, and differentiate into the required cell types in response to injury. Recent technical advances such as global gene expression analysis, RNAi, and small molecule treatment have enabled researchers to identify the signaling cascades involved in the regeneration processes. These include Wnt, BMP, and hedgehog (Hh), all of which are known to be involved in Zic-mediated developmental controls in other animal models and in other developmental contexts. However, neither the involvement of the Zic protein in planarian regeneration nor its relationship to known biological signals has been explained. Here we review the role of planarian Zic in the head regeneration process in two recent independent reports (Vasquez-Doorman and Petersen 2014; Vogg et al. 2014).

There are two planarian *Zic* genes, *zicA* and *zicB*, in both *S. mediterranea* and *D. japonica* (Aruga et al. 2006). *S. mediterranea zicA* and *zicB* are known as *Smed-zic-1* and *Smed-zic-2* (hereafter *zic1* and *zic2*) (Vasquez-Doorman and Petersen 2014), but zic1 is well characterized because of its functional redundancy and clear phenotype of zic1 loss-of-function analysis (Vasquez-Doorman and Petersen 2014). *zic1* is expressed in the anterior pole of intact planarians (Fig. 5.6a) (Vogg et al. 2014; Vasquez-Doorman and Petersen 2014). When the planarians were cut into three pieces to generate anterior (head), middle (trunk), and posterior (tail) pieces, *zic1* expression was enhanced in the anterior-facing side of the head and tail pieces 18 h after amputation, and by 72 h, *zic1* was strongly expressed in the anterior pole and in the surrounding region (Fig. 5.6b). Most *zic1*-expressing cells in the anterior pole were identified as neoblasts at 24 h after amputation. The *zic1*-expressing cells at the anterior pole also expressed notum (secreted Wnt inhibitor), foxD (transcription factor), or follistatin (secreted BMP inhibitor). Double labeling with neoblast markers indicated that some *zic1*-expressing neoblasts form zic1-expressing anterior pole cells.

zic1 knockdown during regeneration resulted in head defects, including cyclopia, absence of the eyes, and head regeneration failure (Fig. 5.6c) (Vasquez-Doorman and Petersen 2014; Vogg et al. 2014). In contrast, tail regeneration was not affected. In the marker analysis, head-defective zic1-suppressed planarians showed the absence of markers for the brain (gpas and chat), head tip (sFRP-1), head region (prep), and

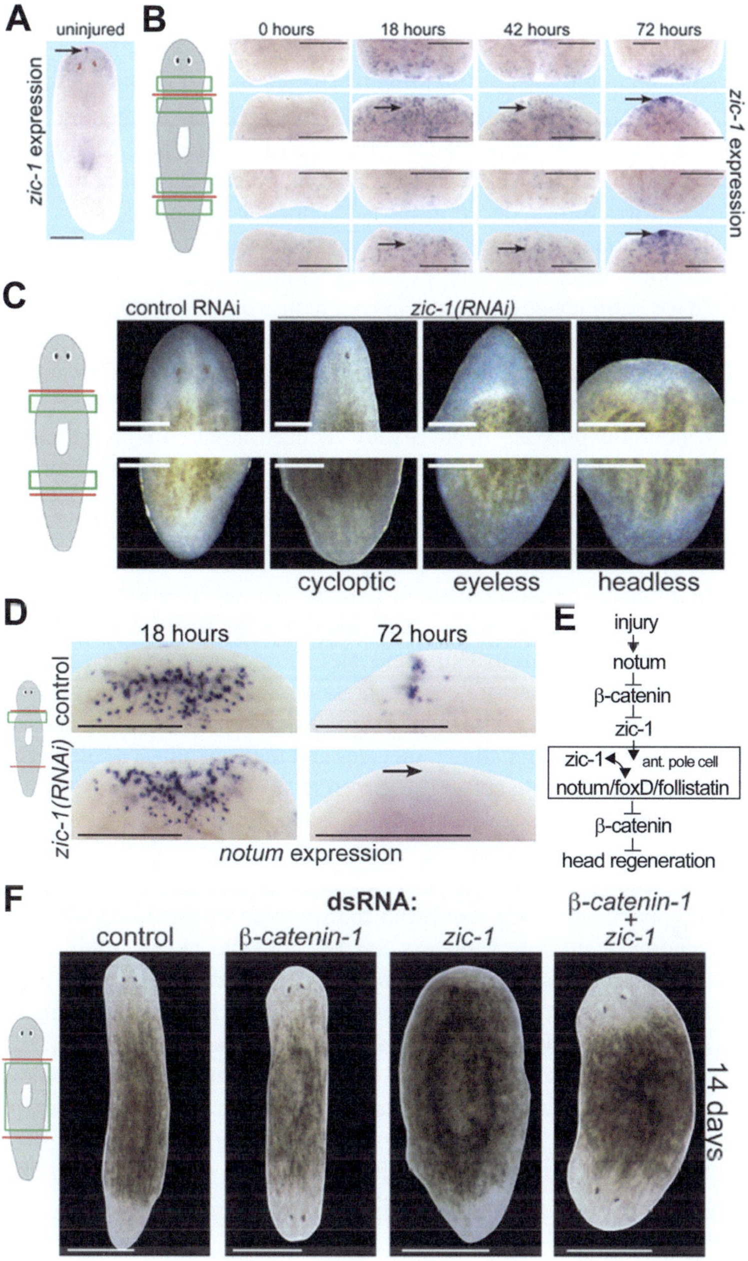

Fig. 5.6 Role of Zic in planarian head regeneration. (**a**) Expression of zic1 in the head region and head tip in uninjured animals (*arrow*). (**b**) Expression of zic-1 in regenerating animals. Zic1 is expressed preferentially near anterior-facing versus posterior-facing amputation (*arrows*). (**c**) Animals 8 days after amputation of heads and tails after zic1 knockdown [zic-1 (RNAi)] and in the control [control (RNAi)]. (**d**) Zic-1 is required for anterior pole formation at 72 h and not for early wound-induced notum expression at 18 h of regeneration. (**e**) Programs underlying zic1-mediated anterior pole cell generation and maintenance. (**f**) Single and double RNAi as indicated to examine interactions between zic-1 and β-catenin (**a**–**d**, **f** Reprinted from Vasquez-Doorman and Petersen 2014)

anterior pole (notum, follistatin, and foxD). The normal tail regeneration and intact neoblast-specific marker expression suggested that zic1 may be required not for the maintenance of neoblasts but for the specification of anterior pole cell types.

Concerning the zic1 downstream targets during head regeneration, ovo, notum, and distalless (dlx) were reduced in the X1 subpopulation of neoblasts from zic1-knockdown planarians (Vasquez-Doorman and Petersen 2014). Ovo is a zinc-finger-type transcription factor required for eye regeneration in planarians and is known to be required for germline development in both mouse and *Drosophila* (Hayashi et al. 2017). Distalless is also required for planarian eye regeneration (Lapan and Reddien 2011). Notum is required for head regeneration and is expressed in the anterior pole (Petersen and Reddien 2011). Accordingly, Vasquez-Doorman and Petersen (2014) found that notum-expressing neoblasts are reduced in zic1-inhibited regenerating heads (Fig. 5.6d), and Vogg et al. (2014) found that notum-expressing anterior pole cells and follistatin-expressing anterior pole cells were reduced in zic1-knockdown regenerating heads.

Interestingly, the regulatory relationship among the zic1, notum, and follistatin is not limited to the regeneration process. This idea comes from the analysis of anterior pole cells in intact planarians. The anterior pole cells in intact head tissue expressed notum, foxD, follistatin, and zic1. Prolonged zic1 RNAi treatments resulted in the disappearance of notum expression in the anterior pole cells, whereas notum RNAi did not affect zic1 expression. These results indicate that zic1 acts upstream of the secreted signaling molecules and that zic1 plays a role in the maintenance of the anterior pole cells. These results suggest the presence of a "zic1→ notum/follistatin" regulatory relationship in both injury-induced regeneration and cell maintenance in intact tissue (Fig. 5.6e).

The relationship between foxD and zic1 is an intriguing issue addressed by the two studies mentioned above. Forkhead transcription factor, foxD, also promotes the specification of midline neoblasts for anterior pole regeneration. Vogg et al. (2014) identified *zic1* as a gene that is downregulated in the foxD1-suppressed regenerating head at 72 h after amputation, and foxD/zic1 suppression showed additive effects on the severity of eye defects. The co-expression of zic1 and foxD was small at an early stage of regeneration, but was extensive in anterior pole cells at 72 h after amputation (Vasquez-Doorman and Petersen 2014). The co-expression profiles of notum-zic1 and follistatin-zic1 were similar to that of foxD-zic1. Therefore, zic1, notum, follistatin, and foxD are not co-expressed early after head amputation, but are later co-expressed at the regenerating anterior pole.

The underlying mechanisms of zic1 induction after wounding have also been reported. Wnt1 and notum are expressed prior to zic1 induction in response to injury. In functional terms, RNAi-mediated notum knockdown reduced zic1 expression on the anterior side, and wnt1 knockdown resulted in ectopic zic1 expression on the posterior side. The involvement of Wnt signaling in the activation of zic1 expression was further confirmed by the knockdown of Wnt signaling downstream components (Vasquez-Doorman and Petersen 2014). This study concluded that Wnt signaling inhibition by early injury-induced notum is necessary and sufficient to activate early zic1 expression by 24 h (Fig. 5.6e) (Vasquez-Doorman and Petersen 2014).

In previous studies, Wnt signaling was shown to play a key role in head regeneration. Functional interaction between zic1 and wnt signaling was further investigated by combining β-catenin (wnt canonical pathway component) knockdown and zic1 knockdown (Fig. 5.6f). When β-catenin was solely suppressed in the trunk pieces, ectopic head regeneration occurred from the posterior side, consistent with its key role in head regeneration. zic1 suppression causes the loss of head structure and intact tail as above. But the combined knockdown treatment abolished the zic1 effect, identical to β-catenin knockdown animals (Fig. 5.6f). This result indicates that β-catenin inhibition can promote head regeneration in the absence of zic1, suggesting a critical role of β-catenin inhibition in zic1-mediated head regeneration (Fig. 5.6e).

Besides the components of Wnt signaling, other candidate genes were examined to determine their requirements for *zic1* expression activation after amputation. Follistatin knockdown resulted in the reduction of the *zic1*-expressing cells at 24 h after amputation. FoxD knockdown resulted in the reduction of *zic1* expression 3 days after amputation (Vogg et al. 2014). In addition, Pbx1 (homeodomain transcription factor) knockdown also reduced the number of *zic1*-expressing cells and *zic1* expression levels 24 days after amputation.

Based on these two studies, the role of planarian Zic in head regeneration can be summarized as below (Fig. 5.6e). At the anterior side of the wounds, *zic1* expression in the stem cells (neoblast) is induced by the secretory wnt signaling inhibitor (notum). zic1 in the stem cells is necessary for the wnt signaling inhibitor and TGFβ (activin) signaling antagonist (follistatin). The anterior-most zic1-expressing stem cells are called anterior pole cells and also express the winged helix-type transcription factor foxD. zic1 and foxD cooperate to establish the anterior pole cell properties.

The method of zic1 involvement in planarian head regeneration is intriguing when compared with the involvement of vertebrate Zic family in the repression of Wnt-β-catenin signaling (Fujimi et al. 2012; Pourebrahim et al. 2011). Xenopus Zic3 suppresses β-catenin signaling to control the organizer, which emanates TGFβ (BMP) signaling antagonists (chordin, follistatin, etc.) (Fujimi et al. 2012). On the other hand, the CNS disorganization caused by planarian zic1 RNAi is similar to brain abnormalities in holoprosencephaly patients caused by ZIC2 loss-of-function mutations (Brown et al. 1998) and in Zic2 knockdown mutant mice (Nagai et al. 2000). In either case, the midline region of the "brain" is impaired. Despite such phenotypic similarities, planarian head regeneration may provide a promising experimental system to clarify the evolutionary conserved gene regulatory network for brain establishment.

5.6 Summary of the Lophotrochozoan *Zic* Genes

Having observed the expression and the role of the lophotrochozoan *Zic* genes, we noticed conservation of their roles in development and regeneration in the neural tissue and mesoderm. The results steadily improved our understanding of bilaterian ancestral characteristics. Furthermore, their possible involvement in novel lophotrochozoan (chaetal sac-associated mesoderm derivative) development and the essential role of planarian *Zic* genes in head regeneration were found based on lophotrochozoan model animals. In addition, Zic in dicyemids, secondarily reduced lophotrochozoans, may shed new light on gametogenesis, a metazoan core developmental process that has not been fully addressed in conventional animal models with complex traits (e.g., muscle and nervous system).

References

Aruga J, Kamiya A, Takahashi H, Fujimi TJ, Shimizu Y, Ohkawa K, Yazawa S, Umesono Y, Noguchi H, Shimizu T, Saitou N, Mikoshiba K, Sakaki Y, Agata K, Toyoda A (2006) A wide-range phylogenetic analysis of Zic proteins: implications for correlations between protein structure conservation and body plan complexity. Genomics 87(6):783–792. https://doi.org/10.1016/j.ygeno.2006.02.011

Aruga J, Odaka YS, Kamiya A, Furuya H (2007) Dicyema *Pax6* and *Zic*: tool-kit genes in a highly simplified bilaterian. BMC Evol Biol 7:201. https://doi.org/10.1186/1471-2148-7-201

Bao Y, Xu F, Shimeld SM (2017) Phylogenetics of lophotrochozoan *bHLH* genes and the evolution of lineage-specific gene duplicates. Genome Biol Evol. https://doi.org/10.1093/gbe/evx047

Biscotti MA, Canapa A, Forconi M, Barucca M (2014) *Hox* and *ParaHox* genes: a review on molluscs. Genesis 52(12):935–945. https://doi.org/10.1002/dvg.22839

Brown SA, Warburton D, Brown LY, Yu CY, Roeder ER, Stengel-Rutkowski S, Hennekam RC, Muenke M (1998) Holoprosencephaly due to mutations in ZIC2, a homologue of Drosophila odd-paired. Nat Genet 20(2):180–183. https://doi.org/10.1038/2484

Dray N, Tessmar-Raible K, Le Gouar M, Vibert L, Christodoulou F, Schipany K, Guillou A, Zantke J, Snyman H, Behague J, Vervoort M, Arendt D, Balavoine G (2010) Hedgehog signaling regulates segment formation in the annelid *Platynereis*. Science 329(5989):339–342. https://doi.org/10.1126/science.1188913

Dunn CW, Hejnol A, Matus DQ, Pang K, Browne WE, Smith SA, Seaver E, Rouse GW, Obst M, Edgecombe GD, Sorensen MV, Haddock SH, Schmidt-Rhaesa A, Okusu A, Kristensen RM, Wheeler WC, Martindale MQ, Giribet G (2008) Broad phylogenomic sampling improves resolution of the animal tree of life. Nature 452(7188):745–749. https://doi.org/10.1038/nature06614

Endo M, Sakai C, Shimizu T (2016) Embryonic expression patterns of *Hox* genes in the oligochaete annelid *Tubifex tubifex*. Gene Expr Patterns: GEP 22(1):1–14. https://doi.org/10.1016/j.gep.2016.09.002

Fujimi TJ, Hatayama M, Aruga J (2012) Xenopus Zic3 controls notochord and organizer development through suppression of the Wnt/beta-catenin signaling pathway. Dev Biol 361(2):220–231. https://doi.org/10.1016/j.ydbio.2011.10.026

Furuya H, Tsuneki K (2003) Biology of dicyemid mesozoans. Zool Sci 20(5):519–532. https://doi.org/10.2108/zsj.20.519

Garcia-Fernandez J, Baguna J, Salo E (1993) Genomic organization and expression of the planarian homeobox genes *Dth-1* and *Dth-2*. Development 118(1):241–253

Goto A, Kitamura K, Shimizu T (1999) Cell lineage analysis of pattern formation in the *Tubifex* embryo. I Segmentation in the mesoderm. Int J Dev Biol 43(4):317–327

Gustus RM, Cloney RA (1972) Ultrastructural similarities between setae of brachiopods and polychaetes. Acta Zool 53:229–233

Halanych KM (2016) Diversification of Lophotrochozoa. In: Kliman RM (ed) Encyclopedia of evolutionary biology. Elsevier Science, Waltham, pp 405–408

Halanych KM, Bacheller JD, Aguinaldo AM, Liva SM, Hillis DM, Lake JA (1995) Evidence from 18S ribosomal DNA that the lophophorates are protostome animals. Science 267(5204):1641–1643

Hayashi M, Shinozuka Y, Shigenobu S, Sato M, Sugimoto M, Ito S, Abe K, Kobayashi S (2017) Conserved role of Ovo in germline development in mouse and Drosophila. Sci Rep 7:40056. https://doi.org/10.1038/srep40056

Kenny NJ, Namigai EK, Dearden PK, Hui JH, Grande C, Shimeld SM (2014) The Lophotrochozoan TGF-beta signalling cassette – diversification and conservation in a key signalling pathway. Int J Dev Biol 58(6–8):533–549. https://doi.org/10.1387/ijdb.140080nk

Kobayashi M, Furuya H, Holland PW (1999) Dicyemids are higher animals. Nature 401(6755):762. https://doi.org/10.1038/44513

Kobayashi M, Furuya H, Wada H (2009) Molecular markers comparing the extremely simple body plan of dicyemids to that of lophotrochozoans: insight from the expression patterns of Hox, Otx, and brachyury. Evol Dev 11(5):582–589. https://doi.org/10.1111/j.1525-142X.2009.00364.x

Kuo JS, Patel M, Gamse J, Merzdorf C, Liu X, Apekin V, Sive H (1998) Opl: a zinc finger protein that regulates neural determination and patterning in Xenopus. Development 125(15):2867–2882

Lapan SW, Reddien PW (2011) dlx and sp6-9 control optic cup regeneration in a prototypic eye. PLoS Genet 7(8):e1002226. https://doi.org/10.1371/journal.pgen.1002226

Layden MJ, Meyer NP, Pang K, Seaver EC, Martindale MQ (2010) Expression and phylogenetic analysis of the *zic* gene family in the evolution and development of metazoans. EvoDevo 1(1):12. https://doi.org/10.1186/2041-9139-1-12

Lee PN, Callaerts P, De Couet HG, Martindale MQ (2003) Cephalopod *Hox* genes and the origin of morphological novelties. Nature 424(6952):1061–1065. https://doi.org/10.1038/nature01872

Lichtneckert R, Reichert H (2005) Insights into the urbilaterian brain: conserved genetic patterning mechanisms in insect and vertebrate brain development. Heredity 94(5):465–477. https://doi.org/10.1038/sj.hdy.6800664

Mallatt J, Winchell CJ (2002) Testing the new animal phylogeny: first use of combined large-subunit and small-subunit rRNA gene sequences to classify the protostomes. Mol Biol Evol 19(3):289–301

Meyer NP, Seaver EC (2009) Neurogenesis in an annelid: characterization of brain neural precursors in the polychaete *Capitella* sp. I. Dev Biol 335(1):237–252. https://doi.org/10.1016/j.ydbio.2009.06.017

Meyer NP, Boyle MJ, Martindale MQ, Seaver EC (2010) A comprehensive fate map by intracellular injection of identified blastomeres in the marine polychaete *Capitella teleta*. EvoDevo 1(1):8. https://doi.org/10.1186/2041-9139-1-8

Nagai T, Aruga J, Minowa O, Sugimoto T, Ohno Y, Noda T, Mikoshiba K (2000) Zic2 regulates the kinetics of neurulation. Proc Natl Acad Sci U S A 97(4):1618–1623

Nakata K, Nagai T, Aruga J, Mikoshiba K (1997) Xenopus Zic3, a primary regulator both in neural and neural crest development. Proc Natl Acad Sci U S A 94(22):11980–11985

Newmark PA, Sanchez Alvarado A (2000) Bromodeoxyuridine specifically labels the regenerative stem cells of planarians. Dev Biol 220(2):142–153. https://doi.org/10.1006/dbio.2000.9645

Ogino K, Tsuneki K, Furuya H (2010) Unique genome of dicyemid mesozoan: highly shortened spliceosomal introns in conservative exon/intron structure. Gene 449(1–2):70–76. https://doi.org/10.1016/j.gene.2009.09.002

Petersen CP, Reddien PW (2011) Polarized notum activation at wounds inhibits Wnt function to promote planarian head regeneration. Science 332(6031):852–855. https://doi.org/10.1126/science.1202143

Pourebrahim R, Houtmeyers R, Ghogomu S, Janssens S, Thelie A, Tran HT, Langenberg T, Vleminckx K, Bellefroid E, Cassiman JJ, Tejpar S (2011) Transcription factor Zic2 inhibits Wnt/beta-catenin protein signaling. J Biol Chem 286(43):37732–37740. https://doi.org/10.1074/jbc.M111.242826

Ruppert EE, Fox RS, Barns RD (2004) Invertebrate zoology. Thomson-Brooks/Cole, Belmonte

Schiemann SM, Martin-Duran JM, Borve A, Vellutini BC, Passamaneck YJ, Hejnol A (2017) Clustered brachiopod *Hox* genes are not expressed collinearly and are associated with lophotrochozoan novelties. Proc Natl Acad Sci U S A 114(10):E1913–E1922. https://doi.org/10.1073/pnas.1614501114

Seaver EC (2014) Variation in spiralian development: insights from polychaetes. Int J Dev Biol 58(6–8):457–467. https://doi.org/10.1387/ijdb.140154es

Shimizu T (1982) Ooplasmic segregation in the Tubifex egg: mode of pole plasm accumulation and possible involvement of microfilaments. Wilehm Roux Arch Dev Biol 191(4):246–256. https://doi.org/10.1007/BF00848412

Steinmetz PR, Kostyuchenko RP, Fischer A, Arendt D (2011) The segmental pattern of otx, gbx, and Hox genes in the annelid *Platynereis dumerilii*. Evol Dev 13(1):72–79. https://doi.org/10.1111/j.1525-142X.2010.00457.x

Struck TH, Wey-Fabrizius AR, Golombek A, Hering L, Weigert A, Bleidorn C, Klebow S, Iakovenko N, Hausdorf B, Petersen M, Kuck P, Herlyn H, Hankeln T (2014) Platyzoan paraphyly based on phylogenomic data supports a noncoelomate ancestry of spiralia. Mol Biol Evol 31(7):1833–1849. https://doi.org/10.1093/molbev/msu143

Suzuki TG, Ogino K, Tsuneki K, Furuya H (2010) Phylogenetic analysis of dicyemid mesozoans (phylum Dicyemida) from innexin amino acid sequences: dicyemids are not related to Platyhelminthes. J Parasitol 96(3):614–625. https://doi.org/10.1645/GE-2305.1

Takahashi H, Shimizu T, Aruga J (2008) Expression pattern of annelid Zic in embryonic development of the oligochaete *Tubifex tubifex*. Dev Genes Evol 218(10):553–560. https://doi.org/10.1007/s00427-008-0252-x

Telford MJ (2006) Animal phylogeny. Curr Biol 16(23):R981–R985. https://doi.org/10.1016/j.cub.2006.10.048

Umesono Y, Watanabe K, Agata K (1997) A planarian orthopedia homolog is specifically expressed in the branch region of both the mature and regenerating brain. Develop Growth Differ 39(6):723–727

Vasquez-Doorman C, Petersen CP (2014) zic-1 expression in planarian neoblasts after injury controls anterior pole regeneration. PLoS Genet 10(7):e1004452. https://doi.org/10.1371/journal.pgen.1004452

Vogg MC, Owlarn S, Perez Rico YA, Xie J, Suzuki Y, Gentile L, Wu W, Bartscherer K (2014) Stem cell-dependent formation of a functional anterior regeneration pole in planarians requires Zic and Forkhead transcription factors. Dev Biol 390(2):136–148. https://doi.org/10.1016/j.ydbio.2014.03.016

Wagner DE, Wang IE, Reddien PW (2011) Clonogenic neoblasts are pluripotent adult stem cells that underlie planarian regeneration. Science 332(6031):811–816. https://doi.org/10.1126/science.1203983

Chapter 6
Ascidian *Zic* Genes

Yutaka Satou and Kaoru S. Imai

Abstract Ascidians are tunicates, which constitute the sister group of vertebrates. The ascidian genome contains two Zic genes, called *Zic-r.a* (also called *Macho-1*) and *Zic-r.b* (*ZicL*). The latter is a multi-copy gene, and the precise copy number has not yet been determined. *Zic-r.a* is maternally expressed, and soon after fertilization *Zic-r.a* mRNA is localized in the posterior pole of the zygote. Zic-r.a protein is translated there and is involved in specification of posterior fate; in particular it is important for specification of muscle fate. *Zic-r.a* is also expressed zygotically in neural cells of the tailbud stage. On the other hand, *Zic-r.b* is first expressed in marginal cells of the vegetal hemisphere of 32-cell embryos and then in neural cells that contribute to the central nervous system during gastrulation. *Zic-r.b* is required first for specification of mesodermal tissues and then for specification of the central nervous system. Their upstream and downstream genetic pathways have been studied extensively by functional assays, which include gene knockdown and chromatin immunoprecipitation assays. Thus, ascidian Zic genes play central roles in specification of mesodermal and neural fates.

Keywords Ascidians · Zic · Macho-1 · Mesoderm specification · Neural fate specification

Y. Satou (✉)
Department of Zoology, Graduate School of Science, Kyoto University, Kyoto, Japan
e-mail: yutaka@ascidian.zool.kyoto-u.ac.jp

K. S. Imai
Department of Biological Sciences, Graduate School of Science, Osaka University, Toyonaka, Japan

J. Aruga (ed.), *Zic family*, Advances in Experimental Medicine and Biology 1046,
https://doi.org/10.1007/978-981-10-7311-3_6

6.1 Two *Zic*-Related Genes in Ascidian Genomes

6.1.1 Ascidians

Ascidians are tunicates, which constitute the sister group of vertebrates. Tunicates and vertebrates, as well as cephalochordates, constitute the phylum Chordata (Satoh et al. 2003; Lemaire 2011). Chordate animals commonly develop a tadpole-type body, in which a notochord is located in the middle of the tail and is flanked laterally by muscle, ventrally by endodermal tissues, and dorsally by the neural tube. Ascidian tadpole larvae metamorphose to the adult form, in which the body plan is totally changed.

In this chapter, we describe *Zic* genes identified in two *Ciona* species and *Halocynthia roretzi*. Although the phylogenetic relationships among tunicates are controversial, most studies agree that *Ciona* and *Halocynthia* are phylogenetically distant within the tunicates (Wada 1998; Tsagkogeorga et al. 2009; Turon and Lopez-Legentil 2004). In the genus *Ciona*, *Ciona savignyi* and *Ciona intestinalis* have been widely studied. *C. intestinalis* has been grouped into two subspecies, type A and type B. It was proposed that these two types are two distinct species, the former *Ciona robusta* and the latter *C. intestinalis* (Brunetti et al. 2015). This proposal has temporarily brought some confusion to *Ciona* biology, mainly because most previous studies did not describe which species or group was used. Although there is genetic evidence that the reproduction systems of these two types of animal are not completely isolated (Nydam and Harrison 2011), it is certain that reproductive isolation between these two types is now under way (Caputi et al. 2007; Nydam and Harrison 2007; Sato et al. 2012). However, because most published papers that study *Zic* genes do not clearly discriminate these two species or types, we refer here to these two types or species simply as *C. intestinalis*.

6.1.2 Zic *Genes*

Two *Zic* genes are encoded in the genomes of *C. intestinalis* and *C. savignyi*. These are *Zic-related.a* (*Zic-r.a*) and *Zic-related.b* (*Zic-r.b*). *Zic-r.a* was originally named *Macho-1*, and *Zic-r.b* was named *ZicL* (*Zic-like*) in *Ciona* and *ZicN* in *Halocynthia*. In this article, we call them *Zic-r.a* and *Zic-r.b* according to the recently published nomenclature rules for tunicate genes (Stolfi et al. 2015).

Molecular phylogenetic analyses have suggested that these two *Zic* genes are duplicated in the ascidian lineage after the divergence of the ascidians and vertebrates (Imai et al. 2002c; Aruga et al. 2006). In other words, these *Zic* genes are

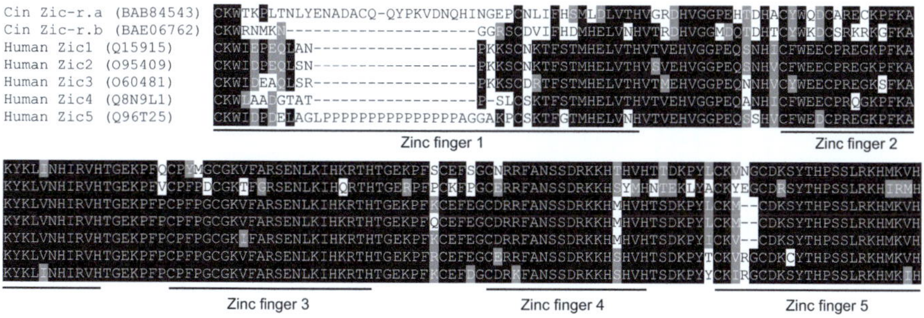

Fig. 6.1 An alignment of zinc finger domains of two *C. intestinalis* Zic proteins and human Zic proteins shows high conservation

equally similar to all *Zic* paralogues in vertebrates. *Zic-r.a* was first identified in *Halocynthia* as a maternal determinant for muscle of the larva (Nishida and Sawada 2001), while *Zic-r.b* was first identified in *Ciona* and *Halocynthia* by two different laboratories (Imai et al. 2002c; Wada and Saiga 2002). *Zic-r.b* was subsequently revealed to be a multi-copy gene in the *C. intestinalis* genome (Yamada et al. 2003). Five and six copies were predicted currently in the *C. intestinalis* and *C. savignyi* genomes, respectively (Dehal et al. 2002; Vinson et al. 2005). In the *C. intestinalis* genome, these copies are encoded in two separate small scaffolds, and, therefore, the exact number might be different. It is possible that the number of *Zic-r.b* copies might be different among individual chromosomes.

In vertebrates, two Zic genes are often encoded as neighbors. In the human genome, *ZIC1*/*ZIC4* and *ZIC2*/*ZIC5* pairs are located on chromosomes 3 and 13, respectively (Aruga 2004). In the *C. intestinalis* genome, the *Zic-r.b* locus has not yet been assigned to any chromosome, while *Zic-r.a* is located on chromosome 1. The neighbors of *Zic-r.a* are homologs of human *HCFC1* and *TRIP12* genes, and thus *Zic-r.a* does not abut *Zic-r.b* in the *Ciona* genome. In the *C. savignyi* genome, these two loci are found on two different scaffolds. The neighbors of *Zic-r.a* in *C. savignyi* are also homologs for human *HCFC1* and *TRIP12*, and, therefore, *Zic-r.a* and *Zic-r.b* are not neighbors in the genome of this species either.

Zic-r.a and Zic-r.b both contain five zinc fingers as do Zic proteins in other animals, and the number of amino acid residues between the two cysteines in the first zinc fingers is large (Fig. 6.1). Indeed, in vitro selection assays have shown that the DNA sequences that bind Zic-r.a and Zic-r.b are highly similar to those that bind vertebrate Zic proteins (Fig. 6.2). On the other hand, these ascidian Zic proteins both lack two conserved domains, Zoc and ZF-NC (Aruga et al. 2006), and, therefore, they are thought to be diverged members of the Zic family.

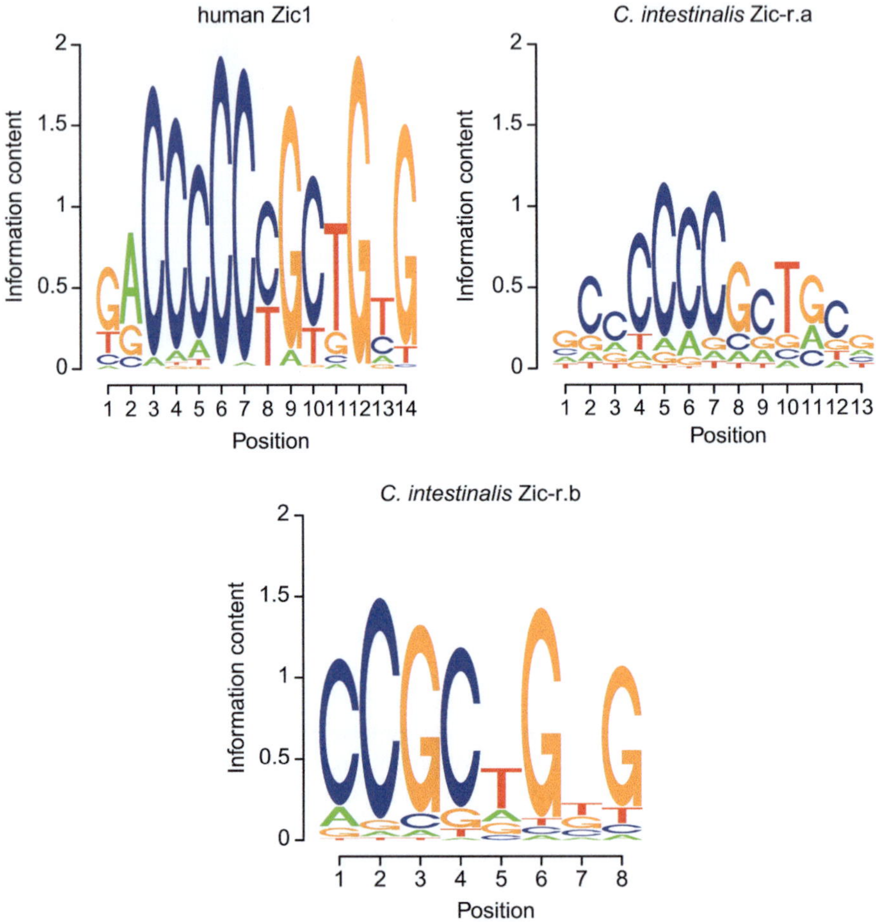

Fig. 6.2 Sequence logos representing binding sequences of human ZIC1, *C. intestinalis* Zic-r.a, and *C. intestinalis* Zic-r.b, which were made using data from in vitro selection (Jolma et al. 2013; Yagi et al. 2004a, b).

6.2 Expression Pattern and Function of *Zic-r.a*

6.2.1 Expression and Function of Maternal Zic-r.a

6.2.1.1 Localization of Maternal mRNA

The number of cells that constitute the ascidian embryo is small, and cell lineages are invariant among individuals. Because of these features, cell lineages are precisely determined, and gene expression patterns are traceable at the single cell level (Fig. 6.3).

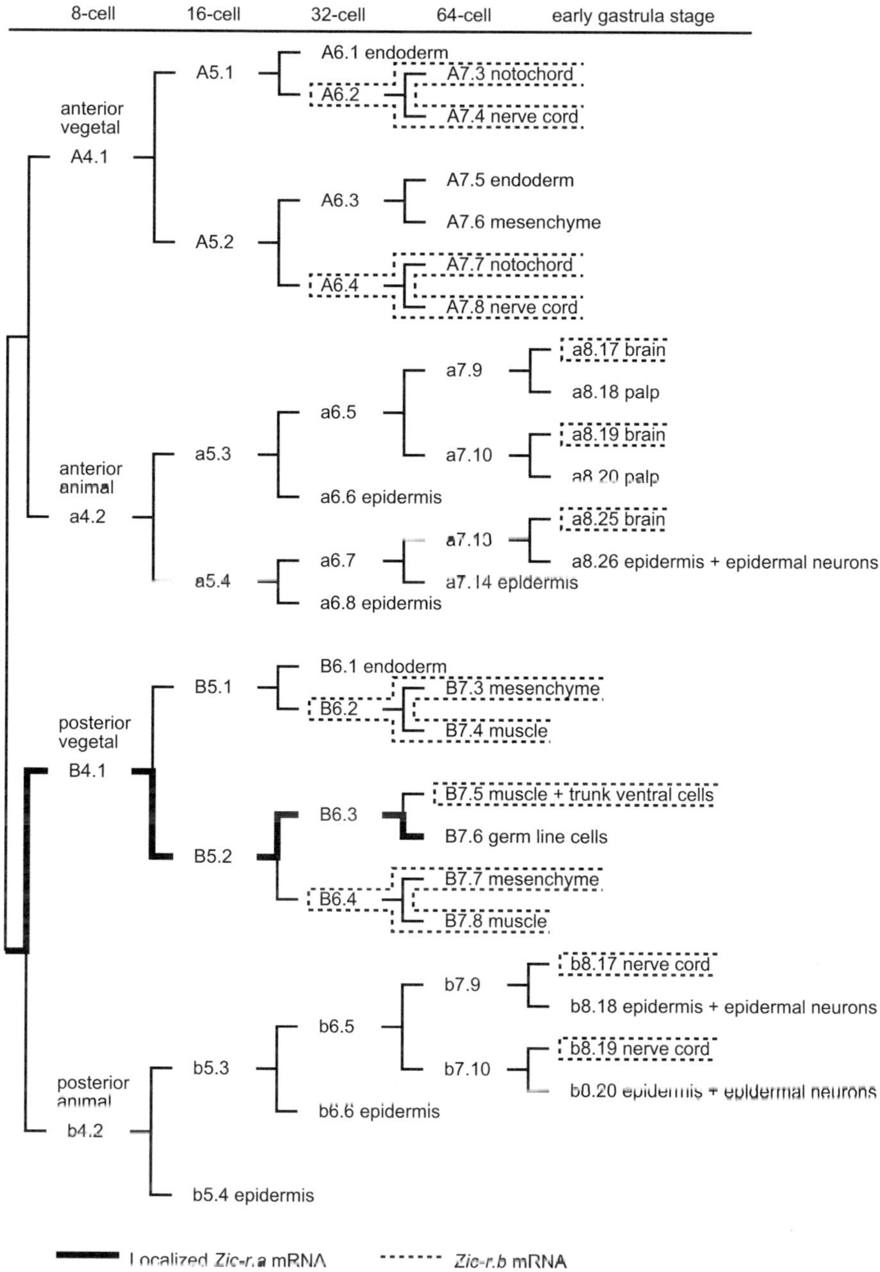

Fig. 6.3 The cell lineage of *Ciona* embryos and the expression pattern of two Zic genes. Note that the ascidian embryo is bilaterally symmetrical, and each pair is designated with unique names

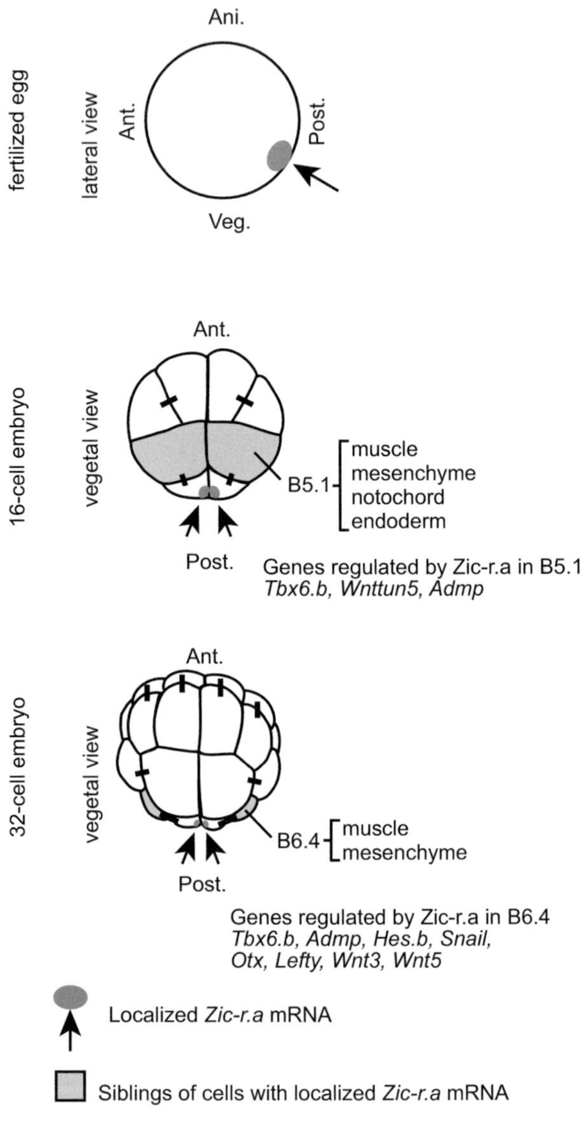

Fig. 6.4 Illustrations showing localization of *Zic-r.a* mRNA in early embryos. In sister cells (gray) of cells in which Zic-r.a is localized, several genes are activated by Zic-r.a

Zic-r.a mRNA is present in unfertilized eggs of *Halocynthia* and *Ciona*. After fertilization, it is prominently localized in the posterior vegetal region (Nishida and Sawada 2001; Satou et al. 2002). This localized mRNA is inherited by a pair of posterior-most cells in subsequent cell divisions (Figs. 6.3 and 6.4). Many mRNAs are similarly localized in the posterior pole (Yoshida et al. 1996; Satou 1999; Satou and Satoh 1997; Paix et al. 2009; Sasakura et al. 1998a, b; Yamada et al. 2005). Such posteriorly localized mRNAs are referred to as postplasmic or posterior-end-mark mRNAs. The posterior-most cells that retain maternal *Zic-r.a* mRNA give rise to the germ line cells. The sister cells (B5.1) of the posterior-most cells at the 16-cell stage contribute to the endoderm, mesenchyme, notochord, and muscle. The sister cells (B6.4) at the 32-cell stage contribute to mesenchyme and muscle. The sister cells (B7.5) at the 64-cell stage contribute to muscle and precursor cells of the adult heart and muscle, which are called trunk ventral cells.

6.2.1.2 Developmental Function of Maternal *Zic-r.a*

Maternal *Zic-r.a* was first identified as a muscle determinant (Nishida and Sawada 2001). Existence of a muscle determinant in ascidian eggs was first shown by a classical embryological experiment in 1905 (Conklin 1905). Its molecular identity was then revealed to be *Zic-r.a* mRNA almost 100 years later (Nishida and Sawada 2001). Prior to this finding, several mRNAs localized in the posterior pole were identified and are called posterior end mark mRNAs (Yoshida et al. 1996; Satou 1999; Satou and Satoh 1997). Among them, only *Zic-r.a* plays a role in specification of muscle fate. *Zic-r.a* was later revealed to also be required for specification of the posterior mesenchyme lineage.

The blastomeres, in which *Zic-r.a* mRNA is localized, give rise to the germ line cells, and transcription is generally suppressed by a protein encoded by another localized mRNA, *pem-1*. Hence, Zic-r.a protein cannot activate its target genes in the posterior-most cells but activates them in sister cells of the posterior-most cells (Fig. 6.4) (Shirae-Kurabayashi et al. 2011; Kumano et al. 2011).

At the 16-cell stage, the posterior-most cells are named B5.2, and their sister cells are B5.1. Although *Zic-r.a* mRNA is localized in B5.2, Zic-r.a protein is present in both of these sister cells and activates its targets in B5.1 (Oda-Ishii et al. 2016). These targets include *Tbx6.b*, which encodes a T-box transcription factor, *Wnttun5*, which encodes a Wnt signaling molecule, and *Admp*, which encodes a signaling molecule of the BMP (bone morphogenetic protein) family. Zic-r.a works with Tcf7 and β-catenin to activate these target genes. Tcf7 is a transcription factor with an HMG-box and acts as an activator by forming a complex with β-catenin. At the 16-cell stage, β-catenin is translocated into nuclei of cells in the vegetal hemisphere. Thus, a collaboration of Tcf7/β-catenin and Zic-r.a activates *Tbx6.b*, *Wnttun5*, and *Admp* in the posterior vegetal blastomeres (Oda-Ishii et al. 2016).

Tcf7/β-catenin activates its targets, including *Foxd* and *Fgf9/16/20*, in the anterior and posterior vegetal cells. How are *Tbx6.b* and *Wnttun5* activated specifically in the posterior vegetal cells? In other words, why are *Tbx6.b* and *Wnttun5* not activated in the anterior vegetal cells? The most likely hypothesis is that Tcf7/β--catenin binding sites are qualitatively different between these two groups; namely, Tcf7/β-catenin binding sites in the *Tbx6.b* and *Wnttun5* enhancers could not activate gene expression alone and would require the help of Zic-r.a. However, this is not the case; when Tcf7 binding sites in the *Tbx6.b* enhancer are replaced with those in the *Fgf9/16/20* enhancer, the chimeric *Tbx6.b* enhancer drives gene expression specifically in the posterior vegetal cells (Oda-Ishii et al. 2016). Instead, there is a repressor element in each of the enhancers of *Tbx6.b* and *Wnttun5*. Without these repressor elements, *Tbx6.b* and *Wnttun5* are expressed in the anterior and posterior vegetal

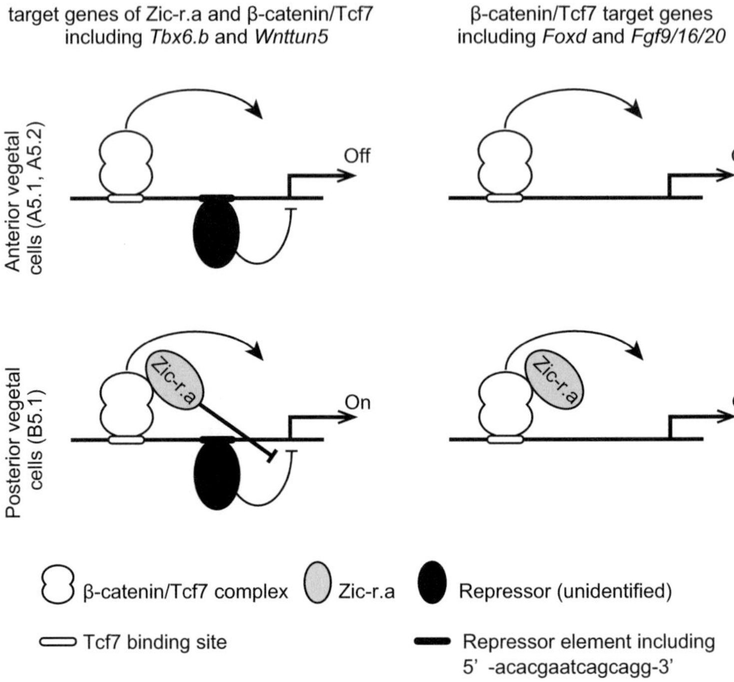

Fig. 6.5 Zic-r.a works together with β-catenin and Tcf7 to activate its targets in posterior vegetal cells. Zic-r.a is thought to suppress the function of a repressor, which has not yet been identified, and for this function, Zic-r.a does not necessarily bind directly to DNA

cells. What then is the function of Zic-r.a? Currently, the most likely hypothesis is that the repressor that binds to the repressor elements is present in the anterior and posterior vegetal cells, and Zic-r.a suppresses the function of the repressor.

Intriguingly, for this function, direct binding of *Zic-r.a* may not necessarily be required (Oda-Ishii et al. 2016). First, there are no clear Zic-r.a binding sites in the enhancers of *Tbx6.b* and *Wnttun5*. Second, no clear binding was found in these enhancers by chromatin immunoprecipitation (ChIP) followed by microarray and deep-sequencing assays; only weak binding was found in the *Tbx6.b* enhancer. Lastly, in the ascidian embryo, Zic-r.a protein can physically interact with Tcf7. Hence, it is possible that Zic-r.a might bind to the enhancer indirectly through Tcf7 (Fig. 6.5). Similar examples, in which Zic protein acts as a cofactor of transcription factors, are known in vertebrates (Koyabu et al. 2001; Sanchez-Ferras et al. 2014).

At the 32-cell stage, genes activated under the control of *Zic-r.a* include *Hes.b*, *Snail*, *Otx*, *Admp*, *Bmp3*, *Lefty*, *Nodal*, *Wnt3*, and *Wnt5*. These genes are all expressed in B6.4 cells, which are sister cells of the posterior-most cells at this stage. *Tbx6.b* and *Wnttun5* are also activated in this pair of cells. The expression of *Tbx6.b* in B6.4 at the 32-cell stage is likely to be regulated differently from the expression in B5.1 at the 16-cell stage, because a different enhancer, which contains putative *Zic-r.a* binding sites, is responsible for the expression in B6.4 (Kugler et al.

2010). Therefore, it is possible that Zic-r.a is bound directly to enhancers of the above genes at the 32-cell stage.

Tbx6.b encodes an essential transcription factor for specification of muscle cells, because it regulates muscle-specific genes including genes encoding myosin light chain, myosin regulatory light chain, muscle actin, troponin I, troponin C, troponin T, and tropomyosin (Yagi et al. 2005). In addition, it regulates a gene encoding a myogenic factor, Mrf, which has a basic helix-loop-helix motif and is the sole ortholog for vertebrate MyoD, Myogenin, MRF4 and Myf5, and another Zic gene, *Zic-r.b* (Yagi et al. 2005). *Mrf* and *Zic-r.b* are also required for proper differentiation of muscle cells. Thus, *Zic-r.a* begins the muscle specification gene pathway.

Zic-r.a is also required for mesenchyme specification (Kobayashi et al. 2003). *Zic-r.a* activates *Otx* in the descendants of B5.1 at the 32-cell stage (Yagi et al. 2004a). Because *Otx* expression does not begin in B5.1 at the 16-cell stage, *Zic-r.a* might activate *Otx* indirectly or might activate it cooperatively with transcription factors activated at the 16-cell stage. At later stages, *Otx* is required for expression of *Twist-r.a*, which is an essential transcription factor for mesenchyme specification (Imai et al. 2003, 2006).

The B5.1 lineage also contributes to endoderm. In *Halocynthia* embryos, Zic-r.a function is suppressed in the posterior endodermal lineage by Fgf and Bmp signaling (Kondoh et al. 2003). The molecular mechanism of this suppression has not been fully elucidated, and it is not clear whether this suppression is required in *Ciona* embryos.

6.2.2 Expression in the Nervous System

In addition to maternal expression, *Zic-r.a* is also expressed in the nervous system of the *Ciona* tailbud embryo (Satou et al. 2002). The central nervous system can be divided morphologically into several regions. The most anterior part is the brain. The next prominent structure is called the visceral ganglion, in which motor neurons are located. The narrow region between the brain and visceral ganglion is called the neck region. The boundary between the neck and visceral ganglion is formed by a mechanism utilizing Fgf8/17/18 signaling, which is evocative of a mechanism in the midbrain-hindbrain boundary (MHB) of vertebrate embryos (Imai et al. 2009). The nerve cord follows the visceral ganglion. This nerve cord is a tube consisting of four rows of cells, the dorsal, ventral, and two lateral (left and right) rows. This nerve cord does not contain neurons but consists of ependymal cells. *Zic-r.a* is expressed in two regions within the central nervous system, a small anterior region within the brain and the dorsal and ventral rows of the nerve cord (Fig. 6.6) (Satou et al. 2002; Imai et al. 2009). The expression in the dorsal row of the nerve cord at the tailbud stage is under the control of *Msx*, which is expressed in their ancestral cells around the gastrula stage (Imai et al. 2009). However, it is not understood how *Zic-r.a* is activated only in a subset of the descendants of

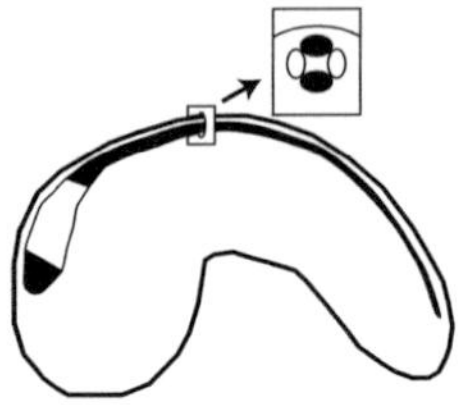

Fig. 6.6 Zic-r.a is expressed in the anterior portion of the brain and in the dorsal and ventral rows of the nerve cord at the tailbud stage, although its function has not been studied

Msx-expression cells or how *Zic-r.a* expression in the anterior part of the brain and the ventral row of the nerve cord is regulated. The function of *Zic-r.a* in the nervous system also remains to be determined.

6.3 Expression Pattern and Function of *Zic-r.b*

6.3.1 Expression and Function at the 32-Cell Stage

6.3.1.1 Upstream Regulatory Mechanisms

Zic-r.b mRNA is not expressed maternally. It is first expressed at the 32-cell stage in marginal cells of the vegetal hemisphere (A6.2, A6.4, B6.2, and B6.4) (Figs. 6.3 and 6.7). Most mesodermal cells and the ventral and lateral rows of the nerve cord cells are derived from these *Zic-r.b* expressing cells.

The expression of *Zic-r.b* in the marginal cells at the 32-cell stage is regulated by Gata.a, Foxd, Fgf9/16/20, and Foxa.a (Hudson et al. 2016; Imai et al. 2016) (Fig. 6.8). Whereas Gata.a is a maternal protein, *Foxd*, *Fgf9/16/20*, and *Foxa.a* are zygotically expressed in all vegetal cells except cells with a germ cell fate at the 16-cell stage. These three genes are among the first genes activated from the zygotic genome. Although it is not known how *Foxa.a* is activated, the expression of *Foxd* and *Fgf9/16/20* is known to be activated directly by β-catenin and Tcf7 (Oda-Ishii et al. 2016; Rothbächer et al. 2007). At the 16-cell stage, β-catenin is translocated to the nuclei of vegetal hemisphere cells (A5.1, A5.2, and B5.1), except the most posterior cells (B5.2) (Hudson et al. 2013), and a complex of β-catenin and Tcf7 directly activates *Foxd* and *Fgf9/16/20* (Imai et al. 2002b; Oda-Ishii et al. 2016). At the same time, this β-catenin/Tcf7 complex suppresses Gata.a binding activity to its target sites through direct interaction between β-catenin/Tcf7 and Gata.a. In this manner, Gata.a activates its target genes specifically in the animal hemisphere, where β-catenin is not translocated to nuclei (Oda-Ishii et al. 2016). At the 32-cell stage, each of the three vegetal cell pairs that express *Foxd* divides into a marginal cell and an endodermal cell. β-catenin continues to be translocated into the nuclei of the endodermal cells, and therefore Gata.a function continues to be suppressed there. Meanwhile, β-catenin is not translocated into the nuclei of the marginal cells, and therefore Gata.a begins to work as a transcriptional activator. Thus, a combination of Gata.a, Foxd, Foxa.a, and Ets1/2 that is activated by Fgf9/16/20 signaling promotes *Zic-r.b* expression specifically in the marginal cells.

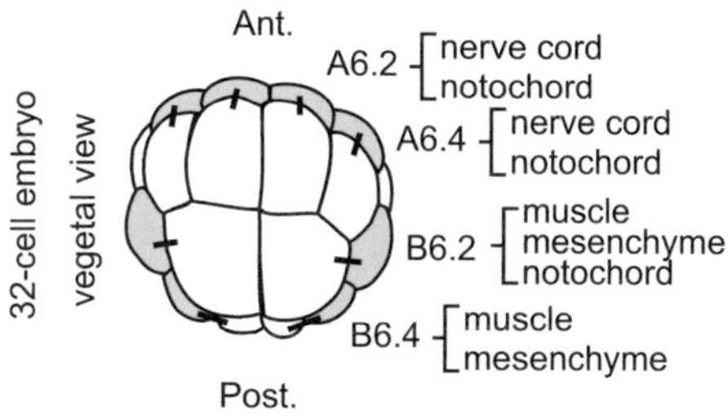

Fig. 6.7 Zic-r.b is expressed in the margin of the vegetal hemisphere of the 32-cell embryo (gray). These cells mainly contribute to mesodermal tissues

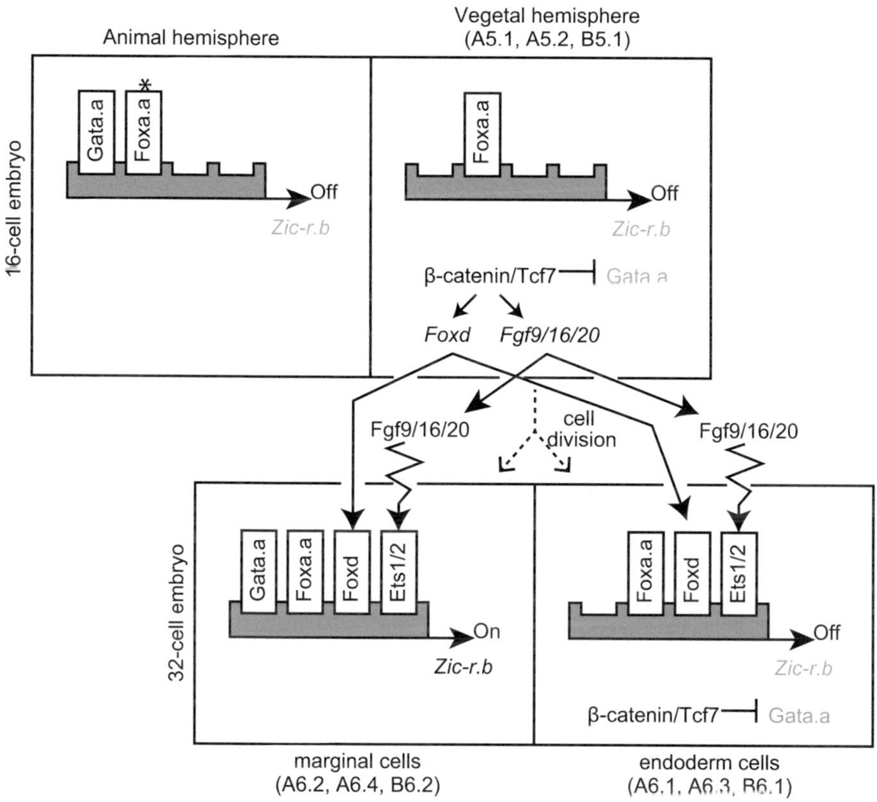

Fig. 6.8 Schematic illustrations for regulation of *Zic-r.b* at the 16-cell and 32-cell stages. Four indicated factors are required for activation of *Zic-r.b*, and these four factors are active only in the marginal cells of the 32-cell embryo. Foxa.a is present in the anterior animal cells (an asterisk) but not in the posterior animal cells. Note that Ets1/2 is the effector of Fgf9/16/20 signaling at the 32-cell stage

Zic-r.b expression in B6.2 is likely regulated by an additional distinct mechanism, in which *Tbx6.b*, expressed under the control of *Zic-r.a*, is involved. The abovementioned mechanism dependent on Gata.a and the *Tbx6.b*-dependent

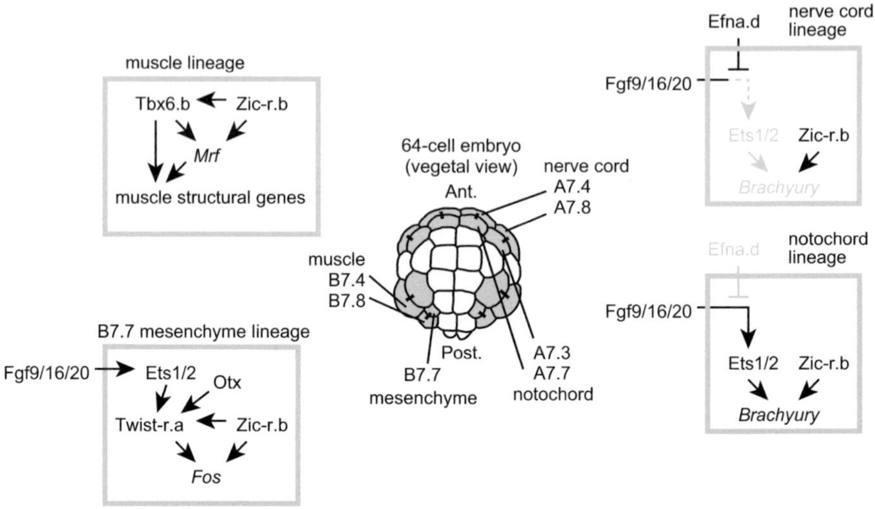

Fig. 6.9 Zic-r.b activates different downstream factors in different mesodermal lineages

mechanism are likely to cooperatively and redundantly regulate *Zic-r.b* expression in B6.2 (Yagi et al. 2005; Imai et al. 2016; Anno et al. 2006).

Intriguingly, the expression of *Zic-r.b* in B6.4, which are sister cells of the most posterior germ line cells, is regulated differently. First, the upstream transcription factors *Foxd* and *Foxa.a* are not expressed in B6.4 or its parental cells (Imai et al. 2002b, 2004; Shimauchi et al. 1997). Second, knockdown or suppression of either β-catenin, *Foxd*, *Fgf9/16/20*, or *Gata.a* does not impair *Zic-r.b* expression in B6.4 (Imai et al. 2002c; Hudson et al. 2013, 2016). However, it has not been revealed how this expression is regulated.

6.3.1.2 Functions in the Anterior Marginal Cells

The notochord and nerve cord are differentiated from the anterior marginal cells of 32-cell embryos (Fig. 6.7). After a cell division between the 32-cell and 64-cell stages, the developmental fate of the anterior descendants is restricted to the nerve cord, and the developmental fate of the posterior descendants is restricted to the notochord. *Brachyury* is then expressed exclusively in the notochord lineage and plays an essential role in specification of notochord fate (Yasuo and Satoh 1993, 1998; Corbo et al. 1998; Chiba et al. 2009). *Zic-r.b* binds to the upstream regulatory region of this gene, and this binding is essential for the activation of *Brachyury* (Imai et al. 2002c; Yagi et al. 2004b; Kumano et al. 2006). Consequently, no notochord cells are differentiated in *Zic-r.b* morphants. Among the daughter cells of cells that express *Zic-r.b* at the 32-cell stage, the posterior cells (A7.3 and A7.7) abut endodermal cells and express *Brachyury* under the control of Fgf9/16/20 signaling (Fig. 6.9) (Imai et al. 2002a; Nakatani and Nishida 1994). Fgf signaling is weakened in the anterior daughter cells (A7.4 and A7.8) by Ephrin signaling. *Ephrina.d* (*Efna.d*) is expressed in the animal hemisphere and encodes a membrane-anchored ligand. Because the contact surface area of A7.4 and A7.8 to cells expressing *Efna.d* is much larger than that of A7.3 and A7.7, A7.4 and A7.8 are expected to receive a stronger Efna.d signal. The activated receptor suppresses Fgf signaling through p120Ras-GAP (Picco et al. 2007; Haupaix et al. 2013). Thus, *Brachyury* is activated only in A7.3 and A7.7 by a combinatorial action of *Zic-r.b* and Fgf signaling. Foxa.a, which activates *Zic-r.b*, also acts cooperatively with these factors for activation of *Brachyury* expression (Imai et al. 2006; Kumano et al. 2006). In addition, *Foxb* is expressed in A7.4 and A7.8 cells of *Halocynthia* embryos, and *Foxb* ensures that *Brachyury* is not activated in A7.4 and A7.8. *Zic-r.b* is required for this *Foxb* expression (Hashimoto et al. 2011).

In addition to *Brachyury*, *Zic-r.b* activates *Chordin*, *Lhx3*, and *Mnx* in A7.3 and A7.7, which was revealed by an observation that *Zic-r.b* knockdown resulted in loss of expression of these genes in A7.3 and A7.7 (Imai et al. 2006). A chromatin immunoprecipitation assay using a Gfp-tagged Zic-r.b suggested that Zic-r.b binds to the upstream regions of *Chordin*, *Lhx3*, and *Mnx* (Kubo et al. 2010).

6.3.1.3 Functions in the Posterior Marginal Cells

In the posterior marginal cells, *Zic-r.b* contributes to specification of muscle and mesenchyme (Imai et al. 2002c, 2006; Kubo et al. 2010), although knockdown of *Zic-r.b* does not completely suppress differentiation of muscle or mesenchyme.

In the muscle lineage, *Zic-r.b* contributes to maintenance of *Tbx6.b* expression in later embryos (Imai et al. 2006). Consequently, *Tbx6.b* expression continues until the neurula stage in *Ciona* and the tailbud stage in *Halocynthia*, although *Zic-r.b* expression in the muscle lineage disappears before the late gastrula stage (Imai et al. 2002c, 2004; Wada and Saiga 2002; Yasuo et al. 1996).

Zic-r.b and *Tbx6.b* activate *Mrf* (the sole ortholog for vertebrate MyoD, Myf5, Myogenin, and Mrf4, as described above) (Imai et al. 2006). Chromatin immunoprecipitation studies have shown that both Zic-r.b and Tbx6.b are bound directly to the upstream region of *Mrf*. Muscle structural genes are expressed under the control of *Zic-r.b*, *Tbx6.b*, and *Mrf*. ChIP assays indicated that most of them are regulated directly by Tbx6.b and Mrf but not by Zic-r.b (Kubo et al. 2010). This observation is consistent with the temporal expression pattern of *Zic-r.b*, because *Zic-r.b* expression disappears earlier than *Mrf* and *Tbx6.b* expression.

Zic-r.b also contributes to expression of *Twist-r.a*, which is a key regulatory gene for specification of mesenchyme fate, because knockdown of *Zic-r.b* reduces the expression of *Twist-r.a* (Imai et al. 2003, 2006). Two distinct cell lineages contribute to the mesenchyme, and *Zic-r.b* is especially important for specification of the mesenchyme that is derived from the posterior pairs of cells (B7.7). Knockdown of *Zic-r.b* resulted in loss of *Fos* expression in the B7.7 lineage mesenchyme (Imai et al. 2006). Zic-r.b is suggested to bind to the upstream region of *Twist-r.a*, and a deletion of the region where Zic-r.b binds impairs expression in the posterior lineage (Kubo et al. 2010).

6.3.2 Expression and Functions at the Gastrula Stage

Zic-r.b is expressed in the neural lineage of the animal hemisphere of gastrula embryos (a8.17, a8.19, a8.25, b8.17, and b8.19) (Fig. 6.3) (Imai et al. 2002c, 2004; Wada and Saiga 2002), while its expression in the vegetal hemisphere is gradually decreased. This *Zic-r.b* expression is important for specification of neural fate, because knockdown of *Zic-r.b* abolishes expression of neuronal marker genes (Imai et al. 2002c; Wada and Saiga 2002).

Fgf9/16/20 signaling is required for activation of *Zic-r.b* in the neural lineage at the gastrula stage, as in the vegetal marginal cells at the 32-cell stage (Ikeda et al. 2013; Imai et al. 2006; Wagner and Levine 2012), while *Zic-r.b* is also regulated by *Dlx.b*, which is activated by *Sox1/2/3* in the animal hemisphere (Imai et al. 2017). Intriguingly, this Fgf signaling acts on the neural lineage of cells from the 32-cell stage, because *Otx* is activated by Fgf signaling (Hudson et al. 2003; Hudson and Lemaire 2001), and the downstream MAP kinase, Erk1/2, is indeed phosphorylated at the 32-cell stage (Ohta and Satou 2013). Three genes encoding transcriptional repressors, *Prdm1-r.a* (*Bz1*), *Prdm1-r.b* (*Bz2*), and *Hes.a*, repress precocious expression of *Zic-r.b* in this lineage (Fig. 6.10) (Ikeda et al. 2013; Ikeda and Satou 2017). These three genes are expressed transiently from the 16-cell and/or 32-cell stages; *Hes.a* is expressed only at the 16-cell stage, and *Prdm1-r.a* and *Prdm1-r.b* expression disappears in the neural lineage before the 64-cell stage. Probably because of a delay between transcription and translation, Prdm1-r.a, Prdm1-r.b, and Hes.a repress *Zic-r.b* at the 32-cell stage, and Prdm1-r.a and Prdm1-r.b repress *Zic-r.b* at the 64-cell stage.

Double knockdown of *Prdm1-r.a* and *Prdm1-r.b* induces precocious expression of *Zic-r.b* in the neural lineage (a7.9 and a7.10, the two pairs of cells marked by asterisks in Fig. 6.10b) at the 64-cell stage (Ikeda et al. 2013). In normal embryos, these cells have developmental potential to become the brain and palps with sensory neurons in normal embryos. After the next division, the posterior daughter cells, which abut the vegetal cells expressing *Fgf9/16/20*, begin to express *Zic-r.b* and are specified to become the brain, while the anterior cells begin to express *Foxc* instead of *Zic-r.b* and are specified to become the palps (Wagner and Levine 2012). The cells with precocious expression of *Zic-r.b* in double *Prdm1-r.a* and *Prdm1-r.b* morphants are precociously specified to become the brain. As a result, the palps are not formed in such morphants (Ikeda et al. 2013).

In triple morphants of *Prdm1-r.a*, *Prdm1-r.b*, and *Hes.a*, *Zic-r.b* expression begins at the 32-cell stage. Because these repressors also repress *Foxa.a*, which is expressed at the 16-cell stage, *Foxa.a* expression does not cease properly in such triple morphants (Ikeda and Satou 2017), and, in the neural lineage, expression of *Foxa.a* and *Zic-r.b* overlaps improperly. As a result, a combination of Foxa.a, Zic-r.b,

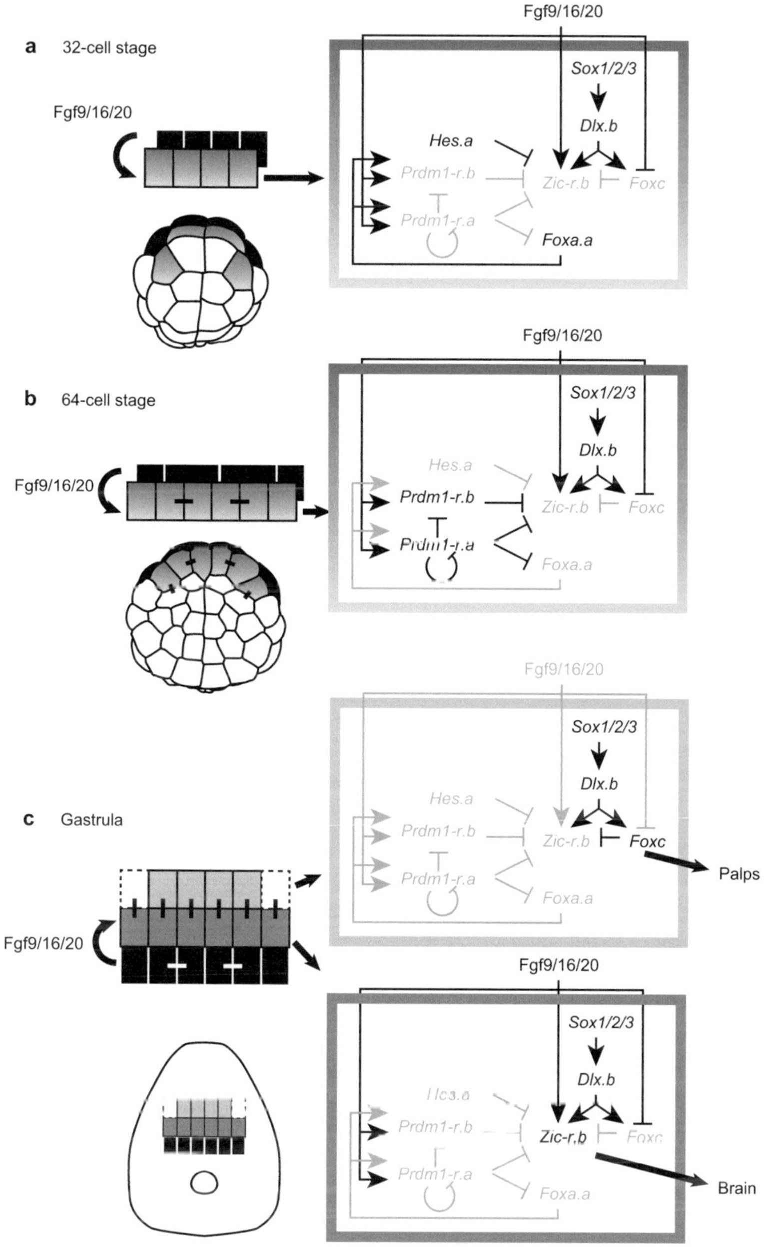

Fig. 6.10 *Zic-r.b* expressed in the animal hemisphere specifies brain fate. Precocious expression of *Zic-r.b* is suppressed by three transcriptional repressors at the 32-cell stage (**a**) and 64-cell stage (**b**). Note that *Prdm1-r.a* and *Prdm1-r.b* are not transcribed at the 64-cell stage, but their protein products are expected to exist and indeed repress *Zic-r.b* expression. This temporal control is important for proper development of the brain and palp lineages (**c**)

and Fgf9/16/20 signaling activates *Brachyury* in the neural lineage, as it does in the notochord lineage. This ectopic *Brachyury* expression leads to ectopic differentiation of notochord. In this way, *Zic-r.b* expression in the neural lineage is strictly controlled temporally by three repressors, and this temporal regulation is important for Zic-r.b to act as a key specifier for the mesodermal fate at the 32-cell stage and the neural fate at the 64-cell stage.

As mentioned, the palp precursors are sister cells of the brain precursors and are located at the anterior end of the neural plate. The palp precursors are often considered to be a rudiment of the anterior placode, which ancestral chordates had and extant vertebrates retain. Because the brain is generally considered to be more ancient than the placode, incorporation of repressors that delay the onset of *Zic-r.b* expression into the gene circuit for differentiation of the brain might give cells in the boundary of the neural plate a chance to take a fate other than that of the brain.

6.4 Conclusions

Two ascidian Zic genes, which are likely to have arisen by duplication in the tunicate lineage, are expressed in mesodermal and neural lineages during embryogenesis. These Zic genes are essential for specification of these cell lineages. Maternally expressed Zic-r.a activates target genes immediately after the zygotic genome is activated at the 16-cell stage. These Zic-r.a target genes are required for specification of mesenchyme and muscle. At the 32-cell stage, *Zic-r.b* begins to be expressed in the marginal cells of the vegetal hemisphere. These cells contribute to mesodermal tissues and part of the nerve cord. In the posterior marginal cells, *Zic-r.b* works cooperatively with targets of Zic-r.a to specify mesenchyme and muscle fate. In the anterior cells, it plays a role in specifying the notochord fate. In the animal hemisphere, *Zic-r.b* expression is repressed until the 64-cell stage. Precocious expression of *Zic-r.b* at the 32-cell stage ectopically activates the notochord fate, and precocious expression at the 64-cell stage expands the brain region in expense of the placode-like region. The expression of *Zic-r.b* is transient and becomes undetectable before differentiation. Thus, *Zic* genes work first for segregating mesodermal cells within endomesodermal cells and then for segregating neural cells within ectodermal cells.

References

Anno C, Satou A, Fujiwara S (2006) Transcriptional regulation of ZicL in the Ciona intestinalis embryo. Dev Genes Evol 216(10):597–605. https://doi.org/10.1007/s00427-006-0080-9

Aruga J (2004) The role of Zic genes in neural development. Mol Cell Neurosci 26(2):205–221. https://doi.org/10.1016/j.mcn.2004.01.004

Aruga J, Kamiya A, Takahashi H, Fujimi TJ, Shimizu Y, Ohkawa K, Yazawa S, Umesono Y, Noguchi H, Shimizu T, Saitou N, Mikoshiba K, Sakaki Y, Agata K, Toyoda A (2006) A wide-range phylogenetic analysis of Zic proteins: implications for correlations between protein structure conservation and body plan complexity. Genomics 87(6):783–792. https://doi.org/10.1016/j.ygeno.2006.02.011

Brunetti R, Gissi C, Pennati R, Caicci F, Gasparini F, Manni L (2015) Morphological evidence that the molecularly determined Ciona intestinalis type A and type B are different species: Ciona robusta and Ciona intestinalis. J Zool Syst Evol Res 53(3):186–193. https://doi.org/10.1111/jzs.12101

Caputi L, Andreakis N, Mastrototaro F, Cirino P, Vassillo M, Sordino P (2007) Cryptic speciation in a model invertebrate chordate. Proc Natl Acad Sci U S A 104(22):9364–9369. https://doi.org/10.1073/pnas.0610158104

Chiba S, Jiang D, Satoh N, Smith WC (2009) Brachyury null mutant-induced defects in juvenile ascidian endodermal organs. Development 136(1):35–39. https://doi.org/10.1242/dev.030981

Conklin EG (1905) Mosaic development in ascidian eggs. J Exp Zool 2(2):145–223. https://doi.org/10.1002/jez.1400020202

Corbo JC, Fujiwara S, Levine M, Di Gregorio A (1998) Suppressor of hairless activates brachyury expression in the Ciona embryo. Dev Biol 203(2):358–368. https://doi.org/10.1006/dbio.1998.9067

Dehal P, Satou Y, Campbell RK, Chapman J, Degnan B, De Tomaso A, Davidson B, Di Gregorio A, Gelpke M, Goodstein DM, Harafuji N, Hastings KE, Ho I, Hotta K, Huang W, Kawashima T, Lemaire P, Martinez D, Meinertzhagen IA, Necula S, Nonaka M, Putnam N, Rash S, Saiga H, Satake M, Terry A, Yamada L, Wang HG, Awazu S, Azumi K, Boore J, Branno M, Chin-Bow S, DeSantis R, Doyle S, Francino P, Keys DN, Haga S, Hayashi H, Hino K, Imai KS, Inaba K, Kano S, Kobayashi K, Kobayashi M, Lee BI, Makabe KW, Manohar C, Matassi G, Medina M, Mochizuki Y, Mount S, Morishita T, Miura S, Nakayama A, Nishizaka S, Nomoto H, Ohta F, Oishi K, Rigoutsos I, Sano M, Sasaki A, Sasakura Y, Shoguchi E, Shin-i T, Spagnuolo A, Stainier D, Suzuki MM, Tassy O, Takatori N, Tokuoka M, Yagi K, Yoshizaki F, Wada S, Zhang C, Hyatt PD, Larimer F, Detter C, Doggett N, Glavina T, Hawkins T, Richardson P, Lucas S, Kohara Y, Levine M, Satoh N, Rokhsar DS (2002) The draft genome of Ciona intestinalis: insights into chordate and vertebrate origins. Science 298(5601):2157–2167. https://doi.org/10.1126/science.1080049

Hashimoto H, Enomoto T, Kumano G, Nishida H (2011) The transcription factor FoxB mediates temporal loss of cellular competence for notochord induction in ascidian embryos. Development 138(12):2591–2600. https://doi.org/10.1242/dev.053082

Haupaix N, Stolfi A, Sirour C, Picco V, Levine M, Christiaen L, Yasuo H (2013) p120RasGAP mediates ephrin/Eph-dependent attenuation of FGF/ERK signals during cell fate specification in ascidian embryos. Development 140(21):4347–4352. https://doi.org/10.1242/dev.098756

Hudson C, Lemaire P (2001) Induction of anterior neural fates in the ascidian Ciona intestinalis. Mech Dev 100(2):189–203

Hudson C, Darras S, Caillol D, Yasuo H, Lemaire P (2003) A conserved role for the MEK signalling pathway in neural tissue specification and posteriorisation in the invertebrate chordate, the ascidian Ciona intestinalis. Development 130(1):147–159

Hudson C, Kawai N, Negishi T, Yasuo H (2013) β-catenin-driven binary fate specification segregates germ layers in ascidian embryos. Curr Biol 23(6):491–495. https://doi.org/10.1016/j.cub.2013.02.005

Hudson C, Sirour C, Yasuo H (2016) Co-expression of Foxa.a, Foxd and Fgf9/16/20 defines a transient mesendoderm regulatory state in ascidian embryos. elife 5:e14692. https://doi.org/10.7554/eLife.14692

Ikeda T, Satou Y (2017) Differential temporal control of Foxa.a and Zic-r.b specifies brain versus notochord fate in the ascidian embryo. Development 144(1):38–43. https://doi.org/10.1242/dev.142174

Ikeda T, Matsuoka T, Satou Y (2013) A time delay gene circuit is required for palp formation in the ascidian embryo. Development 140(23):4703–4708. https://doi.org/10.1242/dev.100339

Imai K, Satoh N, Satou Y (2002a) Early embryonic expression of FGF4/6/9 gene and its role in the induction of mesenchyme and notochord in Ciona savignyi embryos. Development 129(7):1729–1738

Imai KS, Satoh N, Satou Y (2002b) An essential role of a FoxD gene in notochord induction in Ciona embryos. Development 129(14):3441–3453

Imai KS, Satou Y, Satoh N (2002c) Multiple functions of a Zic-like gene in the differentiation of notochord, central nervous system and muscle in Ciona savignyi embryos. Development 129(11):2723–2732

Imai K, Satoh N, Satou Y (2003) A twist-like bHLH gene is a downstream factor of an endogenous FGF and determines mesenchymal fate in the ascidian embryos. Development 130(18):4461–4472. https://doi.org/10.1242/dev.00652

Imai KS, Hino K, Yagi K, Satoh N, Satou Y (2004) Gene expression profiles of transcription factors and signaling molecules in the ascidian embryo: towards a comprehensive understanding of gene networks. Development 131(16):4047–4058. https://doi.org/10.1242/dev.01270

Imai KS, Levine M, Satoh N, Satou Y (2006) Regulatory blueprint for a chordate embryo. Science 312(5777):1183–1187. https://doi.org/10.1126/science.1123404

Imai KS, Stolfi A, Levine M, Satou Y (2009) Gene regulatory networks underlying the compartmentalization of the Ciona central nervous system. Development 136(2):285–293. https://doi.org/10.1242/dev.026419

Imai KS, Hudson C, Oda-Ishii I, Yasuo H, Satou Y (2016) Antagonism between beta-catenin and Gata.a sequentially segregates the germ layers of ascidian embryos. Development 143(22):4167–4172. https://doi.org/10.1242/dev.141481

Imai KS, Hikawa H, Kobayashi K, Satou Y (2017) Tfap2 and Sox1/2/3 cooperatively specify ectodermal fates in ascidian embryos. Development 144(1):33–37. https://doi.org/10.1242/dev.142109

Jolma A, Yan J, Whitington T, Toivonen J, Nitta KR, Rastas P, Morgunova E, Enge M, Taipale M, Wei GH, Palin K, Vaquerizas JM, Vincentelli R, Luscombe NM, Hughes TR, Lemaire P, Ukkonen E, Kivioja T, Taipale J (2013) DNA-binding specificities of human transcription factors. Cell 152(1–2):327–339. https://doi.org/10.1016/j.cell.2012.12.009

Kobayashi K, Sawada K, Yamamoto H, Wada S, Saiga H, Nishida H (2003) Maternal macho-1 is an intrinsic factor that makes cell response to the same FGF signal differ between mesenchyme and notochord induction in ascidian embryos. Development 130(21):5179–5190. https://doi.org/10.1242/dev.00732

Kondoh K, Kobayashi K, Nishida H (2003) Suppression of macho-1-directed muscle fate by FGF and BMP is required for formation of posterior endoderm in ascidian embryos. Development 130(14):3205–3216

Koyabu Y, Nakata K, Mizugishi K, Aruga J, Mikoshiba K (2001) Physical and functional interactions between Zic and Gli proteins. J Biol Chem 276(10):6889–6892. https://doi.org/10.1074/jbc.C000773200

Kubo A, Suzuki N, Yuan X, Nakai K, Satoh N, Imai KS, Satou Y (2010) Genomic cis-regulatory networks in the early Ciona intestinalis embryo. Development 137(10):1613–1623. https://doi.org/10.1242/dev.046789

Kugler JE, Gazdoiu S, Oda-Ishii I, Passamaneck YJ, Erives AJ, Di Gregorio A (2010) Temporal regulation of the muscle gene cascade by Macho1 and Tbx6 transcription factors in Ciona intestinalis. J Cell Sci 123(14):2453–2463. https://doi.org/10.1242/jcs.066910

Kumano G, Yamaguchi S, Nishida H (2006) Overlapping expression of FoxA and Zic confers responsiveness to FGF signaling to specify notochord in ascidian embryos. Dev Biol 300(2):770–784. https://doi.org/10.1016/j.ydbio.2006.07.033

Kumano G, Takatori N, Negishi T, Takada T, Nishida H (2011) A maternal factor unique to ascidians silences the germ line via binding to P-TEFb and RNAP II regulation. Curr Biol 21(15):1308–1313. https://doi.org/10.1016/j.cub.2011.06.050

Lemaire P (2011) Evolutionary crossroads in developmental biology: the tunicates. Development 138(11):2143–2152. https://doi.org/10.1242/dev.048975

Nakatani Y, Nishida H (1994) Induction of notochord during ascidian embryogenesis. Dev Biol 166(1):289–299. https://doi.org/10.1006/dbio.1994.1315

Nishida H, Sawada K (2001) Macho-1 encodes a localized mRNA in ascidian eggs that specifies muscle fate during embryogenesis. Nature 409(6821):724–729. https://doi.org/10.1038/35055568

Nydam ML, Harrison RG (2007) Genealogical relationships within and among shallow-water Ciona species (Ascidiacea). Mar Biol 151(5):1839–1847. https://doi.org/10.1007/s00227-007-0617-0

Nydam ML, Harrison RG (2011) Introgression despite substantial divergence in a broadcast spawning marine invertebrate. Evolution 65(2):429–442. https://doi.org/10.1111/j.1558-5646.2010.01153.x

Oda-Ishii I, Kubo A, Kari W, Suzuki N, Rothbacher U, Satou Y (2016) A maternal system initiating the zygotic developmental program through combinatorial repression in the ascidian embryo. PLoS Genet 12(5):e1006045. https://doi.org/10.1371/journal.pgen.1006045

Ohta N, Satou Y (2013) Multiple signaling pathways coordinate to induce a threshold response in a chordate embryo. PLoS Genet 9(10):e1003818. https://doi.org/10.1371/journal.pgen.1003818

Paix A, Yamada L, Dru P, Lecordier H, Pruliere G, Chenevert J, Satoh N, Sardet C (2009) Cortical anchorages and cell type segregations of maternal postplasmic/PEM RNAs in ascidians. Dev Biol 336(1):96–111. https://doi.org/10.1016/j.ydbio.2009.09.001

Picco V, Hudson C, Yasuo H (2007) Ephrin-Eph signalling drives the asymmetric division of notochord/neural precursors in Ciona embryos. Development 134(8):1491–1497. https://doi.org/10.1242/dev.003939

Rothbächer U, Bertrand V, Lamy C, Lemaire P (2007) A combinatorial code of maternal GATA, Ets and β-catenin-TCF transcription factors specifies and patterns the early ascidian ectoderm. Development 134(22):4023–4032. https://doi.org/10.1242/dev.010850

Sanchez-Ferras O, Bernas G, Laberge-Perrault E, Pilon N (2014) Induction and dorsal restriction of paired-box 3 (Pax3) gene expression in the caudal neuroectoderm is mediated by integration of multiple pathways on a short neural crest enhancer. Biochim Biophys Acta 1839(7):546–558. https://doi.org/10.1016/j.bbagrm.2014.04.023

Sasakura Y, Ogasawara M, Makabe KW (1998a) HrWnt-5: a maternally expressed ascidian Wnt gene with posterior localization in early embryos. Int J Dev Biol 42(4):573–579

Sasakura Y, Ogasawara M, Makabe KW (1998b) Maternally localized RNA encoding a serine/threonine protein kinase in the ascidian, Halocynthia roretzi. Mech Dev 76(1–2):161–163

Sato A, Satoh N, Bishop JDD (2012) Field identification of 'types' A and B of the ascidian Ciona intestinalis in a region of sympatry. Mar Biol 159(7):1611–1619. https://doi.org/10.1007/s00227-012-1898-5

Satoh N, Satou Y, Davidson B, Levine M (2003) Ciona intestinalis: an emerging model for whole-genome analyses. Trends Genet 19(7):376–381. https://doi.org/10.1016/S0168-9525(03)00144-6

Satou Y (1999) Posterior end mark 3 (pem-3), an ascidian maternally expressed gene with localized mRNA encodes a protein with Caenorhabditis elegans MEX-3-like KH domains. Dev Biol 212(2):337–350. https://doi.org/10.1006/dbio.1999.9336

Satou Y, Satoh N (1997) Posterior end mark 2 (pem-2), pem-4, pem-5, and pem-6: maternal genes with localized mRNA in the ascidian embryo. Dev Biol 192(2):467–481. https://doi.org/10.1006/dbio.1997.8730

Satou Y, Yagi K, Imai KS, Yamada L, Nishida H, Satoh N (2002) Macho-1-related genes in Ciona embryos. Dev Genes Evol 212(2):87–92. https://doi.org/10.1007/s00427-002-0218-3

Shimauchi Y, Yasuo H, Satoh N (1997) Autonomy of ascidian fork head/HNF-3 gene expression. Mech Dev 69(1–2):143–154

Shirae-Kurabayashi M, Matsuda K, Nakamura A (2011) Ci-Pem-1 localizes to the nucleus and represses somatic gene transcription in the germ line of Ciona intestinalis embryos. Dev 138(14):2871–2881. https://doi.org/10.1242/dev.058131

Stolfi A, Sasakura Y, Chalopin D, Satou Y, Christiaen L, Dantec C, Endo T, Naville M, Nishida H, Swalla BJ, Volff JN, Voskoboynik A, Dauga D, Lemaire P (2015) Guidelines for the nomenclature of genetic elements in tunicate genomes. Genesis 53(1):1–14. https://doi.org/10.1002/dvg.22822

Tsagkogeorga G, Turon X, Hopcroft RR, Tilak MK, Feldstein T, Shenkar N, Loya Y, Huchon D, Douzery EJP, Delsuc F (2009) An updated 18S rRNA phylogeny of tunicates based on mixture and secondary structure models. BMC Evol Biol 9:187. https://doi.org/10.1186/1471-2148-9-187

Turon X, Lopez-Legentil S (2004) Ascidian molecular phylogeny inferred from mtDNA data with emphasis on the Aplousobranchiata. Mol Phylogenet Evol 33(2):309–320. https://doi.org/10.1016/j.ympev.2004.06.011

Vinson J, Jaffe D, O'Neill K, Karlsson E, Stange-Thomann N, Anderson S, Mesirov J, Satoh N, Satou Y, Nusbaum C, Birren B, Galagan J, Lander E (2005) Assembly of polymorphic genomes: algorithms and application to Ciona savignyi. Genome Res 15(8):1127–1135. https://doi.org/10.1101/gr.3722605

Wada H (1998) Evolutionary history of free-swimming and sessile lifestyles in urochordates as deduced from 18S rDNA molecular phylogeny. Mol Biol Evol 15(9):1189–1194

Wada S, Saiga H (2002) HrzicN, a new Zic family gene of ascidians, plays essential roles in the neural tube and notochord development. Development 129(24):5597–5608. https://doi.org/10.1242/dev.00156

Wagner E, Levine M (2012) FGF signaling establishes the anterior border of the Ciona neural tube. Development 139(13):2351–2359. https://doi.org/10.1242/dev.078485

Yagi K, Satoh N, Satou Y (2004a) Identification of downstream genes of the ascidian muscle determinant gene Ci-macho1. Dev Biol 274(2):478–489. https://doi.org/10.1016/j.ydbio.2004.07.013

Yagi K, Satou Y, Satoh N (2004b) A zinc finger transcription factor, ZicL, is a direct activator of Brachyury in the notochord specification of Ciona intestinalis. Development 131(6):1279–1288. https://doi.org/10.1242/dev.01011

Yagi K, Takatori N, Satou Y, Satoh N (2005) Ci-Tbx6b and Ci-Tbx6c are key mediators of the maternal effect gene Ci-macho1 in muscle cell differentiation in Ciona intestinalis embryos. Dev Biol 282(2):535–549. https://doi.org/10.1016/j.ydbio.2005.03.029

Yamada L, Kobayashi K, Degnan B, Satoh N, Satou Y (2003) A genomewide survey of developmentally relevant genes in Ciona intestinalis. IV. Genes for HMG transcriptional regulators, bZip and GATA/Gli/Zic/snail. Dev Genes Evol 213(5–6):245–253. https://doi.org/10.1007/s00427-003-0316-x

Yamada L, Kobayashi K, Satou Y, Satoh N (2005) Microarray analysis of localization of maternal transcripts in eggs and early embryos of the ascidian, Ciona intestinalis. Dev Biol 284(2):536–550. https://doi.org/10.1016/j.ydbio.2005.05.027

Yasuo H, Satoh N (1993) Function of vertebrate T gene. Nature 364(6438):582–583. https://doi.org/10.1038/364582b0

Yasuo H, Satoh N (1998) Conservation of the developmental role of Brachyury in notochord formation in a urochordate, the ascidian Halocynthia roretzi. Dev Biol 200(2):158–170

Yasuo H, Kobayashi M, Shimauchi Y, Satoh N (1996) The ascidian genome contains another T-domain gene that is expressed in differentiating muscle and the tip of the tail of the embryo. Dev Biol 180(2):773–779. https://doi.org/10.1006/dbio.1996.0345

Yoshida S, Marikawa Y, Satoh N (1996) Posterior end mark, a novel maternal gene encoding a localized factor in the ascidian embryo. Development 122(7):2005–2012

Chapter 7
Amphibian *Zic* Genes

Christa Merzdorf and Jennifer Forecki

Abstract Studies in *Xenopus laevis* have greatly contributed to understanding the roles that the Zic family of zinc finger transcription factors play as essential drivers of early development. Explant systems that are not readily available in other organisms give *Xenopus* embryos a unique place in these studies, facilitated by the recent sequencing of the *Xenopus laevis* genome. A number of upstream regulators of *zic* gene expression have been identified, such as inhibition of BMP signaling, as well as calcium, FGF, and canonical Wnt signaling. Screens using induced ectodermal explants have identified genes that are direct targets of Zic proteins during early neural development and neural crest specification. These direct targets include *Xfeb* (also called *glipr2*; hindbrain development), *aqp3b* (dorsal marginal zone in gastrula embryos and neural folds), *snail* family members (premigratory neural crest), genes that play roles in retinoic acid signaling, noncanonical Wnt signaling, and mesoderm development, in addition to a variety of genes some with and many without known roles during neural or neural crest development. Functional experiments in *Xenopus* embryos demonstrated the involvement of Zic family members in left-right determination, early neural patterning, formation of the midbrain-hindbrain boundary, and neural crest specification. The role of *zic* genes in cell proliferation vs. differentiation remains unclear, and the activities of Zic factors as inhibitors or activators of canonical Wnt signaling may be dependent on developmental context. Overall, *Xenopus* has contributed much to our understanding of how Zic transcriptional activities shape the development of the embryo and contribute to disease.

Keywords Zic genes · *Xenopus*

C. Merzdorf (✉) · J. Forecki
Department of Cell Biology and Neuroscience, Montana State University, Bozeman, MT, USA
e-mail: merzdorf@montana.edu

J. Aruga (ed.), *Zic family*, Advances in Experimental Medicine and Biology 1046,
https://doi.org/10.1007/978-981-10-7311-3_7

7.1 Introduction

The Zic family of zinc finger proteins plays multiple roles during early development. In this chapter, we will examine how studies with *Xenopus laevis* embryos have contributed to our understanding of *zic* genes and their activities. Although complete gene knockout in early developmental stages of *Xenopus* is difficult, partly because maternal mRNAs can persist past MBT (Blum et al. 2015), gene expression levels can easily be altered in *Xenopus* embryos using morpholino oligonucleotides or injection of mRNAs. Further, *Xenopus* embryos readily lend themselves to physical manipulation. Therefore, studies in *Xenopus laevis* have contributed much to our understanding of the functional roles of *zic* genes during neural induction, early neural patterning, and formation of the neural crest. In addition, microarray screens have identified a number of direct targets of Zic proteins, prompting a number of new and ongoing studies. Due to years of study, a large body of knowledge has been amassed on *Xenopus* embryo development, gene regulation, and cell fate mapping, which helps put the roles of *zic* genes into context.

7.1.1 Experimental Approaches Unique to Xenopus

7.1.1.1 Ectodermal Explants (Animal Caps)

Ectodermal explants (animal caps) allow researchers to study gene expression in cells that are competent to respond to neural induction. At the same time, these explants allow the study of gene regulation free from the variety of inductive signals that characterize gastrulation and neural induction. For animal cap experiments, two-cell embryos are typically injected with mRNAs or other molecules into the animal hemisphere of both blastomeres. After maturing to late blastula (stage 9), ectodermal explants are harvested from the animal hemisphere of the embryos. The explants form characteristic balls, which can be aged to gastrula and neurula stages (using intact sibling embryos for staging), at which point they are processed in assays to determine gene expression (Sive et al. 2007). Ectodermal explants have been used extensively to identify gene regulatory relationships between *zic* and other genes, which can then be tested in whole embryos. In addition, ectodermal explants make *Xenopus* embryos uniquely suited to identify or confirm genes that are direct targets of transcription factors active during early development. A hormone-inducible transcription factor is constructed by fusing the glucocorticoid receptor domain (hGR) to the transcription factor, and mRNA for this inducible construct is injected into the embryos. The hGR domain forms a complex with endogenous HSP90, thus retaining the transcription factor in the cytoplasm (Kolm and Sive 1995; Mattioni et al. 1994). Treatment with the hormone dexamethasone allows the hGR-bound transcription factor to detach and enter the nucleus. In order to identify direct transcriptional targets, the hormone-inducible transcription factor is activated in the presence of protein synthesis inhibitors. More detail is provided below in the description of two screens for direct targets of Zic1 (Cornish et al. 2009; Plouhinec et al. 2014).

7.1.1.2 Keller Explants

Keller open-faced explants are derived from the dorsal marginal zone of early gastrula embryos (Keller and Danilchik 1988). They comprise prospective mesoderm and ectoderm and allow powerful studies of the genes involved in regulating convergent extension movements (Keller et al. 1992). With regard to *zic* genes, this system is being used to study the role of *aqp3b*, a direct target of Zic1, in convergent extension (See and Merzdorf unpublished). Keller explants have also been used to study neural induction free from vertical signals, since the signals that pass from the mesoderm to the ectoderm portion of the explant are limited to planar signals. This system demonstrated that calcium transients are required for induction of *zic3* expression (Leclerc et al. 2003). With the identification of direct targets of Zic1 that play roles in noncanonical Wnt signaling (Cornish et al. 2009), Keller explants may help understand the roles that these genes play in convergent extension.

7.2 Zic Family Genes and Their Expression in Xenopus Embryos

7.2.1 *Comparison of* Zic *Genes in the Allotetraploid Genome of* Xenopus laevis

The genomes of both *Xenopus* species are nearly complete (Hellsten et al. 2010; Session et al. 2016). *Xenopus zic* genes show the same chromosomal arrangement as *Zic* genes in mouse and humans, with *zic* genes clustered on the same chromosome in a head-to-head orientation: *zic1* with *zic4*, *zic2* with *zic5*, and *zic3* on a different chromosome (Grinberg and Millen 2005; Aruga et al. 2006). In addition to the zinc finger (ZF) DNA-binding domain, the Zic-Opa (ZOC) and zinc finger-nucleocapsid (ZF-NC) domains (both N-terminal to the zinc fingers) are conserved between *Xenopus* and mammalian *zic* genes (ZOC is present only in *zic1-3*) (Houtmeyers et al. 2013).

Xenopus laevis and *Xenopus tropicalis* both have five *zic* genes, but due to the allotetraploid nature of *X. laevis*, its genome possesses two versions of each *zic* gene, one on either a longer or shorter chromosome. The *zic* genes are therefore named *zic.S* and *zic.L*. Table 7.1 shows the results of comparing nucleotide sequences of S and L *zic* gene-coding regions and the amino acid sequences of S and L Zic proteins. The S and L variants were also compared to the *X. tropicalis* versions of each Zic protein (Table 7.1) (Ricker et al. unpublished). The amino acid sequence identities indicate that the S and L versions of *X. laevis* Zic proteins are about

Table 7.1 Nucleotide and amino acid sequence identity between *Xenopus laevis* Zic.S and Zic.L versions and Zic proteins in *X. tropicalis*

Gene name	Sequence source	CDS nucleotide identity	Amino acid identity between S and L gene versions	Amino acid identity between S and L genes and *X. tropicalis* zics
zic1.S	NM_001090330.1	95%	98%:	98%
zic1.L	Sequence predicted from genome		7 aa substitutions (1 in ZF)	99%
zic2.S	NM_001085959.1	94%	96%:	96%
zic2.L	NM_001087724.1		11 aa substitutions (4 in ZF); 4 gaps in Zic2.S and 4 gaps in Zic2.L	95%
zic3.S	NM_001087619.1	96%	97%:	98%
zic3.L	Sequence predicted from genome		13 aa substitution (2 in ZF)	98%
zic4.S	Sequence predicted from genome	94%	92%:	93%
zic4.L	NM_001127780.1		33 aa substitutions (2 in ZF); 5 gaps in Zic4.S and 4 gaps in Zic4.L	93%
zic5.S	NM_001085657.1	95%	93%:	91%
zic5.L	Sequence predicted from genome		29 aa substitutions (4 in ZF); 3 gaps in Zic5.S and 4 gaps in Zic5.L	93%

The nucleotide sequences of the coding regions and the amino acid sequences of the *X. laevis zic* genes on the S and L chromosomes were compared using Blastn and Blastp, respectively, to determine their sequence identity. The number of amino acid differences is indicated and gaps comprise maximally three consecutive amino acids. The S and L versions in *X. laevis* are as different from each other as they are from the *X. tropicalis* versions of each *zic* gene

equally divergent from each other as they are from the Zic proteins in *X. tropicalis*. The differences include substitutions and small gaps spanning up to three consecutive amino acids. Outside of the coding region, in the 5′UTR and 3′UTR, the sequences are more divergent between the S and L versions. The untranslated regions of the *X. tropicalis zic* genes are significantly different from the UTRs of the *X. laevis zic* genes.

The presence of two versions for each *zic* gene in *X. laevis* allowed each gene to diverge and possibly even perform different functions. For example, the Zic1 direct target gene *aqp3b* is the L version of the X. *laevis aqp3* gene. It is expressed at gastrula and neurula stages, while *aqp3a*, the S version of the gene, is not expressed during early development (Cornish et al. 2009). In adult frogs, the tissues that express the two *aqp3* genes vary, although the composite of the expression patterns is similar to the overall expression pattern of the single *Aqp3* gene in mice (Cornish et al. 2009; King et al. 2004). Thus, the individual roles of the S and L copies of each *zic* gene may vary but, taken together, may perform similar functions as a single copy in other species. Finally, while *X. tropicalis* is a useful model, to date it has not been used to study the roles of *zic* genes.

7.2.2 *Expression of* Zic *Genes in Xenopus Embryos*

The gene expression patterns of *zic* genes in *Xenopus* embryos overlap extensively (Fujimi et al. 2006, 2012), which is also the case for *zic* genes in other vertebrates, for example, mouse and chick embryos (Nagai et al. 1997; Furushima et al. 2000; Gaston-Massuet et al. 2005; McMahon and Merzdorf 2010). Despite the overlap, there are significant differences in the expression domains of *zic* genes.

7.2.2.1 Blastula Embryos

The *zic2* gene is the only maternally expressed *zic* gene in *Xenopus* embryos (Nakata et al. 1998). The expression of *zic1*, *zic3*, and *zic4* begins at stage 9, after midblastula transition, although the expression of *zic4* is initially very low (Fig. 7.1a). The *zic5* gene is not expressed in blastula embryos. In situ hybridization shows that in late blastula embryos (stage 9.5), *zic1*, *zic2*, and *zic3* are expressed in the dorsal marginal zone in both ectoderm and mesoderm (Fig. 7.1b). There does not appear to be significant expression of these *zic* genes in the roof of the blastocoel.

7.2.2.2 Gastrula Embryos

As gastrulation begins, the *zic1-3* genes are strongly expressed in the prospective neural ectoderm and moderately expressed in the mesoderm (stages 10.5 and 11; Fig. 7.1b). *zic4* expression is quite low, and *zic5* expression begins in late gastrula embryos (Figs 7.1a and 7.2). In late gastrula embryos (stage 11.5), *zic1*, *zic2*, *zic3*, and *zic5* are expressed to varying degrees in a broad region of the prospective neural ectoderm (Fig. 7.2), while *zic4* expression is extremely weak. These expression patterns are consistent with the significant roles that the *zic1-3* genes play during early stages of development and show that *zic* genes are among the earliest genes expressed in response to neural induction.

7.2.2.3 Neurula Embryos

During neurula stages, all five *zic* genes are expressed in the neural plate border. Only *zic2* and *zic3* are expressed within the neural plate. *zic3* is found in the midbrain-hindbrain region, and *zic2* is expressed at the midline of the neural plate (Fujimi et al. 2006) and in the progenitor cells located between the stripes of primary neurons (Brewster et al. 1998). At the neural plate border, *zic1-3* are strongly

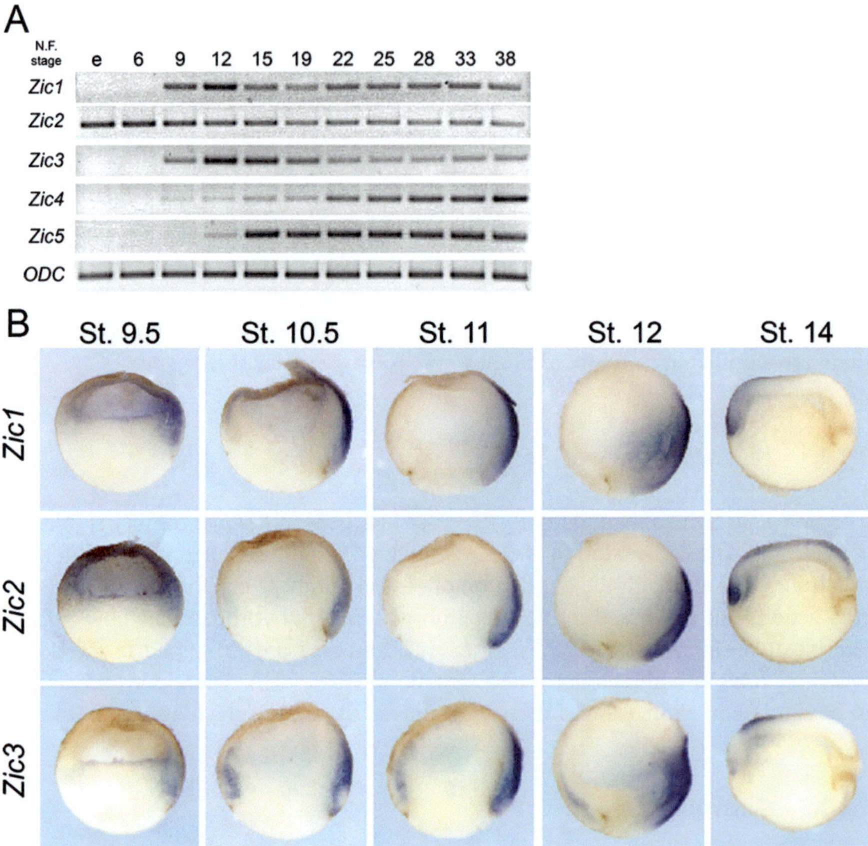

Fig. 7.1 Expression of *zic* genes in *Xenopus* embryos. (**a**) Expression of *zic1-5* determined by RT-PCR in unfertilized eggs (e) and *Xenopus* embryos at different developmental stages, including pre-MBT blastula (stage 6), post-MBT blastula (stage 9), late gastrula (stage 12), mid-neurula (stage 15), and tailbud stages (stage 19 and older). *zic2* is expressed both maternally and throughout early development. *zic1* and *zic3* are first detected at stage 9. The expression of *zic3* peaks in late gastrula/early neurula, while *zic1* expression remains strong. Weak expression of *zic4* is first detected at stage 9 and continues until tailbud (stage 22 and later stages), when it is more strongly expressed. Weak expression of *zic5* is detected by late gastrula (stage 12), and it is strongly expressed in neurula stages and beyond. (**b**) Expression of *zic1-3* by in situ hybridization in whole embryos. Dorsal is to the right. *zic2* mRNAs are more extensively present at stage 9.5, likely due to residual maternal mRNA. During late blastula (stage 9.5) and throughout gastrula (stages 10.5-12), the *zic1-3* genes are expressed in the dorsal ectoderm and in the involuting mesoderm. *zic3* is also expressed in the ventral and lateral involuting mesoderm (Fujimi et al. 2012; Kitaguchi et al. 2000). During neurula (stage 14) *zic1-3* are expressed in the neural plate and to some extent in the notochord (Reproduced from Fujimi et al. 2012 with permission of the publisher)

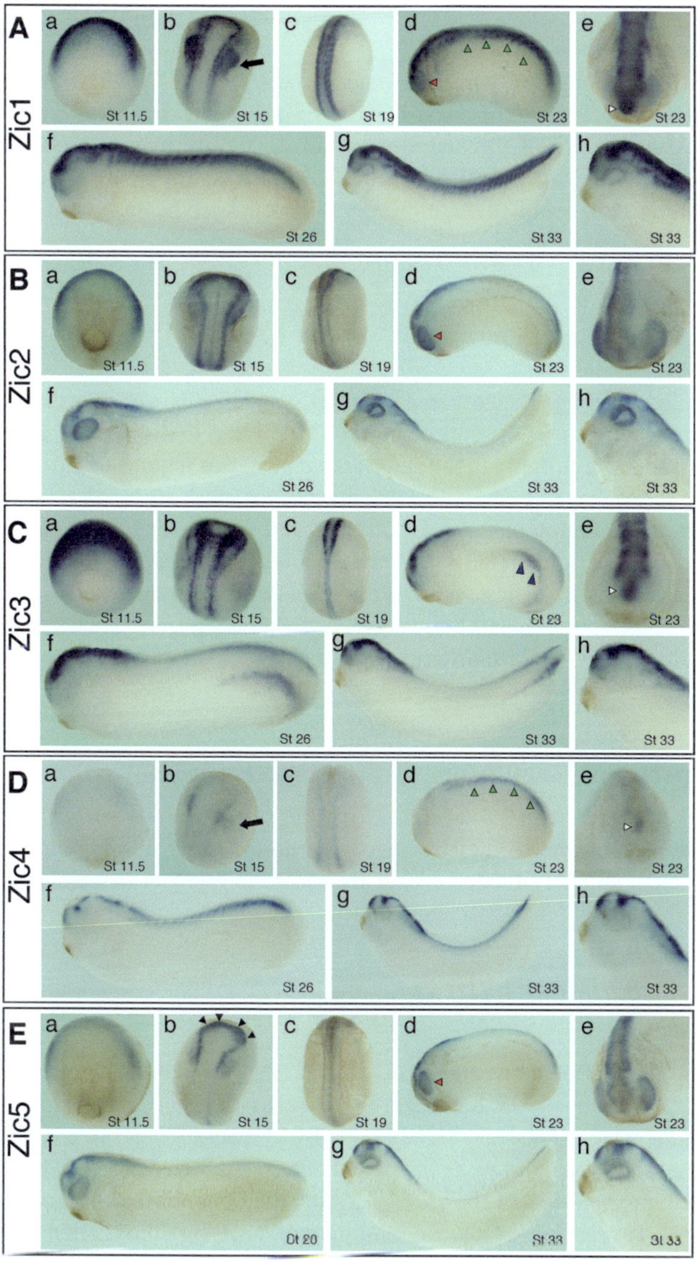

Fig. 7.2 Expression of *zic* genes in gastrula, neurula, and tailbud *Xenopus* embryos. The expression of the *zic1-5* genes was determined by in situ hybridization in whole embryos. During gastrulation (a: stage 11.5), all *zic* genes with the exception of *zic4* are expressed in the presumptive neural plate. During neurula (stages 15 and 19) and tailbud (stages 23 and later stages), all *zic* genes show expression in the dorsal neural tube. Other tissues also show *zic* gene expression, including the hyoid and branchial crest (black arrow), eye (red arrowheads), somites (green arrowheads), lateral mesoderm (blue arrowheads), and olfactory placode (white arrowheads) (Reproduced from Fujimi et al. 2006 with permission of the publisher)

expressed in wide regions, while *zic4* and *zic5* are more restricted to the regions of the neural folds (Fig. 7.2b). *zic1-3* and *zic5* are expressed strongly in anterior regions. After closure of the neural tube, all *zic* family members continue to be expressed in the dorsal neural tube (Fig. 7.2c). Thus, *zic* genes are expressed in areas required for neural patterning, neural crest specification, and neural tube closure.

7.2.2.4 Tailbud Stage Embryos

During tailbud and later stages, *zic1* is strongly and *zic4* is weakly expressed in the dorsal neural tube along the entire embryo (Fig. 7.2A, D d–g). *zic2*, *zic3*, and *zic5* are expressed more strongly in the anterior and posterior regions of the dorsal neural tube (Fig. 7.2B, C, E d–g). In tailbud stage embryos, *zic1*, *zic2*, and *zic5* show some expression in the region of the eye, both *zic1* and *zic4* are expressed in the somites (Nakata et al. 2000), and *zic3* is uniquely expressed in caudal lateral plate mesoderm (Fig. 7.2). Thus, *zic* gene expression patterns overlap extensively but also show unique aspects. Some of these correlate to known differences in *zic* gene function, although many of these differences in expression are not yet understood.

7.3 Upstream Regulators of *Zic* Gene Expression

Xenopus zic genes are expressed extensively during early development (Figs. 7.1 and 7.2), and a number of mechanisms are known to regulate *zic* gene expression.

7.3.1 Inhibition of BMP Signaling

The inhibition of bone morphogenetic protein (BMP) signaling is critical for neural induction (Sasai et al. 1996; Sasai and De Robertis 1997) and plays an early role in regulating *zic* gene expression. Signaling by BMP specifies ventral, non-neural fates and represses neural genes in *Xenopus* and other vertebrates. Thus, BMP signaling represses the expression of the *zic1*, *zic2*, and *zic3* genes (Gamse and Sive 2001; Nakata et al. 1997). Conversely, the inhibition of BMP signaling, often mediated by Noggin and Chordin, is essential for dorsal determination and specification of neural fate. Accordingly, misexpression of *noggin* (Mizuseki et al. 1998; Gamse and Sive 2001) or *FRL-1* (Yabe et al. 2003), which also represses BMP signaling, results in an increase in *zic* gene expression. Indeed, the promoter of *zic1* contains a 215 bp BMP inhibitory response module (BIRM) (−2.7 to −2.5 kb 5′ to the transcription start site). The BIRM is required for transcription of the *zic1* gene in the absence of BMP signaling in animal cap-based reporter assays (Tropepe et al. 2006). The BIRM contains consensus binding sites for several transcription factors, including one Smad binding site and binding sites for the Ets, Oct, Lef/Tcf, and Sox transcription factors. Mutations in most of these putative transcription factor binding sites eliminate the ability of the BIRM to respond to Noggin, suggesting that multiple signals must cooperate to mediate *zic1* transcription in response to BMP inhibition. A dominant interfering Smad is able to induce expression of *zic1* in the absence of translation (Marchal et al. 2009), indicating direct regulation. However, mutation of the putative Smad binding site within the BIRM does not activate reporter gene expression (Tropepe et al. 2006). Thus, the inhibition of BMP signaling is required for *zic* gene expression, but which region of the *zic* gene directly responds to lack of BMP signaling, or exactly how the BIRM is responsive to suppression of BMP signaling remains to be answered.

7.3.2 *Siamois and Twin*

Organizer-specific transcription factors, such as Siamois and Twin, are responsible for the expression of BMP antagonists, including *noggin* and *chordin*. These BMP antagonists are secreted from the organizer (dorsal mesoderm) and block BMP in the neural ectoderm, which results in the upregulation of *zic* genes, as described above. Accordingly, *zic* genes are expressed in a wide domain in the neural ectoderm during gastrula stages (Fig. 7.2). Klein and Moody (2015) examined whether the expression of neural genes could be induced directly by organizer transcription factors, in addition to the indirect induction by BMP inhibitors. They found that in late blastula embryos, ectopic expression of the organizer genes *siamois* and *twin* induced ectopic *zic2* expression directly, in the absence of translation. Later in development, as gastrulation begins, *zic2* is present in the involuting dorsal mesoderm at moderate levels, with stronger expression in the neural ectoderm (Fujimi et al. 2012), while Siamois and Twin are limited to the dorsal mesoderm. This lack of overlap suggests that *zic2* expression in neural ectoderm is now regulated indirectly through induction of BMP inhibitors by Siamois and Twin. The significance for this bimodal regulation needs to be explored further. However, the direct induction of *zic2* by Siamois and Twin may serve to bias the dorsal region of the late blastula/early gastrula toward neural induction, and further studies demonstrated that maternal *zic2* is able to exert this bias as well (Gaur et al. 2016). This is supported by the finding that Zic1 is able to sensitize the future neural ectoderm for neural induction (Kuo et al. 1998).

7.3.3 *Calcium Signaling*

Calcium signaling helps mediate the activity of BMP inhibitors during neural induction. Noggin causes an increase in calcium transients in the prospective neural ectoderm, and experimentally increasing calcium demonstrated that it is a potent inducer of early neural genes and a repressor of epidermal genes (Moreau et al. 2008;

Leclerc et al. 2006). Thus, calcium signaling is required for neural ectoderm formation and is required for *zic3* gene expression. Blocking L-type calcium channels with specific antagonists in gastrula stage embryos and in Keller open-faced explants results in a reduction of *zic3* expression (Leclerc et al. 2000, 2003). The *xPRMT1b* gene, which codes for an arginine methyltransferase, is upregulated by Noggin in a calcium-dependent manner, and xPRMT1b can induce the expression of *zic3* (Batut et al. 2005). Thus, xPRMT1b appears to be a link between early calcium transients resulting from BMP inhibition and the expression of neural genes during neural induction, including *zic* genes.

7.3.4 FGF Signaling

FGF signaling is required for neural ectoderm formation in a variety of vertebrates (Patthey and Gunhaga 2014; Aruga and Mikoshiba 2011). In *Xenopus* embryos, FGF signaling in conjunction with Noggin activates *zic1* gene expression in ectodermal explants (Gamse and Sive 2001). In embryos with blocked BMP signaling, inhibition of FGF signaling only slightly reduced the induction of *zic1* expression but completely abolished the induction of *zic3* expression (Marchal et al. 2009). This suggests that FGF signaling increases the expression of *zic1* from the level established by BMP inhibition, while both FGF signaling and BMP inhibition are required for *zic3* gene expression. Further, *zic3*, but not *zic1*, is upregulated by FGF in the presence of cycloheximide, suggesting a direct mechanism for *zic3* expression. Conversely, *zic1*, but not *zic3*, expression is activated by Noggin in the presence of cycloheximide (Marchal et al. 2009). This suggests different regulatory mechanisms for the induction of these two *zic* genes, and it shows an important involvement for FGF signaling in their regulation.

7.3.5 Wnt Signaling

Wnt signaling contributes to early patterning of the neural ectoderm and promotes the expression of *zic* genes. During early anterior to posterior patterning in *Xenopus* embryos, *wnt* expression (in conjunction with Noggin) activates *zic1* expression in the posterior portion of the presumptive neural plate (Gamse and Sive 2001). Consistent with this finding, the BIRM regulatory element upstream of *zic1* contains a Lef/Tcf binding site. Mutation of this Lef/Tcf site eliminates the ability of the BIRM to respond to Noggin in reporter assays (Tropepe et al. 2006), supporting a requirement for Wnt signaling in *zic* gene regulation in posterior regions of *Xenopus* embryos.

7.3.6 FoxD4 and Other Factors

A regulator of *zic* gene expression in the early neural ectoderm is the forkhead transcription factor FoxD4 (also called FoxD5), which is expressed in tissue destined to become the neural ectoderm (Yan et al. 2009). During early *Xenopus* neural ectoderm formation, inhibiting *foxD4* expression causes a reduction of *zic2* expression but expands the expression domains of both *zic1* and *zic3*. Testing an activator construct of FoxD4 (FoxD4 fused to the VP16-activating domain) and a repressor construct of FoxD4 (FoxD4 fused to the EnR repressor domain) in whole embryos showed that FoxD4 acts as activator to induce *zic2* expression but as repressor to repress *zic1* and *zic3* expression (Yan et al. 2009). The acidic blob region in the N-terminal domain of FoxD proteins is required for induction activity, while interaction with a co-repressor at a site in the C-terminal domain is required for repressive activity (Pohl and Knochel 2005). Structure-function experiments indicate that the activating function of FoxD4 is a direct process, while its ability to inhibit genes requires intermediate factors. Further studies showed that the upregulation of *zic1* and *zic3* expression as a result of inhibiting *foxD4* expression can be rescued by *zic2* mRNA injections (Neilson et al. 2012). Thus, direct induction of *zic2* may contribute to the inhibition of *zic1* and *zic3* expression during the formation of the neural plate. Interestingly, this interaction is different in gastrula embryos, when FoxD4 has an activating effect on *zic1* and *zic3* expression (Yan et al. 2009). Thus, *zic* genes are regulated differently at different times during development, and individual *zic* genes are regulated by independent mechanisms.

Following the broad induction of *zic* gene expression by the inhibition of BMP signaling in conjunction with FGF and Wnt signaling, other factors help refine and limit the expression pattern of *zic* genes. A 5 kb region upstream from the transcription start site of *zic1* (a region containing the BIRM) encompasses additional binding elements that restrict *zic1* expression, since loss of this region caused an expansion of *zic1* expression (Tropepe et al. 2006). Candidate transcriptional repressors that limit the expression of *zic* genes are the Msx1 and Dlx1 transcription factors. Both are direct targets of intermediate levels of BMP signaling and are expressed in the epidermal-neural boundary region. Both repress *zic* gene expression in ectodermal explants, and Dlx1 was shown to repress *zic3* expression in *Xenopus* embryos (Tribulo et al. 2003; Feledy et al. 1999; Yamamoto et al. 2000; Monsoro-Burq et al. 2005). Thus, blocking BMP signaling creates a permissive environment for *zic* gene expression in the presumptive neural plate, while Dlx3 and Msx1 may prevent the expression of *zic* genes beyond the neural plate border region.

The TALE-family homeodomain proteins Pbx1 and Meis1 are important in early neural patterning, and their misexpression causes an increase in *zic3* expression (Maeda et al. 2001, 2002; Kelly et al. 2006). Further analysis showed that Pbx1 and Meis1 synergistically interact with a 3.1 kb region directly upstream of the *zic3* transcription start site (Kelly et al. 2006). In the anterior portion of the neural plate, the Six1, Six3, and Xrx1 transcription factors may promote expression of *zic* genes, since these transcription factors increase the transcription of *zic2* (Brugmann et al.

2004; Gestri et al. 2005; Andreazzoli et al. 2003). These transcription factors help refine *zic* gene expression patterns.

In summary, *zic* genes are expressed in the prospective neural ectoderm during gastrula stages and are among the first genes expressed in the early neural plate. Studies in *Xenopus* have greatly contributed to our understanding of the mechanisms that regulate *zic* gene expression. Inhibition of BMP, the resulting calcium transients, in conjunction with FGF and Wnt signaling are responsible for early zic gene expression. Nodal also counts among the upstream regulators of *zic* gene expression, which has mostly been explored in mouse (Houtmeyers et al. 2016). After initial induction of *zic* genes, their expression patterns are limited and refined by a number of other transcription factors and signaling mechanisms. Among these factors is expression of *shh* in the ventral neural tube, which represses *zic* gene transcription and therefore limits *zic* expression to the dorsal neural tube (Aruga et al. 2002). Overall, the mechanisms that are responsible for regulating *zic* genes individually at different times during development remain to be explored in greater detail.

7.4 Direct Transcriptional Targets of Zic Proteins

The DNA-binding domain of Zic transcription factors consists of five C_2H_2 zinc fingers. While the three-dimensional structure of this domain has not been determined for any of the Zic proteins, the significant similarity between the zinc fingers in Zic and Gli proteins allows the assumption that zinc fingers 2–5 interact with the major groove of the target gene, while zinc finger 1 engages in protein-protein interactions (Pavletich and Pabo 1993). Interestingly, Zic proteins have not been reported to act as homodimers (Brown et al. 2005).

7.4.1 Screens for Zic1 Direct Targets in Xenopus

Zic proteins are involved in the downstream regulation of a wide variety of genes. In *Xenopus*, two screens were conducted for direct target genes that are relevant during early neural development (Cornish et al. 2009) and during neural crest specification (Plouhinec et al. 2014). The unique ability to use ectodermal explants from *Xenopus* embryos makes the identification of direct targets more readily feasible than in other organisms. In these screens, an inducible *zic* construct (*zic1GR*) was used. Zic1GR is a fusion of Zic1 to the ligand-binding domain of the human glucocorticoid receptor, which renders Zic1GR inducible with dexamethasone (Kuo et al. 1998). In order to identify direct targets of Zic1, animal caps injected with *zic1GR* are aged to the desired stage and then first treated with cycloheximide to prevent protein synthesis, followed by treatment with dexamethasone to activate Zic1GR (Fig. 7.3). The animal caps are harvested and assayed for the transcription of new mRNAs, which are direct targets of Zic1. The Cornish et al. (2009) screen

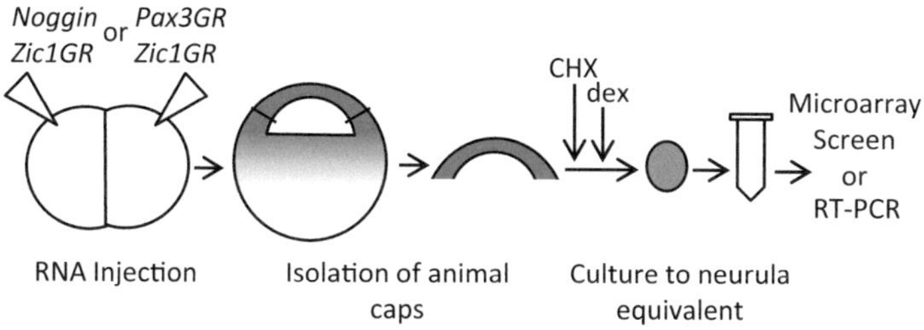

Fig. 7.3 Experimental design for microarray screens to identify direct transcriptional targets of Zic1. Embryos were injected at the two-cell stage into both cells with mRNAs for either *zic1GR/noggin* to induce early neural genes or *zic1GR/pax3GR* to induce neural crest genes. Control embryos were injected with mRNAs for *noggin* only or *pax3GR* only, respectively (Cornish et al. 2009; Plouhinec et al. 2014). Animal caps were dissected at stage 9. At the desired age, the isolated animal caps were treated first with cycloheximide (CHX) to prevent protein synthesis and later with dexamethasone (DEX) to induce the GR-conjugated transcription factors. The caps were then cultured to the correct stage and RNA isolated for microarray analysis and RT-PCR

aimed to identify early neural genes. Therefore, the animal caps were neuralized with a low dose of co-injected *noggin* mRNA. Plouhinec et al. (2014) set out to identify neural crest specifiers. Therefore, the animal caps were co-injected with hormone-inducible *zic1* and *pax3*. Both screens identified a number of genes, which are summarized in Table 7.2. Although both screens used a *zic1GR* construct to induce transcription of direct targets of Zic1, it is likely that the identified genes include direct targets of other Zic proteins, since the zinc finger domains of the Zic1-3 proteins are highly similar (Fujimi et al. 2006).

7.4.2 Direct Targets of Zic1 During Early Neurula Stages

A large number of genes were identified in the screen for direct targets of Zic1 during neural plate development (Cornish et al. 2009). The genes included in Table 7.2 are limited to direct targets that were confirmed by RT-PCR, and many were additionally shown to be regulated by Zic1 in whole embryos by in situ hybridization (Fig. 7.4). The screen was conducted at the equivalent of early neurula stages, and most of these genes are expressed in parts of the neural plate or in the neural plate border, overlapping with the expression patterns of Zic1 (Fig. 7.2).

7.4.2.1 Xfeb (Glipr2)

Among the direct target genes of Zic1, the putative metalloprotease Xfeb (Glipr2) was identified in both screens (Cornish et al. 2009; Plouhinec et al. 2014) and in an earlier spotted array (Li et al. 2006). It is expressed in the hindbrain and represses the expression of both the hindbrain gene *hoxB1* and the *otx2* gene, which is expressed

Table 7.2 Direct targets of Zic1 were identified in two screens

Accession number	Gene name	Confirmation by RT-qPCR	Second confirmation
NM_001095072.1	*Xfeb (glipr2)*[abc]	✓	*dnZic1*/in situ
NM_001094477.1	*aqp3b*[a]	✓	*dnZic1*/in situ
NM_001085780.1	*crabp2*[a]	✓	*dnZic1*/in situ
NM_001088044.1	*ptgds (cpl-1)*[a]	✓	*dnZic1*/in situ
NM_001088263.1	*ncoa3 (SRC-3)*[a]	✓	*ND*
NM_001088688.1	*prickle1*[a]	✓	*ND*
XM_018265023.1	*pkdcc2*[a]	✓	*dnZic1*/in situ
NM_001088196.1	*vegT*[a]	✓	*ND*
NM_001088341.1	*eomesodermin*[a]	✓	*ND*
NM_001085897.1	*myoD1*[a]	✓	*dnZic1*/in situ
NM_001085795.1	*hesx1 (Xanf2)*[a]	✓	*ND*
NM_001172199.1	*sall1*[a]	✓	*ND*
NM_001087226.1	*celf3*[a]	✓	*dnZic1*/in situ
XM_018244664.1	*Sp7 (osterix)*[a]	✓	*dnZic1*/in situ
NM_001088927.1	*lgals4*[a]	✓	*ND*
NM_001088044.1	*dgat2*[a]	✓	*dnZic1*/in situ
Xl.13309.1	*snail1*[b]	✓	*ZicMO1/RT-qPCR*
Xl.3818.1	*snail2*[b]	✓	*ZicMO1/RT-qPCR*
Xl.15393.1	*ets1*[b]	✓	*ZicMO1/RT-qPCR*
Xl.20029.1	*pdgfra*[b]	✓	*ZicMO1/RT-qPCR*
Xl.1946.1	*cyp26c1*[b]	✓	*ZicMO1/RT-qPCR*
Xl.5374.1	*dusp5*[b]	✓	*ZicMO1/RT-qPCR*
Xl.13925.1	*axin2*[b]	✓	*Unconfirmed by ZicMO1/RT-qPCR*

Only direct targets that were verified by quantitative RT-qPCR are included. Several neural targets from the Cornish et al. (2009) screen ([a]) were additionally confirmed by injection of a dominant interfering *zic1* construct (*dnZic1*) into whole embryos and showed decreased target gene expression by in situ hybridization (see Fig. 7.4). Neural crest-specific targets from the Plouhinec et al. (2014) screen ([b]) were additionally confirmed by injection of *zic1MO* and *pax3MO*, which resulted in decreased expression of the target genes, as assayed by RT-qPCR of whole embryos. Li et al. (2006) found *Xfeb* (*glipr2*) in an earlier screen for direct targets of Zic1 ([c]). *Sp7* and *dgat2* were originally identified only by their unigene numbers Xl.8933 and Xl.25952. *pkdcc.2* corresponded to unigene number Xl.73297, which is updated here
ND not determined

anterior to the midbrain-hindbrain boundary (Li et al. 2006). This suggests that Xfeb contributes to patterning the neural plate and may be part of the regulatory mechanism that prevents expression of the *otx2* gene posterior to the midbrain-hindbrain boundary (Fig. 7.4). The identification of Xefb (Glipr2) in the Plouhinec et al. (2014) screen suggests that it also plays a role during neural crest specification. *Xfeb* and *gbx2* are both expressed in the hindbrain (Li et al. 2006; Rhinn and Brand 2001), and Gbx2 has neural crest specifier activity, which is dependent on the presence of Zic1

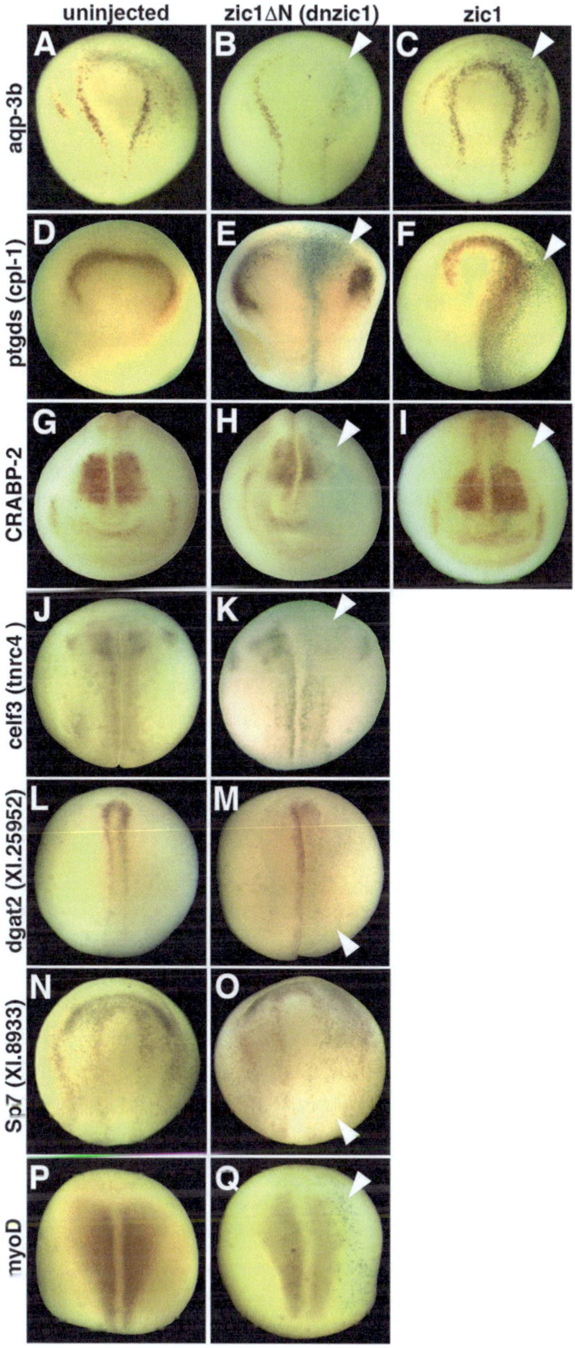

Fig. 7.4 Zic1 regulates the expression of direct target genes in neurula embryos (Cornish et al. 2009). Gene names are listed along the left with the original names or identifiers in parentheses. Shown are in situ hybridization expression patterns for neurula (stage 15–18) embryos that were uninjected (first column), injected with the dominant interfering construct *zic1ΔN* (*dnzic1*; second column), or injected with *zic1* mRNA (third column). Interfering with *zic1* activity reduced the expression levels of all direct target genes shown, indicating that Zic1 is required for their expression. Misexpressing *zic1* resulted in expansion of *aqp-3b*, *ptgds*, and *CRABP-2* expression. *Arrowheads* mark the injected sides (Reproduced from Cornish et al. 2009 with permission from publisher)

activity in ectodermal explants (Li et al. 2009). Thus, the induction of *Xfeb* by Zic1 may be required for the neural crest induction activity by Gbx2.

7.4.2.2 *aqp3b*

The *aqp3b* gene codes for an aquaporin, specifically an aquaglyceroporin. Aquaporins are channel proteins that allow passage of water and other small molecules (like glycerol) across cell membranes along their concentration gradients (Verkman 2005). In *Xenopus* neurula embryos, *aqp3b* is expressed in cells at the tips of the rising neural folds during neural tube closure (Fig. 7.4; Cornish et al. 2009). These cells, called "IS" cells (Schroeder 1970), separate the epidermal ectoderm and the neural ectoderm. During neurulation, the cells of the neural plate apically constrict, which allows the neural folds to rise and the neural tube to close (reviewed in Wallingford 2005). Compromising *aqp3b* expression in *Xenopus* embryos results in loss of apical constriction in neural plate cells and defective neural tube closure (Forecki and Merzdorf unpublished). Neural tube closure defects have been observed with mutations in human or mouse *zic2*, *zic3*, and *zic5* genes (Grinberg and Millen 2005). Thus, Aqp3b may be part of the mechanism that allows *zic* genes to control neural tube closure.

In gastrula embryos, aqp3b is expressed in the marginal zones and in the sensorial layer of the blastocoel roof (Forecki et al. 2018). Thus, *aqp3b* expression overlaps with *zic1-3* expression, which are expressed in the epithelial and sensorial layers of the dorsal marginal zone (Nakata et al. 1998; Fig. 7.1). Disrupting *aqp3b* expression in the dorsal marginal zone of whole embryos results in compromised border integrity between involuted mesendoderm and noninvoluted ectoderm and defective deposition of fibril fibronectin matrix at this boundary (Forecki et al. 2018). Further, inhibiting *aqp3b* expression in explants of the dorsal marginal zone region (Keller explants) interfered with their convergent extension, which was rescued with players in noncanonical Wnt signaling (See and Merzdorf unpublished). Although Zic proteins have not yet been examined for their roles in maintaining border integrities in gastrula embryos between involuted and noninvoluted cells, their expression patterns are consistent with this possibility. Further, involvement of Zic proteins in noncanonical Wnt signaling has not been demonstrated to date. However, identification in this screen of several genes that are involved in noncanonical Wnt signaling pathways suggests that Zic proteins may play such a role.

7.4.2.3 *pkdcc2* and Prickle Act in Noncanonical Wnt Signaling

The *pkdcc2* gene encodes a protein kinase, which regulates JNK-dependent Wnt/PCP signaling. It is important in both blastopore and neural tube closure (Vitorino et al. 2015). Prickle is a cytoplasmic protein that plays a key role in Wnt/PCP signaling as one of the six core components of Wnt/PCP signaling (reviewed in, e.g., Davey and Moens 2017). Accordingly, it is important for cell movements during

Xenopus gastrulation and neural tube closure (Takeuchi et al. 2003). Thus, these direct targets strongly suggest a new role for Zic proteins as regulators of noncanonical Wnt signaling.

7.4.2.4 *crabp2*, *ptgds*, *ncoa3*, and *cyp26cl* Are Genes Related to Retinoic Acid Signaling

The expression domains of the *crabp2* and *ptgds* (also called *cpl-1* or *lpgds)* genes overlap with the *zic1* expression domain (Fig. 7.4; Cornish et al. 2009). Both proteins function in regulating the cellular availability of retinoic acid during development. CRABP2 (cellular retinoic acid-binding protein 2) binds retinoic acid intracellularly and delivers it to the nucleus (Dong et al. 1999; Lepperdinger 2000). PTGDS acts dually as prostaglandin D2 synthase and as a lipocalin carrier for retinoic acid (Urade and Hayaishi 2000). Mutation analysis demonstrated that Zic1 acts only through the lipocalin function of PTGDS (Jaurena et al. 2015). The transcriptional coactivator Ncoa3 (also called SRC-3) activates the RAR/RXR nuclear receptor in response to retinoid binding in *Xenopus* (Kim et al. 1998). The direct target gene *cyp26cl* codes for a retinoic acid metabolizing enzyme, which is involved in anterior/posterior patterning of *Xenopus* embryos (Tanibe et al. 2008). Interestingly, in the pre-placodal ectoderm, Zic1 upregulates both the *cyp26cl* gene and the *raldh2* gene, which codes for a retinoic acid-synthesizing enzyme, although *raldh2* most likely is not a direct target of Zic1 (Jaurena et al. 2015). The authors hypothesize that retinoic acid synthesized by Raldh2 in *zic1*-expressing cells diffuses to and elicits signaling in surrounding cells, while the *zic1*-expressing cells themselves are not subject to signaling by the retinoic acid they produce due to the presence of Cyp26cl. Thus, a sharp boundary of retinoic acid-induced gene expression is created (Jaurena et al. 2015). Therefore, it appears that Zic1 regulates the expression of genes that control multiple aspects of retinoic acid signaling, which includes the synthesis and degradation of retinoic acid and aspects of its transport and availability.

7.4.2.5 VegT, Eomesodermin, and myoD Are Transcription Factors Important for Mesoderm Development

Eomesodermin acts very early in mesoderm development and regulates the expression of the t-box transcription factor VegT (Fukuda et al. 2010). VegT helps organize the paraxial mesoderm in *Xenopus* embryos (Fukuda et al. 2010). Experiments in chick embryos suggest that Zic1 may induce but not maintain *myoD* expression during somite development (Sun Rhodes and Merzdorf 2006). *zic* genes are known to play roles in mesoderm development, which have mostly been studied in other organisms.

7.4.2.6 Other direct targets of Zic1

Additional direct targets of Zic1 include *celf3, sall1, and hesx1,* which are associated with regulating gene expression in the developing nervous system. The *celf3* gene (also called *brunol1*) is broadly and strongly expressed in the neural plate border region. It codes for an RNA-binding protein with roles in regulating splicing events in the nucleus (Wu et al. 2010). The *sall1* transcription factor is expressed in the midbrain and in posterior regions of the neural plate (Hollemann et al. 1996). Sall1 is required for neural tube closure in mice (Böhm et al. 2008). The homeobox transcription factor Hesx1 is expressed in the anterior neural plate, where it promotes differentiation of the neural ectoderm and acts as a repressor of the *xbf-1*, *otx2*, and *pax6* genes (Ermakova et al. 1999).

Interestingly, no genes were identified in this screen, which are directly related to cell cycle control, and the gene most related to cell proliferation or cell differentiation is the *hesx1* gene, described above. Overall, the identified direct targets point to known and new activities for Zic transcription factors during early neural development.

7.4.3 Neural Crest-Specific Direct Targets of Zic1

The screen by Plouhinec et al. (2014) was a multi-step screen designed to limit identification of direct targets to only genes that act during neural crest specification. To this end, inducible *zic1GR* RNA was co-expressed with *pax3GR* RNA in animal caps (Fig. 7.3), and targets of Pax3GR alone were subtracted from the results. A variety of genes were identified, and those that were confirmed by an additional method are included in Table 7.2. Among these targets is the *Xfeb* gene (also called *glipr2*), which was identified in both screens and is discussed above.

The Plouhinec et al. (2014) screen identified the *snail1* and *snail2* (*slug*) genes as direct Zic1 targets, which are known to be expressed in the neural plate border region prior to neural crest migration. Snail1 has also been shown to induce *snail2* and other neural crest markers, including *zic5* and *ets1* (Aybar et al. 2003). Further, there has been indication that Zic1 induces *snail2* acting as a repressor, indicating an indirect regulatory mechanism (a *zic1-EnR* construct activated *snail2* expression; Merzdorf unpublished). Thus, there may be more than one way in which Zic proteins can induce *snail2* expression.

Additional genes identified by Plouhinec et al. (2014) include the *ets1*, *dusp5*, and *pdgfra* genes. The gene for the Ets1 transcription factor is expressed in *Xenopus* premigratory neural crest cells destined to become cardiac tissues and has functions similar to Snail proteins (Nie and Bronner 2015). Dusp5 is a MAP kinase phosphatase and an important regulator of MAPK signaling (Caunt and Keyse 2013). MAPK signaling is essential for neural crest induction (Stuhlmiller and Garcia-Castro 2012a). Pdgfra is a receptor tyrosine kinase for PDGF. It is important for directed migration of cells in *Xenopus* gastrula embryos (Van Stry et al. 2005). During Wnt-induced cell proliferation of osteoblasts, Pdgfra is activated in a disheveled-dependent manner (Caverzaiso et al. 2013).

The number of direct target genes for Zics identified in humans and mouse is relatively small. These targets include *ApoE*, *Math1*, *αCaM kinase II*, *dopamine receptor 1*, and *Pax3* (Salero et al. 2001; Yang et al. 2000; Ebert et al. 2003; Sakurada et al. 2005; Sanchez-Ferras et al. 2014). A ChIP-seq screen for direct targets of zebrafish Zic3 has yielded a large number of regulatory regions that drive a variety of genes involved in early development (Winata et al. 2013).

7.4.4 Interaction with Other Proteins

Zic proteins are transcription factors that bind DNA using C_2H_2 zinc finger domains, as stated earlier. There is some evidence that, like most transcription factors, their activity is regulated by interacting proteins. *Xenopus* Gli proteins, which are also C_2H_2 zinc finger transcription factors, interact with Zic proteins. Zic1, Zic2, and Zic3 and the Gli1, Gli2, and Gli3 proteins interact physically (through zinc fingers 3–5 of both Zic and Gli proteins) (Koyabu et al. 2001). In these Zic/Gli heterodimers, zinc fingers 3–5 would be occupied by binding to each other, thus preventing DNA binding by either protein. Therefore, in cases of co-expression, Zic and Gli proteins may regulate each other's activity as transcription factors. Indeed, in *Xenopus* embryos and in cell culture reporter assays, Zic and Gli proteins are able to reduce each others' activities as transcriptional activators (Brewster et al. 1998; Koyabu et al. 2001; Mizugishi et al. 2001). Zic2 has also been shown to interact with TCF1 and, via its zinc fingers, with TCF4, thereby interfering with Wnt/β-catenin signaling (Fujimi et al. 2012; Pourebrahim et al. 2011). In other organisms, there are not many proteins known to interact with Zic proteins. A yeast two-hybrid screen identified Imfa as a direct binding partner of Zic1, Zic2, and Zic3 in mouse (Mizugishi et al. 2004). In order to understand the activities of Zic factors, it will be important to learn more about proteins that modulate Zic activity by direct protein-protein interactions.

7.5 Biological Roles of Zic Transcription Factors

The *Xenopus* model lends itself to functional studies of genes. Loss and gain of function experiments combine to illustrate the activities of Zic transcription factors during embryonic development, particularly during gastrulation and early neural development.

7.5.1 Role of Maternally Expressed Zic2

Among the *Xenopus zic* genes, only *zic2* maternally expressed (Nakata et al. 1998). The role of maternally expressed *zic2* was studied using the host transfer method, where maternal *zic2* mRNA was depleted in oocytes that were then transferred back into *Xenopus* females for ovulation (Houston and Wylie 2005).

After fertilization and during development, this depletion resulted in exogastrulation, anterior truncations, thickened notochord, and axial abnormalities due to an overall increase in Nodal signaling (Houston and Wylie 2005). Similarly, double *zic2/zic3* morphants had a shortened body axis, smaller heads, and thicker, wider notochords (Fujimi et al. 2012). Thus, maternal expression of *zic2* is essential for early patterning of the embryo.

7.5.2 Zic *Genes During Gastrulation and Early Patterning of* Xenopus *Embryos*

Zygotic expression of the *zic1-4* genes begins shortly after midblastula transition, and all five *zic* genes are expressed during gastrulation, most strongly in the area of the presumptive neural plate (Fig. 7.1). As described above, the expression of *zic* genes appears to bias the ectoderm toward a neural fate in early embryos, since expression of *zic1* in animal cap ectoderm (from late blastula embryos) amplifies the neural inducing effects of Noggin (Kuo et al. 1998). In addition, maternal *zic2* and early zygotically expressed *zic2*, which is induced by the organizer transcription factors Siamois and Twin, also bias the presumptive ectoderm toward neural fate (Klein and Moody 2015; Gaur et al. 2016). The mechanism by which early *zic* gene expression is able to confer this predisposition for neural fate on the future neural ectoderm prior to gastrulation is currently not understood.

Zic3-null mice and *Xenopus zic3* morphants exhibit left-right (L-R) asymmetry defects (Purandare et al. 2002; Ware et al. 2006a; Cast et al. 2012). Of the *Xenopus zic* genes, *zic3* is most widely expressed in gastrula embryos (Fig. 7.2), and it is the only one among the *zic* genes that is involved in L-R asymmetry establishment. There are two prevailing models for the establishment of left-right (L-R) asymmetry in *Xenopus*. Evidence indicates asymmetry establishment either during early cleavage stages via ion flux or during gastrulation by cilia-driven flow (Blum et al. 2014). The result of breaking the symmetry by either mechanism is the asymmetric expression of the TGFβ-type growth factor *nodal* on the left side of the embryo. The *zic3* gene is a direct target of Nodal signaling, most likely via the activin response element found in the first intron of *zic3* (Weber and Sokol 2003). Zic3 then transmits this signal to downstream factors that determine left-sidedness (Kitaguchi et al. 2000). Zic3 has also been shown to regulate *nodal* expression in mice (Ware et al. 2006b). Thus, Zic3 may act upstream and downstream of the Nodal signaling that is required for L-R asymmetry formation.

7.5.3 Zic Proteins and Wnt Signaling

Zic proteins interact with canonical Wnt signaling, although the effects of these interactions appear to be dependent on *Xenopus* developmental stage. In late blastula embryos (stage 9.5), misexpression of *zic3* reduces the expression of the direct Wnt/β-catenin targets *goosecoid* and *siamois*, which are genes expressed in the organizer, resulting in impaired notochord development. Zic3 is hypothesized to act as an early tuner of Wnt/β-catenin signaling in organizer mesoderm, where it is expressed at moderate levels (Fig. 7.1). It is likely that several Zic family members are able to affect Wnt/β-catenin signaling, since all five *zic* genes are able to reduce Wnt/β-catenin transcriptional activity in a luciferase reporter assay in *Xenopus* gastrula embryos and Zic3 was shown to physically interact with TCF1 (Fuijmi et al. 2012). Similarly, Zic2 binds directly to TCF4 and inhibits the ability of the β-catenin/TCF4 complex to activate transcription, thereby reducing the ability of β-catenin to induce Wnt targets in *Xenopus* animal caps (Pourebrahim et al. 2011). Further, the direct Zic1 targets Sp7 (also called Osterix) and Hesx1 (Cornish et al. 2009), both transcription factors, repress Wnt/β-catenin activity, and Hesx1 is expressed during late gastrula stage in the neural ectoderm (Andoniadou et al. 2011; Ermakova et al. 1999). These lines of evidence suggest that during gastrula stages and neural induction, Zic proteins inhibit canonical Wnt signaling.

Later in development, as Zic proteins contribute to patterning the neural plate, the effect of Zic1 on Wnt activity shifts. In neurula embryos, Zic1 acts as an activator of *wnt8b* expression, and it is able to activate *wnt1* and *wnt4* expression in neuralized animal caps. Further, Zic1 requires Wnt signaling to induce expression of the *engrailed-2* gene in ectodermal explants (Merzdorf and Sive 2006), indicating a role for Zic1 in promoting canonical Wnt signaling. Finally, the direct Zic1 targets *pkdcc2* and *prickle* (Cornish et al. 2009) suggest an unexplored role for Zic proteins in noncanonical Wnt signaling.

7.5.4 Zic *Genes During Patterning of the Neural Plate*

During *Xenopus* neurula stages, all five *zic* genes are expressed in overlapping yet distinct domains in the lateral neural plate and in the dorsal region of the closed neural tube (Fig. 7.2). Misexpression of each member of the *Xenopus zic* family expands the neural plate (*zic1*: Kuo et al. 1998; Mizuseki et al. 1998; Nakata et al. 1998), (*zic2*: Brewster et al. 1998; Nakata et al. 1998), (*zic3*: Nakata et al. 1997), (*zic4*: Fujimi et al. 2006), (*zic5*: Nakata et al. 2000). *zic* genes are expressed in relatively broad domains, as are other factors that pattern the neural plate. Combinations of these transcription factors, together with secreted factors, activate the expression of genes that are expressed in more limited domains. These include the *wnt* genes mentioned above (Merzdorf and Sive 2006): *wnt1*, which is expressed at the midbrain-hindbrain boundary, and *wnt4* and *wnt8b*, which are expressed at the forebrain/midbrain boundary and in the midbrain. Additional genes induced by the expression of *zic1* include the dorsal neural marker *pax3*, the hindbrain markers *krox20*, *hoxD1* (Kuo et al. 1998), and *Xfeb (glipr2)* (Li et al. 2006). All *zic* genes induce the midbrain-hindbrain boundary marker *en-2* (Nakata et al. 1997, 1998, 2000; Kuo et al. 1998; Fujimi et al. 2006). *zic1-3* induce the forebrain and midbrain marker *otx2* and the cement gland markers *XAG-1* or *XCG*, while *zic4* and *zic5* are not able to induce these anterior genes (Kuo et al. 1998; Fujimi et al. 2006; Nakata et al. 2000). None of the *zic* genes are able to induce the posterior gene *hoxB9*. Most of these results were obtained in animal cap explants, although the regulation of the *pax3*, *en-2*, *wnt8b*, and *krox20* genes was confirmed in whole embryos (Kuo et al. 1998; Merzdorf and Sive 2006; Gutkovich et al. 2010). Thus, Zic proteins regulate genes in the neural plate regions that give rise to the brain but so far do not appear to be involved in regulation of genes important for spinal cord development.

The *zic1* gene is likely to play a role in the development of the midbrain-hindbrain boundary (MHB). Zic1 is required for expression of the MHB genes *en-2* and *wnt1*. Since Wnt signaling is required for activation of *en-2* expression by Zic1, Zic1 most likely induces *wnt1* transcription, which in turn induces expression of the *en*-2 gene (Merzdorf and Sive 2006). Zic1 may also help maintain the MHB through its direct target gene *Xfeb (glipr2)*. The *Xfeb* gene codes for a putative protease, which represses *otx2* expression (Li et al. 2006). *Xfeb* is expressed in the hindbrain up to the MHB. The transcription factors Otx2 and Gbx2 maintain the MHB by mutual repression (Rhinn and Brand 2001). Xfeb activity may help maintain a posterior limit to *otx2* expression during MHB formation. Thus, Zic1 may play a role in establishing and maintaining the midbrain-hindbrain boundary.

Zic family members appear to be essential for the formation of the hindbrain. Interfering with either *zic1* or *zic5* expression results in the loss of hindbrain cell fates (Gutkovich et al. 2010). Similar defects are observed when the transcription factor Xmeis is knocked down. In fact, defects in *zic1* and *zic5* morphants could be rescued with co-injection of *xmeis* RNA (Gutkovich et al. 2010). *hoxD1*, a gene that contributes to patterning the hindbrain, is a direct target of *Xmeis* and is known to be upregulated by *zic1* (Kuo et al. 1998). This indicates that *zic* genes work upstream of *xmeis* and *hoxD1* to promote formation of the hindbrain in *Xenopus* embryos. In addition, interfering with the expression of the Zic1 direct target gene *Xfeb (glipr2)* resulted in loss of *hoxD1* expression (Li et al. 2006). While it is not known if Xfeb may lie upstream of *xmeis1* or be part of a separate pathway, *zic* genes play an important upstream role during hindbrain development.

7.5.5 Zic *Genes and the Neural Crest*

zic genes act as neural crest specifiers, which has been shown in multiple organisms (reviewed in Merzdorf 2007; Houtmeyers et al. 2013). Neural crest cells are a migratory population of cells that originate from the neural plate border region. Multiple signaling pathways work together to specify the neural crest in two phases.

During phase one, BMP, Wnt, and FGF signaling induces the expression of transcription factors like *pax*, *msx*, and *zic* family members, which are neural border specifiers. During phase two, these neural border specifiers induce the expression of neural crest specifiers, including *snail1*, *snail2*, *ets1*, and *FoxD3* (Stuhlmiller and García-Castro 2012b). Accordingly, mutations in the mouse *Zic2 or Zic5* genes result in a reduction in neural crest cells and deformities in neural crest-derived structures (Inoue et al. 2004; Elms et al. 2003). In *Xenopus*, all *zic* family members are expressed in the neural plate border region (Fig. 7.2; Fujimi et al. 2006) and are important for the formation of neural crest cells. Misexpression of *zic1*, *zic2*, or *zic3* increases the extent of neural crest cell fate in whole embryos, and expression in animal cap explants results in the induction of neural crest markers (Nakata et al. 1997, 1998; Kuo et al. 1998). Similarly, misexpression of *zic4* in *Xenopus* embryos generates ectopic pigment cells, a neural crest-derived cell type (Fujimi et al. 2006). Misexpression of *zic5* in whole embryos causes strong induction of neural crest genes, but, unlike other *zic* family members, *zic5* is not as efficient at inducing neural genes (Nakata et al. 2000). Conversely, interfering with the expression of *zic* genes results in a reduction in the expression of neural crest genes (Hong and Saint-Jeannet 2007; Fujimi et al. 2006; Nakata et al. 2000; Gutkovich et al. 2010). Thus, while having slightly different roles, all members of the *zic* family contribute to induction of the neural crest.

The Zic1 and Pax3 transcription factors work jointly to induce neural crest cell fate in the developing embryo. The expression of *zic1* and *pax3* overlaps in the presumptive neural crest region (Sato et al. 2005; Hong and Saint-Jeannet 2007). Misexpression of either *zic1* or *pax3* alone increases neural crest marker expression only in the ectoderm bordering the neural crest field, while overexpression of both genes together induces ectopic neural crest formation in the ventral ectoderm (Sato et al. 2005) in a Wnt-dependent manner (Monsoro-Burq et al. 2005). When ectopically induced neural crest cells (by activating *zic1* and *pax3* in animal cap explants) are transplanted into embryos, they are able to migrate correctly and form differentiated cell types characteristic of neural crest cell fates. Interestingly, the cooperation between Zic1 and Pax3 is required for these fates, since transplanting cells in which *zic1* alone is activated results in the formation of neural tissue only (Milet et al. 2013). Thus, Zic1 and Pax3 can work together to induce a complete neural crest fate. While physical interaction between the Zic1 and Pax3 proteins was originally elusive (Sato et al. 2005), such an interaction was suggested by expressing these proteins in cultured cells (Himeda et al. 2013).

Since Zic1 and Pax3 together are able to induce a neural crest program in ectodermal explants, this synergy was employed to identify downstream neural crest genes (Plouhinec et al. 2014; Bae et al. 2014). The Plouhinec et al. (2014) screen focused on the identification of direct targets of Zic1/Pax3 and is described above. Bae et al. (2014) used a similar approach but did not limit their screen to direct targets. Both screens identified the *snail1 and snail2* (*slug*) genes. The latter screen identified a variety of additional neural crest genes that may or may not be direct targets. Overall, a variety of familiar and new genes were identified that are activated by Zic1 and Pax3 acting together. Among the neural crest specifiers, the *snail1*, *snail2 (slug)*, and *ets1* genes were identified as direct target genes of the interaction between Zic1 and Pax3 (Plouhinec et al. 2014). Further, Zic1 is required for *snail1*, *snail2*, and *foxD3* expression (Plouhinec et al. 2014; Sasai et al. 2001; Gutkovich et al. 2010). FoxD3, which is not known as a direct target of Zic1 at this time, restricts cells to a neural crest fate and also aids in the migration of neural crest cells (Sasai et al. 2001). The Snail family and Ets1 are among the transcription factors that facilitate the delamination and migration of neural crest cells (Nie and Bronner 2015; Aybar et al. 2003). Both screens also identified the *pdgfra* gene, which codes for the alpha subunit of a platelet-derived growth factor (PDGF) receptor. PDGF receptor is important for migration of neural crest cells in mouse embryos (Soriano 1997), and in *Xenopus* it has been implicated in cell migration during gastrulation (Nagel et al. 2004; Van Stry et al. 2005). Thus, Zic1 is required for stabilizing neural crest fate and for the expression of genes that prepare neural crest cells for delamination and migration. While *zic1* is expressed in premigratory neural crest cells and is essential for the expression of genes required for the transition of neural crest cells to emigrate, studies in chick show that it ceases to be expressed as soon as neural crest cells become migratory (Sun Rhodes and Merzdorf 2006). Since *zic* genes can repress neural differentiation genes, their role may include keeping the premigratory neural crest population in an undifferentiated state until the time of cell migration. Overall, the two screens confirmed that the neural plate border specifier Zic1 acts to induce neural crest specifier genes, with some of these interactions identified as direct. This adds further detail to the role of the Zic transcription factors in the gene regulatory landscape that governs neural crest specification.

7.5.6 Zic *Genes and the Proliferation and Differentiation of Cells in the Nervous System*

Xenopus embryos undergo primary neurogenesis, during which six discrete stripes of N-tubulin-positive primary neurons differentiate in the early neural plate, while the remainder of the neural plate remains as undifferentiated progenitors. *zic2* is expressed in these undifferentiated progenitors between the stripes of primary neurons (Brewster et al. 1998). Misexpression of *zic2* in the regions of primary neuron differentiation resulted in a significant decrease in the number of N-tubulin-positive primary neurons, indicating a role for Zic2 in preventing the differentiation of primary neurons. Consistent with this finding, Zic2 has a repressive effect on transcription of the bHLH gene *neurogenin (ngnr-1)*, a gene that promotes neural differentiation (Brewster et al. 1998). Similarly, Sonic Hedgehog (Shh) signaling upregulates *zic2* expression, and overexpression of *Xenopus shh* during primary neurogenesis causes expanded expression of *zic2* and reduced N-tubulin-positive stripes (Franco et al. 1999). This indicates that *zic2* acts in maintaining progenitors and preventing neurogenesis in certain areas of the neural plate, possibly under the regulation of Shh.

While Zic2 represses transcription of the neural differentiation factor *ngnr-1* (Brewster et al. 1998), Zic1 and Zic3 have inductive effects on the expression of the proneural genes *ngnr-1* and *neuroD* in animal cap explants (Nakata et al. 1997; Mizuseki et al. 1998). Further, *hesx1*, which promotes differentiation, is a direct target of Zic1 (Cornish et al. 2009). However, in mouse and chick embryos, Zic1 represses proneural gene expression (Ebert et al. 2003), misexpression of *Zic1* blocks neuronal differentiation, and mutations in *zic* genes cause a decrease in cell proliferation in the dorsal neural tube (Ebert et al. 2003; Aruga et al. 2002; Nyholm et al. 2007). In addition, *Zic1* promotes proliferation in the cerebellum and *Zic1* and *Zic3* in retinal precursors (Blank et al. 2011; Watabe et al. 2011). Consistent with these results from other organisms, interfering with *btg2* expression in *Xenopus* embryos (Btg2 reduces proliferation and promotes neuronal differentiation) results in increased *zic3* expression (Sugimoto et al. 2007), indicating that Btg2 downregulates *zic3* gene expression to allow neurogenesis to begin. Thus, it appears that Zic2 has a role in maintaining undifferentiated progenitors in the neural ectoderm, while the role of Zic1 and Zic3 in proliferation and differentiation is not completely clear and may be context dependent. Interestingly, the early neural transcription factor FoxD4 regulates these *zic* genes differently. It induces *zic2* transcription directly while indirectly repressing *zic1* and *zic3* expression (Neilson et al. 2012; Yan et al. 2009), which has been interpreted as FoxD4 keeping the neural ectoderm in a proliferative state by promoting *zic2* and repressing *zic1* and *zic3* expression. Since the expression domains of the *zic1*, *zic2*, and *zic3* genes overlap in the neural plate, it will be interesting and important to sort out the potentially opposite and context-dependent influences that these genes exert on neural differentiation.

7.6 *Xenopus* Studies Contribute to Our Understanding of Human Diseases

Xenopus embryos are increasingly employed as a model system in functional studies of human diseases (Kofent and Spagnoli 2016; Lienkamp 2016; Hardwick and Philpott 2015). With the near completion of the *Xenopus* genome and the advent of the TALEN and CRISPR-Cas9 systems of genome editing, such studies have become feasible (Tandon et al. 2016). With regard to diseases caused by mutations in human *ZIC* genes, *Xenopus* embryos were used to examine gene regulatory interactions in human craniosynostosis caused by mutations in the *ZIC1* gene. Craniosynostosis is the premature fusion of skull sutures that leads to abnormalities in brain development and brain function in human patients. Five independent families with a history of coronal craniosynostosis showed four different mutations in the third exon of ZIC1, C-terminal to the zinc finger region (Twigg et al. 2015). These mutations include one point mutation and three nonsense mutations that result in truncations of the ZIC1 protein. The *ZIC1* and *engrailed* (*EN1*) gene expression domains overlap in the developing sutures (Twigg et al. 2015). Using the regulatory relationship between Zic1 and the *engrailed* (*en-2*) gene as a model, misexpression of wild-type *Xenopus zic1* or human *ZIC1* does not change the *engrailed* (*en-2*) expression domain at the midbrain-hindbrain boundary in *Xenopus* embryos (Merzdorf and Sive 2006; Twigg et al. 2015). In contrast, the human mutant *ZIC1* genes elicit increased and/or abnormal *en-2* expression in *Xenopus* embryos, indicating that the mechanism by which these C-terminal *ZIC1* mutations cause craniosynostosis may lie in dysregulation of the *EN1* gene in the developing sutures (Twigg et al. 2015). En1 has been shown to regulate osteogenic differentiation and induction of Osterix (Sp7) during the formation of mouse skull sutures (Deckelbaum et al. 2006). Interestingly, Osterix (Sp7) is a direct target of zic1 (Table 7.2). Thus, ZIC1 appears to participate in a gene regulatory network, which is disturbed by mutations in the C-terminal domain of ZIC1, resulting in abnormal bone development in the coronal sutures and craniosynostosis in human patients.

Xenopus embryos were used to study the mechanism by which a mutation in the first zinc finger of the human *ZIC3* gene causes TGA (transposition of the great arteries), which is a complex heart defect (Chhin et al. 2007). Zic3 plays a role in left-right axis formation and induction of the neural crest (Cast et al. 2012; Kitaguchi et al. 2000; Nakata et al. 1997, 1998), which are processes that may underlie the defects seen in the human patients. Injection of wild-type human *ZIC3* into *Xenopus* embryos induced misexpression of the left lateral plate mesoderm marker *pitx2* and the neural crest marker *snail2*. This induction activity was diminished when *Xenopus* embryos were injected with the mutant *ZIC3* gene (Chhin et al. 2007). Thus, it appears that the mutation in the first zinc finger (which does not bind to DNA but engages in protein-protein interactions) diminishes the overall activity of ZIC3 in both left-right axis formation and neural crest induction. Thus, *Xenopus* embryos have proven useful in studying the interactions of mutant forms of human *ZIC* genes with developmental mechanisms to identify a molecular basis for human disease.

7.7 Conclusion

Work with *Xenopus* embryos has greatly contributed to understanding the role of Zic transcription factors during development. While *zic* gene family members are important players in many developmental processes, much remains to be understood about the molecular mechanisms that govern *zic* gene expression and Zic activities. The screens for direct and indirect targets of Zic transcription factors have yielded a variety of genes that are supporting ongoing and new research and are giving rise to new insights. Important are the advent of new genetic tools, such as new methods for genome editing, and the sequencing of the *Xenopus laevis* genome. Thus, previous studies can now be combined with genomic studies that have long been the strengths of other model organisms to form a more complete understanding of how Zic proteins drive development. Zic gene expression overlaps and their activities are partially redundant. Thus, it will be important to discover how individual *zic* genes are regulated and what distinguishes their functions. These studies will help with understanding the basis for human diseases. Indeed, *Xenopus* embryos have already been used to examine the molecular mechanisms underlying two human diseases caused by mutations in *ZIC* genes.

References

Andoniadou CL, Signore M, Young RM, Gaston-Massuet C, Wilson SW, Fuchs E, Martinez-Barbera JP (2011) HESX1- and TCF3-mediated repression of Wnt/β-catenin targets is required for normal development of the anterior forebrain. Development 138(22):4931–4942. https://doi.org/10.1242/dev.066597

Andreazzoli M, Gestri G, Cremisi F, Casarosa S, Dawid IB, Barsacchi G (2003) Xrx1 controls proliferation and neurogenesis in Xenopus anterior neural plate. Development 130(21):5143–5154. https://doi.org/10.1242/dev.00665

Aruga J, Mikoshiba K (2011) Role of BMP, FGF, calcium signaling, and Zic proteins in vertebrate neuroectodermal differentiation. Neurochem Res 36(7):1286–1292. https://doi.org/10.1007/s11064-011-0422-5

Aruga J, Tohmonda T, Homma S, Mikoshiba K (2002) Zic1 promotes the expansion of dorsal neural progenitors in spinal cord by inhibiting neuronal differentiation. Dev Biol 244(2):329–341. https://doi.org/10.1006/dbio.2002.0598

Aruga J, Kamiya A, Takahashi H, Fujimi TJ, Shimizu Y, Ohkawa K, Yazawa S, Umesono Y, Noguchi H, Shimizu T, Saitou N, Mikoshiba K, Sakaki Y, Agata K, Toyoda A (2006) A wide-range phylogenetic analysis of Zic proteins: implications for correlations between protein structure conservation and body plan complexity. Genomics 87(6):783–792. https://doi.org/10.1016/j.ygeno.2006.02.011

Aybar MJ, Nieto MA, Mayor R (2003) Snail precedes slug in the genetic cascade required for the specification and migration of the Xenopus neural crest. Development 130(3):483–494

Bae CJ, Park BY, Lee YH, Tobias JW, Hong CS, Saint-Jeannet JP (2014) Identification of Pax3 and Zic1 targets in the developing neural crest. Dev Biol 386(2):473–483. https://doi.org/10.1016/j.ydbio.2013.12.011

Batut J, Vandel L, Leclerc C, Daguzan C, Moreau M, Néant I (2005) The Ca2+−induced methyltransferase xPRMT1b controls neural fate in amphibian embryo. Proc Natl Acad Sci U S A 102(42):15128–15133. https://doi.org/10.1073/pnas.0502483102

Blank MC, Grinberg I, Aryee E, Laliberte C, Chizhikov VV, Henkelman RM, Millen KJ (2011) Multiple developmental programs are altered by loss of Zic1 and Zic4 to cause Dandy-Walker malformation cerebellar pathogenesis. Development 138(6):1207–1216. https://doi.org/10.1242/dev.054114

Blum M, Schweickert A, Vick P, Wright CV, Danilchik MV (2014) Symmetry breakage in the vertebrate embryo: when does it happen and how does it work? Dev Biol 393(1):109–123. https://doi.org/10.1016/j.ydbio.2014.06.014

Blum M, De Robertis EM, Wallingford JB, Niehrs C (2015) Morpholinos: antisense and sensibility. Dev Cell 35(2):145–149. https://doi.org/10.1016/j.devcel.2015.09.017

Böhm J, Buck A, Borozdin W, Mannan AU, Matysiak-Scholze U, Adham I, Schulz-Schaeffer W, Floss T, Wurst W, Kohlhase J, Barrionuevo F (2008) Sall1, sall2, and sall4 are required for neural tube closure in mice. Am J Pathol 173(5):1455–1463. https://doi.org/10.2353/ajpath.2008.071039

Brewster R, Lee J, Ruiz i Altaba A (1998) Gli/Zic factors pattern the neural plate by defining domains of cell differentiation. Nature 393(6685):579–583. https://doi.org/10.1038/31242

Brown L, Paraso M, Arkell R, Brown S (2005) In vitro analysis of partial loss-of-function ZIC2 mutations in holoprosencephaly: alanine tract expansion modulates DNA binding and transactivation. Hum Mol Genet 14(3):411–420. https://doi.org/10.1093/hmg/ddi037

Brugmann SA, Pandur PD, Kenyon KL, Pignoni F, Moody SA (2004) Six1 promotes a placodal fate within the lateral neurogenic ectoderm by functioning as both a transcriptional activator and repressor. Development 131(23):5871–5881. https://doi.org/10.1242/dev.01516

Cast AE, Gao C, Amack JD, Ware SM (2012) An essential and highly conserved role for Zic3 in left-right patterning, gastrulation and convergent extension morphogenesis. Dev Biol 364(1):22–31. https://doi.org/10.1016/j.ydbio.2012.01.011

Caunt CJ, Keyse SM (2013) Dual-specificity MAP kinase phosphatases (MKPs): shaping the outcome of MAP kinase signalling. FEBS J 280(2):489–504. https://doi.org/10.1111/j.1742-4658.2012.08716.x

Caverzasio J, Biver E, Thouverey C (2013) Predominant role of PDGF receptor transactivation in Wnt3a-induced osteoblastic cell proliferation. J Bone Miner Res 28(2):260–270. https://doi.org/10.1002/jbmr.1748

Chhin B, Hatayama M, Bozon D, Ogawa M, Schön P, Tohmonda T, Sassolas F, Aruga J, Valard AG, Chen SC, Bouvagnet P (2007) Elucidation of penetrance variability of a ZIC3 mutation in a family with complex heart defects and functional analysis of ZIC3 mutations in the first zinc finger domain. Hum Mutat 28(6):563–570. https://doi.org/10.1002/humu.20480

Cornish EJ, Hassan SM, Martin JD, Li S, Merzdorf CS (2009) A microarray screen for direct targets of Zic1 identifies an aquaporin gene, aqp-3b, expressed in the neural folds. Dev Dyn 238(5):1179–1194. https://doi.org/10.1002/dvdy.21953

Davey CF, Moens CB (2017) Planar cell polarity in moving cells: think globally, act locally. Development 144(2):187–200. https://doi.org/10.1242/dev.122804

Deckelbaum RA, Majithia A, Booker T, Henderson JE, Loomis CA (2006) The homeoprotein engrailed 1 has pleiotropic functions in calvarial intramembranous bone formation and remodeling. Development 133(1):63–74. https://doi.org/10.1242/dev.02171

Dong D, Ruuska SE, Levinthal DJ, Noy N (1999) Distinct roles for cellular retinoic acid-binding proteins I and II in regulating signaling by retinoic acid. J Biol Chem 274(34):23695–23698

Ebert PJ, Timmer JR, Nakada Y, Helms AW, Parab PB, Liu Y, Hunsaker TL, Johnson JE (2003) Zic1 represses Math1 expression via interactions with the Math1 enhancer and modulation of Math1 autoregulation. Development 130(9):1949–1959

Elms P, Siggers P, Napper D, Greenfield A, Arkell R (2003) Zic2 is required for neural crest formation and hindbrain patterning during mouse development. Dev Biol 264(2):391–406

Ermakova GV, Alexandrova EM, Kazanskaya OV, Vasiliev OL, Smith MW, Zaraisky AG (1999) The homeobox gene, Xanf-1, can control both neural differentiation and patterning in the presumptive anterior neurectoderm of the Xenopus laevis embryo. Development 126(20):4513–4523

Feledy JA, Beanan MJ, Sandoval JJ, Goodrich JS, Lim JH, Matsuo-Takasaki M, Sato SM, Sargent TD (1999) Inhibitory patterning of the anterior neural plate in Xenopus by homeodomain factors Dlx3 and Msx1. Dev Biol 212(2):455–464. https://doi.org/10.1006/dbio.1999.9374

Forecki J, Van Antwerp D, Lujan S, Merzdorf C, Antwerp V (2018) Roles for Xenopus aquaporin-3b (aqp3.L) during gastrulation: fibrillar fibronectin and tissue boundary establishment in the dorsal margin. Dev Biol 433(1):3–16. https://doi: 10.1016/j.ydbio.2017.11.001

Franco PG, Paganelli AR, López SL, Carrasco AE (1999) Functional association of retinoic acid and hedgehog signaling in Xenopus primary neurogenesis. Development 126(19):4257–4265

Fujimi TJ, Mikoshiba K, Aruga J (2006) Xenopus Zic4: conservation and diversification of expression profiles and protein function among the Xenopus Zic family. Dev Dyn 235(12):3379–3386. https://doi.org/10.1002/dvdy.20906

Fujimi TJ, Hatayama M, Aruga J (2012) Xenopus Zic3 controls notochord and organizer development through suppression of the Wnt/β-catenin signaling pathway. Dev Biol 361(2):220–231. https://doi.org/10.1016/j.ydbio.2011.10.026

Fukuda M, Takahashi S, Haramoto Y, Onuma Y, Kim YJ, Yeo CY, Ishiura S, Asashima M (2010) Zygotic VegT is required for Xenopus paraxial mesoderm formation and is regulated by nodal signaling and Eomesodermin. Int J Dev Biol 54(1):81–92. https://doi.org/10.1387/ijdb.082837mf

Furushima K, Murata T, Matsuo I, Aizawa S (2000) A new murine zinc finger gene. Opr Mech Dev 98(1–2):161–164

Gamse JT, Sive H (2001) Early anteroposterior division of the presumptive neurectoderm in Xenopus. Mech Dev 104(1–2):21–36

Gaston-Massuet C, Henderson DJ, Greene ND, Copp AJ (2005) Zic4, a zinc-finger transcription factor, is expressed in the developing mouse nervous system. Dev Dyn 233(3):1110–1115. https://doi.org/10.1002/dvdy.20417

Gaur S, Mandelbaum M, Herold M, Majumdar HD, Neilson KM, Maynard TM, Mood K, Daar IO, Moody SA (2016) Neural transcription factors bias cleavage stage blastomeres to give rise to neural ectoderm. Genesis 54(6):334–349. https://doi.org/10.1002/dvg.22943

Gestri G, Carl M, Appolloni I, Wilson SW, Barsacchi G, Andreazzoli M (2005) Six3 functions in anterior neural plate specification by promoting cell proliferation and inhibiting Bmp4 expression. Development 132(10):2401–2413. https://doi.org/10.1242/dev.01814

Grinberg I, Millen KJ (2005) The ZIC gene family in development and disease. Clin Genet 67(4):290–296. https://doi.org/10.1111/j.1399-0004.2005.00418.x

Gutkovich YE, Ofir R, Elkouby YM, Dibner C, Gefen A, Elias S, Frank D (2010) Xenopus Meis3 protein lies at a nexus downstream to Zic1 and Pax3 proteins, regulating multiple cell-fates during early nervous system development. Dev Biol 338(1):50–62. https://doi.org/10.1016/j.ydbio.2009.11.024

Hardwick LJ, Philpott A (2015) An oncologist's friend: how Xenopus contributes to cancer research. Dev Biol 408(2):180–187. https://doi.org/10.1016/j.ydbio.2015.02.003

Hellsten U, Harland RM, Gilchrist MJ, Hendrix D, Jurka J, Kapitonov V, Ovcharenko I, Putnam NH, Shu S, Taher L, Blitz IL, Blumberg B, Dichmann DS, Dubchak I, Amaya E, Detter JC, Fletcher R, Gerhard DS, Goodstein D, Graves T, Grigoriev IV, Grimwood J, Kawashima T, Lindquist E, Lucas SM, Mead PE, Mitros T, Ogino H, Ohta Y, Poliakov AV, Pollet N, Robert J, Salamov A, Sater AK, Schmutz J, Terry A, Vize PD, Warren WC, Wells D, Wills A, Wilson RK, Zimmerman LB, Zorn AM, Grainger R, Grammer T, Khokha MK, Richardson PM, Rokhsar DS (2010) The genome of the western clawed frog Xenopus tropicalis. Science 328(5978):633–636. https://doi.org/10.1126/science.1183670

Himeda CL, Barro MV, Emerson CP (2013) Pax3 synergizes with Gli2 and Zic1 in transactivating the Myf5 epaxial somite enhancer. Dev Biol 383(1):7–14. https://doi.org/10.1016/j.ydbio.2013.09.006

Hollemann T, Schuh R, Pieler T, Stick R (1996) Xenopus Xsal-1, a vertebrate homolog of the region specific homeotic gene spalt of Drosophila. Mech Dev 55(1):19–32

Hong CS, Saint-Jeannet JP (2007) The activity of Pax3 and Zic1 regulates three distinct cell fates at the neural plate border. Mol Biol Cell 18(6):2192–2202. https://doi.org/10.1091/mbc.E06-11-1047

Houston DW, Wylie C (2005) Maternal Xenopus Zic2 negatively regulates nodal-related gene expression during anteroposterior patterning. Development 132(21):4845–4855. https://doi.org/10.1242/dev.02066

Houtmeyers R, Souopgui J, Tejpar S, Arkell R (2013) The ZIC gene family encodes multi-functional proteins essential for patterning and morphogenesis. Cell Mol Life Sci 70(20):3791–3811. https://doi.org/10.1007/s00018-013 1285-5

Houtmeyers R, Tchouate Gainkam O, Glanville-Jones HA, Van den Bosch B, Chappell A, Barratt KS, Souopgui J, Tejpar S, Arkell RM (2016) Zic2 mutation causes holoprosencephaly via disruption of NODAL signalling. Hum Mol Genet 25(18):3946–3959. https://doi.org/10.1093/hmg/ddw235

Inoue T, Hatayama M, Tohmonda T, Itohara S, Aruga J, Mikoshiba K (2004) Mouse Zic5 deficiency results in neural tube defects and hypoplasia of cephalic neural crest derivatives. Dev Biol 270(1):146–162. https://doi.org/10.1016/j.ydbio.2004.02.017

Jaurena MB, Juraver-Geslin H, Devotta A, Saint-Jeannet JP (2015) Zic1 controls placode progenitor formation non-cell autonomously by regulating retinoic acid production and transport. Nat Commun 6:7476. https://doi.org/10.1038/ncomms8476

Keller R, Danilchik M (1988) Regional expression, pattern and timing of convergence and extension during gastrulation of Xenopus laevis. Development 103(1):193–209

Keller R, Shih J, Sater AK, Moreno C (1992) Planar induction of convergence and extension of the neural plate by the organizer of Xenopus. Dev Dyn 193(3):218–234. https://doi.org/10.1002/aja.1001930303

Kelly LE, Carrel TL, Herman GE, El-Hodiri HM (2006) Pbx1 and Meis1 regulate activity of the Xenopus laevis Zic3 promoter through a highly conserved region. Biochem Biophys Res Commun 344(3):1031–1037. https://doi.org/10.1016/j.bbrc.2006.03.235

Kim HJ, Lee SK, Na SY, Choi HS, Lee JW (1998) Molecular cloning of xSRC-3, a novel transcription coactivator from Xenopus, that is related to AIB1, p/CIP, and TIF2. Mol Endocrinol 12(7):1038–1047. https://doi.org/10.1210/mend.12.7.0139

King LS, Kozono D, Agre P (2004) From structure to disease: the evolving tale of aquaporin biology. Nat Rev Mol Cell Biol 5(9):687–698. https://doi.org/10.1038/nrm1469

Kitaguchi T, Nagai T, Nakata K, Aruga J, Mikoshiba K (2000) Zic3 is involved in the left-right specification of the Xenopus embryo. Development 127(22):4787–4795

Klein SL, Moody SA (2015) Early neural ectodermal genes are activated by Siamois and Twin during blastula stages. Genesis 53(5):308–320. https://doi.org/10.1002/dvg.22854

Kofent J, Spagnoli FM (2016) Xenopus as a model system for studying pancreatic development and diabetes. Semin Cell Dev Biol 51:106–116. https://doi.org/10.1016/j.semcdb.2016.01.005

Kolm PJ, Sive HL (1995) Efficient hormone-inducible protein function in Xenopus laevis. Dev Biol 171(1):267–272. https://doi.org/10.1006/dbio.1995.1279

Koyabu Y, Nakata K, Mizugishi K, Aruga J, Mikoshiba K (2001) Physical and functional interactions between Zic and Gli proteins. J Biol Chem 276(10):6889–6892. https://doi.org/10.1074/jbc.C000773200

Kuo JS, Patel M, Gamse J, Merzdorf C, Liu X, Apekin V, Sive H (1998) Opl: a zinc finger protein that regulates neural determination and patterning in Xenopus. Development 125(15):2867–2882

Leclerc C, Webb SE, Daguzan C, Moreau M, Miller AL (2000) Imaging patterns of calcium transients during neural induction in Xenopus laevis embryos. J Cell Sci 113(Pt 19):3519–3529

Leclerc C, Lee M, Webb SE, Moreau M, Miller AL (2003) Calcium transients triggered by planar signals induce the expression of ZIC3 gene during neural induction in Xenopus. Dev Biol 261(2):381–390

Leclerc C, Néant I, Webb SE, Miller AL, Moreau M (2006) Calcium transients and calcium signaling during early neurogenesis in the amphibian embryo Xenopus laevis. Biochim Biophys Acta 1763(11):1184–1191. https://doi.org/10.1016/j.bbamcr.2006.08.005

Lepperdinger G (2000) Amphibian choroid plexus lipocalin, Cpl1. Biochim Biophys Acta 1482(1–2):119–126

Li S, Shin Y, Cho KW, Merzdorf CS (2006) The Xfeb gene is directly upregulated by Zic1 during early neural development. Dev Dyn 235(10):2817–2827. https://doi.org/10.1002/dvdy.20896

Li B, Kuriyama S, Moreno M, Mayor R (2009) The posteriorizing gene Gbx2 is a direct target of Wnt signalling and the earliest factor in neural crest induction. Development 136(19):3267–3278. https://doi.org/10.1242/dev.036954

Lienkamp SS (2016) Using Xenopus to study genetic kidney diseases. Semin Cell Dev Biol 51:117–124. https://doi.org/10.1016/j.semcdb.2016.02.002

Maeda R, Mood K, Jones TL, Aruga J, Buchberg AM, Daar IO (2001) Xmeis1, a protooncogene involved in specifying neural crest cell fate in Xenopus embryos. Oncogene 20(11):1329–1342. https://doi.org/10.1038/sj.onc.1204250

Maeda R, Ishimura A, Mood K, Park EK, Buchberg AM, Daar IO (2002) Xpbx1b and Xmeis1b play a collaborative role in hindbrain and neural crest gene expression in Xenopus embryos. Proc Natl Acad Sci U S A 99(8):5448–5453. https://doi.org/10.1073/pnas.082654899

Marchal L, Luxardi G, Thomé V, Kodjabachian L (2009) BMP inhibition initiates neural induction via FGF signaling and Zic genes. Proc Natl Acad Sci U S A 106(41):17437–17442. https://doi.org/10.1073/pnas.0906352106

Mattioni T, Louvion JF, Picard D (1994) Regulation of protein activities by fusion to steroid binding domains. Methods Cell Biol 43(Pt A):335–352

McMahon AR, Merzdorf CS (2010) Expression of the zic1, zic2, zic3, and zic4 genes in early chick embryos. BMC Res Notes 3:167. https://doi.org/10.1186/1756-0500-3-167

Merzdorf CS (2007) Emerging roles for zic genes in early development. Dev Dyn 236(4):922–940. https://doi.org/10.1002/dvdy.21098

Merzdorf CS, Sive HL (2006) The zic1 gene is an activator of Wnt signaling. Int J Dev Biol 50(7):611–617. https://doi.org/10.1387/ijdb.052110cm

Milet C, Maczkowiak F, Roche DD, Monsoro-Burq AH (2013) Pax3 and Zic1 drive induction and differentiation of multipotent, migratory, and functional neural crest in Xenopus embryos. Proc Natl Acad Sci U S A 110(14):5528–5533. https://doi.org/10.1073/pnas.1219124110

Mizugishi K, Aruga J, Nakata K, Mikoshiba K (2001) Molecular properties of Zic proteins as transcriptional regulators and their relationship to GLI proteins. J Biol Chem 276(3):2180–2188. https://doi.org/10.1074/jbc.M004430200

Mizugishi K, Hatayama M, Tohmonda T, Ogawa M, Inoue T, Mikoshiba K, Aruga J (2004) Myogenic repressor I-mfa interferes with the function of Zic family proteins. Biochem Biophys Res Commun 320(1):233–240. https://doi.org/10.1016/j.bbrc.2004.05.158

Mizuseki K, Kishi M, Matsui M, Nakanishi S, Sasai Y (1998) Xenopus Zic-related-1 and Sox-2, two factors induced by chordin, have distinct activities in the initiation of neural induction. Development 125(4):579–587

Monsoro-Burq AH, Wang E, Harland R (2005) Msx1 and Pax3 cooperate to mediate FGF8 and WNT signals during Xenopus neural crest induction. Dev Cell 8(2):167–178. https://doi.org/10.1016/j.devcel.2004.12.017

Moreau M, Néant I, Webb SE, Miller AL, Leclerc C (2008) Calcium signalling during neural induction in Xenopus laevis embryos. Philos Trans R Soc Lond Ser B Biol Sci 363(1495):1371–1375. https://doi.org/10.1098/rstb.2007.2254

Nagai T, Aruga J, Takada S, Günther T, Spörle R, Schughart K, Mikoshiba K (1997) The expression of the mouse Zic1, Zic2, and Zic3 gene suggests an essential role for Zic genes in body pattern formation. Dev Biol 182(2):299–313. https://doi.org/10.1006/dbio.1996.8449

Nagel M, Tahinci E, Symes K, Winklbauer R (2004) Guidance of mesoderm cell migration in the Xenopus gastrula requires PDGF signaling. Development 131(11):2727–2736. https://doi.org/10.1242/dev.01141

Nakata K, Nagai T, Aruga J, Mikoshiba K (1997) Xenopus Zic3, a primary regulator both in neural and neural crest development. Proc Natl Acad Sci U S A 94(22):11980–11985

Nakata K, Nagai T, Aruga J, Mikoshiba K (1998) Xenopus Zic family and its role in neural and neural crest development. Mech Dev 75(1–2):43–51

Nakata K, Koyabu Y, Aruga J, Mikoshiba K (2000) A novel member of the Xenopus Zic family, Zic5, mediates neural crest development. Mech Dev 99(1–2):83–91

Neilson KM, Klein SL, Mhaske P, Mood K, Daar IO, Moody SA (2012) Specific domains of FoxD4/5 activate and repress neural transcription factor genes to control the progression of immature neural ectoderm to differentiating neural plate. Dev Biol 365(2):363–375. https://doi.org/10.1016/j.ydbio.2012.03.004

Nie S, Bronner ME (2015) Dual developmental role of transcriptional regulator Ets1 in Xenopus cardiac neural crest vs. heart mesoderm. Cardiovasc Res 106(1):67–75. https://doi.org/10.1093/cvr/cvv043

Nyholm MK, Wu SF, Dorsky RI, Grinblat Y (2007) The zebrafish zic2a-zic5 gene pair acts downstream of canonical Wnt signaling to control cell proliferation in the developing tectum. Development 134(4):735–746. https://doi.org/10.1242/dev.02756

Patthey C, Gunhaga L (2014) Signaling pathways regulating ectodermal cell fate choices. Exp Cell Res 321(1):11–16. https://doi.org/10.1016/j.yexcr.2013.08.002
Pavletich NP, Pabo CO (1993) Crystal structure of a five-finger GLI-DNA complex: new perspectives on zinc fingers. Science 261(5129):1701–1707
Plouhinec JL, Roche DD, Pegoraro C, Figueiredo AL, Maczkowiak F, Brunet LJ, Milet C, Vert JP, Pollet N, Harland RM, Monsoro-Burq AH (2014) Pax3 and Zic1 trigger the early neural crest gene regulatory network by the direct activation of multiple key neural crest specifiers. Dev Biol 386(2):461–472. https://doi.org/10.1016/j.ydbio.2013.12.010
Pohl BS, Knochel W (2005) Of fox and frogs: fox (fork head/winged helix) transcription factors in Xenopus development. Gene 344:21–32. https://doi.org/10.1016/j.gene.2004.09.037
Pourebrahim R, Houtmeyers R, Ghogomu S, Janssens S, Thelie A, Tran HT, Langenberg T, Vleminckx K, Bellefroid E, Cassiman JJ, Tejpar S (2011) Transcription factor Zic2 inhibits Wnt/β-catenin protein signaling. J Biol Chem 286(43):37732–37740. https://doi.org/10.1074/jbc.M111.242826
Purandare SM, Ware SM, Kwan KM, Gebbia M, Bassi MT, Deng JM, Vogel H, Behringer RR, Belmont JW, Casey B (2002) A complex syndrome of left-right axis, central nervous system and axial skeleton defects in Zic3 mutant mice. Development 129(9):2293–2302
Rhinn M, Brand M (2001) The midbrain – hindbrain boundary organizer. Curr Opin Neurobiol 11(1):34–42
Sakurada T, Mima K, Kurisaki A, Sugino H, Yamauchi T (2005) Neuronal cell type-specific promoter of the alpha CaM kinase II gene is activated by Zic2, a Zic family zinc finger protein. Neurosci Res 53(3):323–330. https://doi.org/10.1016/j.neures.2005.08.001
Salero E, Pérez-Sen R, Aruga J, Giménez C, Zafra F (2001) Transcription factors Zic1 and Zic2 bind and transactivate the apolipoprotein E gene promoter. J Biol Chem 276(3):1881–1888. https://doi.org/10.1074/jbc.M007008200
Sanchez-Ferras O, Bernas G, Laberge-Perrault E, Pilon N (2014) Induction and dorsal restriction of paired-box 3 (Pax3) gene expression in the caudal neuroectoderm is mediated by integration of multiple pathways on a short neural crest enhancer. Biochim Biophys Acta 1839(7):546–558. https://doi.org/10.1016/j.bbagrm.2014.04.023
Sasai Y, De Robertis EM (1997) Ectodermal patterning in vertebrate embryos. Dev Biol 182(1):5–20. https://doi.org/10.1006/dbio.1996.8445
Sasai Y, Lu B, Piccolo S, De Robertis EM (1996) Endoderm induction by the organizer-secreted factors chordin and noggin in Xenopus animal caps. EMBO J 15(17):4547–4555
Sasai N, Mizuseki K, Sasai Y (2001) Requirement of FoxD3-class signaling for neural crest determination in Xenopus. Development 128(13):2525–2536
Sato T, Sasai N, Sasai Y (2005) Neural crest determination by co-activation of Pax3 and Zic1 genes in Xenopus ectoderm. Development 132(10):2355–2363. https://doi.org/10.1242/dev.01823
Schroeder TE (1970) Neurulation in Xenopus laevis. An analysis and model based upon light and electron microscopy. J Embryol Exp Morpholog 23(2):427–462
Session AM, Uno Y, Kwon T, Chapman JA, Toyoda A, Takahashi S, Fukui A, Hikosaka A, Suzuki A, Kondo M, van Heeringen SJ, Quigley I, Heinz S, Ogino H, Ochi H, Hellsten U, Lyons JB, Simakov O, Putnam N, Stites J, Kuroki Y, Tanaka T, Michiue T, Watanabe M, Bogdanovic O, Lister R, Georgiou G, Paranjpe SS, van Kruijsbergen I, Shu S, Carlson J, Kinoshita T, Ohta Y, Mawaribuchi S, Jenkins J, Grimwood J, Schmutz J, Mitros T, Mozaffari SV, Suzuki Y, Haramoto Y, Yamamoto TS, Takagi C, Heald R, Miller K, Haudenschild C, Kitzman J, Nakayama T, Izutsu Y, Robert J, Fortriede J, Burns K, Lotay V, Karimi K, Yasuoka Y, Dichmann DS, Flajnik MF, Houston DW, Shendure J, DuPasquier L, Vize PD, Zorn AM, Ito M, Marcotte EM, Wallingford JB, Ito Y, Asashima M, Ueno N, Matsuda Y, Veenstra GJ, Fujiyama A, Harland RM, Taira M, Rokhsar DS (2016) Genome evolution in the allotetraploid frog Xenopus laevis. Nature 538(7625):336–343. https://doi.org/10.1038/nature19840

Sive HL, Grainger RM, Harland RM (2007) Animal cap isolation from Xenopus laevis. CSH Protoc 2007:pdb.prot4744

Soriano P (1997) The PDGF alpha receptor is required for neural crest cell development and for normal patterning of the somites. Development 124(14):2691–2700

Stuhlmiller TJ, García-Castro MI (2012a) FGF/MAPK signaling is required in the gastrula epiblast for avian neural crest induction. Development 139(2):289–300. https://doi.org/10.1242/dev.070276

Stuhlmiller TJ, García-Castro MI (2012b) Current perspectives of the signaling pathways directing neural crest induction. Cell Mol Life Sci 69(22):3715–3737. https://doi.org/10.1007/s00018-012-0991-8

Sugimoto K, Okabayashi K, Sedohara A, Hayata T, Asashima M (2007) The role of XBtg2 in Xenopus neural development. Dev Neurosci 29(6):468–479. https://doi.org/10.1159/000097320

Sun Rhodes LS, Merzdorf CS (2006) The zic1 gene is expressed in chick somites but not in migratory neural crest. Gene Expr Patterns 6(5):539–545. https://doi.org/10.1016/j.modgep.2005.10.006

Takeuchi M, Nakabayashi J, Sakaguchi T, Yamamoto TS, Takahashi H, Takeda H, Ueno N (2003) The prickle-related gene in vertebrates is essential for gastrulation cell movements. Curr Biol 13(8):674–679

Tandon P, Conlon F, Furlow JD, Horb ME (2016) Expanding the genetic toolkit in Xenopus: approaches and opportunities for human disease modeling. Dev Biol. https://doi.org/10.1016/j.ydbio.2016.04.009

Tanibe M, Michiue T, Yukita A, Danno H, Ikuzawa M, Ishiura S, Asashima M (2008) Retinoic acid metabolizing factor xCyp26c is specifically expressed in neuroectoderm and regulates anterior neural patterning in Xenopus laevis. Int J Dev Biol 52(7):893–901. https://doi.org/10.1387/ijdb.082683mt

Tribulo C, Aybar MJ, Nguyen VH, Mullins MC, Mayor R (2003) Regulation of Msx genes by a Bmp gradient is essential for neural crest specification. Development 130(26):6441–6452. https://doi.org/10.1242/dev.00878

Tropepe V, Li S, Dickinson A, Gamse JT, Sive HL (2006) Identification of a BMP inhibitor-responsive promoter module required for expression of the early neural gene zic1. Dev Biol 289(2):517–529. https://doi.org/10.1016/j.ydbio.2005.10.004

Twigg SR, Forecki J, Goos JA, Richardson IC, Hoogeboom AJ, van den Ouweland AM, Swagemakers SM, Lequin MH, Van Antwerp D, SJ MG, Westbury I, Miller KA, Wall SA, van der Spek PJ, Mathijssen IM, Pauws E, Merzdorf CS, Wilkie AO, Consortium W (2015) Gain-of-function mutations in ZIC1 are associated with coronal craniosynostosis and learning disability. Am J Hum Genet 97(3):378–388. https://doi.org/10.1016/j.ajhg.2015.07.007

Urade Y, Hayaishi O (2000) Biochemical, structural, genetic, physiological, and pathophysiological features of lipocalin-type prostaglandin D synthase. Biochim Biophys Acta 1482(1–2):259–271

Van Stry M, Kazlauskas A, Schreiber SL, Symes K (2005) Distinct effectors of platelet-derived growth factor receptor-alpha signaling are required for cell survival during embryogenesis. Proc Natl Acad Sci U S A 102(23):8233–8238. https://doi.org/10.1073/pnas.0502885102

Verkman AS (2005) More than just water channels: unexpected cellular roles of aquaporins. J Cell Sci 118(Pt 15):3225–3232. https://doi.org/10.1242/jcs.02519

Vitorino M, Silva AC, Inácio JM, Ramalho JS, Gur M, Fainsod A, Steinbeisser H, Belo JA (2015) Xenopus Pkdcc1 and Pkdcc2 are two new tyrosine kinases involved in the regulation of JNK dependent Wnt/PCP signaling pathway. PLoS One 10(8):e0135504. https://doi.org/10.1371/journal.pone.0135504

Wallingford JB (2005) Neural tube closure and neural tube defects: studies in animal models reveal known knowns and known unknowns. Am J Med Genet C: Semin Med Genet 135C(1):59–68. https://doi.org/10.1002/ajmg.c.30054

Ware SM, Harutyunyan KG, Belmont JW (2006a) Zic3 is critical for early embryonic patterning during gastrulation. Dev Dyn 235(3):776–785. https://doi.org/10.1002/dvdy.20668

Ware SM, Harutyunyan KG, Belmont JW (2006b) Heart defects in X-linked heterotaxy: evidence for a genetic interaction of Zic3 with the nodal signaling pathway. Dev Dyn 235(6):1631–1637. https://doi.org/10.1002/dvdy.20719

Watabe Y, Baba Y, Nakauchi H, Mizota A, Watanabe S (2011) The role of Zic family zinc finger transcription factors in the proliferation and differentiation of retinal progenitor cells. Biochem Biophys Res Commun 415(1):42–47. https://doi.org/10.1016/j.bbrc.2011.10.007

Weber JR, Sokol SY (2003) Identification of a phylogenetically conserved activin-responsive enhancer in the Zic3 gene. Mech Dev 120(8):955–964

Winata CL, Kondrychyn I, Kumar V, Srinivasan KG, Orlov Y, Ravishankar A, Prabhakar S, Stanton LW, Korzh V, Mathavan S (2013) Genome wide analysis reveals Zic3 interaction with distal regulatory elements of stage specific developmental genes in zebrafish. PLoS Genet 9(10):e1003852. https://doi.org/10.1371/journal.pgen.1003852

Wu J, Li C, Zhao S, Mao B (2010) Differential expression of the Brunol/CELF family genes during Xenopus laevis early development. Int J Dev Biol 54(1):209–214. https://doi.org/10.1387/ijdb.082685jw

Yabe S, Tanegashima K, Haramoto Y, Takahashi S, Fujii T, Kozuma S, Taketani Y, Asashima M (2003) FRL-1, a member of the EGF-CFC family, is essential for neural differentiation in Xenopus early development. Development 130(10):2071–2081

Yamamoto TS, Takagi C, Ueno N (2000) Requirement of Xmsx-1 in the BMP-triggered ventralization of Xenopus embryos. Mech Dev 91(1–2):131–141

Yan B, Neilson KM, Moody SA (2009) FoxD5 plays a critical upstream role in regulating neural ectodermal fate and the onset of neural differentiation. Dev Biol 329(1):80–95. https://doi.org/10.1016/j.ydbio.2009.02.019

Yang Y, Hwang CK, Junn E, Lee G, Mouradian MM (2000) ZIC2 and Sp3 repress Sp1-induced activation of the human D1A dopamine receptor gene. J Biol Chem 275(49):38863–38869. https://doi.org/10.1074/jbc.M007906200

Chapter 8
Zic Genes in Teleosts: Their Roles in Dorsoventral Patterning in the Somite

Kota Abe, Toru Kawanishi, and Hiroyuki Takeda

Abstract The medaka contains seven *zic* genes, two of which, *zic1* and *zic4*, have been studied extensively. The analyses are mainly based on the *double anal fin* (*Da*) mutant, which was isolated from the wild. *Da* is an enhancer mutant of *zic1/zic4*, and the expression of *zic1/zic4* is specifically lost in the dorsal half of the somites, which leads to a mirror-image duplication of the ventral half across the lateral midline from larva to adult. The studies of medaka *Da* give us important insights into the function of *zic1/zic4* in mesodermal tissues and also the mechanism of dorsoventral patterning in the vertebrate trunk region occurring during late development, which is a long-standing mystery in developmental biology. In this chapter, we introduce genomic organization of medaka *zic* genes and discuss their function, mainly focusing on *zic1* and *zic4* in dorsoventral patterning of the trunk region and possible connections to human congenital disorders.

Keywords Zic · Dorsoventral patterning · Somite · Epigenetics · Evolution · Spina bifida occulta · *Oryzias latipes*

8.1 Genomic Organization and Protein Structure of Medaka *zic* Genes

Medaka has seven *zic* genes: *zic1*, *zic2*, *zic3*, *zic4*, *zic5a*, *zic5b*, and *zic6*. The genomic organization of medaka *zic* genes is similar to that of other vertebrates (Fig. 8.1a). *zic1* and *zic4* and *zic2* and *zic5* are located in a paired configuration in the same way as in tetrapods (Houtmeyers et al. 2013). Medaka also has a fish-specific *zic* gene, *zic6*, adjacent to *zic3*, as do the other teleosts including zebrafish (Keller and Chitnis 2007). Additionally, protein structures of medaka Zic1–6 are

K. Abe (✉) · H. Takeda
Department of Biological Sciences, Graduate School of Science, University of Tokyo, Tokyo, Japan
e-mail: kabe@bs.s.u-tokyo.ac.jp

T. Kawanishi
Department of Systems Biology, Harvard Medical School, Boston, MA, USA

J. Aruga (ed.), *Zic family*, Advances in Experimental Medicine and Biology 1046,
https://doi.org/10.1007/978-981-10-7311-3_8

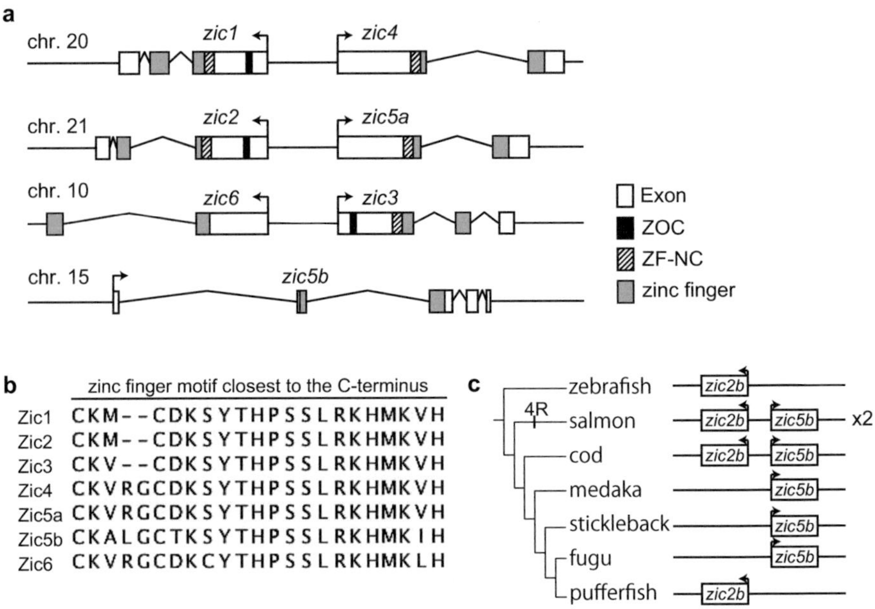

Fig. 8.1 Medaka *zic* genes. (**a**) Genomic organization of medaka *zic* genes as shown in the Ensembl genome browser medaka (HdrR) assembly, Oct 2005. Medaka has seven *zic* genes, three tandem pairs (*zic1/4*, *zic2/5a*, *zic3/6*) and *zic5b*. (**b**) Amino acid sequences of the zinc-finger motif closest to the C-terminus of medaka *zic* genes. Zic proteins have two amino acids between the first and the second cysteine residues, with the exception of Zic4, Zic5a, Zic5b, and Zic6, which have four. (**c**) *zic2b* and *zic5b* in teleosts

almost the same as those of other vertebrates (Fig. 8.1a) (Houtmeyers et al. 2013). Furthermore, medaka and zebrafish have a seventh *zic* gene, *zic5b* and *zic2b*, respectively (Toyama et al. 2004). Medaka Zic5b has lost the *N*-terminal portion of the protein and has only three zinc-finger motifs unlike other Zic proteins that have five. The medaka *zic5b* gene was annotated based on the synteny conservation with the zebrafish genome and the amino acid sequence homology of the zinc finger domain closest to the C-terminus. In this domain Zic4, Zic5a, Zic5b, and Zic6 contain four amino acids between the first and the second cysteine residues, unlike other Zic proteins that have two (Fig. 8.1b) (Aruga et al. 2006).

The presence of *zic* genes in other teleost species was examined to understand the evolution of *zic* genes in teleosts. Medaka (*Oryzias latipes*), zebrafish (*Danio rerio*), cod (*Gadus morhua*), stickleback (*Gasterosteus aculeatus*), fugu (*Takifugu rubripes*), and pufferfish (*Tetraodon nigroviridis*) have one copy of *zic1*, *zic3*, *zic4*, and *zic6* and one or two *zic2* and *zic5* (Fig. 8.1c). In contrast, salmon (*Salmo salar*) has two *zic3* genes and four pairs of *zic2*/*zic5*, in addition to single *zic1*, *zic4*, and *zic6*. These observations suggest that the common ancestor of teleost may have had eight *zic* genes: *zic1*, *zic2a/b*, *zic3*, *zic4*, *zic5a/b*, and *zic6*. Subsequently, *zic2b* or *zic5b* might have been lost independently in each species. It is also suggested that salmon acquired another *zic3* and additional *zic2/zic5* pairs as a result of the fourth-round whole genome duplication (Fig. 8.1c) (Lien et al. 2016). Because the genome assembly and gene annotation are rather incomplete in many teleosts, this is one possible interpretation of the currently available data, and the picture could be revised in the future. Only zebrafish *zic2b* has been reported to contribute to the neural crest and craniofacial development together with *zic2a* (TeSlaa et al. 2013). Considering that the structures of most of the Zic2b and Zic5b are relatively diverged especially in the N-terminal region, while that of zebrafish and salmon *zic2b* are highly conserved, functions of those *zic* genes, if any, could be divergent.

8.2 The Expression Patterns of Medaka *zic* Genes

To date, expression patterns of medaka *zic1*, *zic3*, and *zic4* have been investigated. The expression patterns of *zic3* were examined in stage 31 embryo (gill blood vessel-forming stage) by whole-mount in situ hybridization (WISH). *zic3* is expressed in the roof plate, rhombomeres (N4-8), cerebellum, tectum marginal zone, epithalamus, and telencephalon. The expression pattern of *zic3* is based on Medaka Expression Pattern Database (MEPD). These expression patterns almost coincide with those of zebrafish *zic3* at the corresponding developmental stage (Thisse et al. 2001).

The expression patterns of *zic1/zic4* in medaka embryo were examined by WISH (Fig. 8.2a, c, and e). The early expression pattern of *zic1* in the medaka head region is almost the same as that in zebrafish (Rohr et al. 1999; Ohtsuka et al. 2004). During somitogenesis, *zic1* is expressed dorsally in all parts of the brain (Fig. 8.2a). In the trunk and tail regions, *zic1* is expressed in the dorsal neural tube and the somites (Fig. 8.2a, c). The expression in the somites can be detected between the two-somite stage and six-somite stage. In later stages of development, the expression in the dorsal somites continues and intensifies in the dorsal sclerotome, more than other parts of the somites (Ohtsuka et al. 2004; Kawanishi et al. 2013). The expression pattern was confirmed by making a transgenic medaka line in which a bacterial artificial chromosome (BAC) construct encoding *zic1*- and *zic4*-responsive reporter genes was introduced (Kawanishi et al. 2013) (Fig. 8.2g). The transgenic medaka allowed analysis of the expression patterns in later developmental stages. First, it was revealed that all somite derivatives, dermomyotome, myotome, and sclerotome, maintain the dorsal expression and share the sharp and straight ventral expression boundary at the level of the notochord. The position of the expression boundary finally coincides with horizontal myoseptum, a structure that divides dorsal and ventral body wall muscle masses in fish. Surprisingly, the dorsal expression and the clear expression boundary are persisted even in adult medaka (Kawanishi et al. 2013) (Fig. 8.2g–i). The dorsal expression in the somite derivatives in adult was

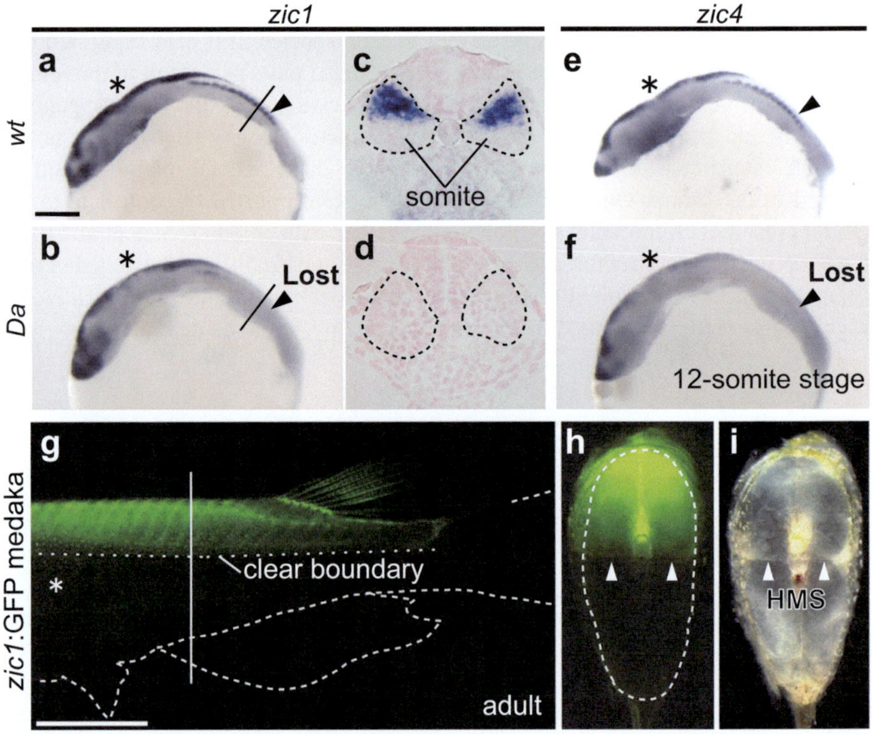

Fig. 8.2 Expression of *zic1* and *zic4* in the embryo and adult of *wt* and *Da* mutant medaka. (**a–f**) *zic1* and *zic4* expression in the 12-somite stage embryo. *zic1* and *zic4* are expressed in the neural tube (asterisks) and dorsal somite (arrowheads) in the wild-type medaka embryo (**a**, **e**). In contrast, expression in the dorsal somite is lost in *Da* (**b**, **f**). Transverse sections at the level of solid lines in (**a**, **b**) confirm the expression in the dorsal somite in wt (**c**) and loss of expression in *Da* (**d**) (**c**, **d**, dashed lines indicate the somites). (**g–i**) Fluorescent images of the adult *zic1*:GFP transgenic medaka. A dotted line in **g** indicates the clear ventral boundary of GFP expression. Transverse section of the transgenic medaka at the level indicated by the solid line in **g** shows that the location of the expression boundary coincides with the horizontal myoseptum (HMS) (arrowheads in **h**, **i**). An asterisk in **g** indicates autofluorescence of pigment cells. Scale bars: 100 μm for **a**, **b**, **e**, and **f**; 5 mm for **g** (**a–f** are modified from Kawanishi et al. 2013)

confirmed by RT-PCR. The expression pattern of *zic4* is almost the same as that of *zic1*, although the signal is weaker (Fig. 8.2e). This lower expression level of *zic4* was confirmed by transcriptome analysis using the myotome (Nakamura et al. 2014). The expression patterns of *zic1/zic4* in the brain and the somite are highly conserved among vertebrates, such as mouse, *Xenopus laevis*, and chick (Merzdorf 2007; Mcmahon and Merzdorf 2010).

8.3 The Function of *zic1/zic4* in the Somite: Insights from Medaka *Double anal fin* (*Da*) Mutant

zic1 and *zic4* are expressed in both the neural tissue and somite. While the function of *zic1/zic4* is well known in the neural tissues (Aruga et al. 1998, 2002; Grinberg et al. 2004), reports about their function in the somites are relatively few. Aruga et al. reported that the deletion of *Zic1* caused abnormalities in dorsal structures of the vertebrae in mouse (Aruga et al. 1999). However, the function in other somite-derived tissues was largely unknown. In this section, we introduce the function of *zic1/zic4* in the somite revealed by analyses of a medaka spontaneous mutant *double anal fin* (*Da*).

Da is an enhancer mutant of *zic1/zic4* in which a large DNA fragment is inserted 8.6 kb downstream of *zic4* (Moriyama et al. 2012) (Fig. 8.3a). This fragment was found to be a DNA-based transposable element, and its size (180 kb) is larger than any other transposon reported to date (Inoue et al. 2017). The transposon is named *Teratorn* after the name of one of the extinct gigantic birds of prey. In *Da*, the expression of *zic1/zic4* in the dorsal somite is lost, while the expression in the neural tissues is less affected, suggesting that the somite enhancer of *zic1/zic4* is specifically disrupted (Ohtsuka et al. 2004) (Fig. 8.2b, d, and f). In *Da*, the dorsal morphology in the trunk region is ventralized, and the dorsal half of the trunk appears to be a mirror image of the ventral half across the lateral midline (Fig. 8.3b, c). Thus, *zic1/zic4* was revealed to play an important role in dorsoventral patterning of the vertebrate (at least teleost) trunk during late development, which was a long-standing mystery in developmental biology. Interestingly, the ventral duplication phenotypes are seen not only in the somite-derived tissues but also in the ectoderm- and the neural crest-derived tissues surrounding the somites. Transplantation of the wild-type somites to *Da* demonstrated that the loss of *zic1/zic4* expression in the somites is responsible for the *Da* phenotypes (Kawanishi et al. 2013). The following detailed phenotypes of *Da* are mainly based on Ohtsuka et al. 2011.

8.3.1 *Embryonic Phenotypes*

The phenotypes of the *Da* embryo can be classified into five features: (1) localization of the pigment cells, (2) shape of the dorsal fin fold, (3) shape of the dorsal myotome, (4) caudal structures, and (5) positioning of the neuromast.

In the wild-type medaka embryo, the dorsal melanocytes of the trunk and tail region are arranged in a single line along the dorsal midline, whereas those on the ventral side are arranged in two lines along the ventral midline. In contrast, the dorsal melanocytes in *Da* are arranged in two lines in the same way as the ventral melanocytes (Fig. 8.3d, e). In addition to the melanophores, the location of other chromatophores, namely, xanthophores and leucophores, is altered in *Da*. Numerous

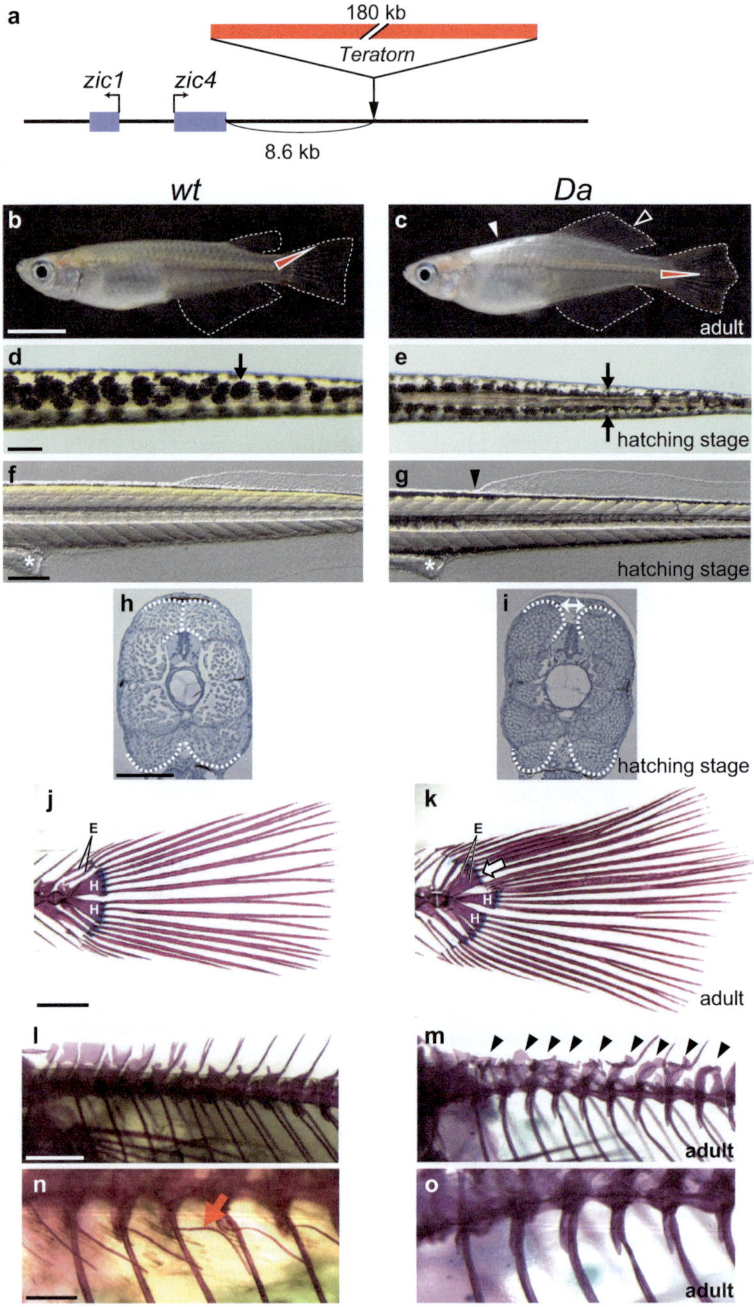

Fig. 8.3 Phenotypes of the *Da* mutant. (**a**) A huge transposon, *Teratorn*, is inserted 8.6 kb downstream of *zic4* in the *Da* genome. Adult wild-type (**b**) and *Da* mutant (**c**) medaka. *Da* shows a ventralized pigmentation (white arrowhead) and dorsal fin (black arrowhead), as well as a rhombic

xanthophores and leucophores appear in a single row along the dorsal midline of the wild-type, while only a few of these chromatophores are seen in *Da*.

The median fins, including the dorsal, caudal, and anal fins, develop directly from the continuous epidermal fold surrounding the trunk and tail region. In the wild-type medaka embryo, the anterior end of the dorsal fin fold is situated seven somites posterior to that of the ventral fin fold (Fig. 8.3f). However, the anterior end of the dorsal fin fold in *Da* has shifted anteriorly toward the position of the ventral fin fold (Fig. 8.3g).

In the wild-type medaka embryo, the tops of the dorsal myotomes elongate and come in contact at the dorsal midline. Subsequently, the top of the dorsal myotome thickens, covering the neural tube (Fig. 8.3h). In contrast, in *Da*, the top of the dorsal myotomes remain at both sides of the neural tube, and each myotome grows to protrude in the dorsal direction without covering the neural tube (Fig. 8.3i) (Tamiya et al. 1997).

Teleosts including medaka have a homocercal caudal fin. Although it is superficially symmetric along DV axis, the caudal part of the vertebral column tilts strongly dorsally so that the fin expanse is purely a ventral structure. The notochord terminus of a wild-type embryo starts to bend dorsally at stage 32 (somite-completion stage), whereas it extends straight into the tail in *Da* (Moriyama et al. 2012). This leads to the rhombic caudal fin of adult *Da* (Fig. 8.3b, c).

Neuromasts, which make up a small cluster of sensory cells that are part of the lateral line system, allow the sensing of mechanical changes in water. These cells are normally localized in the lateral and ventral regions in the wild-type medaka. However, they are mislocated at the dorsal region of *Da*.

8.3.2 Adult Phenotypes

The phenotypes of the adult *Da* mutant include (1) shapes of the anal fin and caudal fin, (2) mislocalization of iridophores, (3) an altered body shape, and (4) an irregular axial skeleton.

Fig. 8.3 (continued) caudal fin (red arrowhead) and teardrop body shape. (**d**, **e**) Distribution of melanocytes in the tail region of hatching stage embryo (dorsal view). In wild-type, melanocytes are arranged in a single line along the dorsal midline (arrow in **d**). In contrast, those in *Da* are arranged in two lines in the same manner as ventral melanocytes (arrows in **e**). (**f**, **g**) Position of the dorsal fin fold at hatching stage (lateral view). The anterior limit of the dorsal fin fold (arrowhead) is shifted anteriorly in *Da*. (**h**, **i**) Myotomal morphology at hatching stage. Dashed lines delineate the edge of the myotome. A double-headed arrow indicates the gap between the dorsal myotomes in *Da*. (**j**, **k**) Caudal skeleton morphology of adult wild-type and *Da* medaka. An arrow indicates enlarged epurals in *Da*. *E* epural, *H* hypural. (**l**–**o**) Phenotypes of the trunk axial skeleton. The neural spines of *Da* are also deformed and larger (arrowheads in **m**) than those of wild-type medaka (**l**). The epipleurals, which are normally attached to the proximal region of the ribs in the wild-type (arrow in **n**), are missing or truncated in *Da* (**o**). Scale bars: 1 cm for **b**, **c**, **j**, and **k**; 200 μm for **d**–**i**; 10 mm for **l**, **m**; 1 mm for **n** and **o** (**b**–**i** are modified from Kawanishi et al. 2013)

The dorsal fin of adult *Da* is enlarged and similar in shape to the anal fin as if this mutant has two anal fins (Fig. 8.3b, c). The *Da* (*double anal fin*) nomenclature refers to this severe phenotype. In the case of the caudal fin, the shape becomes rhombic instead of triangular. The triangular caudal fin of wild-type medaka, called the homocercal caudal fin, exhibits dorsoventrally asymmetric morphology; the urostyle bends dorsally, and the hypurals are formed on the ventral side to support the caudal fin rays. However, in *Da*, the urostyle does not bend, and both the hypurals and epurals become equally larger in size. The fin rays articulate with both hypurals and epurals, which results in the dorsoventrally symmetric morphology of the caudal fin of *Da* (Moriyama et al. 2012) (Fig. 8.3j, k).

The iridophores, which are normally found in the abdominal region of wild-type medaka, are abundant at the dorsal side as well as the ventral side of the *Da* trunk region (Fig. 8.3b, c). Furthermore, *Da* exhibits a teardrop body shape, instead of a dorsally flattened one. The body shape can be mostly attributed to the shape of the myotome. The dorsal myotome of *Da* grows abnormally without filling the gap over the neural tube.

The axial skeleton is also morphologically altered in *Da* (Fig. 8.3l–o). The trunk neural arches of *Da* are irregularly shaped, and, in some cases, the midline fusion of several neural arches is disturbed, resulting in spina bifida occulta. The neural spines of *Da* are deformed and larger than those of wild-type medaka (Fig. 8.3l, m). The epipleurals, which are normally attached to the proximal region of the ribs in the wild-type, are missing or truncated in *Da* (Fig. 8.3n, o). However, the centrums, ribs, and hemal spines appear normal in this mutant. These axial skeleton phenotypes are quite similar to those reported in the *Zic1*$^{-/-}$ mouse (Aruga et al. 1999).

8.4 Transcriptional Regulation of *zic1/zic4*

8.4.1 Regulatory Elements of zic1/zic4

The specific loss of *zic1/zic4* expression in the somites suggests that insertion of the transposon disrupts the function of a somite enhancer downstream of *zic1/zic4*. Transgenic medaka lines using modified BAC clones containing *zic1/zic4* genes were established to test this possibility (Moriyama et al. 2012). In the resulting transgenic medaka, the reporter signals recapitulated the endogenous expression patterns of *zic1/zic4* both in the neural tube and the somites, indicating that the BAC clone contains all of the cis-regulatory elements required for *zic1/zic4* expression (Fig. 8.2g). In contrast, the transgenic medaka harboring a BAC construct that lacks the entire region downstream of the *Teratorn* insertion point in *Da* shows a dramatic reduction in reporter gene expression in the somites, whereas that in the neural tissue was only weakly reduced. This expression pattern is almost identical to that of *zic1/zic4* in *Da*. In the genomic region around the insertion point, four conserved noncoding elements (CNEs) were found by comparison of genomic sequences from

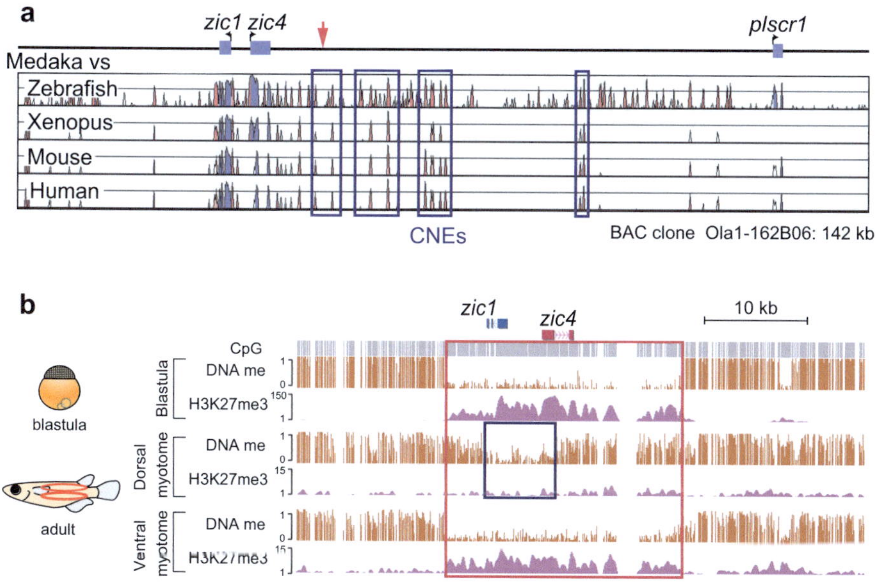

Fig. 8.4 Transcriptional regulation of *zic1* and *zic4*. (**a**) Homology comparison of the *zic1/zic4* region. The location of the *Teratorn* transposon insertion in the *Da* mutant genome is indicated by a red arrow. (**b**) Active-gene-expression-state-dependent K27HMD shortening at the *zic1/zic4* locus in the adult myotome. Patterns of DNA methylation and H3K27me3 are shown for the blastula embryo, and for the adult dorsal and ventral myotome. The red box indicates a large K27HMD identified in the blastula embryo, and the blue box indicates a shortened HMD in the adult dorsal myotome (**b** is modified from Nakamura et al. 2014)

distantly related vertebrates including human (Fig. 8.4a). Transgenic medaka, which only contains these four CNEs, shows a reporter signal in the neural tube and the somite, indicating that the somite and the neural enhancers reside in the region of the CNEs.

Furthermore, because the state of epigenetic modifications, namely, histone modifications and DNA methylation, around *zic1/zic4* locus is not altered in *Da* (Nakamura et al. 2014), it is suggested that physical interactions between the promoter and somite enhancer of *zic1/zic4* may be disrupted in *Da*. Another *Da* mutant, which exhibits almost identical phenotypes to those of *Da*, was also isolated and is named *Da-2* (Moriyama et al. 2012). *Da-2* was genetically mapped to the region (74 kb) containing *zic1/zic4*. *Teratorn* is also associated with this allele, but the insertion point is within the intergenic region between *zic1* and *zic4* in this mutant. Despite the different insertion point of *Teratorn*, the expressions of both *zic1* and *zic4* in somite are significantly reduced like original *Da*. These results suggest that *Teratorn* may affect the transcriptional regulation of *zic1/zic4* not simply by extending the distance between the regulatory elements and the promoter but by affecting the higher-order structure of chromatin and disrupting the regulatory landscape around the *zic1/zic4* locus. Indeed, the distance between the promoter region of *zic1/zic4* and the most distant CNE is about 50 kb in medaka and about 400 kb in human, suggesting the existence of some mechanisms that mediate such long-range interaction. This needs to be addressed in the future.

8.4.2 Two Distinct Modes of Regulation of zic1/zic4 Transcription

The dorsal-specific expression of *zic1/zic4* in somites is established in a cell-nonautonomous manner (Kawanishi et al. 2013). However, as embryos grow rapidly, the signaling relationship between the somite and surrounding tissues could be changed, and so could be the case for gene regulation. Notably, expression of *zic1/zic4* is maintained even in somite-derived tissues of adult, suggesting that *zic1/zic4* expression becomes less dependent on external signals as development proceeds (Fig. 8.2g). This idea is supported by in vitro culture experiments of somite-derived tissues (Kawanishi et al. 2013). While the cells derived from the somite at the segmentation stage lose *zic1* expression within 1 day, those derived from later embryonic stages and adult tissues maintained the expression at least for 1 week. This indicates that the regulation of the *zic1* expression is changed from a signal-dependent induction by surrounding tissues to cell-autonomous maintenance.

Epigenetic modification is thought to play a role in the maintenance of *zic1/zic4* expression. DNA methylation is one of the essential epigenetic modifications in vertebrates, and a small fraction of the genome is hypomethylated. Importantly, key developmental genes including *zic1/zic4* are marked by especially large hypomethylated domains harboring the repressive H3K27me3 histone modification (large K27HMD), before activation (Nakamura et al. 2014) (Fig. 8.4b). The large K27HMD, including *zic1/zic4*, is shortened in the adult dorsal myotome, but not in ventral, suggesting that the large K27HMD strictly represses the activation of *zic1/zic4* in a pluripotent state, whereas its shortening consolidates long-term gene expression in adult differentiated cells (Fig. 8.4b). This is also the case with other key developmental genes (Nakamura et al. 2014).

Taken together, *zic1/zic4* expression in dorsal somite is initially established by signals derived from surrounding tissues; however, it is later maintained in a cell-autonomous manner. Changes in epigenetic modifications may be involved in the long-term maintenance of gene expression.

8.5 *zic1/zic4* Function in DV Patterning Is Conserved in Vertebrates

The conserved dorsal-specific expression of *zic1/zic4* in vertebrate somites suggests that the function of *zic1/zic4* in the somite is also conserved. A search for other species that have altered trunk morphology similar to the *Da* mutant uncovered one variant of *Betta* that has been established during domestication (Kawanishi et al. 2013). This variant, known as "double tail," exhibits typical *Da* phenotypes in terms of fin morphology when compared with the common type *Betta*. The shape and position of the dorsal fin are transformed into those of the anal fin, and the caudal-most vertebrae do not bend dorsally, similar to what is observed in medaka *Da*, which leads to duplicated caudal fin lobes in this variant (Fig. 8.5a, b). The distribution of the lateral line is also ventralized. When the expression of *zic1/zic4* was compared between the common type and double-tail *Betta* embryos, *zic1/zic4* were expressed in the dorsal part of somites and neural tissues but lost specifically in double-tail somites (Fig. 8.5c–h). These results suggest that the function of Zic1/Zic4 to orchestrate trunk morphology is conserved in *Betta* somites, although the cause of altered *zic1/zic4* expression in double tail remains elusive.

Other than teleost, the function of *zic1* in somite has only been investigated in mouse. In *Zic1*$^{-/-}$ mice, skeletal abnormalities are found only in dorsal sclerotome derivatives, such as vertebral arches and ribs (Aruga et al. 1999). This implies that Zic1 also regulates dorsal-specific morphology of the trunk region in mouse. However, roles of *Zic1* in the patterning of other phenotypes of the dorsal trunk, such as the shape of myotome and the coloration of the skin, remain largely unknown. Further analyses including somite-specific knockout of *Zic1* and *Zic4* may be required.

8.6 Evolution of *zic* Dorsal Expression in the Vertebrate Somite

To understand the evolutionary aspects of dorsal expression of *zic* in the vertebrate somite, we summarized the expression of *zic* among chordate (Fig. 8.6). Somites are present in the basal chordate amphioxus, in which *zic* is expressed in the dorsal somite (Gostling and Shimeld 2003). This suggests that dorsal expression of *zic* in the somite was acquired very early on in vertebrate evolution. However, the ventral expression boundary of *zic* is not sharp, and the expression appears to expand ventrally in amphioxus. This is also the case with lampreys and Chondrichthyes (Freitas et al. 2006; Kusakabe et al. 2011). In tetrapods, *zic* is expressed in dorsal somites,

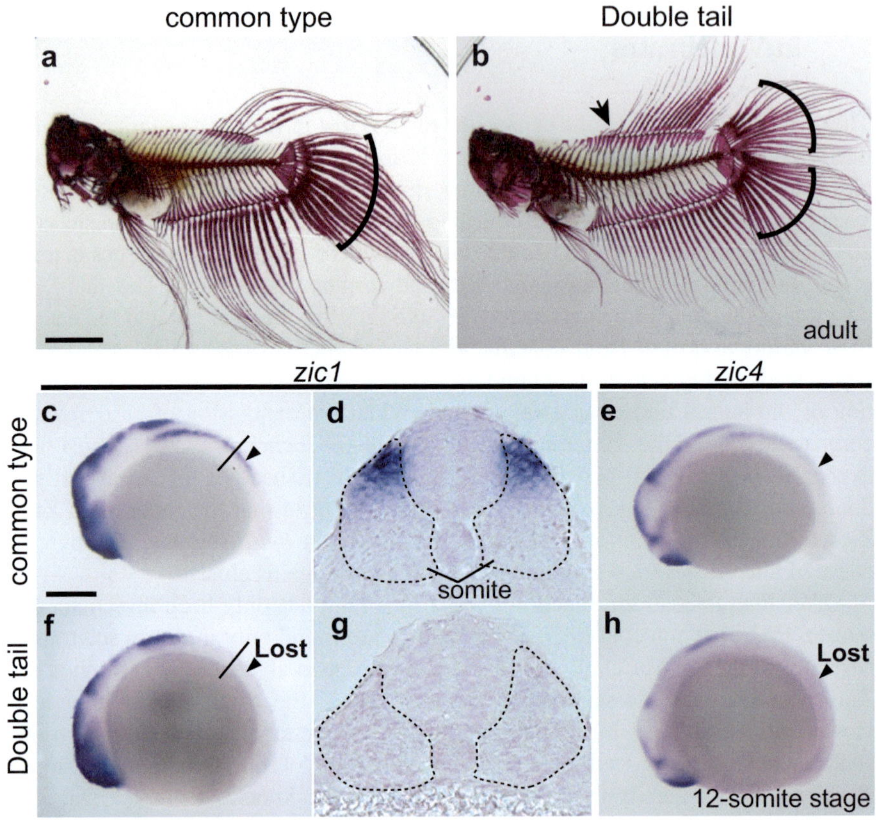

Fig. 8.5 Conserved function of *zic1* and *zic4* in *Betta splendens*. Skeletal staining of common type (**a**) and double-tail (**b**) adult male *Betta splendens* with alizarin red. The dorsal fin of the double tail is expanded anteriorly like an anal fin (arrow). The caudal fin lobe is duplicated in double tail (brackets). (**c**, **h**) Expression patterns of *zic1* and *zic4* in common type and double tail during embryogenesis (12 somite). *zic1* and *zic4* are expressed in the dorsal somite in common type (arrowheads in **c**, **f**), but they are lost in double tail (arrowheads in **f**, **h**). The dashed lines in **d** and **g** delineate somites in transverse sections at the level indicated by solid lines in (**d**, **f**), respectively. Scale bars: 2 cm for **a** and **b**; 200 μm for **c**, **e**, **f**, and **h** (**c**–**h** modified from Kawanishi et al. 2013)

and the ventral boundary appears sharp like in teleosts (Merzdorf 2007; McMahon and Merzdorf 2010). The detailed expression patterns of *zic* in tetrapods in terms of the existence of the sharp ventral expression boundary are now under examination. We speculate that the evolution of the sharp expression boundary of *zic* might accelerate the functional differentiation of the dorsal and ventral trunk.

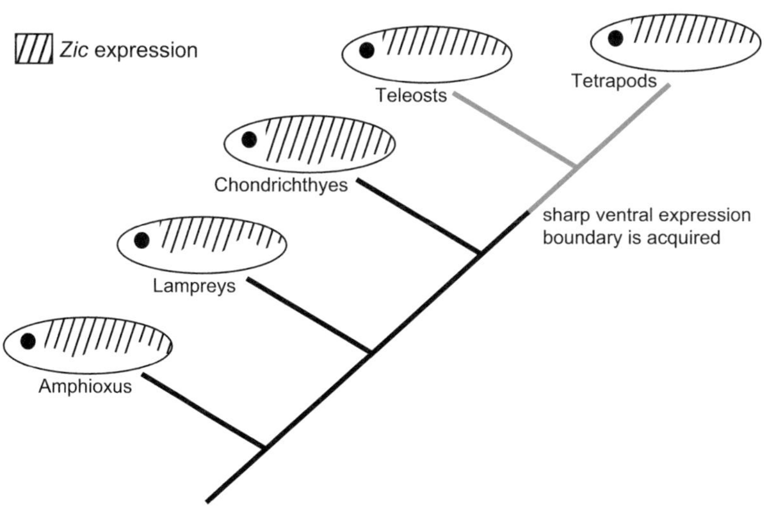

Fig. 8.6 Evolution of the *zic* expression patterns in chordate somites. Various expression patterns of *zic* in the somite are depicted in the schematic phylogenetic tree of chordates. Black stripes indicate the expression of *zic* in the somites

8.7 *zic1* and *zic4* in the Trunk and Implications for Human Congenital Defects

Neural tube defects (NTDs) are one of the most common human congenital malformations. One frequent form of NTD is spina bifida, a malformation of the spinal cord and/or vertebral arches characterized by herniation or exposure of the spinal cord through incompletely closed vertebrae (Wong et al. 2008; Cabrera et al. 2012). The medaka *Da* mutant develops spina bifida occulta, the mildest form of spina bifida, in which midline fusion of some neural arches is disturbed but the spinal cord does not protrude (Ohtsuka et al. 2004). The *Zic1*$^{-/-}$ mouse also exhibits spina bifida occulta (Aruga et al. 1999). However, unlike *Da*, spinous processes, probably corresponding to the neural spines of the medaka, are missing in the knockout mouse. This can be ascribed to the absence of *Zic1* expression in both the neural tube and somite in *Zic1*$^{-/-}$ mouse, whereas *Da* loses *zic1* expression specifically in the somite. Since *zic1* expression in the neural tube seems to be required for the formation of the spinous processes, the *Da* phenotype in neural spines (Fig. 8.3l, m) suggests that *zic1* expression in the somite derivatives is required for late patterning of the spinous processes (Ohtsuka et al. 2004). Although medaka *Da* and *Zic1*$^{-/-}$ mouse develop only mild spina bifida, it is reported that *Zic1* expression in the posterior part of the embryo is downregulated in mouse NTD models that exhibit severe spina bifida (Cabrera et al. 2012). This suggests that the downregulation of *zic1* may indeed be one of the causes of severe spina bifida.

Dandy-Walker malformation (DWM) is a common congenital cerebellar malformation in humans. It has been shown that *ZIC1* and *ZIC4* on human chromosome

3q24 are DWM-causative genes, and heterozygous loss of both *Zic1* and *Zic4* in mouse results in a DWM-like phenotype (Grinberg et al. 2004). DWM is characterized by cerebellar hypoplasia and other malformations in the brain. Although the responsible genes of DWM are the same as those of the *Da* mutant, no brain phenotypes are evident in *Da*. This is probably due to the almost intact expression of *zic1* and *zic4* in the brain of *Da*. However, DWM is occasionally associated with spina bifida (Takahashi et al. 1999; Tohyama et al. 2011) which may be related to the abnormal expression of *zic1* and *zic4* in the trunk region.

8.8 Conclusions and Future Perspectives

Studies using medaka revealed that *zic1/zic4* in the dorsal somites regulate the dorsal-specific external morphology of the trunk region. Importantly, establishment of a transgenic medaka line that reproduces the expression patterns of *zic1/zic4* revealed that *zic1/zic4* are exclusively expressed in the dorsal part of somite derivatives with a sharp and straight boundary at the level of the notochord, suggesting that the trunk region of medaka is compartmentalized into dorsal and ventral domains (Fig. 8.7) (Kawanishi et al. 2013). The dorsal domain defined by the

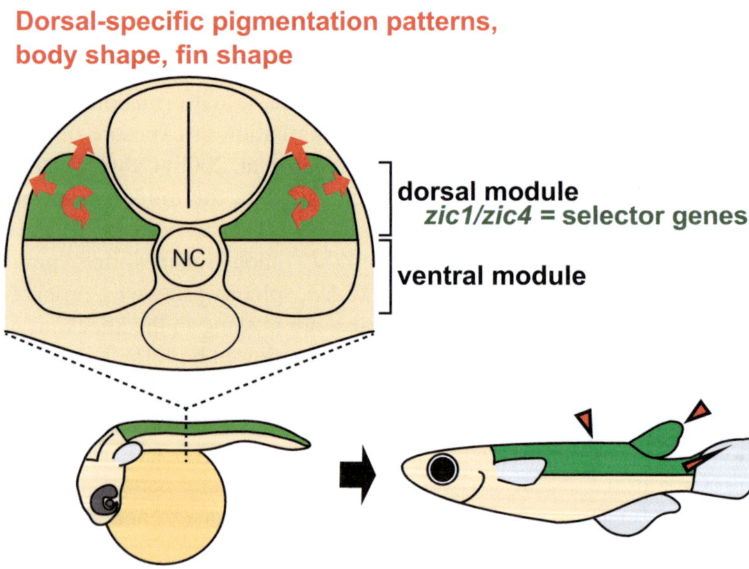

Fig. 8.7 A model for the modular organization of the medaka trunk region. The somite is compartmentalized into dorsal and ventral domains with a sharp and straight ventral expression boundary of *zic1/zic4* at the level of the notochord (NC). The dorsal domain could be a developmental module, and *zic1/zic4* could serve as selector genes, regulating the global morphology of the dorsal trunk such as pigmentation patterns, body shape, and fin shape

zic1/zic4 expression could be a developmental module—a quasi-independent developmental unit—because the loss of *zic1/zic4* activity does not affect the ventral part of the trunk. Since *zic1/zic4* globally determine the fates of various organs on the dorsal side, they serve as selector genes in the dorsal module. This function of *zic1/zic4* should be conserved among teleosts and potentially among vertebrates.

The modular organization of the animal body could promote diversification in morphology during evolution; one module can adopt a novel phenotype without affecting the others (Wagner et al. 2007). In general, vertebrates exhibit a variety of color patterns and structures on the dorsal side, whereas those on the ventral side are relatively conserved. This could be achieved through the modular organization and recruitment of various target genes of the selector gene, *zic1/zic4*, during adaptation to ever-changing environmental conditions (Kawanishi et al. 2013).

The specific functions of *zic1/zic4* in the dorsal module are still largely unknown. Roughly, there could be two functions: regulation of somite-derived tissue morphology and regulation of ectodermal tissue patterning mediated by signaling molecules produced by somite cells. The search for factors downstream of *zic1/zic4* would lead to further understanding of morphogenesis in the vertebrate trunk region and human developmental defects such as spina bifida.

References

Aruga J, Minowa O, Yaginuma H et al (1998) Mouse Zic1 is involved in cerebellar development. J Neurosci 18:284–293

Aruga J, Mizugishi K, Koseki H et al (1999) Zic1 regulates the patterning of vertebral arches in cooperation with Gli3. Mech Dev 89:141–150. https://doi.org/10.1016/S0925-4773(99)00220-8

Aruga J, Tohmonda T, Homma S, Mikoshiba K (2002) Zic1 promotes the expansion of dorsal neural progenitors in spinal cord by inhibiting neuronal differentiation. Dev Biol 244:329–341. https://doi.org/10.1006/dbio.2002.0598

Aruga J, Kamiya A, Takahashi H et al (2006) A wide-range phylogenetic analysis of Zic proteins: implications for correlations between protein structure conservation and body plan complexity. Genomics 87:783–792. https://doi.org/10.1016/j.ygeno.2006.02.011

Cabrera RM, Finnell RH, Zhu H et al (2012) Transcriptional analyses of two mouse models of spina bifida. Birth Defects Res Part A – Clin Mol Teratol 94:782–789. https://doi.org/10.1002/bdra.23081

Freitas R, Zhang G, Cohn MJ (2006) Evidence that mechanisms of fin development evolved in the midline of early vertebrates. Nature 442:1033–1037. https://doi.org/10.1038/nature04984

Gostling NJ, Shimeld SM (2003) Protochordate Zic genes define primitive somite compartments and highlight molecular changes underlying neural crest evolution. Evol Dev 5:136–144. https://doi.org/10.1046/j.1525-142X.2003.03020.x

Grinberg I, Northrup H, Ardinger H et al (2004) Heterozygous deletion of the linked genes ZIC1 and ZIC4 is involved in Dandy-Walker malformation. Nat Genet 36:1053–1055. https://doi.org/10.1038/ng1420

Houtmeyers R, Souopgui J, Tejpar S, Arkell R (2013) The ZIC gene family encodes multi-functional proteins essential for patterning and morphogenesis. Cell Mol Life Sci 70:3791–3811. https://doi.org/10.1007/s00018-013-1285-5

Inoue Y, Saga T, Aikawa T, Kumagai M, Shimada A, Kawaguchi Y, Naruse K, Morishita S, Koga A, Takeda H (2017) Complete fusion of a transposon and herpesvirus created the Teratorn mobile element in medaka fish. Nat Commun 8(1):551. https://doi:10.1038/s41467-017-00527-2

Kawanishi T, Kaneko T, Moriyama Y et al (2013) Modular development of the teleost trunk along the dorsoventral axis and zic1/zic4 as selector genes in the dorsal module. Development 140:1486–1496. https://doi.org/10.1242/dev.088567

Keller MJ, Chitnis AB (2007) Insights into the evolutionary history of the vertebrate zic3 locus from a teleost-specific zic6 gene in the zebrafish, Danio rerio. Dev Genes Evol 217:541–547. https://doi.org/10.1007/s00427-007-0161-4

Kusakabe R, Kuraku S, Kuratani S (2011) Expression and interaction of muscle-related genes in the lamprey imply the evolutionary scenario for vertebrate skeletal muscle, in association with the acquisition of the neck and fins. Dev Biol 350:217–227. https://doi.org/10.1016/j.ydbio.2010.10.029

Lien S, Koop BF, Sandve SR et al (2016) The Atlantic salmon genome provides insights into rediploidization. Nature 533:200–205. https://doi.org/10.1038/nature17164

Mcmahon AR, Merzdorf CS (2010) Expression of the zic1, zic2, zic3 and zic4 genes in early chick embryos

Merzdorf CS (2007) Emerging roles for zic genes in early development. Dev Dyn 236:922–940. https://doi.org/10.1002/dvdy.21098

Moriyama Y, Kawanishi T, Nakamura R, Tsukahara T, Sumiyama K, Suster ML, Kawakami K, Toyoda A, Fujiyama A, Yasuoka Y, Nagao Y, Sawatari E, Shimizu A, Wakamatsu Y, Hibi M, Taira M, Okabe M, Naruse K, Hashimoto H, Shimada A, Takeda H (2012) The Medaka zic1/zic4 mutant provides molecular insights into teleost caudal fin evolution. Curr Biol 22(7):601–607

Nakamura R, Tsukahara T, Qu W et al (2014) Large hypomethylated domains serve as strong repressive machinery for key developmental genes in vertebrates. Dev 141:2568–2580. https://doi.org/10.1242/dev.108548

Ohtsuka M, Kikuchi N, Yokoi H et al (2004) Possible roles of zic1 and zic4, identified within the medaka Double anal fin (Da) locus, in dorsoventral patterning of the trunk-tail region (related to phenotypes of the Da mutant). Mech Dev 121:873–882. https://doi.org/10.1016/j.mod.2004.04.006

Ohtsuka M, Takeda H, Shimada A (2011) In: Naruse K, Tanaka M, Takeda H (eds) Double anal fin (Da): a medaka mutant exhibiting a mirror-image pattern duplication of the dorsal-ventral axis. Medaka, Springer Japan. https://doi.org/10.1007/978-4-431-92691-7_13

Rohr KB, Schulte-Merker S, Tautz D (1999) Zebrafish zic1 expression in brain and somites is affected by BMP and hedgehog signalling. Mech Dev 85:147–159. https://doi.org/10.1016/S0925-4773(99)00044-1

Takahashi S, Sekido K, Ohtsubo Y et al (1999) Dandy-Walker syndrome meningocele and associated spinal with lipoma occipital case report & mdash. Neurol Chir 39:544–547

Tamiya G, Wakamatsu Y, Ozato K (1997) An embryological study of ventralization of dorsal structures in the tail of medaka (Oryzias latipes) Da mutants. Develop Growth Differ 39:531–538

TeSlaa JJ, Keller AN, Nyholm MK, Grinblat Y (2013) Zebrafish Zic2a and Zic2b regulate neural crest and craniofacial development. Dev Biol 380:73–86. https://doi.org/10.1016/j.ydbio.2013.04.033

Thisse, B, Pflumio, S, Fürthauer, M, et al (2001) Expression of the zebrafish genome during embryogenesis (NIH R01 RR15402). ZFIN Direct Data Submission (http://zfin.org)

Tohyama J, Kato M, Kawasaki S et al (2011) Dandy-Walker malformation associated with heterozygous ZIC1 and ZIC4 deletion: report of a new patient. Am J Med Genet Part A 155:130–133. https://doi.org/10.1002/ajmg.a.33652

Toyama R, Gomez DM, Mana MD, Dawid IB (2004) Sequence relationships and expression patterns of zebrafish zic2 and zic5 genes. Gene Expr Patterns 4:345–350. https://doi.org/10.1016/j.modgep.2003.09.011

Wagner GP, Pavlicev M, Cheverud JM (2007) The road to modularity. Nat Rev Genet 8:921–931. https://doi.org/10.1038/nrg2267

Wong RLY, Wlodarczyk BJ, Min KS et al (2008) Mouse Fkbp8 activity is required to inhibit cell death and establish dorso-ventral patterning in the posterior neural tube. Hum Mol Genet 17:587–601. https://doi.org/10.1093/hmg/ddm333

Chapter 9
Zebrafish Zic Genes Mediate Developmental Signaling

Cecilia Lanny Winata and Vladimir Korzh

Abstract The introduction of genomics into the field of developmental biology led to a vast expansion of knowledge about developmental genes and signaling mechanisms they are involved in. Unlike mammals, the zebrafish features seven Zic genes. This provides an interesting insight into Zic gene evolution. In addition, an unprecedented bioimaging capability of semitransparent zebrafish embryos turns to be a crucial factor in medium- to large-scale analysis of the activity of potential regulatory elements. The Zic family of zinc finger proteins plays an important, relatively well-established, role in the regulation of stem cells and neural development and, in particular, during neural fate commitment and determination. At the same time, some Zic genes are expressed in mesodermal lineages, and their deficiency causes a number of developmental defects in axis formation, establishing body symmetry and cardiac morphogenesis. In stem cells, Zic genes are required to maintain pluripotency by binding to the proximal promoters of pluripotency genes (Oct4, Nanog, Sox2, etc.). During embryogenesis, the dynamic nature of Zic transcriptional regulation is manifested by the interaction of these factors with distal enhancers and other regulatory elements associated with the control of gene transcription and, in particular, with the Nodal and Wnt signaling pathways that play a role in establishing basic organization of the vertebrate body. Zic transcription factors may regulate development through acting alone as well as in combination with other transcription factors. This is achieved due to Zic binding to sites adjacent to the binding sites of other transcription factors, including Gli. This probably leads to the formation of multi-transcription factor complexes associated with enhancers.

Keywords Zebrafish · Enhancer · Promoter · Transcription · Stem cells · Gastrulation · Left right asymmetry · Neurogenesis · Developmental signaling

C. L. Winata (✉)
International Institute of Molecular and Cell Biology, Warsaw, Poland

Max-Planck Institute for Heart and Lung Research, Bad Nauheim, Germany
e-mail: cwinata@iimcb.gov.pl

V. Korzh (✉)
International Institute of Molecular and Cell Biology, Warsaw, Poland
e-mail: vkorzh@iimcb.gov.pl

J. Aruga (ed.), *Zic family*, Advances in Experimental Medicine and Biology 1046,
https://doi.org/10.1007/978-981-10-7311-3_9

9.1 Introduction

The patterning of the embryo is achieved through a process involving the determination of embryonic body axes and defining which cell types develop in each embryonic coordinate. The Zic family proteins are known to be involved in such process (Nagai et al. 1997; Aruga 2004). They are characterized by the presence of a C2H2-type Zn finger DNA-binding domain, which is highly similar to that of the Gli family proteins (Aruga et al. 1994). The ability of Zic proteins to activate transcription has been demonstrated through reporter assays (Kuo et al. 1998; Salero et al. 2001; Sakurada et al. 2005), as well as identification of its binding sites near promoters of putative target genes (Lim et al. 2010). Their physical interactions with Gli proteins, as well as their ability to bind Gli consensus motif (Mizugishi et al. 2001), suggested an interaction with the Hh signaling pathway. However, the range of such interactions with signaling mechanisms seems to be much broader. Genome-wide profiling of Zic3-binding sites in zebrafish provided preliminary insight into a rather complex regulatory network, including components of Nodal, Hh, Wnt, and Notch signaling pathways (Winata et al. 2013).

9.2 The Zebrafish Zic Family of Transcription Factors

The Zic family proteins are known for their involvement in multiple aspects of embryonic patterning (Aruga 2004; Ware et al. 2006; Merzdorf 2007). Their study dates back more than 20 years ago, when the first gene in the family, murine *Zic1*, expressed abundantly in the granule cells of the cerebellum, was cloned (Aruga et al. 1994). Comparisons of DNA sequences and gene structures of the Zic genes revealed their homology to the *odd-paired* gene of *Drosophila* known to specify the anterior-posterior identity of the embryonic body segments (Benedyk et al. 1994). Additional vertebrate *Zic* genes were subsequently identified and characterized (Nakata et al. 2000; Ishiguro et al. 2004) making a total of five in frog, chicken, and mammals. Two additional *zic* genes are present in zebrafish. *zic2b* arose from the teleost-specific gene duplication (Toyama et al. 2004), and *zic6* (Parinov et al. 2004; Keller and Chitnis 2007) represents a molecular evidence of existence in the evolution of the third pair of Zic genes (Zic3-6) similar to that of the Zic1-4 and Zic2-5 pairs. To date, no evidence exists of the presence of Zic2b in tetrapods, latimeria, and sharks, which suggests that it never evolved outside of the teleost lineage; on the other hand Zic6 remains only in teleosts. Given the fact that this book contains several reviews on Zic genes in other species, we will limit our analysis to the zebrafish Zic genes.

In all vertebrates *Zic* genes are located on opposite DNA strands as head-to-head pairs. For instance, the *zic1*-*zic4* pair is located in this configuration on zebrafish chromosome 24, the *zic2*-*zic5* pair on zebrafish chromosome 3, and the *zic3* paired with *zic6* on zebrafish chromosome 14 (Parinov et al. 2004) (Fig. 9.1). Such close

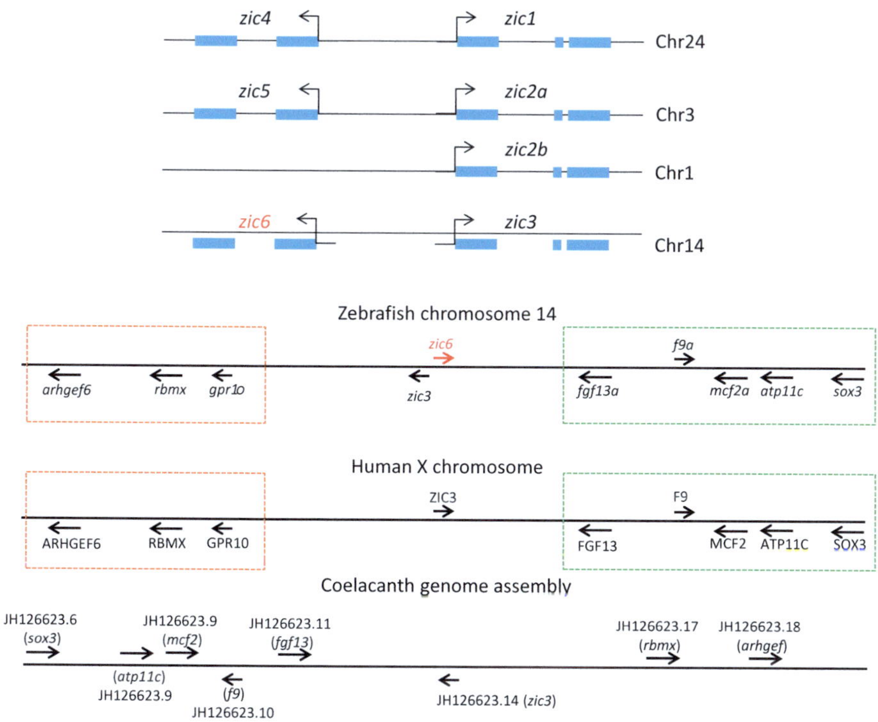

Fig. 9.1 The pairwise arrangement of *zic* genes in the zebrafish genome. The zebrafish genome contains seven *zic* genes, an additional two from that of higher vertebrates. These additions consist of *zic2b* located in chromosome 1 and *zic6* located in pair with *zic3* on chromosome 14. These two *zic* genes arose through chromosomal duplication in *Euteleostomi*, possibly excluding the *Coelacanthiformes*, and were subsequently lost in higher vertebrates. Nevertheless, the fragment of chromosome 14 containing *zic3* retains the syntenic relationship with that of the human X chromosome and the ancient *Coelacanth* (*Latimeria*) genome

proximity of pairs of Zic genes has been proposed to facilitate the sharing of regulatory regions. This idea is supported by the similarities in spatiotemporal expression patterns and somewhat overlapping functions between pairs of *zic* genes (Inoue et al. 2004; Ohtsuka et al. 2004; Merzdorf 2007; Nyholm et al. 2007). Members of the Zic family share many features of the characteristic expression pattern. In general, *zic* genes are expressed in the neural plate and later on in the dorsal neural tube. The role of *Zic* genes during neural development is conserved in all organisms that possess the nervous system (Lindgens et al. 2004), suggesting that these genes play an important role in the development and evolution of the nervous system. In particular, members of the Zic family positively regulate the proliferation of neural progenitors and negatively regulate the expression of proneural factors driving neural differentiation (Aruga 2004; Toyama et al. 2004; Nyholm et al. 2007). Based on the expression of a transgene in the ET33 transgenics that faithfully recapitulate the expression of *zic3-zic6* pair, these genes may also play a role in the development of the brain meninges (Parinov et al. 2004; Kondrychyn et al. 2013; Chap. 12 of this book). Some of Zic genes are expressed in the mesoderm/mesendoderm, where they play an important role in the development of the body axis, etc.

Comparative analysis across different metazoan phyla revealed that *zic* genes probably evolved from an ancestral gene of the *gli/glis/nk*-like family that existed in the last common ancestor of the placozoans, cnidarians, and bilaterians. In these basal metazoans, *zic* genes are expressed in the endomesodermal tissues and highly neuralized developing tentacles, indicating that their function has likely been conserved since the early stages of metazoan evolution (Wada and Saiga 2002; Layden et al. 2010). Hence, a detailed study of regulation even of a single Zic gene could be rather informative in the context of regulation of all Zic genes.

9.3 Zic3 as a Transcription Factor

Application of next-generation sequencing (NGS) allows an unbiased genome-wide assessment of transcription factor (TF)-binding sites by ChIP-seq (Lim et al. 2010; Winata et al. 2013). This approach was applied to study the transcriptional activity of Zic3 in the developing zebrafish embryo at two different developmental stages. At 8 hpf Zic3 is expressed in the neural plate and mesoderm upon formation of germ layers (ectoderm, mesoderm, endoderm) and commencement of neural induction (Parinov et al. 2004). At 24 hpf Zic3 is expressed during neural differentiation in the dorsal neural tube (Winata et al. 2013). This analysis revealed that only a relatively small fraction of Zic3-binding events (8–9%) were associated with promoters. Most of these events were mapped to distant genomic locations. This is in line with the idea that Zic3, similar to other TFs, regulates gene activity through long-distance regulatory elements (Carroll et al. 2005; Wederell et al. 2008; Sanyal et al. 2012). Hence, the results of these studies led to the formulation of novel hypotheses regarding Zic3 function.

First, a difference in localization of Zic3-binding sites in stem cells and during embryogenesis possibly reflects changes in the role of this TF during different developmental periods. In stem cells that are in a relatively stable pluripotent state, Zic3 often acts as a general TF that binds to the core transcription machinery (Lim et al. 2010). This seems to be a common feature among TFs known to regulate ES cell pluripotency in mouse, such as Oct4, Stat3, and Klf4, all of which often bind sites within promoter regions (Chen et al. 2008). In contrast, during later development, when cells begin to differentiate, Zic3 binding to distal elements prevails. Such shift in site specificity of Zic3 suggests an acquisition of cellular functions specific for differentiating cells. A precise mechanism of this phenomenon remains unknown. At chromatin level it could be due to a decrease in the availability of binding sites in promoters or increase in the availability of distant binding sites. Both explanations suggest major epigenetic changes taking place during transition from a period of extensive cell proliferation to a period of cell fate determination and differentiation. Epigenetic rearrangements in the form of changes in genome-wide histone methylation pattern on gene promoters have been well documented during the midblastula transition (Lindeman et al. 2010, 2011; Vastenhouw et al. 2010) and could thus support a model of TF-binding site accessibility. Equally important are changes at transcriptome level, which in principle could be both a cause and outcome of transcriptional regulation. A shift in Zic3 site specificity also correlates with a replacement of maternal transcripts by zygotic ones (Giraldez et al. 2006). Future studies could focus on investigating the relationship between these two events.

Second, a consensus-binding motif of Zic3 in zebrafish is highly similar to that found in mouse ES cells (Lim et al. 2010; Hong et al. unpublished). In sharp contrast, most of the surrounding regions appear to be poorly conserved in evolution. It is well documented that the evolution of divergent traits mostly involves modifications of regulatory elements rather than structural or functional changes in effector molecules, as the latter may impose dramatic changes in the gene regulatory network (GRN) controlling development (Carroll 2008; Wittkopp and Kalay 2012). In accordance with this idea, the binding sites of Zic3 diverge greatly, while their core structure and, possibly, their binding specificity remain largely conserved across metazoans (Layden et al. 2010; Winata et al. 2013).

Lastly, a large group of Zic3-binding sites fails to induce an expression of reporter. These sites could be nonfunctional or perform functions differently compared to regulatory elements that stimulate transcription. Analysis of such sites requires experimental output other than an increase in transient expression of reporter during embryogenesis in vivo used in this study. Possibly such sites could become functional at postembryonic stages or during adulthood. Zic3 may require interacting partners to induce transcription at these sites. This possibility is especially attractive since binding motifs of other TFs often are identified in proximity to Zic3 motifs of TF-binding sites, and enhancer studies have demonstrated that multiple TF-binding sites tend to co-localize with enhancers (Winata et al. 2013). Some of these enhancers form large regulatory regions up to 50 kb previously termed as "super enhancers" (Whyte et al. 2013). Co-binding of a particular TF with different partners has been shown to cause transcriptional outcome distinct from the one brought about by a single TF. The presence of other TF-binding sites nearby Zic3 peaks therefore suggests that Zic3 may act in multi-TF complexes. Among possible candidates for Zic3-binding partners are Gli proteins. These effectors of Hh signaling are structurally similar to Zic (Aruga et al. 1994). Gli-Zic physical interactions as well as Zic ability to bind Gli consensus motif (Mizugishi et al. 2001) suggested an interaction with the Hh signaling pathway. This is further supported by the fact that a deficiency of Zic2 has been linked to holoprosencephaly, the phenotype also connected to defects in Hh signaling (Brown et al. 1998; Sanek and Grinblat 2008). Finally, genome-wide analysis of Zic3-binding sites showed that almost half of all Zic3-binding sites contain both Zic3 and Gli motifs (Winata et al. 2013). This provided additional support for Zic-Gli interaction in the regulation of gene activity. Interestingly, the Hh signaling pathway is activated as a result of zygotic transcription, i.e., after a shift toward Zic3 regulation of enhancers. The

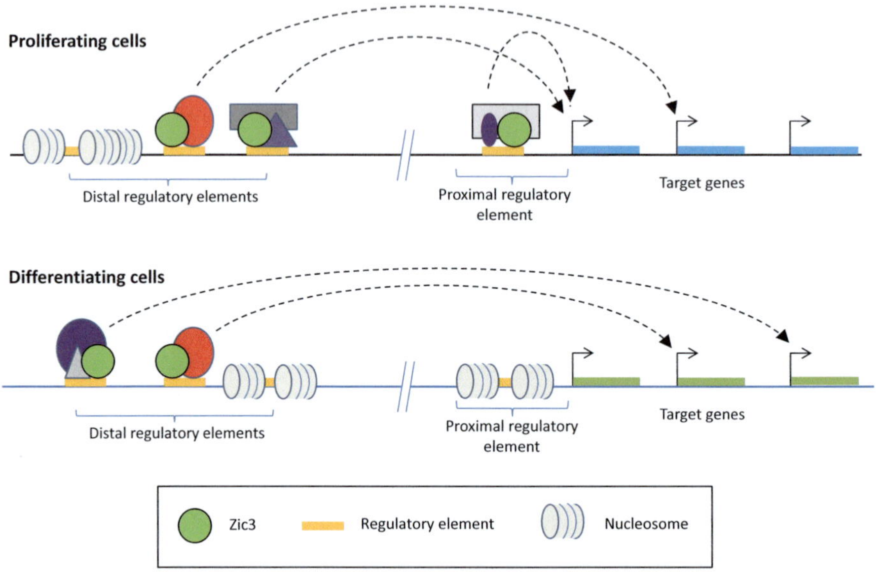

Fig. 9.2 Proposed model of Zic3 regulatory mechanism. In the developing embryo, Zic3 binds mainly to distal enhancer elements to regulate tissue-specific expression of target genes. The binding to different enhancer elements is regulated spatiotemporally through epigenetic mechanisms or recruitment by different transcription factors as binding partners

same could be true regarding other conserved binding sites detected in proximity of Zic3 motifs (Winata et al. 2013), which may become functional later on. At least for now, without detailed study of these potential interacting partners, it is difficult to determine the exact nature of their interaction with Zic3. Given the developmental shift from promoter-driven transcriptional regulation by Zic3 to enhancer-driven regulation and possible interaction with some other TFs, a mechanism involving Zic3-mediated transcriptional regulation in different spatiotemporal contexts could be illustrated as in Fig. 9.2.

9.4 Zic3 and Global Regulation of Development

The role of Zic3 in multiple, disparate aspects of development reflects its "mosaic pleiotropism" (Hadorn 1956; Carroll 2008). This property is exemplified by its involvement in the patterning of at least two different germ layers (ectoderm and mesoderm) and its role in activating different pathways at different developmental stages. The ability of TFs to perform multiple functions in different spatiotemporal contexts could be achieved through interactions with different partners which confer spatiotemporal specificity of its function (Carroll 2008). In the case of Zic3, the presence of this mechanism is supported by the co-localization of binding sites of different TFs and Zic3, as well as evidence of possible physical interactions between Zic3 and Gli proteins. Within the wider context, comparative studies of metazoan evolution showed that the conserved Zic protein is repeatedly utilized in developmental processes. This is compatible with an idea of evolutionary "bricolage" (Wilkins 2007), which manifests itself as usage of an existing set of molecules during the evolution of new GRNs that eventually acquire novel developmental functions (Carroll 2008). The novel features generated from re-usage of a conserved TF often result in changes in the sequences of cis-regulatory elements (CREs) in the form of addition or deletion of a TF-binding site or modification of TF affinity or the strength of regulatory effects through changes in the number of binding sites (Carroll 2008). The case of Zic3 illustrates this principle – a majority of Zic3-binding sites are surrounded by poorly conserved regions, which may suggest distinct compositions of multiprotein complexes binding to the target CREs, resulting in evolutionary diversification. It is possible that an additional round of genome duplication in teleosts further contributed into relaxing selection pressure on CREs as it led to an even greater diversification of regulatory elements. This could be seen not only due to the genome-wide shift in a mode of Zic3 binding. It also correlates with a shift in Zic3 functionality, which is evident due to a difference in GO enrichment of associated genes during development. Importantly, the recognition motif of Zic3 involved in two developmental GRNs (8 and 24 hpf) remains the same, which highlights the importance of its pleiotropism.

In this context, it is worthwhile to mention *embryonic competence*, an actively acquired ability of developing cells to respond to an inductive signal. The developmental regulation of accessibility of regulatory regions, i.e., enhancers and promoters, in cells that respond to developmental signaling may be determined by the epigenetic state of chromatin, which reveals itself via embryonic competence. An analysis of developmental regulation of genetic activity by Zic3 revealed an important genome-wide switch from the regulation of the promoter-driven cellular functions during pluripotency state to the enhancer-driven regulation of functions associated with progressing development – cell migration, commitment and determination during gastrulation, as well as neural cell differentiation (Lim et al. 2010; Winata et al. 2013) (Fig. 9.3). Given the role of *Zic* genes in brain tumors (Aruga et al. 2010), it is easy to imagine that under pathological conditions such as dedifferentiation, a reversal from enhancer-driven regulation to promoter-driven general cellular activities such as cell proliferation may take place. When supported by experimental evidence, this emerging knowledge may help to formulate a novel paradigm of searching druggable targets.

9.5 Zic3 in Gastrulation and Left-Right (L-R) Patterning

Zic3 is distinguished from other Zic family members by its involvement in the L-R patterning (Aruga 2004). It was shown that Zic3 loss-of-function (LOF) causes laterality defects in *Xenopus* and zebrafish in support of the conserved role of Zic3 in

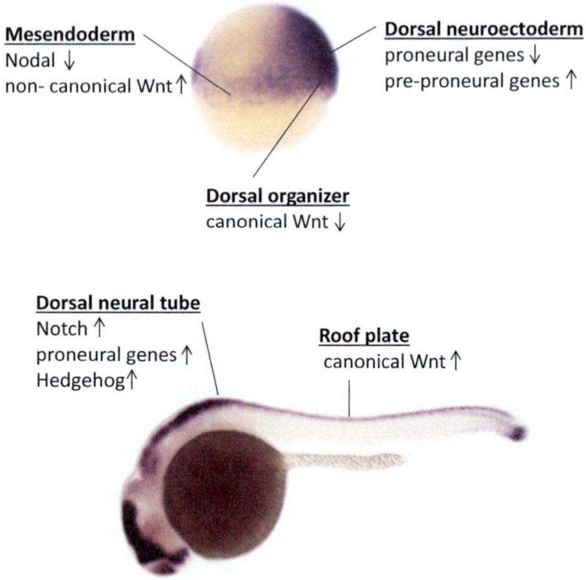

Fig. 9.3 Signaling pathways downstream of Zic3 in different spatiotemporal contexts in zebrafish development. Expression of *zic3* in 8 hpf and 24 hpf zebrafish embryo. Different signaling pathways within specific expression domains are subjected to regulation by Zic3. Spatiotemporal regulation of multiple developmental processes by Zic3 (right panel) is achieved by regulating distinct sets of target genes through binding different sets of regulatory elements at different developmental stages

regulating the L-R specification in vertebrates (Cast et al. 2012). Despite its role as a determinant of the L-R asymmetry, *Zic3* is not expressed unilaterally like some other determinants of this process (Nakata et al. 1997; Kitaguchi et al. 2000; Grinblat and Sive 2001). Moreover, organs, in which laterality is affected by Zic3 LOF, do not express *Zic3*. This raises a question as to how Zic3 confers the L-R patterning. Some studies suggested that an action of Zic3 in L-R asymmetry is an early developmental event, in which Zic3 regulates the formation of the dorsal organizer and therefore the midline structures (Fujimi et al. 2012), through its suppression of the canonical Wnt signaling (Winata et al. 2013). Defects of the midline structures are associated with aberrations in the L-R patterning (Danos and Yost 1996; Bisgrove et al. 2000). Furthermore, it was shown that Zic3 LOF defects in convergence extension (CE) correlate with subsequent defects in the L-R patterning (Cast et al. 2012). Therefore, Zic3 participates in establishing the L-R asymmetry by means of binary involvement as a regulator of the canonical Wnt signaling in the embryonic shield (organizer) and the noncanonical Wnt signaling during CE movements. ChIP-seq study in zebrafish suggested that this is indeed the case (Winata et al. 2013). Zic3 downstream targets include genes acting in the Nodal and canonical Wnt pathways that regulate early midline development. Zic3 also regulates genes directly implicated in the L-R patterning including members of the noncanonical Wnt (or planar cell polarity) signaling pathway, such as *dvl2*, *invs*, and *vangl2*, known to regulate ciliogenesis in zebrafish Kupffer's vesicle, which acts as an "organizer" (Morgan et al. 1998; Okada et al. 2005; Hashimoto et al. 2010; Wang et al. 2011). Of note is a differential tissue-specific activity of Zic3 in respect of canonical Wnt signaling, where, in the dorsal mesoderm, Zic3 inhibits Wnt signaling and, in the neuroectoderm, it activates the same branch of this signaling pathway. This example serves as an illustration of the multiplicity of Zic3 developmental functions (Winata et al. 2013).

9.6 Zic Genes in Neural Development

Zic genes are some of the earliest TFs expressed in the neuroectoderm, where their expression precedes that of the proneural genes or starts at the same time as the expression of proneural genes (Nakata et al. 1997, 1998; Gamse and Sive 2001). In the zebrafish embryo, the earliest expression of Zic genes is represented by genes expressed during gastrula in the mesoderm – *zic2b* and *zic3* (Grinblat and Sive 2001; Toyama et al. 2004). Next, a domain of *zic3* expression appears in the posterior ectoderm. The expression of *zic1-zic4* is initiated at around mid-gastrula in the anterior neural plate (Elsen et al. 2008). By this time, *zic3* expression is higher in the posterior dorsal neuroectoderm in contrast to *zic1* and *zic2*, whose expression is higher anteriorly (Grinblat and Sive 2001). Later on, several *zic* genes are expressed in highly overlapping domains along the dorsal part of the neural tube and mesoderm – *zic1* is expressed throughout the brain, optic stalk, and dorsal somites (Rohr et al. 1999); *zic2* in the diencephalon, cerebellum, and optic stalk and weakly in the tail bud; *zic3* in the diencephalon, tectum, and optic stalk and strongly in the tail bud; and *zic5* and *zic6* at the beginning of neurulation (Toyama et al. 2004; Keller and Chitnis 2007; Kondrychyn et al. 2013). Hence, based upon expression patterns, the Zic genes of zebrafish seem to fall into at least three categories. The first one consists of genes that have only neural expression (*zic4*, *zic6*). The second is represented by their paired counterparts (*zic1*, *zic3*) expressed not only in neural tissues but also in the mesoderm and/or mesendoderm. In the third category, the Zic2a-Zic5 genes share expression in neural tissues, neural crest cells (NCC), and the apical ectodermal ridge (AER). Based on expression pattern, *zic2b* belongs to the same group of genes. Hence, it is possible to predict that the hypothetical counterpart of *zic2b* – *zic5b* – probably had similar expression, which could be one of the reasons for its elimination in evolution. The second prediction is based on a fact that since one of the genes with neural expression only (*zic6*) was lost in evolution, at some point another Zic gene with neural expression only (*zic4*) may follow the same fate.

Zic genes are negatively regulated by the BMP signaling and positively regulated by signaling emanating from the dorsal mesoderm. An inhibition of the ventralizing BMP activity induces expression of *zic1-3* in the dorsal neuroectoderm. This marks the earliest event in the determination of neural fate (Nakata et al. 1997; Grinblat and Sive 2001; Marchal et al. 2009). In the Bmp2b mutant *swirl*, *zic1* expression during late gastrulation expands significantly. In addition, expression in somites expands ventrally to encompass the whole somite. The opposite effect was observed upon Bmp gain-of-function (GOF) achieved by injection of *bmp2b* mRNA or upon decrease of dorsal mesoderm signaling in *chordino* mutant; both situations cause significant reduction of *zic1* expression (Rohr et al. 1999). Similarly, *zic3* expression in the neural plate of dorsalized *swirl* (*bmp7*) and *snakehead* (*atp1a1a.1*) mutants expands into the ventral ectoderm. This supports the idea that BMP activity is necessary for restriction of *zic3* expression. Correspondingly, in the ventralized *chordino* mutant, *zic3* expression in the neuroectoderm was abolished (Grinblat and Sive 2001). The genome-wide analysis of Zic3-binding sites using ChIP-seq combined with gene expression profiling of Zic3 LOF demonstrated that Zic3 positively regulates genes essential to maintain neural progenitors during neuroectodermal specification (Winata et al. 2013). Some of these are direct targets of Zic3 (*dlx4b* and *msxe*), whereas others could be regulated indirectly (*msxc*, *irx1a*, *irx7*). Targets of Zic3 include also several Her genes implicated in the Notch signaling (*her4.2*, *her6*, *her9*). These genes are expressed in the neural plate, in which the marginal zone contains proliferating progenitors that contribute into dorsal neural tube and neural crest derivatives (Ekker et al. 1997; Thisse et al. 2001; Woda et al. 2003; Lecaudey et al. 2004; Thisse et al. 2004; Lecaudey et al. 2005). The identification of neural pre-pattern genes as downstream targets of Zic3, along with the repressive action of Zic3 on proneural genes (Winata et al. 2013), suggests that Zic3 acts to maintain a certain level of proliferation of neural progenitors. Thus, Zic3 establishes a particular number of neurons. It looks like Zic3 acts by maintaining an undifferentiated state of neural progenitors by positive regulation of repressors of the neural fate and, possibly, negatively regulating proneural genes. In support of this interpretation, Zic3 LOF caused an increase of neural differentiation markers, such as *neurog1* and *her9*. This indicates the repressive action of Zic3 on neural differentiation. Interestingly, binding sites of Zic3 were also found within 100 kb of *neurog1*, *neurod4*, and *ncam1a* promoters, which suggests that genes involved in neural differentiation could be direct targets of Zic3 also (Winata et al. 2013). Such mode of action is consistent with a model according to which Zic3 acts in parallel to Notch (Fig. 9.3).

9.6.1 Zic and Neural Crest

It has been demonstrated that several Zic proteins are expressed in and promote differentiation of the neural crest cells (NCC) (Teslaa et al. 2013). These cells originate from precursors located during gastrulation at the lateral edge of neural plate. As a result of neurulation, they converge at the dorsal neural tube together with precursors of the roof plate (see below). Subsequently, NCC undergo epithelial-to-mesenchymal transition (EMT), delaminate from the neural tube, and migrate out to differentiate into various cell types (Eisen and Weston 1993). Precursors of the NCC are specified through the combined action of Wnt, Fgf, and Bmp signaling. Wnt and Fgf signals along with Bmp antagonists are generated by the neural plate, while Bmp signals are generated by the adjacent epidermal ectoderm (Bang et al. 1999; Garcia-Castro et al. 2002). Taken together with the signaling activity of the dorsal mesoderm (i.e., Bmp antagonists), these signaling pathways combined result in a dorso-ventral gradient of Bmp which increases ventrally and induces the expression of *zic* and *pax3/7* genes sufficient to induce NCC fate (Sato et al. 2005). In the zebrafish it was shown that *zic3* expression in the lateral neural plate depends upon Wnt signaling and an intermediate level of Bmp signaling. These signaling pathways seem to operate by regulating the two evolutionarily conserved enhancers located upstream and downstream of the *zic3-zic6* locus (Garnett et al. 2012). *Zic* genes are also involved in the subsequent migration and differentiation of the NCC. In zebrafish, *foxd3* and *pax3a* involved in NCC induction and migration (Bang et al. 1999; Stewart et al. 2006) were downregulated upon *zic3* knockdown. These genes are downstream targets of Zic3 (Winata et al. 2013), which acts as a positive transcriptional regulator. Although this result derives from analysis at 24 hpf, i.e., much later than when the NCC specification and migration from the dorsal neural tube take place (Raible et al. 1992; Kimmel et al. 1995), one needs to consider that *zic3* is constantly expressed in the NCC starting from gastrula. Its role in NCC migration can therefore be extrapolated based on this evidence.

9.6.2 Zic and Neurulation

Neurulation is a process in which the neural plate folds to form the neural tube. It involves apical constriction driven by the actin-based cytoskeletal microfilament network, which localizes at the subapical surface of the neuroepithelium to drive the folding movement (van Straaten et al. 2002; Sawyer et al. 2010; Korzh 2014). As a result of this movement, the neural folds are bent at a point known as the dorso-lateral hinge point (DLHP). This structure is crucial during neurulation as it determines the formation of the neural tube lumen and its shape and, eventually, the proper closure of the neural tube. It is also a landmark dividing the two lateral plates of the neural tube – the basal and alar plates. At the very beginning of neurulation, *zic2a* is expressed at the lateral edges of the neural plate giving rise to the future dorsal neural tube. As the neural plate folds to form the neural tube, *zic2a* and *zic5* expression continues to be dorsally restricted. Even prior to cavitation of the neural tube, the ventral border of the *zic2a* and *zic5* expression domain determines the position of the DLHP at the border of the dorsal alar plate and the ventral basal plate (Nyholm et al. 2007). The same two genes, Zic2a and Zic5, are required for the apical localization of F-actin and myosin II in the dorsal midbrain, particularly along the future luminal surface. The enrichment of myosin II, in turn, further ensures the structural integrity of the neuroepithelium. Interestingly, the canonical Wnt signaling is similarly required for the apical myosin II accumulation, further enhancing a link between Zic2a-Zic5 and Wnt signaling in the regulation of this process.

9.6.3 Zic and Roof Plate

Upon migration of the NCC out of a neural tube, the roof plate cells move dorsally to become the most dorsal cell lineage of the neural tube (Krispin et al. 2010a, b). Zic3 negatively regulates several proneural bHLH genes (*neurog1*, *neurod4*, and *her9*) (Winata et al. 2013). Probably this prevents differentiation of these cells as neurons and lets the roof plate cells to maintain their properties as signaling glia. In the zebrafish, Zic1, Zic3, and Zic4 control the expression of roof plate determinant *lmx1b* (Chizhikov and Millen 2004a, b; Elsen et al. 2008; Winata et al. 2013). In addition, Zic3-related Zic6 has been implicated in the regulation of cell adhesion in the dorsal neural tube during elongation of the roof plate. It is accepted that cell specification in the dorsal spinal cord depends mostly on Gli3-independent Wnt signaling. Hence, it comes as no surprise that several genes of the Wnt signaling pathway expressed in the dorsal neural tube are targets of Zic3 (Grinblat and Sive 2001; Fujimi et al. 2006; Winata et al. 2013). This in turn illustrates that at least Zic3 and, given Zic redundancy, other members of this family could act as dorsal developmental regulators alternative to Hh-independent Gli3. The developmental regulation mediated by Zic may play a role in a major morphogenetic rearrangement that prospective roof plate cells undergo between 24 hpf and 36 hpf. Being initially polarized along the medial-lateral axis, these cells rearrange polarity along the dorso-ventral axis, and a deficiency in the Zic genes affects this process (Kondrychyn et al. 2013). *lgl2* and *dlg2* are Zic3 targets expressed in the roof plate, where they regulate cell polarity at the level of cell adhesion (Bilder et al. 2003; Harris and Peifer 2004; Sonawane et al. 2005; Sabherwal et al. 2009; Sonawane et al. 2009; Winata et al. 2013). Hence, it is possible that Zic3 regulation of *lgl2* and *dlg2* plays an essential part in changes in cell adhesion necessary for reorientation of the prospective roof plate cells during stretching morphogenesis (Sonawane et al. 2009; Kondrychyn et al. 2013).

Subsequent stages of dorso-ventral patterning of the neural tube involve both Gli3-dependent and Gli3-independent mechanisms that mediate Wnt action in the intermediate and ventral neural tube. In the ventral neural tube, Wnts expressed in the floor plate contribute into the development of motor neurons (Ungar et al. 1995; Liu et al. 2000; Agalliu et al. 2009). The mechanisms by which the dorsal Wnts pattern the neural tube in a Gli3-dependent manner lack a few important details. It was proposed that Wnts acting in parallel with Bmps directly control the expression of homeodomain and basic helix-loop-helix (bHLH) TFs (Ulloa and Marti 2010). But in the absence of a mechanism for delivery of Wnts expressed dorsally into the ventral neural tube, this model remains incomplete. This is of particular importance since, unlike some other morphogens, the hydrophobic Wnts do not diffuse efficiently and act only at a short distance from Wnt-producing cells (Logan and Nusse 2004). In *Drosophila* the long-distance transport of the Wnt-related Wg is achieved by specialized cell extensions (cytonemes) (Roy et al. 2011) or transcytosis (Strigini and Cohen 2000; Marois et al. 2006; Gallet et al. 2008). Morphogens are known to be secreted by highly polarized cells such as the roof plate and floor plate. As a matter of fact, the end feet of roof plate cells are tightly aligned with *glyt1*-positive stemlike cells adjacent to the primitive lumen/central canal prior to, during, and after stretching morphogenesis of the roof plate resulting in them spanning at least two-thirds of the diameter of the neural tube from cell bodies located in the dorsal neural tube. Such elongation of the roof plate may allow a long-distance transport of dorsal Wnts (Kondrychyn et al. 2013). Furthermore, it has been shown that the secreted Frizzled-related proteins enhance the diffusion of Wnt ligands to expand their signaling range (Mii and Taira 2009). Since Zic3 negatively regulates *sfrp1a* in the roof plate (Winata et al. 2013), this could be a mechanism to restrict a spread of Wnt signaling to a vicinity of a small apical footprint that the roof plate cells maintain at the dorsal aspect of the primitive lumen/central canal. Zic3 also negatively regulates the expression of ventral *wnt4b* [40], which could be a mechanism to restrict the effect of this ventral Wnt to the ventral-most aspect of the neural tube in close vicinity to the floor plate, where *wnt4b* is expressed (Liu et al. 2000). To summarize this part, it seems that the long-distance Wnt signaling could be regulated by the Zic3-based inbuilt transcriptionally regulated molecular system controlling Wnt signaling at multiple levels. It stimulates canonical Wnt signaling (by promoting *axin1*, *dvl2*, *lef1*) over noncanonical Wnt signaling (by blocking *vangl2* expression) as well as regulates expression of the Wnt receptors (Fzd8) and soluble Wnt-binding modulators (*sfrp1*) (Winata et al. 2013). In more general context, Zic expression is itself a subject of Wnt signaling regulation. Hence, Zic genes emerge as an important positive regulatory node enhancing Wnt signaling. Given a well-known role of Wnts as oncogenes and an activation of Zic expression in brain tumors (Chap. 16 of this book), the regulation of Zic genes and their targets in tumors should be explored further in search for anticancer therapy.

9.6.4 Zic and Midbrain

The regulation of Zic2a and Zic5 by the canonical Wnt signaling is well documented in the developing dorsal midbrain, which gives rise to the optic tectum. These two Zic genes are located in tandem with their transcription start sites separated by 4.6 kb of intergenic sequences. This suggests that they likely share regulatory elements. The expression of *zic5* does not initiate as early as that of *zic2a*. However, starting from late gastrula, the two genes are co-expressed at the future neural tube and retain very similar expression patterns in the dorsal brain. This sets another example, where a pair of Zic genes located in tandem could be transcriptionally co-regulated to some extent by the same *cis*- and *trans*-regulatory elements. Further analysis reveals that the *zic2a* and *zic5* locus contains transcriptional enhancers with the ability to drive expression in the midbrain. These enhancers are recognized by the canonical Wnt effectors, Tcf/Lef transcription factors, which activate the expression of *zic2a* and *zic5*. Although lacking any patterning defects, Zic2a and Zic5 knockdown resulted in a reduction of cell proliferation in the tectum likely through the activation of *cyclinB1* transcription. Therefore, the canonical Wnt pathway activates *zic2a* and *zic5* expression during dorsal midbrain development to promote cell proliferation (Nyholm et al. 2007).

9.6.5 Zic and Hindbrain

In the hindbrain, a similar regulatory module operates downstream of Zic2a and Zic2b to regulate the patterning of the hindbrain motor neurons (Drummond et al. 2013). Here Zic2a and Zic2b act upstream of two enzymes acting in opposite manner. *aldh1a2* catalyzes the production of RA and activates the retinoic acid (RA) signaling (Niederreither et al. 2000). *cyp26a1* catalyzes the degradation of RA and blocks the RA signaling. Unlike *zic1*, two other Zic genes (*zic2a* and *zic2b*) are expressed early, starting from 5 hpf in the prospective anterior neural ectoderm. By 8 hpf, *zic1* expression became restricted to the anterior, while *zic2a* and *zic2b* exhibit a broader and more posterior expression pattern overlapping that of both *aldh1a2* and *cyp26a1*. Intriguingly, depletion of both Zic2 paralogs results in a reduction of the expression level and size of expression domains of both *aldh1a2* and *cyp26a1*. This reduction in expression occurs as early as at 7 hpf stage, i.e., during neuroectodermal specification. Nevertheless the overall effect is a reduction in RA signaling within the posterior hindbrain and spinal cord, which persists up to 26 hpf. Ultimately this results in the loss of vagal motor neurons. Interestingly, *zic3* is also expressed in a domain largely overlapping with that of *zic2a* and *zic2b*. However, in one set of Zic3 knockdown experiments, no obvious phenotype was noted, and authors did not pursue their analysis further (Drummond et al. 2013). This is seemingly at odds with our observation, according to which Zic3 morpholino caused prominent convergence extension and gastrulation defects (Winata et al. 2013). Moreover, Zic3-binding peaks were detected at 24 hpf upstream (−14 kb and −52 kb) of the *aldh1a2* locus, as well as upstream (−73 kb) of the locus encoding a *cyp26a1* paralog, *cyp26b1*. Hence, a combination of these two observations makes it likely that Zic3 may perform a role overlapping with those defined for Zic2a and Zic2b in hindbrain patterning. To resolve a difference in observations, this notion awaits more detailed analyses.

9.6.6 Zic and Disease

Dandy-Walker Syndrome (DWS) correlates with the reduction and displacement of the cerebellum and hydrocephalus of the fourth ventricle. Humans affected may have defects of motor development, hypotonia, and ataxia as well as mental retardation (http://omim.org/entry/220200; Chap. 13 of this book). This genetic defect was mapped to the Zic1-Zic4 pair of genes, and an analysis in animal models, including the zebrafish (Elsen et al. 2008), supported an idea that DWS could be due to deficiency of this pair of genes. The expression patterns of *zic1* and *zic4* largely overlap, with *zic1* expression initiated at mid-gastrula and that of *zic4* initiated at late gastrula in the anterior neuroectoderm and at the lateral edges of the future neural plate. The expression of these two genes subsequently extends posteriorly to encompass the dorsal-most portion of the neural tube, including the roof plate at the level of the midbrain and hindbrain. Loss of Zic1-Zic4 function results in an improper opening of the hindbrain ventricle. As a result, the ventricle is fused from the level of rhombomere 2 toward the posterior. In addition, proliferation of progenitor cells of the dorsal hindbrain is reduced. The roof plate is disrupted or absent. This is particularly clear in the regions of fused ventricle. Zic1 and Zic4 positively regulate expression of genes known to play a role in the development of the roof plate – *lmx1b-1* and *lmx1b-2* and their downstream targets (*gdf6a*). Zic1-Zic4 deficiency also affects the dorsal aspect of *wnt1* expression at the rhombomere boundaries, but not its more dorsal expression, or that at the ventral aspect of rhombomere boundaries. Therefore, Zic1-Zic4 contribute in the development of some aspects of the roof plate and the dorsal aspect of the rhombomere boundaries. These processes seem to have some bearing on the canonical Wnt signaling. Importantly, disturbances of these processes are significant enough to cause defects in the development of the dorsal neural tube resulting in hydrocephalus.

In the hindbrain ventricular zone, Zic5 is involved in neurogenesis by promoting progenitor maintenance and proliferation and preventing commitment and cell-cycle exit resulting in the birth of a neuron. *zic5* expression is negatively controlled by miR-9, which promotes cell-cycle exit. This poises neural progenitors to be responsive to alternative signals, which promote commitment toward neural fate (Coolen et al. 2012). It remains unclear whether the expression of other zebrafish Zic genes is also controlled by miRNA, but given many similarities in Zic gene biology, this possibility may be not that far-fetched. This idea is supported by the fact that miR-564 negatively regulates the expression of Zic3 in human lung cancer (Yang et al. 2015). So it seems that Zic genes play various, albeit to some extent redundant, roles in neurulation and, in particular, the development of the dorsal neural tube.

Holoprosencephaly (HPE) is one of the prominent defects associated with the impairment of Zic function (Chap. 14 of this book). HPE is an outcome of the deficient axial mesoderm (notochord) and ventral neural midline tissues (floor plate) manifested in mammals by an incomplete separation of the bilateral hemispheres of the telencephalon and deficiency of ventral diencephalon including the eye fields (Fernandes and Hebert 2008). Mutations of Zic2 are known to cause HPE. This could be due to a role of Zic2 in directing events of early gastrulation prior to formation of the prechordal plate (Warr et al. 2008). In the zebrafish anterior neural tissue, Zic2a and Zic1 seem to have overlapping expression and function as their combined LOF results in the forebrain midline defect (Grinblat and Sive 2001). However, while the loss of Zic2a alone did not cause any midline defects, a single Zic1 knockdown strongly affects midline development and causes partial cyclopia. Unlike *zic2a*, the expression of *zic1* starts later during mid-gastrula, which excludes its involvement in the formation of the prechordal plate. Zic1 LOF affects the optic stalk and diencephalon derivatives. The optic stalk cells convert into a more dorsal fate of pigmented cells, and optic vesicles ventralize as shown by the dorsally retracted darkly stained retinal pigmented epithelium (RPE). These phenotypes could be caused also by a decreased Nodal and Hedgehog (Hh) signaling, which similarly reduces the optic stalk markers, or by an increased RA signaling, which ventralizes the optic vesicle. It has been suggested that in regulating neural midline development, Zic1 acts upstream of these signaling pathways by controlling the Nodal signaling at the level of *cyc* expression, the Hedgehog signaling at the level of *shh* expression, and the RA signaling at the level of the RA-degrading enzyme *cyp26a1* expression (Abu-Abed et al. 2001). In support of these results, it was shown that Zic genes control Nodal signaling at multiple levels with Zic3 acting as the transcriptional inhibitor of Nodal co-receptor Oep (Winata et al. 2013).

Zic1 expression is under negative regulation of the Hedgehog (Hh) pathway. Shha GOF causes a reduction of *zic1* expression, resulting in the reduction of the anterior neuroectoderm expression domain and telencephalon later on and total absence of expression elsewhere (Rohr et al. 1999). The LOF experiments are more tedious due to inherent redundancy of genes encoding Hh ligands. As a result, changes in Zic expression in a single mutant of any of the Hh genes could be relatively mild. No wonder thus that *zic1* expression was unaffected in the *shha* mutant, which could be due to maintenance of an effect of Hh signaling by two other Hh-related genes (*shhb* and *ihhb*) expressed in dorsal mesodermal derivatives. Indeed, in mouse Shh mutants, where there is no such redundancy of Hh genes expressed in axial mesodermal derivatives, the midbrain expression of *zic1* decreased (Fogel et al. 2008). In this context, the analysis of zebrafish mutants affecting Hh receptors or mediators of Hh signaling could be more informative. Given that Zic3 acts as a negative regulator of Hh signaling at the level of transcription of *shha* and *shhb* (Winata et al. 2013), this may illustrate an autoregulatory feedback loop between Hh signaling and Zic proteins.

9.7 Conclusions

From the point of view of the evolution of Zic gene family in vertebrates, zebrafish having seven Zic genes is of particular interest. These genes could be segregated into categories that play roles in neural, mesodermal, and neural crest development with at least one gene in each pair combining roles in two germ layers. In many instances, these genes seem to perform redundant developmental roles. Analysis of Zic genes in zebrafish turned to be rather useful for rapid evaluation of functional and tissue-specific activity of regulatory elements driving the expression of these genes and identification of their transcriptional targets. Combining genomic studies (ChIP-seq) with bioimaging revealed developmental changes in the molecular regulatory mechanism involving the interaction of one of the Zic genes – Zic3 – with regulatory regions. This dynamic is reflected in the shift from a proximal promoter-based regulation in proliferating cells to a distal enhancer-based regulation during the determination of cell fate. Such shift of Zic regulatory mode of action may involve epigenetic regulation at the chromatin level. In the near future, characterization of genetic and epigenetic factors behind the changes in spatiotemporal specificity of Zic transcriptional activity may provide more details about the molecular mechanism of the recruitment of different Zic proteins across developmental stages and cell lineages. These will contribute to the information necessary to understand the general mechanism of developmental regulation of pleiotropic transcription factors such as Zic.

References

Abu-Abed S, Dolle P, Metzger D, Beckett B, Chambon P, Petkovich M (2001) The retinoic acid-metabolizing enzyme, CYP26A1, is essential for normal hindbrain patterning, vertebral identity, and development of posterior structures. Genes Dev 15:226–240

Agalliu D, Takada S, Agalliu I, McMahon AP, Jessell TM (2009) Motor neurons with axial muscle projections specified by Wnt4/5 signaling. Neuron 61:708–720

Aruga J (2004) The role of Zic genes in neural development. Mol Cell Neurosci 26:205–221

Aruga J, Yokota N, Hashimoto M, Furuichi T, Fukuda M, Mikoshiba K (1994) A novel zinc finger protein, zic, is involved in neurogenesis, especially in the cell lineage of cerebellar granule cells. J Neurochem 63:1880–1890

Aruga J, Nozaki Y, Hatayama M, Odaka YS, Yokota N (2010) Expression of ZIC family genes in meningiomas and other brain tumors. BMC Cancer 10:79

Bang AG, Papalopulu N, Goulding MD, Kintner C (1999) Expression of Pax-3 in the lateral neural plate is dependent on a Wnt-mediated signal from posterior nonaxial mesoderm. Dev Biol 212:366–380

Benedyk MJ, Mullen JR, DiNardo S (1994) Odd-paired: a zinc finger pair-rule protein required for the timely activation of engrailed and wingless in Drosophila embryos. Genes Dev 8:105–117

Bilder D, Schober M, Perrimon N (2003) Integrated activity of PDZ protein complexes regulates epithelial polarity. Nat Cell Biol 5:53–58

Bisgrove BW, Essner JJ, Yost HJ (2000) Multiple pathways in the midline regulate concordant brain, heart and gut left-right asymmetry. Development 127:3567–3579

Brown SA, Warburton D, Brown LY, Yu CY, Roeder ER, Stengel-Rutkowski S, Hennekam RC, Muenke M (1998) Holoprosencephaly due to mutations in ZIC2, a homologue of Drosophila odd-paired. Nat Genet 20:180–183

Carroll SB (2008) Evo-devo and an expanding evolutionary synthesis: a genetic theory of morphological evolution. Cell 134:25–36

Carroll JS, Liu XS, Brodsky AS, Li W, Meyer CA, Szary AJ, Eeckhoute J, Shao W, Hestermann EV, Geistlinger TR et al (2005) Chromosome-wide mapping of estrogen receptor binding reveals long-range regulation requiring the forkhead protein FoxA1. Cell 122:33–43

Cast AE, Gao C, Amack JD, Ware SM (2012) An essential and highly conserved role for Zic3 in left-right patterning, gastrulation and convergent extension morphogenesis. Dev Biol 364:22–31

Chen X, Vega VB, Ng HH (2008) Transcriptional regulatory networks in embryonic stem cells. Cold Spring Harb Symp Quant Biol 73:203–209

Chizhikov VV, Millen KJ (2004a) Control of roof plate development and signaling by Lmx1b in the caudal vertebrate CNS. J Neurosci 24:5694–5703

Chizhikov VV, Millen KJ (2004b) Mechanisms of roof plate formation in the vertebrate CNS. Nat Rev Neurosci 5:808–812

Coolen M, Thieffry D, Drivenes O, Becker TS, Bally-Cuif L (2012) miR-9 controls the timing of neurogenesis through the direct inhibition of antagonistic factors. Dev Cell 22:1052–1064
Danos MC, Yost HJ (1996) Role of notochord in specification of cardiac left-right orientation in zebrafish and Xenopus. Dev Biol 177:96–103
Drummond DL, Cheng CS, Selland LG, Hocking JC, Prichard LB, Waskiewicz AJ (2013) The role of Zic transcription factors in regulating hindbrain retinoic acid signaling. BMC Dev Biol 13:31
Eisen JS, Weston JA (1993) Development of the neural crest in the zebrafish. Dev Biol 159:50–59
Ekker M, Akimenko MA, Allende ML, Smith R, Drouin G, Langille RM, Weinberg ES, Westerfield M (1997) Relationships among msx gene structure and function in zebrafish and other vertebrates. Mol Biol Evol 14:1008–1022
Elsen GE, Choi LY, Millen KJ, Grinblat Y, Prince VE (2008) Zic1 and Zic4 regulate zebrafish roof plate specification and hindbrain ventricle morphogenesis. Dev Biol 314:376–392
Fernandes M, Hebert JM (2008) The ups and downs of holoprosencephaly: dorsal versus ventral patterning forces. Clin Genet 73:413–423
Fogel JL, Chiang C, Huang X, Agarwala S (2008) Ventral specification and perturbed boundary formation in the mouse midbrain in the absence of hedgehog signaling. Dev Dyn 237:1359–1372
Fujimi TJ, Mikoshiba K, Aruga J (2006) Xenopus Zic4: conservation and diversification of expression profiles and protein function among the Xenopus Zic family. Dev Dyn 235:3379–3386
Fujimi TJ, Hatayama M, Aruga J (2012) Xenopus Zic3 controls notochord and organizer development through suppression of the Wnt/beta-catenin signaling pathway. Dev Biol 361:220–231
Gallet A, Staccini-Lavenant L, Therond PP (2008) Cellular trafficking of the glypican Dally-like is required for full-strength Hedgehog signaling and wingless transcytosis. Dev Cell 14:712–725
Gamse JT, Sive H (2001) Early anteroposterior division of the presumptive neurectoderm in Xenopus. Mech Dev 104:21–36
Garcia-Castro MI, Marcelle C, Bronner-Fraser M (2002) Ectodermal Wnt function as a neural crest inducer. Science 297:848–851
Garnett AT, Square TA, Medeiros DM (2012) BMP, Wnt and FGF signals are integrated through evolutionarily conserved enhancers to achieve robust expression of Pax3 and Zic genes at the zebrafish neural plate border. Development 139:4220–4231
Giraldez AJ, Mishima Y, Rihel J, Grocock RJ, Van Dongen S, Inoue K, Enright AJ, Schier AF (2006) Zebrafish MiR-430 promotes deadenylation and clearance of maternal mRNAs. Science 312:75–79
Grinblat Y, Sive H (2001) Zic gene expression marks anteroposterior pattern in the presumptive neurectoderm of the zebrafish gastrula. Dev Dyn 222:688–693
Hadorn E (1956) Patterns of biochemical and developmental pleiotropy. Cold Spring Harb Symp Quant Biol 21:363–373
Harris TJ, Peifer M (2004) Adherens junction-dependent and -independent steps in the establishment of epithelial cell polarity in Drosophila. J Cell Biol 167:135–147
Hashimoto M, Shinohara K, Wang J, Ikeuchi S, Yoshiba S, Meno C, Nonaka S, Takada S, Hatta K, Wynshaw-Boris A et al (2010) Planar polarization of node cells determines the rotational axis of node cilia. Nat Cell Biol 12:170–176
Inoue T, Hatayama M, Tohmonda T, Itohara S, Aruga J, Mikoshiba K (2004) Mouse Zic5 deficiency results in neural tube defects and hypoplasia of cephalic neural crest derivatives. Dev Biol 270:146–162
Ishiguro A, Inoue T, Mikoshiba K, Aruga J (2004) Molecular properties of Zic4 and Zic5 proteins: functional diversity within Zic family. Biochem Biophys Res Commun 324:302–307
Keller MJ, Chitnis AB (2007) Insights into the evolutionary history of the vertebrate zic3 locus from a teleost-specific zic6 gene in the zebrafish, Danio rerio. Dev Genes Evol 217:541–547
Kimmel CB, Ballard WW, Kimmel SR, Ullmann B, Schilling TF (1995) Stages of embryonic development of the zebrafish. Dev Dyn 203:253–310
Kitaguchi T, Nagai T, Nakata K, Aruga J, Mikoshiba K (2000) Zic3 is involved in the left-right specification of the Xenopus embryo. Development 127:4787–4795

Kondrychyn I, Teh C, Sin M, Korzh V (2013) Stretching morphogenesis of the roof plate and formation of the central canal. PLoS One 8:e56219
Korzh V (2014) Stretching cell morphogenesis during late neurulation and mild neural tube defects. Dev Growth Differ 56:425–433
Krispin S, Nitzan E, Kalcheim C (2010a) The dorsal neural tube: a dynamic setting for cell fate decisions. Dev Neurobiol 70:796–812
Krispin S, Nitzan E, Kassem Y, Kalcheim C (2010b) Evidence for a dynamic spatiotemporal fate map and early fate restriction of premigratory avian neural crest. Development 137:585–595
Kuo JS, Patel M, Gamse J, Merzdorf C, Liu X, Apekin V, Sive H (1998) Opl: a zinc finger protein that regulates neural determination and patterning in Xenopus. Development 125:2867–2882
Layden MJ, Meyer NP, Pang K, Seaver EC, Martindale MQ (2010) Expression and phylogenetic analysis of the zic gene family in the evolution and development of metazoans. EvoDevo 1:12
Lecaudey V, Anselme I, Rosa F, Schneider-Maunoury S (2004) The zebrafish Iroquois gene iro7 positions the r4/r5 boundary and controls neurogenesis in the rostral hindbrain. Development 131:3121–3131
Lecaudey V, Anselme I, Dildrop R, Ruther U, Schneider-Maunoury S (2005) Expression of the zebrafish Iroquois genes during early nervous system formation and patterning. J Comp Neurol 492:289–302
Lim LS, Hong FH, Kunarso G, Stanton LW (2010) The pluripotency regulator Zic3 is a direct activator of the Nanog promoter in ESCs. Stem Cells 28:1961–1969
Lindeman LC, Winata CL, Aanes H, Mathavan S, Alestrom P, Collas P (2010) Chromatin states of developmentally-regulated genes revealed by DNA and histone methylation patterns in zebrafish embryos. Int J Dev Biol 54:803–813
Lindeman LC, Andersen IS, Reiner AH, Li N, Aanes H, Ostrup O, Winata C, Mathavan S, Muller F, Alestrom P et al (2011) Prepatterning of developmental gene expression by modified histones before zygotic genome activation. Dev Cell 21:993–1004
Lindgens D, Holstein TW, Technau U (2004) Hyzic, the Hydra homolog of the zic/odd-paired gene, is involved in the early specification of the sensory nematocytes. Development 131:191–201
Liu A, Majumdar A, Schauerte HE, Haffter P, Drummond IA (2000) Zebrafish wnt4b expression in the floor plate is altered in sonic hedgehog and gli-2 mutants. Mech Dev 91:409–413
Logan CY, Nusse R (2004) The Wnt signaling pathway in development and disease. Annu Rev Cell Dev Biol 20:781–810
Marchal L, Luxardi G, Thome V, Kodjabachian L (2009) BMP inhibition initiates neural induction via FGF signaling and Zic genes. Proc Natl Acad Sci U S A 106:17437–17442
Marois E, Mahmoud A, Eaton S (2006) The endocytic pathway and formation of the wingless morphogen gradient. Development 133:307–317
Merzdorf CS (2007) Emerging roles for zic genes in early development. Dev Dyn 236:922–940
Mii Y, Taira M (2009) Secreted frizzled-related proteins enhance the diffusion of Wnt ligands and expand their signalling range. Development 136:4083–4088
Mizugishi K, Aruga J, Nakata K, Mikoshiba K (2001) Molecular properties of Zic proteins as transcriptional regulators and their relationship to GLI proteins. J Biol Chem 276:2180–2188
Morgan D, Turnpenny L, Goodship J, Dai W, Majumder K, Matthews L, Gardner A, Schuster G, Vien L, Harrison W et al (1998) Inversin, a novel gene in the vertebrate left-right axis pathway, is partially deleted in the inv mouse. Nat Genet 20:149–156
Nagai T, Aruga J, Takada S, Gunther T, Sporle R, Schughart K, Mikoshiba K (1997) The expression of the mouse Zic1, Zic2, and Zic3 gene suggests an essential role for Zic genes in body pattern formation. Dev Biol 182:299–313
Nakata K, Nagai T, Aruga J, Mikoshiba K (1997) Xenopus Zic3, a primary regulator both in neural and neural crest development. Proc Natl Acad Sci U S A 94:11980–11985
Nakata K, Nagai T, Aruga J, Mikoshiba K (1998) Xenopus Zic family and its role in neural and neural crest development. Mech Dev 75:43–51
Nakata K, Koyabu Y, Aruga J, Mikoshiba K (2000) A novel member of the Xenopus Zic family, Zic5, mediates neural crest development. Mech Dev 99:83–91

Niederreither K, Vermot J, Schuhbaur B, Chambon P, Dolle P (2000) Retinoic acid synthesis and hindbrain patterning in the mouse embryo. Development 127:75–85
Nyholm MK, Wu SF, Dorsky RI, Grinblat Y (2007) The zebrafish zic2a-zic5 gene pair acts downstream of canonical Wnt signaling to control cell proliferation in the developing tectum. Development 134:735–746
Ohtsuka M, Kikuchi N, Yokoi H, Kinoshita M, Wakamatsu Y, Ozato K, Takeda H, Inoko H, Kimura M (2004) Possible roles of zic1 and zic4, identified within the medaka Double anal fin (Da) locus, in dorsoventral patterning of the trunk-tail region (related to phenotypes of the Da mutant). Mech Dev 121:873–882
Okada Y, Takeda S, Tanaka Y, Izpisua Belmonte JC, Hirokawa N (2005) Mechanism of nodal flow: a conserved symmetry breaking event in left-right axis determination. Cell 121:633–644
Parinov S, Kondrichin I, Korzh V, Emelyanov A (2004) Tol2 transposon-mediated enhancer trap to identify developmentally regulated zebrafish genes in vivo. Dev Dyn 231:449–459
Raible DW, Wood A, Hodsdon W, Henion PD, Weston JA, Eisen JS (1992) Segregation and early dispersal of neural crest cells in the embryonic zebrafish. Dev Dyn 195:29–42
Rohr KB, Schulte-Merker S, Tautz D (1999) Zebrafish zic1 expression in brain and somites is affected by BMP and hedgehog signalling. Mech Dev 85:147–159
Roy S, Hsiung F, Kornberg TB (2011) Specificity of Drosophila cytonemes for distinct signaling pathways. Science 332:354–358
Sabherwal N, Tsutsui A, Hodge S, Wei J, Chalmers AD, Papalopulu N (2009) The apicobasal polarity kinase aPKC functions as a nuclear determinant and regulates cell proliferation and fate during Xenopus primary neurogenesis. Development 136:2767–2777
Sakurada T, Mima K, Kurisaki A, Sugino H, Yamauchi T (2005) Neuronal cell type-specific promoter of the alpha CaM kinase II gene is activated by Zic2, a Zic family zinc finger protein. Neurosci Res 53:323–330
Salero E, Perez-Sen R, Aruga J, Gimenez C, Zafra F (2001) Transcription factors Zic1 and Zic2 bind and transactivate the apolipoprotein E gene promoter. J Biol Chem 276:1881–1888
Sanek NA, Grinblat Y (2008) A novel role for zebrafish zic2a during forebrain development. Dev Biol 317:325–335
Sanyal A, Lajoie BR, Jain G, Dekker J (2012) The long-range interaction landscape of gene promoters. Nature 489:109–113
Sato T, Sasai N, Sasai Y (2005) Neural crest determination by co-activation of Pax3 and Zic1 genes in Xenopus ectoderm. Development 132:2355–2363
Sawyer JM, Harrell JR, Shemer G, Sullivan-Brown J, Roh-Johnson M, Goldstein B (2010) Apical constriction: a cell shape change that can drive morphogenesis. Dev Biol 341:5–19
Sonawane M, Carpio Y, Geisler R, Schwarz H, Maischein HM, Nuesslein-Volhard C (2005) Zebrafish penner/lethal giant larvae 2 functions in hemidesmosome formation, maintenance of cellular morphology and growth regulation in the developing basal epidermis. Development 132:3255–3265
Sonawane M, Martin-Maischein H, Schwarz H, Nusslein-Volhard C (2009) Lgl2 and E-cadherin act antagonistically to regulate hemidesmosome formation during epidermal development in zebrafish. Development 136:1231–1240
Stewart RA, Arduini BL, Berghmans S, George RE, Kanki JP, Henion PD, Look AT (2006) Zebrafish foxd3 is selectively required for neural crest specification, migration and survival. Dev Biol 292:174–188
Strigini M, Cohen SM (2000) Wingless gradient formation in the Drosophila wing. Curr Biol 10:293–300
Teslaa JJ, Keller AN, Nyholm MK, Grinblat Y (2013) Zebrafish Zic2a and Zic2b regulate neural crest and craniofacial development. Dev Biol 380:73–86
Thisse B, Pflumio S, Fürthauer M, Loppin B, Heyer V, Degrave A, Woehl R, Lux A, Steffan T, Charbonnier XQ, Thisse C (2001) Expression of the zebrafish genome during embryogenesis. ZFIN direct data submission

Thisse B, Heyer V, Lux A, Alunni V, Degrave A, Seiliez I, Kirchner J, Parkhill JP, Thisse C (2004) Spatial and temporal expression of the zebrafish genome by large-scale in situ hybridization screening. Methods Cell Biol 77:505–519

Toyama R, Gomez DM, Mana MD, Dawid IB (2004) Sequence relationships and expression patterns of zebrafish zic2 and zic5 genes. Gene Expr Patterns 4:345–350

Ulloa F, Marti E (2010) Wnt won the war: antagonistic role of Wnt over Shh controls dorso-ventral patterning of the vertebrate neural tube. Dev Dyn 239:69–76

Ungar AR, Kelly GM, Moon RT (1995) Wnt4 affects morphogenesis when misexpressed in the zebrafish embryo. Mech Dev 52:153–164

van Straaten HW, Sieben I, Hekking JW (2002) Multistep role for actin in initial closure of the mesencephalic neural groove in the chick embryo. Dev Dyn 224:103–108

Vastenhouw NL, Zhang Y, Woods IG, Imam F, Regev A, Liu XS, Rinn J, Schier AF (2010) Chromatin signature of embryonic pluripotency is established during genome activation. Nature 464:922–926

Wada S, Saiga H (2002) HrzicN, a new Zic family gene of ascidians, plays essential roles in the neural tube and notochord development. Development 129:5597–5608

Wang G, Cadwallader AB, Jang DS, Tsang M, Yost HJ, Amack JD (2011) The Rho kinase Rock2b establishes anteroposterior asymmetry of the ciliated Kupffer's vesicle in zebrafish. Development 138:45–54

Ware SM, Harutyunyan KG, Belmont JW (2006) Zic3 is critical for early embryonic patterning during gastrulation. Dev Dyn 235:776–785

Warr N, Powles-Glover N, Chappell A, Robson J, Norris D, Arkell RM (2008) Zic2-associated holoprosencephaly is caused by a transient defect in the organizer region during gastrulation. Hum Mol Genet 17:2986–2996

Wederell ED, Bilenky M, Cullum R, Thiessen N, Dagpinar M, Delaney A, Varhol R, Zhao Y, Zeng T, Bernier B et al (2008) Global analysis of in vivo Foxa2-binding sites in mouse adult liver using massively parallel sequencing. Nucleic Acids Res 36:4549–4564

Whyte WA, Orlando DA, Hnisz D, Abraham BJ, Lin CY, Kagey MH, Rahl PB, Lee TI, Young RA (2013) Master transcription factors and mediator establish super-enhancers at key cell identity genes. Cell 153:307–319

Wilkins AS (2007) Between "design" and "bricolage": genetic networks, levels of selection, and adaptive evolution. Proc Natl Acad Sci U S A 104(Suppl 1):8590–8596

Winata CL, Kondrychyn I, Kumar V, Srinivasan KG, Orlov Y, Ravishankar A, Prabhakar S, Stanton LW, Korzh V, Mathavan S (2013) Genome wide analysis reveals Zic3 interaction with distal regulatory elements of stage specific developmental genes in zebrafish. PLoS Genet 9:e1003852

Wittkopp PJ, Kalay G (2012) Cis-regulatory elements: molecular mechanisms and evolutionary processes underlying divergence. Nat Rev Genet 13:59–69

Woda JM, Pastagia J, Mercola M, Artinger KB (2003) Dlx proteins position the neural plate border and determine adjacent cell fates. Development 130:331–342

Yang B, Jia L, Guo Q, Ren H, Hu D, Zhou X, Ren Q, Hu Y, Xie T (2015) MiR-564 functions as a tumor suppressor in human lung cancer by targeting ZIC3. Biochem Biophys Res Commun 467:690–696

Chapter 10
Overview of Rodent *Zic* Genes

Koula E. M. Diamand, Kristen S. Barratt, and Ruth M. Arkell

Abstract The five murine *Zic* genes encode multifunctional transcriptional regulator proteins important for a large number of processes during embryonic development. The genes and proteins are highly conserved with respect to the orthologous human genes, an attribute evidently mirrored by functional conservation, since the murine and human genes mutate to give the same phenotypes. Each ZIC protein contains a zinc finger domain that participates in both protein-DNA and protein-protein interactions. The ZIC proteins are capable of interacting with the key transcriptional mediators of the SHH, WNT and NODAL signalling pathways as well as with components of the transcriptional machinery and chromatin-modifying complexes. It is possible that this diverse range of protein partners underlies characteristics uncovered by mutagenesis and phenotyping of the murine *Zic* genes. These features include redundant and unique roles for ZIC proteins, regulatory interdependencies amongst family members and pleiotropic *Zic* gene function. Future investigations into the complex nature of the *Zic* gene family activity should be facilitated by recent advances in genome engineering and functional genomics.

Keywords Mouse · Zinc finger · Intrinsic disorder · Transcription factor · Cofactor · Pleiotropy · SHH · WNT · NODAL

10.1 Discovery of the Murine *Zic* Genes

In July 1952 Ed Garber reported a new dominant mutation in the mouse, Bent-tail (*Bn*) (Garber 1952). This spontaneous mutant was isolated when a single Bent-tail male was found in a litter of seven mice resulting from a mating between a normal female of the Nanru strain and a bald, hr^{b4}, male. Upon mating this male with a normal sibling, multiple female Bent-tail progeny were produced along with normal males. This, and other initial breeding data, indicated the mutation was dominant and sex-linked and represented just the third example of a sex-linked rodent locus.

K. E. M. Diamand · K. S. Barratt · R. M. Arkell (✉)
Early Mammalian Development Laboratory, John Curtin School of Medical Research, The Australian National University, Canberra, ACT, Australia
e-mail: ruth.arkell@anu.edu.au

J. Aruga (ed.), *Zic family*, Advances in Experimental Medicine and Biology 1046,
https://doi.org/10.1007/978-981-10-7311-3_10

Further breeding experiments enabled Garber to correctly conclude that the *Bn* locus is X-linked (rather than Y-linked) and that the *Bn* mutation had a partial lethal effect in hemizygous males and homozygous females. Although the *Bn* mouse provided the first glimpse into the importance of the *Zic* gene family, it was 48 years before this became apparent, when two teams showed that *Zic3* is the causative gene in the *Bn* mouse (Carrel et al. 2000; Klootwijk et al. 2000). By this stage, the single *Drosophila* orthologue (odd paired; *opa*) of the *Zic*s had been identified by its mutant phenotype and cloned (Benedyk et al. 1994), murine *Zic1* had been identified by cDNA screening for genes expressed in the cerebellum (Aruga et al. 1994), and *Zic2-4* had been identified in the mouse on the basis of their homology to *Zic1* (Aruga et al. 1996a, b). We now know that in rodents, as in man, there are five *Zic* genes (Furushima et al. 2000), each of which encodes a transcriptional regulator protein containing a zinc finger domain.

10.2 The Genomic Arrangement and Sequence of the Rodent *Zic* Genes

Genome sequencing of a large number of vertebrate organisms has revealed that the *Zic* genes are generally arranged in tandem pairs. In the mouse genome, the five *Zic* genes reside at three locations. *Zic1* and *Zic4* are arranged as a divergently transcribed tandem gene pair on mouse chromosome 9 (MMu9) as are *Zic2* and *Zic5* on MMu14 and *Zic3* is an X-linked singleton. Phylogenetic analysis of genome sequence and intron/exon position has led to a model of *Zic* gene evolution (Aruga et al. 2006; reviewed in Houtmeyers et al. 2013). Briefly, an ancestral single-copy *Zic* gene, containing a single intron and a full set of conserved domains, underwent tandem duplication to produce two divergently transcribed genes. Structural variation arose between the tandem pair with changes in the protein sequence of one tandem copy and the gain of an intron in the other copy. Subsequently, the genome duplication events proposed to have taken place early in the vertebrate lineage resulted in eight *Zic* genes, arranged in four bi-gene clusters. Each gene pair contained one gene with two introns and sequences most closely related to the ancestral single-copy gene along with one single intron gene with sequence divergence. During vertebrate evolution, gene loss has occurred, and there have been instances of intron gain and individual gene duplications. The mouse, like most (but not all) vertebrate organisms, has five *Zic* genes. *Zic1*, *2* and *3* each have 2 introns, and the murine *Zic4* gene has four introns, whereas the *Zic5* gene still contains a single intron.

The relatively recent compilation of partial or complete genome sequences from rodents other than the laboratory mouse (*Mus musculus*) provides an opportunity to examine *Zic* gene arrangement and sequence divergence across the rodents. Orthologues have been identified in the rat (*Rattus norvegicus*), Chinese hamster

(*Cricetulus griseus*), guinea pig (*Cavia porcellus*), long-tailed chinchilla (*Chinchilla lanigera*), naked mole rat (*Heterocephalus glaber*), prairie vole (*Microtus ochrogaster*) and ground squirrel (*Ictidomys tridecemlineatus*). Amongst these species, the rat is the only genome in which five *Zic* genes are currently annotated. Presumably, this reflects incomplete genome assembly rather than widespread differences in the number of *Zic* genes amongst rodents. Table 10.1 shows a percent

Table 10.1 Percent identity comparison of mammalian *Zic* genes

	Intron number	5′UTR	3′UTR	Introns	Coding DNA	Protein
Zic1						
Mouse (NM_009573)	2	100.00	100.00	100.00	100.00	100.00
Human (NM_003412)	2	81.74	83.73	54.33	90.85	99.33
Rat (ENSRNOG00000014644)	2	96.04	94.93	84.49	98.07	100.00
Chinese hamster (ENSCGRG00001015672)	2	88.89	85.86	73.75	96.13	100.00
Ground squirrel (ENSSTOG00000000952)	2	87.02		55.99	92.56	99.11
Naked mole rat (ENSHGLG00000019340)	2	82.27		54.43	92.02	98.43
Long-tailed chinchilla (ENSCLAG00000006845)	2	82.85		52.45	92.26	99.11
Prairie vole (ENSMOCG00000015344)	2	87.80	89.25	74.12	96.13	100.00
Zic2						
Mouse (NM_009574)	2	100.00	100.00	100.00	100.00	100.00
Human (NM_007129)	2	76.95	81.88	72.68	91.32	98.68
Rat (NM_001108392)	2	93.78	91.90	90.98	97.36	99.81
Guinea pig (XM_003477600)	2	75.46	69.16	69.74	89.51	98.08
Chinese hamster (ENSCGRG00001009292)	2	90.77	84.90	82.79	94.45	99.43
Naked mole rat (ENSHGLG00000019583)	2	77.44	71.48	68.98		
Long-tailed chinchilla (ENSCLAG00000003273)	2		69.49	69.46		
Prairie vole (ENSMOCG00000005659)	2	88.79	85.59	85.10	94.20	99.05
Zic3						
Mouse (NM_009575)	2	100.00	100.00	100.00	100.00	100.00
Human (NM_003413)	2	81.33	85.67	71.83	93.22	98.50
Rat (ENSRNOG00000000861)	2	96.91	92.82	92.64	97.73	99.57
Guinea pig (ENSCPOG00000002374)	2		87.46	68.70	83.18	
Ground squirrel (ENSSTOG00000024153)	2	80.44	87.91	71.90	92.91	98.07
Zic4						
Mouse (NM_009576)	4	100.00	100.00	100.00	100.00	100.00
Human (ENSG00000174963)	4	NA	74.74	64.8	87.06	87.43
Rat (ENSRNOG00000014871)	3	NA	89.60		97.66	98.16
Guinea pig (ENSCPOG00000010981)	3	NA	95.27	59.94	86.57	88.32
Ground squirrel (ENSSTOG00000015977)	3	91.45	78.26	62.66	86.77	87.72
Chinese hamster (ENSCGRG00001024222)	2	NA	99.00	69.63	92.84	93.11
Naked mole rat (ENSHGLG00000019306)	5	37.70	75.27	57.07	69.03	54.46

(continued)

Table 10.1 (continued)

	Intron number	5′UTR	3′UTR	Introns	Coding DNA	Protein
Zic5						
Mouse (NM_022987)	1	100.00	100.00	100.00	100.00	100.00
Human (NM_033132)	1	90.52	92.49	53.29	89.24	88.73
Rat (ENSRNOG00000014358)	1	97.07	95.38	80.49	96.36	97.27
Ground squirrel (ENSSTOG00000027399)	1	86.38	88.55	55.52	90.03	89.27
Chinese hamster (ENSCGRG00001022535)	1	93.55	94.22	71.55	94.09	94.02
Naked mole rat (ENSHGLG00000019584)	1	87.02	90.70	53.39	87.71	85.81
Long-tailed chinchilla (ENSCLAG00000003432)	1	86.47	89.53	54.41	79.91	88.96
Prairie vole (ENSMOCG00000003890)	1	94.13	91.33	70.92	93.88	94.70

Available DNA and protein sequences for the human (*Homo sapiens*), rat (*Rattus norvegicus*), Chinese hamster (*Cricetulus griseus*), guinea pig (*Cavia porcellus*), long-tailed chinchilla (*Chinchilla lanigera*), naked mole rat (*Heterocephalus glaber*), prairie vole (*Microtus ochrogaster*) and ground squirrel (*Ictidomys tridecemlineatus*) were compared to the laboratory mouse (*Mus musculus*). *Blank cells* indicate species where only a partial sequence was available. *NA* indicates sequences that could not be sufficiently aligned to the mouse

identity matrix for the currently annotated rodent *Zic* genes and for the human *Zic* genes. It shows that *Zic4* and *Zic5* coding sequences vary more amongst rodent species than those of *Zic1-3*. For *Zic2* and *Zic3*, the intron sequences are relatively conserved, whereas in *Zic4* not only are the intron sequences variable amongst rodent species, but so is the number of introns. When analysed for conservation between mouse and human sequences, a similar trend is seen. For each of *Zic1-3*, the coding DNA is at least 90% conserved and protein sequences greater than 98% conserved, whereas, for *Zic4* and *Zic5*, the coding DNA is 87% and 89% conserved, respectively, and the protein sequence is ~88% conserved for each protein. All five *Zic* genes are, however, above average for mouse-human conservation of coding DNA sequence (~85%) and protein sequence (70.1% for all orthologues and 90.3% for orthologous disease causing proteins) (Chinwalla et al. 2002). In general the *Zic* UTR conservation is higher than the average 5′ and 3′ UTR mouse-human conservation (~75%), whereas only *Zic2* and *Zic3* have above average mouse-human intron conservation (69%) (Chinwalla et al. 2002).

10.3 The Structure and Conservation of the Murine ZIC Proteins

The defining feature of all ZIC proteins is the inclusion of a zinc finger domain (ZFD) composed of five tandem Cys2His2-type zinc fingers (Fig. 10.1). This domain is most closely related to the ZFD of the GLI, GLIS and NKL families and is highly conserved between ZIC family members both within an individual species

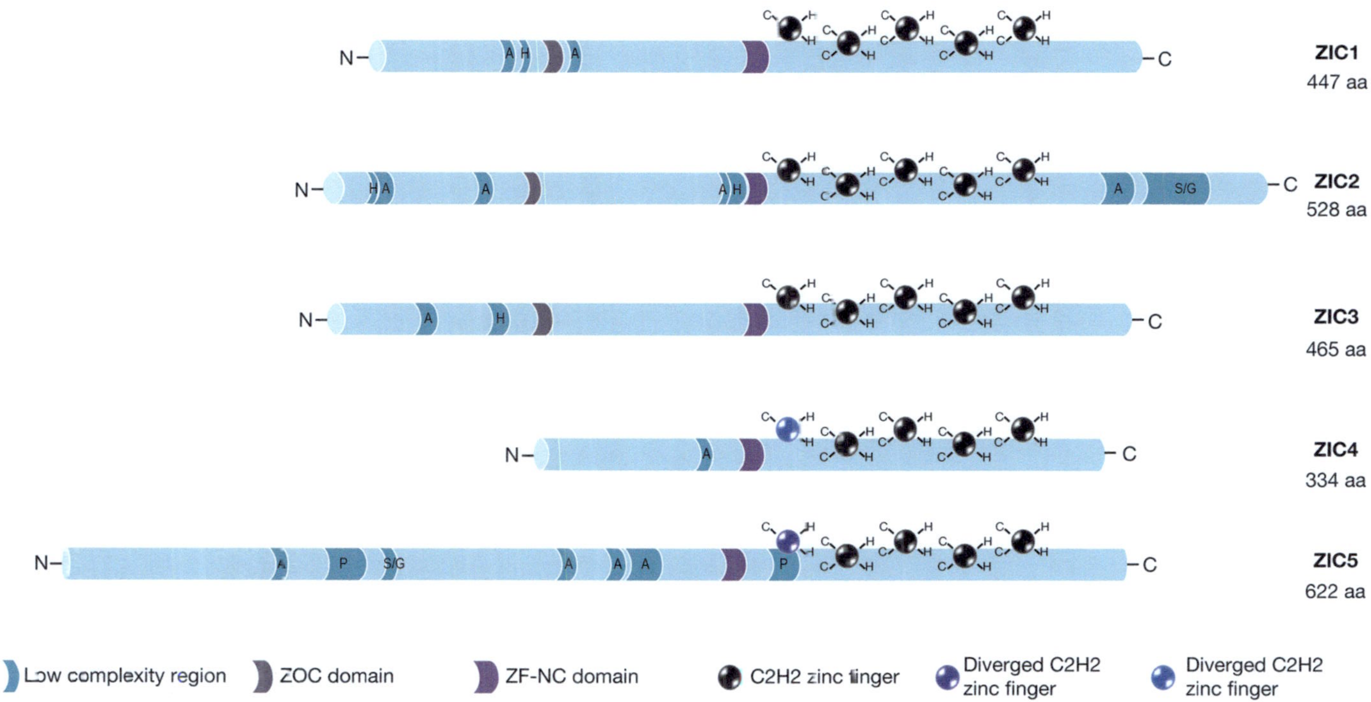

Fig. 10.1 Structural features of the five murine ZIC proteins. All mouse ZIC proteins contain a zinc finger domain (ZFD) that consists of five tandem C2H2-type zinc fingers. This domain is highly conserved, with only the first zinc fingers of ZIC4 and ZIC5 showing some divergence. Additionally, all five proteins also contain a short (14–21 amino acids) highly conserved domain immediately upstream of the zinc fingers, called the ZF-NC domain, as well as low-complexity regions with the major amino acid found at each low-complexity region shown by the associated letter (A alanine, H histidine, P proline, S/G serine/glycine). ZIC1, ZIC2 and ZIC3 share a small domain (9–10 amino acids) of homology towards the N-terminal of the protein, termed the ZOC motif. The ZIC proteins can be divided into two distinct structural subclasses on the basis of the presence or absence of the ZOC motif and the degree of conservation within the first zinc finger domain

and across species. The ZIC ZFD is, however, distinguished by an atypical first zinc finger. Generally one to five amino acid residues separate the two cysteines of a C2H2 zinc finger, whereas, in the first zinc finger of ZIC proteins, this number is both increased and highly variable, ranging from 6 to 38 in species so far examined (Aruga et al. 2006; Layden et al. 2010). Additionally, structural analysis of the ZIC3 protein indicates that the first two zinc finger domains may not be canonical (DNA binding) C2H2-type zinc fingers. Instead these fingers may form a single structural unit called the tandem CWCH2 motif, the hallmark of which is a tryptophan residue located between the two canonical cysteines of each zinc finger (Hatayama et al. 2008). This motif is conserved across a wide range of metazoan species (Hatayama and Aruga 2010; Aruga et al. 2006), indicative of biological significance, as is the identification of a missense mutation of the tryptophan in the first zinc finger of human ZIC3 in association with congenital heart malformations. This mutation reduces protein stability and perturbs the nuclear localization of the protein (Chhin et al. 2007). The ZFD is intimately involved in ZIC protein function, mediating DNA binding and protein-protein interactions. Additionally, none of the ZIC proteins contain a canonical nuclear localization signal, but the human ZIC3 zinc finger domain (fingers 2–5) has been shown to be essential for this function and presumably harbours an interspersed nuclear localization signal (Bedard et al. 2007; Hatayama et al. 2008).

The regions outside of the ZFD exhibit greater sequence variability between the five murine ZIC proteins, but evolutionary conserved domains are present (Fig. 10.1). The ZF-NC (zinc finger N-flanking conserved) domain, located directly prior to the ZFD, is a small (14–21 aa) conserved region that is SUMOylated in human ZIC3 (Aruga et al. 2006; Chen et al. 2013). Additionally, a small (9–10 aa), N-terminally located domain called the *Zic-opa* conserved (ZOC) domain is conserved in ZIC1-3. This domain is one of two shown to be involved in transcriptional activation (the other being the ZFD) and has also been shown to be involved in protein-protein interactions, such as binding the myogenic repressor protein, I-mfa (Mizugishi et al. 2004). As reviewed in Houtmeyers et al. (2013), phylogenetic comparison of the ZIC protein sequences has led to the classification of ZICs into two distinct structural subclasses based on the presence of the ZOC domain and the variation within the first zinc finger. The structural subclass A contains the more highly conserved ZIC1, 2 and 3 proteins, and subclass B contains the less conserved ZIC4 and ZIC5 proteins. Notably, the subclass division is reflected in the genome arrangement and evolution of the *Zic* genes, since each gene pair (i.e. *Zic1/Zic4* and *Zic2/Zic5*) contains one subclass A and one subclass B protein.

Each murine ZIC protein contains several low-complexity regions (including poly-alanine, poly-histidine, poly-proline and poly-serine/poly-glycine tracts), with the number and type of sequence varying between the five ZIC proteins (Fig. 10.1). Low-complexity regions can undergo expansion and contraction mutations that alter protein function, and poly-alanine tract expansions in ZIC2 and ZIC3 are associated with human disease pathogenesis. Expansion of the alanine tract in *ZIC2* is implicated in Holoprosencephaly, whilst expansion mutations of this domain in *ZIC3* are associated with Heterotaxy (Brown et al. 2001; Wessels et al. 2010). In vitro cell-based assays have shown that expansion of the ZIC2 alanine tract from 15As to 25As results in a near-complete loss of transactivation ability, dependent on the promoter that is used (Brown et al. 2005), suggesting a role for the alanine tract in the modulation of ZIC transactivation. It is unclear, however, if poly-alanine tract expansions affect the functionality of other ZIC family members. Previous analysis of ZIC3 found no significant change in transactivation ability upon modification of the alanine tract (Cowan et al. 2014), a result which may be attributed to the differences in promoters and experimental conditions in the two studies.

10.4 Intrinsic Disorder in the ZIC Proteins

Low-complexity regions such as those found in the ZIC proteins are characteristic of intrinsic disorder, a term used to describe proteins or regions of proteins that lack a conventional, well-defined tertiary structure in their native state. Instead, disordered proteins, or regions of disorder, are unstructured but will fold upon binding to their target or in response to posttranslational modifications (as reviewed in Dyson 2016). This flexibility in structure appears to confer multiple advantages including the ability to interact with multiple binding partners. Additionally, it enables interactions that are highly specific but have low affinity, which facilitates the rapid disassembly of the complex (Dyson and Wright 2005; Dyson 2016). This structural polymorphism allows disordered proteins to be involved in a vast array of biological processes and protein functions, such as accessibility to posttranslational modifications, involvement in cell signalling, control of the cell cycle and transcriptional activation (reviewed in Dyson and Wright 2005).

Computer algorithms can predict disordered proteins or regions of disorder from sequence composition since disorder is characterized by high concentration of charged and polar amino acids and low prevalence of hydrophilic residues (Dyson and Wright 2005). Analysis of the murine ZICs identifies intrinsically disordered regions in the residues adjacent to the ZFD, with only minimal disorder predicted within the ZFD itself (Fig. 10.2). For example, in all mouse ZIC proteins, the C-terminus is predicted to be almost completely disordered as is the first 223 amino acids of the ZIC5 N-terminus. Despite many functions mapping to the ZIC ZFD, the presence of disordered regions flanking the ZFD suggests these regions play a currently underappreciated role in ZIC protein function. Moreover, the plasticity afforded by the combining regions of disorder which vary amongst the ZIC proteins with a highly conserved ZFD domain is consistent with each ZIC having multiple biological roles as well as both unique and shared proposed binding partners (Ishiguro et al. 2007; Fujimi et al. 2012; Pourebrahim et al. 2011; Luo et al. 2015; Houtmeyers et al. 2016). Comparative studies generally focus on conserved domains within the ZIC protein family, whereas it is possible that differences in the sequence composition of the family members underlie the incredible functional diversity of the ZIC proteins.

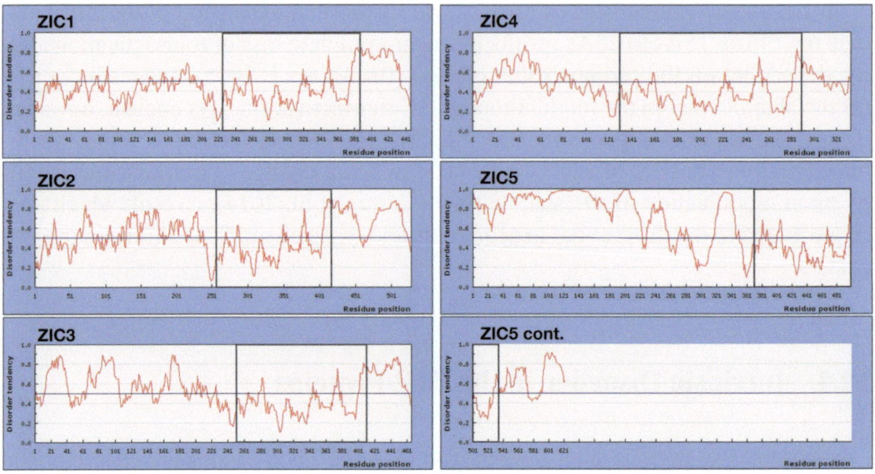

Fig. 10.2 Murine ZIC proteins are predicted to have intrinsically disordered regions. IUPred (Dosztányi et al. 2005) was used to predict disorder within each of the five mouse ZIC proteins. A score between 0 and 0.5 indicated a predicted ordered region, whilst a score of 0.5–1 indicates a predicted disordered region. The *grey* box denotes the zinc finger domain in each ZIC protein

10.5 Gene Expression

The temporal-spatial distribution of murine *Zic* gene transcripts has been documented across a range of developmental stages, and (as reviewed in Houtmeyers et al. 2013) the murine *Zic* genes can be divided into two groups based on their overlapping expression patterns during early embryo development. *Zic2*, *Zic3* and *Zic5* are the first of the genes to be expressed, with transcription of each initiating prior to implantation. During gastrulation, *Zic2*, *Zic3* and *Zic5* are co-expressed in the ectoderm and some of the newly formed mesoderm that has ingressed through the primitive streak, whilst *Zic1* and *Zic4* are not yet detected (Fig. 10.3) (Furushima et al. 2000; Elms et al. 2004; Houtmeyers et al. 2013). By the early head-fold stage, all three expressed genes are restricted to the neurectoderm and later become limited to the most dorsal neurectoderm that will form the future neural crest cells and dorsal neurons, as well as pre-somitic and somatic populations of the lateral mesoderm. In contrast, *Zic1* and *Zic4* expression initiates during early organogenesis in the neurectoderm and somitic mesoderm. By 9.5 dpc, all five murine *Zic* genes are co-expressed in the dorsal spinal cord, in the dorsal cranial neural tube and in the somites, with small differences in expression domains such as the absence of *Zic4* expression in the roof-plate of the neural tube (Fig. 10.3) (Nagai et al. 1997; Furushima et al. 2000; Elms et al. 2004). One day later, the expression of *Zic2*, *Zic3* and *Zic5* is initiated in newly developing eye and in the limb buds. At later stages of development, *Zic* gene expression has not been assayed in the entire embryo, but brain expression patterns have been reported. The progenitor cells of the thalamus,

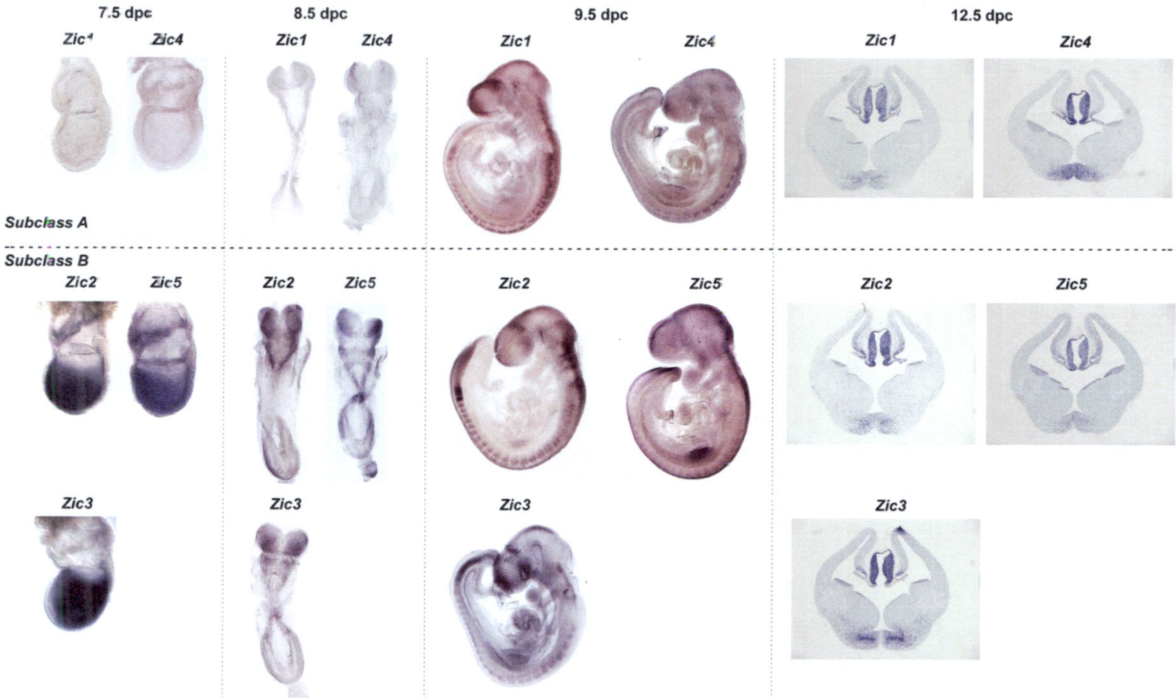

Fig. 10.3 Co-expression of the murine *Zic* genes during mouse gastrulation and organogenesis. The murine *Zic* gene expression subclasses reflect their genomic arrangement. The expression subclass A genes are expressed prior to and throughout gastrulation, during neurulation and in organogenesis. The expression of the subclass B genes does not initiate until after neurulation and is absent from the optic cup, the forelimb bud and the pre-somitic mesoderm at early organogenesis. (**a**) Lateral view of 7.5 dpc mouse embryos (anterior is to the left), (**b**) dorsal view of 8.5 dpc mouse embryos (anterior is to the top) and (**c**) lateral view of 9.5 dpc mouse embryos (anterior is to the left) following whole-mount in situ hybridization to *Zic1-5* (courtesy of Kristen Barratt). (**d**) Coronal sections of 12.5 dpc mouse forebrains through the hippocampus and habenular nucleus following in situ hybridization to *Zic1-5* (courtesy of Jun Aruga)

the preoptic area, the septum, the cortical hem and the retina all express the *Zic* genes in largely overlapping domains at mid-gestation. *Zic* expression in the central nervous system becomes more restricted by embryonic day 12.5 at which time all five *Zic* genes are expressed in the dorsal midline neural tissues (including the cortical hem, the septum and the ventricular and sub-ventricular zone) and ventral neural tube (Fig. 10.3). Interestingly, as shown in Fig. 10.3, these expression groups mirror the genomic arrangement of the genes. The *Zic1/4* bi-gene pair form one expression group, whilst the single gene *Zic3* and the *Zic2/5* bi-gene pair form another. It is possible that the bi-gene pairs share regulatory regions as a direct result of their evolved tandem gene arrangement, resulting in similar gene expression patterns (Houtmeyers et al. 2013).

10.6 Biological Function of the Murine *Zic* Genes

Mutation and phenotype analysis of the murine *Zic* genes provides one of the most comprehensive insights into *Zic* gene function. As can be seen from Table 10.2, multiple alleles for each of the murine *Zic* genes exist and have been produced from a variety of mutagenesis strategies. The majority of alleles, however, are severe loss-of-function with partial loss-of-function alleles (aka hypomorphic alleles) available only for *Zic2*. Moreover, *Zic3* is the only gene for which a conditional allele is available. Together this means that for many of the genes, the available alleles are primarily useful for studying the earliest function of that gene. Despite this, the analysis of these alleles has yielded an extensive list of processes, tissues and organs that require the function of one or more murine *Zic* genes for normal development. Processes affected by the absence of *Zic* gene function include axis formation, as well as the patterning and morphogenesis of the central nervous, visual, musculoskeletal and vascular systems (Table 10.3). Mutation of the murine *Zic* genes gives rise to the same conditions associated with *Zic* gene loss-of-function in humans. The phenotypic consequences of *Zic* gene loss-of-function have been summarized before (Houtmeyers et al. 2013), and many are covered in depth in other chapters of this book. Here, we review some features of *Zic* gene and ZIC protein activity that emerge when one considers the combined information from *Zic* gene phenotyping studies.

10.6.1 Redundant Zic *Gene Function*

Genetic redundancy means that two or more genes are performing the same function and that inactivation of one gene has little or no effect on the biological phenotype. Full functional redundancy is demonstrated when single mutants have no phenotype but the double mutant does. In contrast, partial redundancy is said to occur when one or both of the single mutants exhibits a phenotype that is enhanced

Table 10.2 Murine *Zic* alleles

MGI allele ID	Allele symbol	Allele name	Synonyms	Allele type	Allele attributes	Human disease models	References
Zic1							Roderick (1983)
MGI:3789699	*In(9)26Rk*	Inversion	In26Rk	CI			Aruga et al. (2002b)
MGI:3759850	*Tg(Zic1,-lacZ)1Jaru*	Transgene insertion 1	Z19K	Tr	Inserted expressed sequence		Aruga et al. (2002b)
MGI:3759865	*Tg(Zic1tr,-lacZ)1Jaru*	Transgene insertion 1	TR9K	Tr	Inserted expressed sequence		Aruga et al. (2002b)
MGI:3510680	*Zic1/Zic4*^{tm1Kjm}	Targeted mutation 1	$Zic1^{-}Zic4^{-}$	T	Null	Dandy-Walker syndrome	Grinberg et al. (2004)
MGI:2156822	*Zic1*^{tm1Jaru}	Targeted mutation 1		T	Null	Joubert syndrome	Aruga et al. (1998)
Zic2							
MGI:5566852	*Zic2*^{em1Bend}	Endonuclease-mediated mutation 1	BL6–1	E	Null		Davies et al. (2013)
MGI:5566853	*Zic2*^{em2Bend}	Endonuclease-mediated mutation 2	C3H-5	E	Null		Davies et al. (2013)
MGI:5566854	*Zic2*^{em3Bend}	Endonuclease-mediated mutation 3	C3H-8	E	Null		Davies et al. (2013)
MGI:5566855	*Zic2*^{em4Bend}	Endonuclease-mediated mutation 4	C3H-10	E	Null		Davies et al. (2013)
MGI:5566856	*Zic2*^{em5Bend}	Endonuclease-mediated mutation 5	C3H-19	E	Null		Davies et al. (2013)
MGI:1852004	*Zic2*^{Ku}	Kumba	Gena29, Ku	ENU		Holoprosencephaly	Brown et al. (2000)
MGI:4943187	*Zic2*^{m1Nisw}	Mutation 1	Line7–12	ENU	Hypomorph		Zhang and Niswander (2013)
MGI:2156825	*Zic2*^{tm1Jaru}	Targeted mutation 1	$Zic2^{Kd}$	T	Hypomorph	Holoprosencephaly, schizophrenia	Nagai et al. (2000)

(continued)

Table 10.2 (continued)

MGI allele ID	Allele symbol	Allele name	Synonyms	Allele type	Allele attributes	Human disease models	References
Zic3							
MGI:1856679	*Zic3*^{Bn}	Bent-tail	Bn	S		Heterotaxy	Garber (1952)
MGI:3043028	*Zic3*^{Ka}	Katun	Ka	S		Heterotaxy	Bogani et al. (2004)
MGI:5476827	*Zic3*$^{tm1.1Smwa}$	Targeted mutation 1.1	Zic3^{flox}	T			Sutherland et al. (2013)
MGI:2180720	*Zic3*^{tm1Bca}	Targeted mutation 1	Zic3^{-}, Zic3^{null}	T	Null	Heterotaxy	Purandare et al. (2002)
MGI:3698161	*Zic3*^{tm1Jwb}	Targeted mutation 1	Zic3^{neo}, Zic3^{null}	T	Null	Goldenhar syndrome	Zhu et al. (2007)
MGI:5470150	*Zic3*$^{tm2.1Jwb}$	Targeted mutation 2.1	Zic3^{flox}	T	No functional change		Jiang et al. (2013)
Zic4							
MGI:3789699	*In(9)26Rk*	Inversion	In26Rk	CI			Roderick (1983)
MGI:3510681	*Zic1/Zic4*^{tm1Kjmi}	Targeted mutation 1	Zic1^{-}Zic4^{-}	T	Null	Dandy-Walker syndrome	Grinberg et al. (2004)
MGI:3510678	*Zic4*^{tm2Kjmi}	Targeted mutation 2	Zic4^{-}	T	Null		Grinberg et al. (2004)
Zic5							
MGI:3052154	*Zic5*^{tm1Jaru}	Targeted mutation 1	Deltaneo	T	Null		Inoue et al. (2004)
MGI:3574814	*Zic5*^{tm1Sia}	Targeted mutation 1	Opr^{-}	T	Null		Furushima et al. (2005)

Tr transgenic, *T* targeted, *E* endonuclease-mediated, *CI* chemically induced (non-ENU), *S* spontaneous

Table 10.3 Murine *Zic* loss-of-function phenotypes

Zic phenotypes	Genotypes	References
Gastrulation		
Primitive streak dysgenesis	$Zic3^{/-}$, $Zic3^{-/Y}$	Ware et al. (2006)
Failure of anterior notochord production	$Zic2^{Ku/Ku}$	Warr et al. (2008)
Left-right axis defects	$Zic2^{Ku/Ku}$	Carrel et al. (2000), Purandare et al. (2002), Ahmed et al. (2013), and Barratt et al. (2014)
	$Zic3^{Bn/Bn}$, $Zic3^{Bn/Y}$, $Zic3^{Bn/+}$, $Zic3^{-/-}$, $Zic3^{-/Y}$	
	$Zic3^{Ka/Ka}$, $Zic3^{Ka/Y}$	
Node cilia defects	$Zic2^{Ku/Ku}$	Barratt et al. (2014)
Patterning and morphogenesis		
Exencephaly	$Zic2^{Ku/Ku}$, $Zic2^{kd/kd}$, $Zic2^{m1Nisw/m1Nisw}$	Zhang and Niswander (2013), Nagai et al. (2000), Klootwijk et al. (2000), Purandare et al. (2002), Elms et al. (2003), Inoue et al. (2004), and Furushima et al. (2005)
	$Zic3^{Bn/Bn}$, $Zic3^{Bn/Y}$	
	$Zic3^{-/-}$, $Zic3^{-}/Y$	
	$Zic5^{-/-(a)}$, $Zic5^{-/-(b)}$	
Spina bifida (and curled tail)	$Zic2^{Ku/+}$, $Zic2^{kd/kd}$, $Zic2^{m1Nisw/m1Nisw}$	(Zhang and Niswander (2013), Garber (1952), Nagai et al. (2000), Klootwijk et al. (2000), and Elms et al. (2003)
	$Zic3^{Bn/Bn}$, $Zic3^{Bn/Y}$	
Omphalocele	$Zic3^{Bn/Bn}$, $Zic3^{Bn/Y}$	Klootwijk et al. (2000)
Cleft lip	$Zic3^{Bn/Bn}$, $Zic3^{Bn/Y}$	Klootwijk et al. (2000)
Cranial neural crest deficit	$Zic2^{Ku/Ku}$, $Zic2^{kd/kd}$	Nagai et al. (2000), Elms et al. (2003), Inoue et al. (2004), and Furushima et al. (2005)
	$Zic5^{-/-(a)}$, $Zic5^{-/-(b)}$	
Trunk neural crest deficit	$Zic2^{Ku/Ku}$	Elms et al. (2003)
Hindbrain patterning	$Zic2^{Ku/Ku}$	Elms et al. (2003)
Skeletal defects		
Craniofacial	$Zic5^{-/-(a)}$, $Zic5^{-/-(b)}$	Inoue et al. (2004) and Furushima et al. (2005)
Sternum/ribs	$Zic5^{-/-(a)}$, $Zic1^{-/-}$	Aruga et al. (1999) and Inoue et al. (2004)
Vertebral arches	$Zic1^{-/-}$	GrüNeberg (1963), Aruga et al. (1999), Nagai et al. (2000), Purandare et al. (2002), and Inoue et al. (2004, 2007)
	$Zic2^{kd/kd}$, $Zic2^{kd/+}$;$Zic3^{Bn/Y}$	
	$Zic3^{Bn/X}$, $Zic3^{Bn/Bn}$, $Zic3^{Bn/Y}$, $Zic3^{-/+}$	
	$Zic3^{-/-}$, $Zic3^{-/Y}$	
	$Zic5^{-/-(a)}$	
Tail (reduced number and or dysmorphic caudal vertebrae)	$Zic2^{kd}/^{kd}$	GrüNeberg (1963), Purandare et al. (2002), Inoue et al. (2004), and Ahmed et al. (2013)
	$Zic3^{Bn/X}$, $Zic3^{Bn/Bn}$, $Zic3^{Bn}/^{Y}Zic3^{-/+}$	
	$Zic3^{-/-}$, $Zic3^{-/Y}$	
	$Zic3^{Ka/+}$	
	$Zic5^{-/-(a)}$	
Limb	$Zic2^{kd/kd}$	Nagai et al. (2000)
Classical HPE	$Zic2^{Ku/Ku}$	Warr et al. (2008)

(continued)

Table 10.3 (continued)

Zic phenotypes	Genotypes	References
Roof-plate abnormality/MIHV HPE	*Zic2*$^{kd/kd}$	Nagai et al. (2000) and Furushima et al. (2005)
	Zic5$^{-/-(b)}$	
Dorsal spinal cord hypoplasia	*Zic1*$^{-/-}$	Aruga et al. (1998)
Organogenesis		
Forebrain		
Dysgenesis of medial structures	*Zic1*$^{-/-}$, *Zic1*$^{-/-}$;*Zic3*$^{Bn/Y}$	Purandare et al. (2002), Inoue et al. (2004), and Furushima et al. (2005)
	Zic3$^{-/+}$, *Zic3*$^{-/-}$, *Zic3*$^{-/Y}$, *Zic3*$^{Bn/Bn}$	
	Zic5$^{-/-(a)}$	
Enlarged ventricles	*Zic5*$^{-/-(b)}$	Furushima et al. (2005)
Microcephaly	*Zic2*$^{kd/kd}$, *Zic5*$^{-/-(a)}$	Nagai et al. (2000) and Furushima et al. (2005)
Cerebellum: reduced size	*Zic1*$^{-/-}$	Aruga et al. (1998), Aruga (2004), Grinberg et al. (2004), and Blank et al. (2011)
	Zic1$^{+/-}$; *Zic4*$^{+/-}$	
	Zic1$^{-/-}$;*Zic4*$^{-/-}$	
	Zic3$^{Bn/Y}$, *Zic3*$^{Bn/Bn}$	
Cerebellum: folia patterning	*Zic1*$^{-/-}$,	Aruga et al. (1998, 2002b), Aruga (2004), Grinberg et al. (2004), and Blank et al. (2011)
	Zic1$^{+/-}$;*Zic2*$^{kd/+}$	
	Zic1$^{+/-}$;*Zic4*$^{+/-}$	
	Zic1$^{-/-}$;*Zic4*$^{-/-}$	
	Zic1$^{+/-}$;*Zic2*$^{Kd/+}$	
	Zic3$^{Bn/Y}$, *Zic3*$^{Bn/Bn}$	
Non-forebrain		
Hydrocephalus	*Zic1*$^{+/-}$;*Zic2*$^{kd/+}$	Aruga et al. (2002b) and Inoue et al. (2004)
	Zic5$^{-/-(a)}$	
Eye: micro- and anophthalmia	*Zic2*$^{kd/kd}$	Klootwijk et al. (2000), Herrera et al. (2003), and Furushima et al. (2005)
	Zic3$^{Bn/Bn}$, *Zic3*$^{Bn/Y}$	
	Zic5$^{-/-(b)}$	
Eye: optic chiasm: morphology	*Zic2*$^{kd/kd}$	Herrera et al. (2003)
Eye: aberrant ipsilateral projection of retinal ganglion cells	*Zic2*$^{kd/+}$, *Zic2*$^{kd/kd}$	Herrera et al. (2003)
Limb	*Zic2*$^{kd/kd}$	Garber (1952)
	Zic3$^{Bn/Y}$, *Zic3*$^{Bn/Bn}$	
Small size at birth	*Zic1*$^{+/-}$;*Zic4*$^{+/-}$	Garber (1952), Nagai et al. (2000), Inoue et al. (2004), and Grinberg et al. (2004)
	Zic1$^{-/-}$; *Zic4*$^{-/-}$	
	Zic2$^{kd/kd}$	
	Zic3$^{Bn/Bn}$, *Zic3*$^{Bn/Y}$	
	Zic5$^{-/-}$	

(continued)

Table 10.3 (continued)

Zic phenotypes	Genotypes	References
Abnormal intestine morphology	$Zic2^{m1Nisw/m1Nisw}$	Zhang and Niswander 2013
Abnormal enteric neural crest morphology	$Zic2^{m1Nisw/m1Nisw}$	Zhang and Niswander 2013
Abnormal neurite morphology	$Zic2^{m1Nisw/m1Nisw}$	Zhang and Niswander 2013
Post-birth		
Embryonic lethality	$Zic2^{Ku/Ku}$,	Garber (1952), Purandare et al. (2002), and Elms et al. (2003)
	$Zic3^{Bn/Bn}$, $Zic3^{Bn/Y}$, $Zic3^{-/-}$, $Zic3^{-/Y}$	
Perinatal lethality	$Zic2^{kd/kd}$, $Zic2^{m1Nisw/m1Nisw}$	Zhang and Niswander 2013, Nagai et al. (2000), and Inoue et al. (2004)
	$Zic5^{-/-(a)}$	
Juvenile lethality	$Zic1^{+/-}$;$Zic2^{kd/+}$*(4 weeks)*	Nagai et al. (2000), Aruga et al. (2002b), Inoue et al. (2004), Grinberg et al. (2004), and Blank et al. (2011)
	$Zic1^{+/-}$;$Zic4^{+/-}$*(4 weeks)*	
	$Zic1^{-/-}$;$Zic4^{-/-}$	
	$Zic1^{-/-}$, $Zic2^{kd/kd}$	
	$Zic5^{-/-}$*(4 weeks)*$^{(a)}$	
Behavioural		
Prepulse inhibition	$Zic2^{+/kd}$	Ogura et al. (2001)
Motor control	$Zic1^{+/-}$, $Zic1^{-/-}$	Ogura et al. (2001), Aruga (2004), and Grinberg et al. (2004)
	$Zic1^{+/-}$;$Zic4^{+/-}$	
	$Zic1^{-/-}$;$Zic4^{-/-}$	
	$Zic3^{Bn/Y}$, $Zic3^{Bn/Bn}$	
Abnormal gait and posture	$Zic1^{+/-}$	Aruga et al. (2002b) and Inoue et al. (2004)
	$Zic2^{kd/+}$	
	$Zic5^{-/-(a)}$	

$^{(a)}$*$Zic5^{Tm1Jaru}$ strain*
$^{(b)}$*$Zic5^{Tm1Sia}$ strain*

in the double mutant. Functional redundancy is often proposed as the mechanism underlying the absence of a phenotype predicted on the basis of gene expression pattern and/or gain-of-function phenotypes. For example, the *Zic* genes are implicated in the early stages of neural crest cell development, as overexpression of each *Zic* gene in *Xenopus* embryos (*Xenopus zic1 5*) induces neural crest (Nakata et al. 1997, 1998, 2000; Fujimi et al. 2006). In the mouse, *Zic2*, *3* and *5* are all expressed in the neural crest precursor population from early on in its ontogeny, and *Zic1* and *4* expression in the dorsal neurectoderm domain from which the neural crest arises initiates early in crest migration (Nagai et al. 1997; Furushima et al. 2000; Elms et al. 2004; Gaston-Massuet et al. 2005). Despite these data clearly predicting a role for each of the murine *Zic* genes in neural crest development, as shown in Table 10.3, only *Zic2* or *Zic5* loss-of-function has been associated with cranial neural crest defects in mammals, and only *Zic2* severe loss (i.e. $Zic2^{Ku/Ku}$) has been associated with trunk neural crest deficits (Nagai et al. 2000; Elms et al. 2003; Inoue et al. 2004; Furushima et al. 2005). Moreover, the *Zic2*$^{Ku/Ku}$ trunk neural crest cell phenotype is mild, with a decrease in crest production attributed to a delay in neural crest cell differentiation, rather than a total loss of neural crest cell production (Elms et al. 2003). It is possible that this occurs because of functional redundancy and that inactivating other *Zic* genes in the *Zic2* loss-of-function background would precipitate a more severe phenotype. In particular, it may be predicted that *Zic2* and *Zic5* are partially redundant during neural crest cell development and that combined loss of *Zic2* and *Zic5* would enhance the neural crest phenotype relative to the individual mutations. This experiment is difficult to conduct, however, because these two genes are closely linked (being separated by ~10 Kb of genomic DNA) and the chance of finding a recombination event between these two genes (as required to make compound homozygotes) is very small. The most sensible means of generating a compound *Zic2/Zic5* mutant would be to make a deletion which encompasses both genes. Bi-gene arrangement and the concomitant close linkage of highly homologous sequences have, however, been associated with inefficient homologous recombination, and this may have contributed to the paucity of alleles produced via targeted mutation in ES cells at the *Zic* bi-gene pairs (Table 10.3).

Despite these difficulties, alleles that allow the consequence of individual or combined loss of the *Zic1/Zic4* bi-gene pair have been produced and analysed, demonstrating that *Zic1* and *Zic4* are partially redundant during cerebellum development (Grinberg et al. 2004; Blank et al. 2011). The null mutation of *Zic1* alone is associated with small size and abnormal foliation of the cerebellum, whereas animals null for *Zic4* exhibit no detectable cerebellar phenotype. Both the size and foliation cerebellar defects are enhanced (relative to the *Zic1* null) in the double mutant. The data demonstrate that ZIC1 and ZIC4 both play a role in cerebellar development and that ZIC1 function can compensate for the loss of ZIC4 activity but ZIC4 cannot fully replace the function of ZIC1.

10.6.2 Interdependent Zic *Gene Function*

One feature commonly associated with compound mutant analysis in the mouse is the presentation of the individual homozygous phenotype in animals trans-heterozygous for the two mutations. This phenomenon is known as non-allelic non-complementation (because when trans-heterozygous mutations at the *same* gene precipitate a known homozygous phenotype, the alleles are said to 'fail to complement' each other). The combined trans-heterozygous loss of *Zic1* and *Zic4* leads to cerebellar size and foliation defects in both man and mouse (Grinberg et al. 2004; Blank et al. 2011). Similarly, animals homozygous for the hypomorphic allele of *Zic2* (*Zic2*$^{kd/kd}$) have no cerebellar phenotype, yet one copy of this allele in combination with heterozygous loss of *Zic1* produces a phenotype similar to that of the *Zic1* homozygous null (Aruga et al. 2002a). It is considered that non-allelic non-complementation occurs because the two genes involved are both redundant and dose-dependent (Pérez-Pérez et al. 2009). This combined haploinsufficiency implies that the amalgamated gene products fall below a threshold level necessary to achieve the wild-type phenotype and is consistent with a situation in which the expression or activity of the paralogs depends upon each other's expression. Here the heterozygous null mutation in one or both of the genes causes decreased activity of the other leading to apparent haploinsufficiency. The *Zic1/Zic4* and *Zic1/Zic2* data imply a regulatory interdependency between these two factors, the specific nature of which is unknown. It could arise because the mRNA expression of one gene is dependent on the other or it could be that the proteins interact in a complex in which ZIC protein stoichiometry potentiates function. Perhaps surprisingly, the ZIC proteins have never been reported to form homomeric or heteromeric complexes. The presence of homomeric ZIC2 complexes has been specifically searched for, but not found, by complex immunoprecipitation following overexpression in HEK293T cells of differentially epitope-tagged ZIC2 proteins and by a yeast two-hybrid analysis to directly assay ZIC2 physical interaction with itself (Brown et al. 2005). Additionally, a ZIC3 variant protein containing only the amino acids N-terminal of the ZFD does not dominantly interfere with full length ZIC3, ZIC2 or ZIC5 proteins in a HEK293T cell-based transactivation assay. This suggests these proteins do not bind one another via their N-terminus (Ahmed et al. 2013).

10.6.3 Unique Zic *Gene Function*

It is tempting to speculate, given the extensive overlap in *Zic* gene expression patterns, that most if not all *Zic* gene activity will exhibit some level of redundancy. It is clear that some ZIC activity is nonredundant but this would be expected if the particular requirement is traced to a unique site of *Zic* gene expression, as occurs, for example, in one class of *Zic2*-associated Holoprosencephaly (HPE). As described in Chap. 14, loss-of-function alleles of *Zic2* result in a structural defect of forebrain development known as HPE. The term HPE refers to a wide spectrum of structural defects that can affect the entire dorsal-ventral forebrain axis (classic HPE; due to aberrant ventral patterning) or only the dorsal forebrain (middle interhemispheric variant HPE; due to abnormal dorsal patterning). Severe *Zic2* loss-of-function alleles give rise to classic HPE, whereas hypomorphic alleles are associated with middle interhemispheric variant (MIHV) HPE in man and mouse (Brown et al. 1998, 2001; Nagai et al. 2000; Warr et al. 2008; Solomon et al. 2010; Dubourg et al. 2016). The cause of classic HPE following *Zic2* mutation has been traced to the node of the mid-gastrulation stage mouse embryo (Warr et al. 2008), a site at which *Zic2* is the only *Zic* transcript detected by whole-mount in situ hybridization studies (Elms et al. 2004). This unique *Zic2* phenotype therefore does not arise due to a lack of redundancy. The *Zic2* transcripts are, however, not restricted to the node but are broadly distributed throughout the ectoderm and mesoderm germ layers of the mid-gastrula embryo, and yet loss of activity in these tissues has no apparent phenotypic consequence. Two other *Zic* genes (*Zic5* and *Zic3*) share these other sites of *Zic2* expression (Elms et al. 2004; Furushima et al. 2000), and it seems likely that these proteins compensate for ZIC2 activity in these cells. It remains to be tested whether *Zic3* or *Zic5* ablation on the *Zic2* mutant background would evoke the additional defects expected if redundancy is occurring.

Zic2-associated MIHV HPE seems to arise because of a requirement for ZIC2 function in the neural plate border and/or neural tube roof-plate (as described in Chap. 14) (Nagai et al. 2000). Both tissues are sites of multiple *Zic* gene expression, yet this phenotype is uniquely associated with *Zic2* mutation. In contrast to classic HPE, the MIHV *Zic2*-associated HPE phenotype evidently occurs despite the normal co-expression of other *Zic* genes, indicating that they are not able to fully compensate for decreased *Zic2* activity in this tissue. A similar situation also seems to occur in the case of *Zic3*-associated Heterotaxy, a defect traced to the role of *Zic3* in the nascent mesoderm of the mid-gastrula (Sutherland et al. 2013). Both *Zic2* and *Zic5* are expressed in this region of the embryo yet are evidently unable to fully compensate for the loss of ZIC3 activity. In both cases, it remains to be seen whether partial redundancy could be revealed by compound mutant analysis.

10.6.4 Pleiotropic Zic *Gene Function*

The catalogue of *Zic*-related phenotypes shown in Table 10.3 makes clear the remarkable extent of *Zic* gene pleiotropy, a term which refers to the production by a single gene of two or more apparently unrelated effects. Pleiotropy may arise for a variety of reasons (for review, see Paaby and Rockman 2013), and disparate causes are evident for different cases of *Zic* pleiotropy. For example, the $Zic2^{Ku/Ku}$ allele exhibits both classical HPE and defects of heart development. At first glance, these may appear unrelated events, but detailed phenotypic analysis reveals that both phenotypes can be traced to defective node function in the mid-gastrulation stage embryo (Warr et al. 2008; Barratt et al. 2014). Here, the two phenotypes arise because the mid-gastrula node has multiple functions: cells that move through the node give rise to the anterior notochord, which in turns ensures development of the prechordal plate (PrCP) cells. These PrCP cells signal to the overlying neurectoderm to impart ventral neural character, whilst fluid flow across the node leads to asymmetric signalling that imparts left character to the left lateral plate mesoderm, directly influencing heart looping and morphogenesis. The node is crucial to the development of multiple tissues; hence, a single gene defect that perturbs node function will give rise to multiple phenotypes. This condition is termed developmental pleiotropy by Paaby and Rockman (2013). The $Zic2^{Ku/Ku}$ allele generates phenotypes in many other developmental processes. For example, defects in neurulation, in neural crest cell production and in hindbrain patterning have also been documented (Table 10.3). These processes are not dependent upon the activity of the node nor on some other common precursor cell, but rather this pleiotropy appears to reflect the fact that most proteins play a role in distinct cell types, and any genetic change that alters gene expression or function can therefore affect a variety of tissues. This is referred to as molecular-gene pleiotropy by Paaby and Rockman (2013).

Pleiotropy is also evident upon mutation of other *Zic* genes: *Zic1* and *Zic4* mouse mutants are small at birth and have cerebellar abnormalities (Aruga et al. 1998; Grinberg et al. 2004), whilst mutation of *Zic5* affects diverse systems or processes such as neurulation, neural crest development and skeletal and eye development (Inoue et al. 2004; Furushima et al. 2005). *Zic3* mutation, like *Zic2*, affects a broad range of tissues or processes, some of which are linked by their ontogeny (e.g. the limb and tail defects may both reflect a role in skeletal development) and others which arise from a role in multiple tissues (such as craniofacial, cerebellar, neural tube and ventral closure defects) (Garber 1952; Purandare et al. 2002; Ahmed et al. 2013; Klootwijk et al. 2000). Overall, the extent of *Zic* gene pleiotropy implies a multitude of roles for the Zic proteins during embryonic development. Furthermore, as discussed above, *Zic* gene redundancy means that many phenotypes may not have been uncovered by the current analyses that are largely restricted to the characterization of null alleles of individual genes. One question is whether the striking *Zic* molecular-gene pleiotropy is attributable to the iterative use of the same molecular function or whether the ZIC proteins have an unusually large number of protein-protein interaction partners.

The later possibility is supported by the evident ability of ZIC proteins to control target gene transcription by direct binding to DNA or by physical interaction with other transcription factors. As shown in Fig. 10.4, the ZIC proteins have been shown to physically interact with key transcriptional mediators of three widely used embryonic signalling pathways (GLI mediators of HH signals, TCF/LEF mediators of canonical WNT signals and SMAD mediators of NODAL signals) (Mizugishi et al. 2001; Koyabu et al. 2001; Pourebrahim et al. 2011; Fujimi et al. 2012; Houtmeyers et al. 2016). Indeed, the demonstration that the ZIC proteins can bind GLI proteins via their highly homologous ZFDs provided one of the first demonstrations that C2H2 zinc fingers could bind proteins as well as DNA (Koyabu et al. 2001). The binding of the subclass A ZIC proteins (ZICs 1–3) to each of the GLI proteins (GLI1–3) was demonstrated via Co-IP following overexpression of tagged proteins in several transformed cell lines. Binding of the subclass B proteins to GLI molecules has not been assessed. The ZIC/GLI interaction facilitated the nuclear localization of the GLI proteins and altered GLI-dependent transcription activity in cell-based luciferase assays independent of ZIC DNA binding activity. GLI-dependent transcription was variably augmented or suppressed in different cell lines (Mizugishi et al. 2001; Koyabu et al. 2001). There is evidence that ZIC function intersects the SHH pathway during mouse development. For example, a null allele of *Zic3*, whilst not exhibiting limb defects, can rescue the heterozygous *Gli3* loss-of-function phenotype of the extra-toes mutant (Quinn et al. 2012). GLI3 protein can activate SHH target genes or can be processed into a C-terminally truncated repressor form. ZIC3 appears to influence the balance between the repressor and activator forms such that in the extra-toes mutant, ZIC3 deficiency increases the amount of the GLI3 repressor form, thereby abrogating SHH ectopic expression and preventing the resulting preaxial polydactyly (Quinn et al. 2012). The role of ZIC1

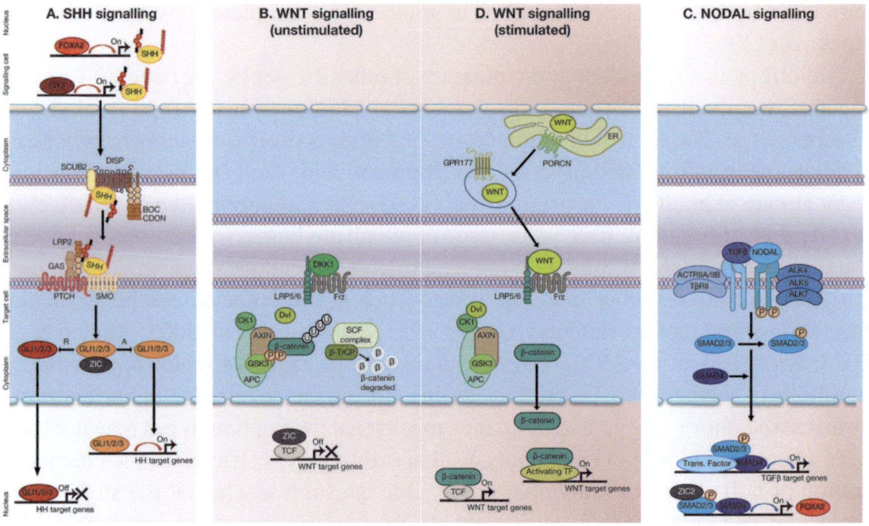

Fig. 10.4 The role of the ZIC proteins in signalling pathways. (**a**) *SHH signalling*. In the absence of ligand, the transmembrane domain protein Patched1 (PTCH) inhibits the activity of Smoothened (SMO) in the target cell. In a signalling cell, SHH expression is initiated by the transcription factor FOXA2 or SIX3 binding to enhancers. In the extracellular space, the binding of secreted SHH to PTCH releases SMO, allowing for regulation of HH target genes by the activator (A) form of the transcription factors GLI1, GLI2 and GLI3. GLI3 protein can activate SHH target genes or can be processed into a C-terminally truncated repressor form (R). Subclass A ZIC proteins (ZICs 1–3) can bind to each of the GLI proteins (GLI1–3), facilitating nuclear localization of the GLI proteins and altered GLI-dependent transcription activity (Mizugishi et al. 2001; Koyabu et al. 2001). Evidence from extra-toes mutants found that ZIC3 appears to influence the balance between the repressor and activator forms, whereby ZIC3 deficiency increases the amount of the GLI3 repressor form (Quinn et al. 2012). (**b**) *WNT signalling*. In the absence of ligand/presence of inhibitors such as DKK1 (unstimulated), β-catenin is phosphorylated (P) by the kinase activity of the destruction complex (consisting of Axin, APC, GSK3 and CK1), polyubiquitylated (U) by the SCF (SKP1, Cullin, F-box)/BTrCP complex and degraded by the proteasome. ZICs act as transcriptional corepressors, complexing with TCF proteins to prevent transcription of WNT target genes (Pourebrahim et al. 2011; Fujimi et al. 2012). In the presence of WNT ligands binding to the receptor (Frz)/co-receptor (LRP5/6) complex (stimulated), a cascade of cytoplasmic events culminating in β-catenin nuclear entry occur. Nuclear β-catenin mediates derepression of targets, potentially via the phosphorylation (P) of TCF7L1. β-catenin also mediates the activation of target genes, including some WNT pathway components. Processing and secretion of WNT ligand is regulated by PORCN and GPR177, whereby WNT ligand is bound and lipidated by PORCN in the endoplasmic reticulum (ER) before transport to the Golgi where GPR177 binds to WNT and escorts it to the plasma membrane. (**c**) *NODAL signalling*. Mature NODAL ligands complex with the EGF-CFC cofactor TDGF1 (Cripto), type I receptors (ALK4/5/7) and type II receptors (ActRII or ActRIIB). Receptor activation leads to the phosphorylation of the type I receptor by the type II kinase, as well as phosphorylation of SMAD2 or SMAD3, which dimerize with SMAD4. The SMAD2/3-4 complex translocates to the nucleus and interacts with the transcription factor FOXH1 and promotes transcription of TGFβ target genes, or with ZIC2 to promote transcription of FOXA2 (Houtmeyers et al. 2016)

and ZIC4 for controlling cerebellar size also intersects the SHH pathway, since the reduction in cerebellar size seen in *Zic1* and *Zic4* mutants results from decreased proliferation of the granule cell progenitor population, a SHH-dependent process. *Shh* itself is expressed normally in the developing cerebellum of the compound mutant embryos, suggesting that ZIC1 and ZIC4 proteins are directly or indirectly involved in receiving or transducing the SHH signal (Blank et al. 2011).

The ZIC proteins have also been shown to physically interact with members of the TCF family of proteins, which are the transcriptional mediators of canonical WNT signals. As shown in Fig. 10.4b, c and reviewed in Arkell et al. 2013, in a low WNT signalling environment, the TCF proteins complex with corepressors to dampen transcription at TCF binding sites. In a high WNT environment, stabilized β-catenin enters the nucleus and complexes with TCF proteins to activate expression at WNT target genes. The ZIC proteins have been shown to physically interact with TCF proteins via their ZFDs. Binding was demonstrated via Co-IP following overexpression of tagged proteins in transformed mammalian cell lines (HEK293T and COS7 cells) (Pourebrahim et al. 2011; Fujimi et al. 2012). In the presence of ZIC proteins, TCF-dependent transcription is inhibited in HEK293T cell-based transcription assays (Pourebrahim et al. 2011) and in heterologous reporter assays conducted in the *Xenopus* embryo (Fujimi et al. 2012). The ZIC zinc finger mediates the physical interaction with the TCF proteins, and transcription inhibition occurs independent of ZIC DNA binding, indicating a cofactor role for ZIC proteins at TCF binding sites (Pourebrahim et al. 2011). Currently, genetic interaction studies between ZIC proteins and the WNT pathway have not been conducted in the mouse, but there is evidence that ZIC proteins can inhibit endogenous WNT signalling in *Xenopus* embryos. For example, *zic3* overexpression in *Xenopus* embryos decreases expression of endogenous WNT target genes; conversely morpholino knockdown of zic2 protein causes increased expression of a TCF-dependent reporter transgene. Overexpression of β-catenin in the ventral side of *Xenopus* embryos results in the formation of secondary axes (Funayama et al. 1995), but co-injection of either zic2 (Pourebrahim et al. 2011) or zic3 (Fujimi et al. 2012) RNA with β-catenin in this experiment prevents secondary axis formation and confirms the ability of the ZIC proteins to inhibit the canonical WNT pathway.

Recently the ZIC2 protein has been shown to interact with SMAD2 and SMAD3, transcriptional mediators which respond to receptor activation by some ligands of the TGF-β superfamily (including TGF-βs and NODAL) (Fig. 10.4d). This binding was demonstrated via Co-IP following overexpression of tagged proteins in a transformed cell line (HEK293T cells), and ZIC2 was also precipitated in a complex with the endogenous phosphorylated SMAD2 protein following stimulation of the HEK293T cells with recombinant TGF-β (Houtmeyers et al. 2016). The presence of ZIC2 antagonized SMAD activity in HEK293T cell-based transactivation assays and at endogenous SMAD-dependent loci in the A549 lung adenocarcinoma cell line. For example, in A549 cells, SMAD-dependent activation of the SERPINE1 locus was converted to repression in the presence of ZIC2, and SMAD-dependent repression of the FOXA2 locus was converted to activation in the presence of ZIC2. The N-terminal 140 aa of the ZIC2 protein is required for SMAD binding. Variant forms of ZIC2 protein in which the N-terminus is intact, but the ZFD disrupted, are unable to oppose SMAD activity, despite still binding SMAD protein, suggesting a bipartite mechanism of ZIC2/SMAD interaction and antagonism. SMAD antagonism occurred in the absence of ZIC-binding sites, suggesting that the ZIC-mediated antagonism of SMAD activity occurs in a DNA binding-independent manner and that, similar to the GLI and TCF interactions, ZIC2 acts as a cofactor to antagonize SMAD-dependent transcription control. There is evidence that ZIC2 intersects the NODAL pathway during murine gastrulation since halving the dose of NODAL activity on the $Zic2^{Ku/Ku}$ background precipitates anterior truncation, a known consequence of loss of NODAL signalling, in ~80% of $Zic2^{Ku/Ku}$ embryos (whereas $Nodal^{+/-}$ embryos exhibit no phenotype) (Houtmeyers et al. 2016). This apparent *Nodal* haploinsufficiency in the absence of ZIC2 function implies that ZIC2 potentiates the NODAL signal during murine gastrulation.

Transcription cofactors (i.e. proteins that influence transcription without binding template DNA) are generally thought to mediate protein-protein interactions between regulatory transcription factors and the basal transcription machinery or chromatin modifiers. Given the evidence that the ZIC proteins can act as cofactors in a range of embryonic signalling pathways, it may be expected that interactions between ZIC and other classes of transcriptional proteins occur. Experiments in which ZIC2-containing complexes are purified and analysed by mass spectrometry following overexpression in HEK293T cells or embryonic stem cells have identified a range of molecules involved in the transcription machinery and chromatin modification. For example, ZIC2-containing complexes purified from HEK293T cells contained an RNA helicase and molecules involved in DNA repair (Ishiguro et al. 2007). Similarly, ZIC2-containing complexes purified from embryonic stem cells contained all of the core components of the Mbd3-containing NuRD complex, which has nucleosome remodelling and histone deacetylase activities (Luo et al. 2015). In each case, the interaction with ZIC2 was confirmed by Co-IP experiments. Whether these interactions are required for ZIC function in vivo, however, has not yet been examined.

10.6.5 Unified *Zic* Gene Function

Despite the widespread use of the ZIC proteins in multiple signalling pathways to activate or repress transcription in a context-dependent manner and the resulting pleiotropy, it is possible that the ongoing phenotype analysis may reveal a unifying biological function for these proteins at the organelle and cellular level. One hypothesis that emerges from the current phenotype data is that ZIC proteins may be required for function of the primary cilium, an organelle that emanates from the cell surface of most mammalian cells and appears to coordinate a number of signalling pathways (reviewed in Murdoch and Copp 2010). This is supported by the finding that the node dysfunction of the $Zic2^{Ku/Ku}$ embryos occurs alongside visibly abnormal node cilia and is associated with loss of expression of ciliogenesis transcription factors such as *Foxj1* and *Rfx3* (Barratt et al. 2014). Additionally, the phenotypes exhibited in some *Zic* mouse mutants can potentially be attributed to an underlying ciliopathy defect. For example, $Zic5^{-/-}$ mutants and $Zic1^{+/-};Zic2^{Kd/+}$ compound heterozygotes exhibit hydrocephalus (Aruga et al. 2002a; Inoue et al. 2004) (Table 10.3), an established phenotype of primary ciliary dyskinesia (Norris and Grimes 2012). Hydrocephaly develops when cerebral spinal fluid (CSF) accumulates within the cerebral ventricles. Motile cilia are required to move CSF through the cerebral aqueduct and ventricles, with cilia defects resulting in a build-up of CSF in the brain (reviewed in Sotak and Gleeson 2012; Ibañez-Tallon et al. 2004). It is currently unknown whether $Zic5^{-/-}$ mutants and $Zic1^{+/-};Zic2^{Kd/+}$ exhibit abnormal cilia development in the brain or spinal cord.

Hydrocephalus cases frequently co-occur with spina bifida aperta (Greene and Copp 2009), suggesting a common underlying cause. Spina bifida, one of the most common forms of neural tube defect (Au et al. 2010), results from failure of the posterior neuropore to completely close (Greene and Copp 2009). Interestingly, spina bifida aperta is regularly detected in murine alleles of *Zic2*, with abnormal folding of the posterior neuropore at 9.5 dpc the origin of the defect (Nagai et al. 2000; Elms et al. 2003) (Table 10.3). Currently, the underlying molecular cause of this spina bifida in *Zic2* mutants is unknown. It is known, however, that cilia line the ventricular zone of the neural tube and extend into the lumen (Huangfu and Anderson 2005; Bay and Caspary 2012). Here, the cilia are thought to transduce the SHH signalling cascade in the dorsal neural tube via the interactions of SMO, a transmembrane receptor and the transcription factors GLI2 and GLI3. The cilia here may also play a role in transduction of WNT and BMP signalling in the ventral neural tube (reviewed in Murdoch and Copp 2010); however, the mechanisms behind signal transduction in the neural tube are still largely unknown, and it is undetermined whether neural tube cilia are motile or if they can sense secreted ligands in the neural tube lumen. Whilst there is currently no evidence explicitly linking spina bifida to cilia defects, the link between cilia defects and neural tube defects via abrogated signalling pathways is strong. Conditional *Arl13b* mutants have been shown to exhibit spina bifida, most likely due to abnormal SHH signalling in the neural tube. ARL13B, a GTPase that localizes to the cilia, leads to significantly shorter cilia when mutated. This in turn results in low levels of GLI activators in the neural tube compared to wild-type embryos and thus aberrant SHH signalling (Caspary et al. 2007). Whether the spina bifida in *Arl13b* mutants develops secondary to other defects such as exencephaly is still unknown; however, a link between defective ciliogenesis and spina bifida is still presented. Furthermore, mutant embryos for the cilia-localized polycystin PKD1 exhibit abnormally sized renal cilia, as well as spina bifida (Lu 2001; Nikonova et al. 2014), providing further evidence for a possible link between the two phenotypes.

At the cellular level, one function repeatedly ascribed to the *Zic* genes is control of pluripotency. *Zic* mutant end phenotypes are often associated with small size (of the affected organ or of the entire animal), and, where investigated, the reduced size has been attributed to the premature differentiation of a progenitor population. This has the effect of ultimately decreasing the size of the organ since the progenitor differentiation occurs at the cost of increasing the size of the progenitor pool via self-renewal. Investigations in murine ES cells have ascribed the task of promoting pluripotency to *Zic3* (Lim et al. 2007) and of controlling exit from pluripotency to *Zic2* (Kalkan and Smith 2014). These studies have been conducted in embryonic stem cells, and it is not clear how the proposed roles of *Zic2* and *Zic3* in pluripotency relate to the loss-of-function phenotypes in mouse embryos. It is possible that functional redundancy has obscured these roles during early embryogenesis in murine *Zic* mutants.

10.7 Conclusion

Rodent *Zic* genes share many features conserved across vertebrate lineages and show high levels of homology to the human genes at both the DNA and protein levels. When murine *Zic* genes are mutated, they give rise to the same disorders as those associated with human *ZIC* gene dysfunction (Nagai et al. 2000; Garber 1952; Bogani et al. 2004; Purandare et al. 2002; Grinberg et al. 2004; Brown et al. 2000), indicating functional conservation. This structural and functional conservation relative to the human sets the rodent *Zic* genes apart from other vertebrate model organisms in which these features are less conserved. The mouse therefore serves as a key model organism for the comprehensive investigation of the role of *ZIC* genes in health and disease. Investigations to date have shown that the murine *Zic* genes encode multifunctional transcription regulators. They appear to interact with DNA and with a range of protein partners to connect DNA with positive or negative components of the transcription apparatus and higher-order chromatin complexes. In vitro overexpression studies suggest their protein partners include the transcriptional mediators of several embryonic signalling pathways (HH, WNT and TGF-β) as well as binding partners from transcription and chromatin complexes (Mizugishi et al. 2001; Koyabu et al. 2001; Pourebrahim et al. 2011; Fujimi et al. 2012; Houtmeyers et al. 2016; Ishiguro et al. 2007; Luo et al. 2015). Notably, cell-based experiments suggest that interaction of ZIC proteins alters the behaviour of the transcriptional mediators, converting activation to repression or vice versa in a protein- and context-dependent manner (Pourebrahim et al. 2011; Fujimi et al. 2012). This suggests that the ZIC proteins are able to work as hub proteins, integrating signals from multiple signal transduction pathways to change the balance between gene repression and activation of a broad range of targets.

Support for this hypothesis includes the ZIC protein structure, which in general consists of a ZFD flanked by unstructured regions. Both of these domain types can bind proteins (Mizugishi et al. 2001, 2004; Pourebrahim et al. 2011), supporting the unusually broad range of proposed binding partners (Ishiguro et al. 2007; Pourebrahim et al. 2011; Luo et al. 2015; Houtmeyers et al. 2016). Additionally, an ability to interact with multiple signalling pathways is implied by *Zic* gene pleiotropy and by the fact that *Zic* gene mutant phenotypes do not phenocopy those of a particular signalling pathway, instead exhibiting the hallmarks of many. Moreover, different propensities of each family member to interact with a particular signalling pathway in a given context could underlie the unique family member functions which are sometimes apparent even in tissues where other members are co-expressed. The challenge for Zic research is to validate the hub protein hypothesis in an in vivo mammalian setting. Recent technical advances should enable these previously intractable issues to be resolved. CRISPR-mediated mutagenesis will facilitate the production of refined *Zic* alleles which should include hypomorphic alleles, conditional alleles, alleles that mimic putative pathogenic human variants and tagged alleles. In particular, there is a pressing need to determine which of the many protein partners identified via overexpression in transformed or primary cell lines represent bona fide in vivo partners. In this respect, alleles which enable the in vivo documentation of protein/protein and protein/DNA complexes will be crucial, as will the use of improved resolution of 'omics' techniques to enable the transcriptional consequences of ZIC protein dysfunction to be assessed on a cell-by-cell basis.

References

Ahmed JN, Ali RG, Warr N et al (2013) A murine Zic3 transcript with a premature termination codon evades nonsense-mediated decay during axis formation. Dis Model Mech 6:755–767. https://doi.org/10.1242/dmm.011668\rdmm.011668 [pii]

Arkell RM, Fossat N, Tam PPL (2013) Wnt signalling in mouse gastrulation and anterior development: new players in the pathway and signal output. Curr Opin Genet Dev 23:454–460. https://doi.org/10.1016/j.gde.2013.03.001

Aruga J (2004) The role of Zic genes in neural development. Mol Cell Neurosci 26:205–221. https://doi.org/10.1016/j.mcn.2004.01.004

Aruga J, Yokota N, Hashimoto M et al (1994) A novel zinc finger protein, zic, is involved in neurogenesis, especially in the cell lineage of cerebellar granule cells. J Neurochem 63:1880–1890. https://doi.org/10.1046/j.1471-4159.1994.63051880.x

Aruga J, Nagai T, Tokuyama T et al (1996a) The mouse zic gene family: homologues of the drosophila pair-rule gene odd- paired. J Biol Chem 271:1043–1047

Aruga J, Yozu A, Hayashizaki Y et al (1996b) Identification and characterization Zic gene family of Zic4, a new member of the mouse. Gene 172:291–294

Aruga J, Minowa O, Yaginuma H et al (1998) Mouse Zic1 is involved in cerebellar development. J Neurosci 18:284

Aruga J, Mizugishi K, Koseki H et al (1999) Zic1 regulates the patterning of vertebral arches in cooperation with Gli3. Mech Dev 89:141–150

Aruga J, Inoue T, Hoshino J, Mikoshiba K (2002a) Zic2 controls cerebellar development in cooperation with Zic1. J Neurosci 22:218–225

Aruga J, Tohmonda T, Homma S, Mikoshiba K (2002b) Zic1 promotes the expansion of dorsal neural progenitors in spinal cord by inhibiting neuronal differentiation. Dev Biol 244:329–341. https://doi.org/10.1006/dbio.2002.0598

Aruga J, Kamiya A, Takahashi H et al (2006) A wide-range phylogenetic analysis of Zic proteins: implications for correlations between protein structure conservation and body plan complexity. Genomics 87:783–792. https://doi.org/10.1016/j.ygeno.2006.02.011

Au KS, Ashley-Koch A, Northrup H (2010) Epidemiologic and genetic aspects of spina bifida and other neural tube defects. Dev Disabil Res Rev 16:6–15

Barratt KS, Glanville-Jones HC, Arkell RM (2014) The Zic2 gene directs the formation and function of node cilia to control cardiac situs. Genesis 52:626–635. https://doi.org/10.1002/dvg.22767

Bay SN, Caspary T (2012) What are those cilia doing in the neural tube? Cilia 1:19. https://doi.org/10.1186/2046-2530-1-19

Bedard JEJ, Purnell JD, Ware SM (2007) Nuclear import and export signals are essential for proper cellular trafficking and function of ZIC3. Hum Mol Genet 16:187–198. https://doi.org/10.1093/hmg/ddl461

Benedyk MJ, Mullen JR, DiNardo S (1994) Odd-paired: a zinc finger pair-rule protein required for the timely activation of engrailed and wingless in Drosophila embryos. Genes Dev 8:105–117

Blank MC, Grinberg I, Aryee E et al (2011) Multiple developmental programs are altered by loss of Zic1 and Zic4 to cause Dandy-Walker malformation cerebellar pathogenesis. Development 138:1207–1216. https://doi.org/10.1242/dev.054114

Bogani D, Warr N, Elms P et al (2004) New semidominant mutations that affect mouse development. Genesis 40:109–117. https://doi.org/10.1002/gene.20071

Brown S, Warburton D, Brown LY et al (1998) Holoprosencephaly due to mutations in ZIC2, a homologue of drosophila odd-paired. Nat Genet 20:180–183. https://doi.org/10.1038/2484

Brown SDM, Nolan PM, Peters J et al (2000) A systematic, genome-wide, phenotype-driven mutagenesis programme for gene function studies in the mouse. Nat Genet 25:440–443. https://doi.org/10.1038/78140

Brown LY, Odent S, David V et al (2001) Holoprosencephaly due to mutations in ZIC2: alanine tract expansion mutations may be caused by parental somatic recombination. Hum Mol Genet 10:791–796

Brown L, Paraso M, Arkell R, Brown S (2005) In vitro analysis of partial loss-of-function ZIC2 mutations in holoprosencephaly: alanine tract expansion modulates DNA binding and transactivation. Hum Mol Genet 14:411–420. https://doi.org/10.1093/hmg/ddi037

Carrel T, Purandare SM, Harrison W et al (2000) The X-linked mouse mutation bent tail is associated with a deletion of the Zic3 locus. Hum Mol Genet 9:1937–1942

Caspary T, Larkins CE, Anderson KV (2007) The graded response to sonic hedgehog depends on cilia architecture. Dev Cell 12:767–778. https://doi.org/10.1016/j.devcel.2007.03.004

Chen L, Ma Y, Qian L, Wang J (2013) Sumoylation regulates nuclear localization and function of zinc finger transcription factor ZIC3. Biochim Biophys Acta, Mol Cell Res 1833:2725–2733. https://doi.org/10.1016/j.bbamcr.2013.07.009

Chhin B, Hatayama M, Bozon D, Ogawa M, Schön P, Tohmonda T, Sassolas F, Aruga J, Valard A-G, Chen S-C, Bouvagnet P (2007) Elucidation of penetrance variability of a mutation in a family with complex heart defects and functional analysis of mutations in the first zinc finger domain. Hum Mutat 28(6):563–570

Chinwalla AT, Cook LL, Delehaunty KD et al (2002) Initial sequencing and comparative analysis of the mouse genome. Nature 420:520–562. https://doi.org/10.1038/nature01262

Cowan J, Tariq M, Ware SM (2014) Genetic and functional analyses of ZIC3 variants in congenital heart disease. Hum Mutat 35:66–75. https://doi.org/10.1002/humu.22457

Davies B, Davies G, Preece C et al (2013) Site specific mutation of the Zic2 locus by microinjection of TALEN mRNA in mouse CD1, C3H and C57BL/6J oocytes. PLoS One 8:e60216. https://doi.org/10.1371/journal.pone.0060216

Dosztányi Z, Csizmok V, Tompa P, Simon I (2005) IUPred: Web server for the prediction of intrinsically unstructured regions of proteins based on estimated energy content. Bioinformatics 21:3433–3434. https://doi.org/10.1093/bioinformatics/bti541

Dubourg C, Carré W, Hamdi-Rozé H et al (2016) Mutational spectrum in holoprosencephaly shows that FGF is a new major signaling pathway. Hum Mutat 37:1329–1339

Dyson HJ (2016) Making sense of intrinsically disordered proteins. Biophys J 110:1013–1016. https://doi.org/10.1016/j.bpj.2016.01.030

Dyson HJ, Wright PE (2005) Intrinsically unstructured proteins and their functions. Nat Rev Mol Cell Biol 6:197–208. https://doi.org/10.1038/nrm1589

Elms P, Siggers P, Napper D et al (2003) Zic2 is required for neural crest formation and hindbrain patterning during mouse development. Dev Biol 264:391–406

Elms P, Scurry A, Davies J et al (2004) Overlapping and distinct expression domains of Zic2 and Zic3 during mouse gastrulation. Gene Expr Patterns 4:505–511. https://doi.org/10.1016/j.modgep.2004.03.003

Fujimi TJ, Mikoshiba K, Aruga J (2006) Xenopus Zic4: conservation and diversification of expression profiles and protein function among the Xenopus Zic family. Dev Dyn 235:3379–3386. https://doi.org/10.1002/dvdy.20906

Fujimi TJ, Hatayama M, Aruga J (2012) Xenopus Zic3 controls notochord and organizer development through suppression of the Wnt/Beta-catenin signaling pathway. Dev Biol 361:220–231. https://doi.org/10.1016/j.ydbio.2011.10.026

Funayama N, Fagotto F, McCrea P, Gumbiner BM (1995) Embryonic axis induction by the armadillo repeat domain of beta-catenin: evidence for intracellular signaling. J Cell Biol 128:959–968

Furushima K, Murata T, Matsuo I, Aizawa S (2000) A new murine zinc finger gene. Opr Mech Dev 98:161–164. https://doi.org/10.1016/S0925-4773(00)00456-1

Furushima K, Murata T, Kiyonari H, Aizawa S (2005) Characterization of Opr deficiency in mouse brain: subtle defects in dorsomedial telencephalon and medioventral forebrain. Dev Dyn 232:1056–1061. https://doi.org/10.1002/dvdy.20253

Garber ED (1952) "Bent-tail," a dominant, sex-linked mutation in the mouse. Proc Natl Acad Sci U S A 38:876

Gaston-Massuet C, Henderson DJ, Greene NDE, Copp AJ (2005) Zic4, a zinc-finger transcription factor, is expressed in the developing mouse nervous system. Dev Dyn 233:1110–1115. https://doi.org/10.1002/dvdy.20417

Greene NDE, Copp AJ (2009) Development of the vertebrate central nervous system: formation of the neural tube. Prenat Diagn 29:303–311. https://doi.org/10.1002/pd.2206

Grinberg I, Northrup H, Ardinger H et al (2004) Heterozygous deletion of the linked genes ZIC1 and ZIC4 is involved in Dandy-Walker malformation. Nat Genet 36:1053–1055. https://doi.org/10.1038/ng1420

GrüNeberg H (1963) The pathology of development: a study of inherited skeletal disorders in animals. Blackwell Scientific Publications, Oxford

Hatayama M, Aruga J (2010) Characterization of the tandem CWCH2 sequence motif: a hallmark of inter-zinc finger interactions. BMC Evol Biol 10(1):53

Hatayama M, Tomizawa T, Sakai-Kato K et al (2008) Functional and structural basis of the nuclear localization signal in the ZIC3 zinc finger domain. Hum Mol Genet 17:3459–3473. https://doi.org/10.1093/hmg/ddn239

Herrera E, Brown L, Aruga J et al (2003) Zic2 patterns binocular vision by specifying the uncrossed retinal projection. Cell 114:545–557

Houtmeyers R, Souopgui J, Tejpar S, Arkell R (2013) The ZIC gene family encodes multifunctional proteins essential for patterning and morphogenesis. Cell Mol Life Sci 70:3791–3811. https://doi.org/10.1007/s00018-013-1285-5

Houtmeyers R, Gainkam OT, Glanville-Jones HA et al (2016) Zic2 mutation causes holoprosencephaly via disruption of NODAL signalling. Hum Mol Genet 25:3946–3959. https://doi.org/10.1093/hmg/ddw235

Huangfu D, Anderson KV (2005) Cilia and hedgehog responsiveness in the mouse. Proc Natl Acad Sci U S A 102:11325–11330. https://doi.org/10.1073/pnas.0505328102

Ibañez-Tallon I, Pagenstecher A, Fliegauf M et al (2004) Dysfunction of axonemal dynein heavy chain Mdnah5 inhibits ependymal flow and reveals a novel mechanism for hydrocephalus formation. Hum Mol Genet 13:2133–2141. https://doi.org/10.1093/hmg/ddh219

Inoue T, Hatayama M, Tohmonda T et al (2004) Mouse Zic5 deficiency results in neural tube defects and hypoplasia of cephalic neural crest derivatives. Dev Biol 270:146–162. https://doi.org/10.1016/j.ydbio.2004.02.017

Inoue T, Ota M, Mikoshiba K, Aruga J (2007) Zic2 and Zic3 synergistically control neurulation and segmentation of paraxial mesoderm in mouse embryo. Dev Biol 306:669–684. https://doi.org/10.1016/j.ydbio.2007.04.003

Ishiguro A, Ideta M, Mikoshiba K et al (2007) ZIC2-dependent transcriptional regulation is mediated by DNA-dependent protein kinase, poly(ADP-ribose) polymerase, and RNA helicase A. J Biol Chem 282:9983–9995. https://doi.org/10.1074/jbc.M610821200

Jiang Z, Zhu L, Hu L et al (2013) Zic3 is required in the extra-cardiac perinodal region of the lateral plate mesoderm for left-right patterning and heart development. Hum Mol Genet 22:879–889. https://doi.org/10.1093/hmg/dds494

Kalkan T, Smith A (2014) Mapping the route from naive pluripotency to lineage specification. Philos Trans R Soc Lond Ser B Biol Sci 369:20130540–20130540. https://doi.org/10.1098/rstb.2013.0540

Klootwijk R, Franke B, van der Zee CE et al (2000) A deletion encompassing Zic3 in bent tail, a mouse model for X-linked neural tube defects. Hum Mol Genet 9:1615–1622

Koyabu Y, Nakata K, Mizugishi K et al (2001) Physical and functional interactions between Zic and Gli proteins. J Biol Chem 276:6889–6892. https://doi.org/10.1074/jbc.C000773200

Layden MJ, Meyer NP, Pang K et al (2010) Expression and phylogenetic analysis of the zic gene family in the evolution and development of metazoans. EvoDevo 1:12. https://doi.org/10.1186/2041-9139-1-12

Lim LS, Loh Y-H, Zhang W et al (2007) Zic3 is required for maintenance of pluripotency in embryonic stem cells. Mol Biol Cell 18:1348–1358. https://doi.org/10.1091/mbc.E06-07-0624

Lu W (2001) Comparison of Pkd1-targeted mutants reveals that loss of polycystin-1 causes cystogenesis and bone defects. Hum Mol Genet 10:2385–2396. https://doi.org/10.1093/hmg/10.21.2385

Luo Z, Gao X, Lin C et al (2015) Zic2 is an enhancer-binding factor required for embryonic stem cell specification. Mol Cell 57:685–694. https://doi.org/10.1016/j.molcel.2015.01.007

Mizugishi K, Aruga J, Nakata K, Mikoshiba K (2001) Molecular properties of Zic proteins as transcriptional regulators and their relationship to GLI proteins. J Biol Chem 276:2180–2188. https://doi.org/10.1074/jbc.M004430200

Mizugishi K, Hatayama M, Tohmonda T et al (2004) Myogenic repressor I-mfa interferes with the function of Zic family proteins. Biochem Biophys Res Commun 320:233–240. https://doi.org/10.1016/j.bbrc.2004.05.158

Murdoch JN, Copp AJ (2010) The relationship between sonic Hedgehog signaling, cilia, and neural tube defects. Birth Defects Res A Clin Mol Teratol 88:633–652. https://doi.org/10.1002/bdra.20686

Nagai T, Aruga J, Takada S et al (1997) The expression of the mouse Zic1, Zic2, and Zic3 gene suggests an essential role for Zic genes in body pattern formation. Dev Biol 182:299–313. https://doi.org/10.1006/dbio.1996.8449

Nagai T, Aruga J, Minowa O et al (2000) Zic2 regulates the kinetics of neurulation. Proc Natl Acad Sci U S A 97:1618–1623

Nakata K, Nagai T, Aruga J, Mikoshiba K (1997) Xenopus Zic3, a primary regulator both in neural and neural crest development. Proc Natl Acad Sci U S A 94:11980–11985

Nakata K, Nagai T, Aruga J, Mikoshiba K (1998) Xenopus Zic family and its role in neural and neural crest development. Mech Dev 75:43–51

Nakata K, Koyabu Y, Aruga J, Mikoshiba K (2000) A novel member of the Xenopus Zic family, Zic5, mediates neural crest development. Mech Dev 99:83–91

Nikonova AS, Plotnikova OV, Serzhanova V et al (2014) Nedd9 restrains renal cystogenesis in Pkd1−/− mice. Proc Natl Acad Sci U S A 111:12859–12864. https://doi.org/10.1073/pnas.1405362111

Norris DP, Grimes DT (2012) Cilia discern left from right. Science 338:206–207. https://doi.org/10.1126/science.1230401

Ogura H, Aruga J, Mikoshiba K (2001) Behavioral abnormalities of Zic1 and Zic2 mutant mice: implications as models for human neurological disorders. Behav Genet 31:317–324. https://doi.org/10.1023/A:1012235510600

Paaby AB, Rockman MV (2013) The many faces of pleiotropy. Trends Genet 29:66–73. https://doi.org/10.1016/j.tig.2012.10.010

Pérez-Pérez JM, Candela H, Micol JL (2009) Understanding synergy in genetic interactions. Trends Genet 25:368–376. https://doi.org/10.1016/j.tig.2009.06.004

Pourebrahim R, Houtmeyers R, Ghogomu S et al (2011) Transcription factor Zic2 inhibits Wnt/Beta-catenin protein signaling. J Biol Chem 286:37732–37740. https://doi.org/10.1074/jbc.M111.242826

Purandare SM, Ware SM, Kwan KM et al (2002) A complex syndrome of left-right axis, central nervous system and axial skeleton defects in Zic3 mutant mice. Development 129:2293–2302

Quinn ME, Haaning A, Ware SM (2012) Preaxial polydactyly caused by Gli3 haploinsufficiency is rescued by Zic3 loss of function in mice. Hum Mol Genet 21:1888–1896. https://doi.org/10.1093/hmg/dds002

Roderick TH (1983) Using inversions to detect and study recessive lethals and detrimentals in mice. In: Utilization of mammalian specific locus studies in hazard evaluation and estimation of genetic risk. Springer US, Boston, pp 135–167

Solomon BD, Lacbawan F, Mercier S et al (2010) Mutations in ZIC2 in human holoprosencephaly: description of a novel ZIC2 specific phenotype and comprehensive analysis of 157 individuals. J Med Genet 47:513–524. https://doi.org/10.1136/jmg.2009.073049

Sotak BN, Gleeson JG (2012) Can't get there from here: cilia and hydrocephalus. Nat Med 18:1742–1743. https://doi.org/10.1038/nm.3011

Sutherland MJ, Wang S, Quinn ME et al (2013) Zic3 is required in the migrating primitive streak for node morphogenesis and left-right patterning. Hum Mol Genet 22:1913–1923. https://doi.org/10.1093/hmg/ddt001

Ware SM, Harutyunyan KG, Belmont JW (2006) Heart defects in X-linked heterotaxy: evidence for a genetic interaction of Zic3 with the Nodal signaling pathway. Dev Dyn 235:1631–1637. https://doi.org/10.1002/dvdy.20719

Warr N, Powles-Glover N, Chappell A et al (2008) Zic2-associated holoprosencephaly is caused by a transient defect in the organizer region during gastrulation. Hum Mol Genet 17:2986–2996. https://doi.org/10.1093/hmg/ddn197

Wessels MW, Kuchinka B, Heydanus R, Smit BJ, Dooijes D, de Krijger RR, Lequin MH, de Jong EM, Husen M, Willems PJ, Casey B (2010) Polyalanine expansion in the ZIC3 gene leading to X-linked heterotaxy with VACTERL association: a new polyalanine disorder? J Med Genet 47(5):351–355

Zhang Y, Niswander L (2013) Zic2 is required for enteric nervous system development and neurite outgrowth: a mouse model of enteric hyperplasia and dysplasia. Neurogastroenterol Motil 25(6):538–541

Zhu L, Peng JL, Harutyunyan KG et al (2007) Craniofacial, skeletal, and cardiac defects associated with altered embryonic murine Zic3 expression following targeted insertion of a PGK-NEO cassette. Front Biosci 12:1680–1690

Chapter 11
Rodent Zic Genes in Neural Network Wiring

Eloísa Herrera

Abstract The formation of the nervous system is a multistep process that yields a mature brain. Failure in any of the steps of this process may cause brain malfunction. In the early stages of embryonic development, neural progenitors quickly proliferate and then, at a specific moment, differentiate into neurons or glia. Once they become postmitotic neurons, they migrate to their final destinations and begin to extend their axons to connect with other neurons, sometimes located in quite distant regions, to establish different neural circuits. During the last decade, it has become evident that *Zic* genes, in addition to playing important roles in early development (e.g., gastrulation and neural tube closure), are involved in different processes of late brain development, such as neuronal migration, axon guidance, and refinement of axon terminals. ZIC proteins are therefore essential for the proper wiring and connectivity of the brain. In this chapter, we review our current knowledge of the role of *Zic* genes in the late stages of neural circuit formation.

Keywords Neuronal migration · Axon pathfinding · Axon terminal refinement · Guidance receptors · Bilateral circuits wiring · Axonal laterality

11.1 Overview of Neural Circuit Development

The central nervous system (CNS) derives from a simple epithelial cell sheet called the neural plate, which is generated from the dorsal ectoderm early in development, in a process known as neural induction. During the late stages of gastrulation, this plate closes to form the neural tube that subsequently regionalizes according to the anteroposterior and dorsoventral axes. As the embryo develops, the anterior part of the neural tube forms a series of bulges called vesicles, which become the primary anatomical regions of the brain: the forebrain (prosencephalon), midbrain (mesencephalon), and hindbrain (rhombencephalon). These simple vesicles enlarge and further divide into the telencephalon (future cerebral cortex and basal ganglia),

E. Herrera (✉)
Instituto de Neurociencias de Alicante, Consejo Superior de Investigaciones Científicas Universidad Miguel Hernández, CSIC-UMH, Alicante, Spain
e-mail: e.herrera@umh.es

J. Aruga (ed.), *Zic family*, Advances in Experimental Medicine and Biology 1046,
https://doi.org/10.1007/978-981-10-7311-3_11

diencephalon (future thalamus and hypothalamus), mesencephalon (future colliculi), metencephalon (future pons and cerebellum), and myelencephalon (future spinal cord). Gradually, some of the cells stop dividing and start to differentiate into neurons or glial cells. After their final mitosis, most immature neurons or neuroblasts migrate substantial distances from the ventricular zones to different parts of the developing brain to self-organize into the various brain structures. Once the neurons have reached their regional positions, they extend axons and dendrites, which allow them to communicate with other neurons through the establishment of synapses. Synaptic communication between neurons leads to the establishment of functional neural circuits that mediate sensory and motor processing and underlie behavior (Puelles and Rubenstein 2003; Wilson and Houart 2004; Lim and Golden 2007; Scholpp and Lumsden 2010; Robertshaw and Kiecker 2012; Beccari et al. 2013; Andoniadou and Martinez-Barbera 2013). Therefore, together with the early mechanisms of regionalization and cellular specification, the late steps of neural development – such as the migration of immature neurons, axon growth, guidance, and tissue targeting – are all critical for the proper establishment of brain connectivity. In this chapter, we will provide an overview of the role of *Zic* genes and related molecules in these late processes of brain development.

11.2 ZIC Proteins in Neuronal Migration

Several populations of migrating neuroblasts express ZIC proteins in the forebrain and hindbrain. In all the cases, ZIC-positive neural precursors seem to delaminate from a neuroepithelium and start to migrate through stereotyped paths to reach their final destinations where they will finish their maturation and exert their function. Next, we describe some of these ZIC-positive migrating cell populations and what is currently known about the role of the ZIC family in these immature neurons.

11.2.1 ZIC Proteins in the Migration of Telencephalic Cell Populations

In the telencephalic vesicles, at approximately embryonic day 10 (E10) in the mouse, the earliest cortical neurons form a transient structure known as the preplate. This primordial layer consists of Cajal-Retzius (CR) cells and the first cohort of pyramidal neurons, which will eventually populate the subplate. CR cells arise from discrete pallial sources, the hem, the septum (ST), and the pallial-subpallial border (PBS), and they colonize the entire surface of the cortex through tangential migration (Takiguchi-Hayashi et al. 2004; Bielle et al. 2005; Yoshida et al. 2006). The next cohort of pyramidal cells forms the cortical plate by intercalating in the preplate and splitting this primitive structure into a superficial layer, the marginal zone (MZ or layer I), and a deep layer, the subplate. The development of the neocortex

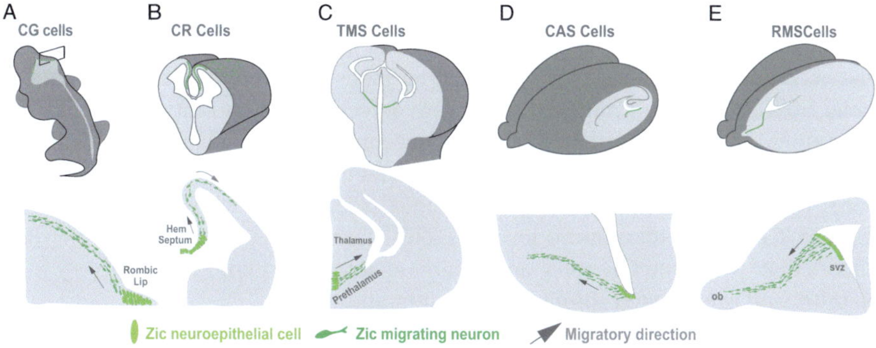

Fig. 11.1 Migration of several populations of ZIC neuroblasts through different paths. (**a**) Cerebellar granule (CG) progenitors positive for ZIC proteins are generated in the rhombic lip of the rhombencephalon. At approximately E11–E12, cells from the neuroepithelial rhombic lip detach and repolarize to initiate migration and reach the external granule layer. (**b**) Cajal-Retzius (CR) cells derived from the hem/septum start to migrate at E11 to cover the entire telencephalic surface by E13. The hem and septum also express Zic genes at low levels, whereas CR cells express high levels of ZIC2 while migrating. (**c**) Near E12, a restricted population of ZIC2 neuroepithelial cells originate in the wall of the third ventricle at the level of the *zona limitans interthalamic*. These cells detach from the epithelium and start to travel through the thalamic prethalamic boundary to ultimately populate the ventral lateral geniculate nucleus. (**d**) Neuroepithelial cells expressing low levels of ZIC2 are located in the most ventrocaudal part of the ventricular zone. At approximately E13, when ZIC2 levels increase, these neuroepithelial cells detach and begin to navigate rostrally, traversing the caudal amygdaloid stream to end in layer 2 of the amygdaloid nucleus. (**e**) In the adult mouse brain, newly generated type A neuroblasts in the subventricular zone (SVZ) detach and repolarize to start their migration through the rostral migratory stream (RMS) and finalize their trip in the olfactory bulb (ob). At least a subpopulation of these type A cells express ZIC proteins

progresses with new waves of neurons that occupy progressively more superficial positions within the cortical plate (Marín and Rubenstein 2003). Therefore, the different layers of the cerebral cortex are generated in an "inside-out" sequence.

CR cells from the different sources intermingle in a manner such that by the end of the process, the entire cortical surface contains CR cells from the different origins (Griveau et al. 2010), although the majority of them derive from the hem (Subramanian and Tole 2009). It has long been thought that the proper migration and positioning of CR cells on the telencephalic surface influences cortical lamination, and although some recent studies have challenged this view (Bielle et al. 2005; Yoshida et al. 2006; Meyer et al. 2002), it seems that once CR cells completely cover the cortical surface, they stimulate the radial migration of late-born cortical neurons via secretion of the extracellular protein REELIN, (D'Arcangelo et al. 1995; Ogawa et al. 1995; Frotscher 1998; Rice and Curran 2001; Shinozaki et al. 2002; Tissir and Goffinet 2003; Gil-Sanz et al. 2013).

The hem, the septum, and the CR cells arising from these two structures are highly positive for ZIC genes (Inoue et al. 2008; Murillo et al. 2015) (Fig. 11.1). Mice mutant for *Zic2* (*Zic2* $^{kd/kd}$) or *Zic1/3* (*Zic1*$^{-/-}$ *Zic3*$^{-/-}$) exhibit hypoplastic development of the septum and the hem, suggesting that ZIC proteins are important for the formation of these structures (Inoue et al. 2007). In addition, live imaging of flattened telencephalic vesicles from reporter mouse embryos has revealed that CR cells from *Zic2* mutant mice exhibit severe defects in cell motility while they are migrating and being distributed on the telencephalic surface (Murillo et al. 2015). *Zic2* mutant CR cells exhibit an uneven distribution across the telencephalic surface, which is in contrast with homogeneous spreading in wild-type embryos. In culture, CR cells with reduced levels of ZIC2 form aggregates, cover shorter distances than wild-type CR cells, and exhibit aberrant morphologies with very long neurites and exacerbated sprouting while migrating.

In agreement with the idea that *Zic* genes are involved in cell migration, it has been reported that ZIC2 controls the expression of EPH receptors, which are tyrosine kinase membrane proteins (García-Frigola et al. 2008; Escalante et al. 2013). These guidance proteins facilitate the distribution and dispersion of CR cells by a mechanism known as cell contact inhibition or repulsion (Villar-Cerviño et al. 2013). The role for the rest of the ZIC family members in CR cell migration has not been studied, but it is likely that, at least in *Zic2* mutant mice, the aggregates observed in the developing telencephalic vesicles result from the lack of repulsion among CR cells mediated by EPHB/EPHRINB signaling during their dispersion in the cortical surface. Also supporting a role for ZIC proteins in the migration of CR cells, it has been shown that *Zic* mutant mice exhibit a strongly reduced expression of CXCR4 (Inoue et al. 2008), the receptor for CXCL12, which is an attractive chemokine expressed in meningeal cells. CR cell migration depends on CXCR4/CXCLl12 signaling (Borrell and Marín 2006), and therefore, a reduction in the levels of CXCR4 in the mutants may compromise CR cell migration.

Cortical defects observed in *Zic* mutants may have however a different origin. As discussed in the next chapter, *Zic* genes are also highly expressed in meningeal cells (Inoue et al. 2008; Murillo et al. 2015). As CR cells, meningeal cells seem to be important for cortical lamination; therefore, the cortical alterations observed in *Zic* mutant mice may result from the lack of ZIC proteins in the CR cells, in the meninges, or in both. Further experiments are required to clarify this issue.

Another group of forebrain migratory neurons that express at least one of the *Zic* genes is a group of cells that will integrate the nucleus of the lateral olfactory tract (nLOT) in the amygdala (Fig. 11.1). The nLOT is connected to the olfactory bulb and piriform cortex and is composed of three layers (Santiago and Shammah-Lagnado 2004). Layer 2 (nLOT2) cells are generated between E11.5 and E12.5 in the neuroepithelium of the most caudal part of the dorsal telencephalon and then migrate rostrally via the caudal amygdaloid stream (CAS) to reach the amygdala (Remedios et al. 2007). ZIC2 is modestly expressed in the neuroepithelium that gives rise to the CAS cells and strongly in the migrating CAS cells. In hypomorphic *Zic2* mutant mice, CAS cells occupy a wider area and do not form the typical curve-shaped stream observed in control mice, indicating that ZIC2 is also necessary for the proper migration of these immature nLOT2 amygdaloid neurons. Whether other *Zic* genes exert any influence on the migration of CAS cells is unknown.

11.2.2 ZIC Proteins in the Migration of Diencephalic Cell Populations

In the diencephalon, a member of the ZIC family has also been detected in a population of migratory neurons travelling along the boundary between the thalamus and the prethalamus, the thalamic-prethalamic stream (TPS) (Murillo et al. 2015) (Fig. 11.1). TPS cells originate in the pallial wall of the third ventricle and move adjacently to the zona limitans, a compartment and signaling center located between the thalamus and the prethalamus, to reach the ventral lateral geniculate nucleus (vLGN) in the thalamus.

Whether TPS migrating cells express other members of the ZIC family is unknown, but at least ZIC2 is first expressed at basal levels in a restricted population of neuroepithelial cells facing the third ventricle located at the frontier between the prospective thalamus and prethalamus, and then it is highly upregulated as soon as these cells exit the cell cycle and start moving. The number of migrating cells in the TPS is significantly reduced in *Zic2* mutant mice compared to the controls, a phenotype that seems to be due to two phenomena. First, cells from the pallial wall do not delaminate properly to start migration and instead they form aggregations in the ventricular zone. Second, time-lapse experiments show that many $Zic2^{kd/kd}$ TPS cells escape from the stereotypical pathway at different points of their journey and never reach the vLGN. In addition, those TPS cells abandoning the regular trajectory exhibit an altered morphology in the *Zic2* mutant mice with many protrusions and longer processes than in control embryos, a phenotype that resembles the morphological defects observed in $Zic2^{kd/kd}$ CR and CAS cells. ZIC2-positive TPS cells express interneuron markers such as the transcription factor OTX2, suggesting that they will become vLGN interneurons. While travelling to the vLGN, these cells also express the tyrosine kinase receptor EPHB1. EPHRINBs, the ligands for EPHB receptors, are expressed in the thalamic and prethalamic areas on both sides of the TPS. EPHRINB2 is expressed near the ventricular area of the thalamus and very mildly in the prethalamus. EPHRINB3 is expressed at basal levels in the rest of the thalamus and strongly in the prethalamic area. Together, EPHRINB2 and EPHRINB3 expressions delineate a complementary pattern to the expression of EPHB1 throughout the entire TPS pathway (Murillo et al. 2015). Because *Zic2* mutant mice exhibit a significant reduction in EPHB1 expression, the lack of EPHB1-/EPHRINB-mediated repulsion in *Zic2* mice could partially explain why an elevated number of TPS cells escape from the stereotypical path.

11.2.3 ZIC Proteins in Rhombencephalic Structures

The cerebellum, a region of the brain that plays an important role in motor control and motor learning, comprises three major types of neurons: Purkinje cells (PCs), granule cells, and interneurons, including basket, stellate, and Golgi cells. PCs and interneurons are GABA-releasing inhibitory neurons, whereas granule neurons are glutamatergic. These cerebellar cell types are organized into distinct neuronal layers: the outermost molecular layer, the Purkinje cell monolayer, the densely populated internal granule layer, and the innermost white matter.

The cerebellum develops from the hindbrain (rhombencephalon) region. Cerebellar development occurs during embryogenesis and early postnatal periods. The diverse cell types that make up the different layers originate from two distinct germinal centers in the early cerebellum: the ventricular zone, a monolayer of cells lining the fourth ventricle on the ventral surface of the cerebellar anlage, and the rhombic lip, a transient neuroepithelial structure in the most posterior part of the cerebellar primordium. In mice, PCs arise at the ventricular zone at E11–E13 and migrate radially to form the PC layer. However, granule cell progenitors migrate tangentially from the lower rhombic lip to form the external granule cell layer. Later, at postnatal stages, external granule cell precursors expand dramatically to ultimately migrate inward and form the granule cell layer (Hatten 1999; Sotelo 2004; Machold and Fishell 2005; Wang et al. 2005). ZIC1, ZIC2, and ZIC3 are all expressed during cerebellar development (Fig. 11.1), with ZIC2 and ZIC3 expressions being considerably lower than that of ZIC1 (Aruga et al. 1996a, b). ZIC1 is expressed at the rhombic lip and external granule cell layer in an abundant and very restricted manner that begins as early as the formation of the neuroectoderm, and it persists during granule cell differentiation and migration (Aruga et al. 1994).

The cerebellum of *m*ice carrying one mutated *Zic1* allele together with one mutated *Zic2* allele *(Zic1*$^{+/-}$*Zic2*$^{+/kd}$ mice) is poorly developed. The number of granule cells is reduced in these mice, the lobule in the anterior vermis is missing, and the most posterior lobule has a truncation. These phenotypes are not evident however in *Zic1*$^{-/-}$ mice. The differences in the phenotypes between the two mouse lines suggest that *ZIC1* and *ZIC2* have similar but also different functions in the regulation of cerebellar development (Aruga et al. 2002).

Both *Zic* mutant lines exhibit a reduction in cell proliferation in the anterior external germinal layer, a reduction in *CyclinD1* expression, and enhanced expression of the mitotic inhibitors *p27* and *p16*. These observations led to the idea that the main function of *Zic* genes in cerebellar development relates to cell proliferation. However, in light of new results attributing a role for *Zic* genes in neuroblast migration and taking into account that cerebellar granule precursors suffer several embryonic and postnatal rounds of migratory processes, it seems reasonable to speculate that defects in the migration of granule precursors also contribute to the cerebellar phenotypes observed in the *Zic* mutant mice. In support of this idea, it is worth mentioning that the chemokine CXCL12 is a strong chemoattractant to upper rhombic lip cells, preventing migration of CXCR4-expressing premature granule cells away from the external granule layer (Zou et al. 1998; Ozawa et al. 2016). As discussed above, *Zic* mutant mice exhibit reduced expression of CXCR4 in the developing cortex, and although this has not yet been reported, it is possible that ZIC proteins also regulate CXCR4 expression in migrating cerebellar granule cells to control their correct migratory pathway. Also, the expression patterns of EPHs and EPHRINs in the cerebellar anlage of the chick embryo, in particular EPHA4, suggest multiple roles of these proteins in cerebellar formation, including restriction of the early phase of granule cell migration to ribbons (Karam et al. 2000). As *Zic* genes seem to be regulating EPH receptors, it is likely that they are also linked during the development of this tissue.

11.3 ZIC Proteins in the Wiring of Bilateral Circuits

ZIC proteins have also been implicated in axon pathfinding. Indeed, the role of ZIC proteins in postmitotic neurons was first described in neurons extending their axons and navigating through the developing brain to reach their targets. In particular, ZIC2 was reported as the first transcription factor controlling the laterality of optic axons in species with binocular vision (Herrera et al. 2003). Later, the role of this transcription factor as a determinant of axonal laterality was confirmed in bilateral circuits other than the visual system (Escalante et al. 2013).

Throughout evolution, the cephalization of different sensory organs and the emergence of bilateral symmetry imposed the need for robust lines of communication between brain hemispheres (Colamarino and Tessier-Lavigne 1995). Consequently, the formation of contralateral (crossing the midline) or ipsilateral (not crossing the midline) tracts during development became essential to distribute and integrate sensory information from both sides of the body and subsequently generate coordinated motor responses (Engle 2010; Izzi and Charron 2011; Nugent et al. 2012). Bilateral symmetry has been so successful in nature that today, 98% of the animal species on Earth have this body disposition. Two well-characterized examples of bilateral circuits are the visual system of mammals and the ascending and descending connections between the brain and the spinal cord, which are essential for binocular vision and motor coordination, respectively. In the next section, we explain in detail some of the molecular details related to the role of *Zic* genes in the establishment of bilateral wiring.

11.3.1 ZIC Proteins in the Development of the Visual System

Zic1, *Zic2*, and *Zic3* mRNAs are all expressed in the optic cup during very early eye development. All three genes are also expressed at basal levels in the ciliary margin zone (CMZ) through retinal development (Nagai et al. 1997), although their role in this niche of retinal stem cells remains unknown. At E10, *Zic1* and *Zic2* (but not *Zic3*) mRNAs are detected in the inner neural layer and continue to be expressed in the neuroblastic layer until at least E13.5. *Zic1*$^{-/-}$ *Zic3*$^{-/-}$ mutant mice have normal eyes (Nagai et al. 1997). However, some *Zic2*$^{kd/kd}$ embryos have small eyes, although they exhibit a variable phenotype (Herrera et al. 2003). This suggests a role for ZIC2, but not other ZIC members, in eye formation. Despite this obvious eye defect, the precise function of ZIC2 in retinal morphogenesis has not been uncovered so far.

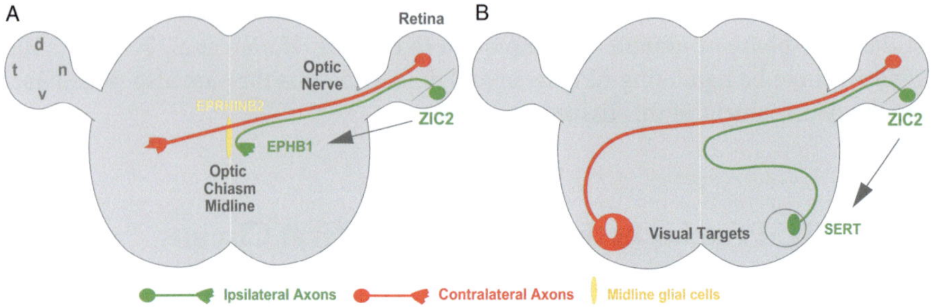

Fig. 11.2 ZIC2 function in the visual system. (**a**) Retinal ganglion cell neurons located at the peripheral ventrotemporal crescent of the retina express high levels of ZIC2. In turn, ZIC2 induces the expression of EPHB1. EPHB1 expressed at the plasma membrane of the growth cone will bind to its ligand EPHRINB2, which is expressed in midline glial cells at the optic chiasm region. EPHB1/EPHRINB2 signaling mediates a repulsive response that obligates ipsilateral axons to repolarize and continue its trip into the same hemisphere. (**b**) Once retinal ipsilateral and contralateral ganglion cell axons leave the optic chiasm region, they navigate to reach the visual targets. Contralateral and ipsilateral axon terminals project in a complementary pattern in the visual targets as a result of a refinement process. Ipsilateral axons occupy discrete patches in the visual targets, represented by green circles, whereas contralateral axons fill the rest of the target tissue (red circles). ZIC2 induces the expression of the serotonin transporter (SERT) only in ipsilaterally projecting axons to facilitate their refinement and segregation from the contralaterally projecting axon terminals

The adult retina is integrated by six different types of neurons: cone and rod photoreceptors and horizontal, bipolar, amacrine, and ganglion cells. Visual information is conveyed in the eye from cones and rods, and they send sensory input to the retinal ganglion cells (RGCs) through amacrine, bipolar, and horizontal cells. RGCs are the only neurons in the retina that send axons to the brain. RGC axons exit the retina through the optic disk and gather to form the optic nerve. Then, visual axons reach the midline at the optic chiasm, and there they decide whether to cross it or not. In mice, retinal axons that do not cross the midline and project to the same (ipsilateral) brain hemisphere originate in the ventrotemporal retina. However, axons that cross the midline and project to the opposite side of the brain (contralateral) arise from RGCs located across the entire retina (Fig. 11.2). After reaching the optic chiasm, retinal axons from each eye regroup in the optic tract before projecting to their main targets, the lateral geniculate nucleus in the thalamus and the superior colliculus in the brainstem. The proper establishment of this circuit is crucial for binocular vision and all related visually driven behaviors.

ZIC2, but not ZIC1 or ZIC3, is transiently expressed in the ventrotemporal peripheral retina of mouse embryos, in recently differentiated RGCs that project ipsilaterally, but not in those projecting contralaterally (Fig. 11.2). *Zic2*$^{kd/kd}$ mice exhibit a severe reduction in the number of uncrossed axons, and conversely, ectopic expression of ZIC2 in RGCs that normally project contralaterally switches their laterality to ipsilateral. These functional experiments demonstrated that ZIC2 determines the ipsilateral projection in the visual system. Strikingly, ectopic expression of ZIC1 is also able to produce an ectopic ipsilateral projection (EH unpublished observations) despite ZIC1 not being expressed in the ventrotemporal retina (Herrera et al. 2003).

The function of ZIC2 as a determinant of axonal laterality in the visual system seems to have been conserved throughout evolution because the number of ZIC2-positive cells in the developing retina of different species correlates with the number of ipsilateral axons, which in turn matches the extent of binocular vision (Herrera et al. 2003). For instance, the number of cells expressing ZIC2 in the developing ferret retina is larger than in the mouse retina, matching with a greater proportion of uncrossed axons and enhanced binocular vision. In *Xenopus*, ZIC2 expression is upregulated at metamorphosis, coinciding with the late development of the uncrossed component in this species. Chicken and zebrafish lack an ipsilateral projection because they have panoramic instead of binocular vision. Accordingly, ZIC2 is not expressed in their developing retinas (Herrera et al. 2003; Seth et al. 2006). In addition, agreeing to its role as a determinant of the uncrossed RGC component, the expression of ZIC2 in albino mice, which for unknown reasons have fewer ipsilateral axons than pigmented mice, is reduced (Rice et al. 1995).

The molecular mechanisms orchestrating the induction of ZIC2 expression in ipsilateral RGCs are largely unknown; some observations may help us start to understand this event. Knockout mice for *Foxd1*, a gene that encodes for a forkhead box transcription factor specifically expressed in progenitors of the temporal part of the retinal cup, lack ZIC2 expression, in addition to many other genes expressed in the temporal retina, such as EPHB1, EPHA5/6, or the serotonin transporter. Thus, FOXD1 is needed for the acquisition of temporal identity, which seems to be essential to later induce ZIC2 expression (Carreres et al. 2011). Another transcription factor that influences ZIC2 expression in the retina is the homeobox protein ISL2. ISL2 is expressed in 40% of the contralateral RGCs, excluding ZIC2-positive cells. In the absence of ISL2, the ZIC2 ventrotemporal domain expands, thus suggesting that ISL2 is able to repress ZIC2 expression (Pak et al. 2004).

Very recent results have evidenced that the most peripheral part of the developing retina, the CMZ, is also critical for the generation of ipsilateral RGCs. The CMZ is a neuroepithelial germinal zone at the interface between the neural retina and the prospective ciliary epithelium (Fischer et al. 2014). During development, a number of CMZ cells that express low levels of ZIC proteins and high levels of CyclinD2 detach from the CMZ and move laterally to generate a subset of ipsilateral RGCs (Bélanger et al. 2016; Marcucci et al. 2016). Further investigations are needed to clarify whether this population of ipsilateral RGCs derived from the CMZ and the ipsilateral population arised from the neural retina activate ZIC2 via similar or different mechanisms.

11.3.2 ZIC2 in the Wiring of the Spinal Cord and Thalamus

The ascending and descending projections that connect the spinal cord and the brain integrate another circuit that contains ipsilaterally and contralaterally projecting neurons in the vertebrate nervous system. The spinal cord is the most caudal part of the central nervous system, and it arises from the neural tube. Three different populations of cells differentiate after neural tube closure: (1) dorsal midline cells that result in the roof plate and constitutes a signaling center essential for proper patterning of dorsal interneurons, (2) neural crest cells that delaminate and migrate out of the neural tube to form diverse structures, and (3) the rest of the cells that will become neural progenitors in the future spinal cord. As development proceeds, the roof plate cells start to secrete different families of dorsalizing molecules, such as TGF-b, WNTs, and FGFs. This complex molecular combination regulates the expression of several cross-inhibitory transcription factors and causes the subdivision of the precursor area in six dorsal domains of interneuron progenitors (dP1-6) that are organized dorsoventrally. Subsequent activation of LIM-HD factors further specifies these domains of progenitors to yield six postmitotic populations of dorsal interneurons (dl1-6) (Jessell 2000; Gross et al. 2002; Müller et al. 2002). Following differentiation of the dl1-6, a late-born dorsal interneuron (dIL) population originates from the dorsal subventricular zone in a second round of differentiation (Helms and Johnson 2003). dIL neurons, which include inhibitory (dIL_A) and excitatory (dIL_B) cell types, will populate the superficial dorsal horns (Altman 1969; Gross et al. 2002; Müller et al. 2002). Each subpopulation of spinal cord interneurons is characterized by specific patterns of axonal projections. dI1 neurons are a mixed population, integrated by both ipsi- and contralaterally projecting neurons (Wilson et al. 2008; Avraham et al. 2009; Ding et al. 2012), whereas dI2 neurons are mostly contralateral (Avraham et al. 2009). dI3 interneurons are exclusively ipsilateral. Some of these ipsilateral neurons project ventrally toward the ventrolateral fascicle, whereas some others project dorsally toward the dorsal root entry zone (Avraham et al. 2010). dI4 neurons seem to be ipsilaterally projecting association neurons that target proprioceptive afferents during their course to motoneurons (Alaynick et al. 2011). The projection pattern of dI5 neurons remains unknown, and dI6 cells are premotor neurons with both ipsi- and contralateral axons (Vallstedt and Kullander 2013). Late-born dIL_A interneurons are locally or contralaterally projecting neurons, whereas dIL_B neurons are long-range ipsilateral projection neurons (Petkó and Antal 2012).

Tracing experiments using a Tg(*Zic2*^{EGFP}) reporter mouse line have revealed that dIL_B spinal neurons, which ascend ipsilaterally through the dorsal funiculus (DF) or the dorsolateral funiculus (DLF), are positive for ZIC2 (Escalante et al. 2013). Mature dorsal horn neurons, as adult retinal ganglion cells, do not express ZIC2, which demonstrates that ZIC2 is needed only during embryonic and early postnatal stages when axons are growing and reaching their target tissues and is not required for mature neuronal function (Escalante et al. 2013).

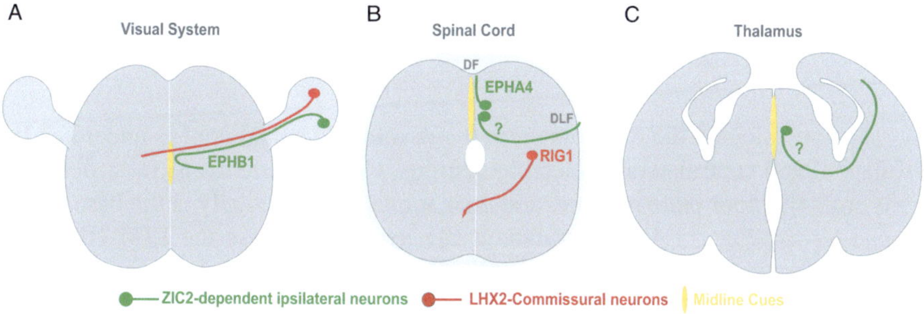

Fig. 11.3 ZIC2 is a general determinant of axon midline repulsion in the CNS. ZIC2 is expressed in different types of ipsilateral neuron populations when their axons grow toward the midline. Spinal contralateral neurons (red) express receptors essential for midline crossing, such as RIG1. In contrast, ipsilateral retinal ganglion cells in the visual system (**a**), dorsal spinal cord neurons that exit the cord through the dorsal funiculus (DF) or the dorsolateral funiculus (DLF) (**b**), and thalamocortical axons (**c**) all express ZIC2 to induce the expression of midline repulsion receptors. In the visual system, ZIC2 upregulates the expression of EPH1, and in the spinal cord, ZIC2 induces the expression of EPHA4 in the cells growing into the DF. The receptors upregulated by ZIC2 in the dorsal spinal cord neurons that exit the cord through DLF or in thalamocortical neurons to promote midline avoidance have not been described

In utero electroporation approaches combined with efficient delivery of RNA interference for ZIC2 at late stages of mouse embryogenesis led to downregulation of ZIC2 in dorsal spinal neurons once the neural tube is closed and neural crest cells have migrated. Using this approach the disruption of ZIC2 does not interfere with earlier developmental stages and therefore, these loss-of-function experiments demonstrated that expression of ZIC2 in dorsal horns spinal neurons is essential for preventing axon midline crossing in dorsal spinal neurons. Interestingly, downregulation of ZIC2 in these cells also caused aberrant midline crossing in a number of neuron cell bodies. This result points to the expression of ZIC2 as a mechanism for controlling the positioning of the cell bodies of dorsal spinal neurons away from the midline and could somehow be related to the role of ZIC2 in cell migration.

In addition to promoting axon midline avoidance in visual and spinal neurons, ZIC2 establishes proper axonal laterality in thalamocortical neurons. Thalamocortical neurons are located in the thalamus and project to the ipsilateral cortex. Sense organs provide input to one or more nuclei in the thalamus, and thalamic neurons transmit sensory information to the same cortical hemisphere where it will be processed (López-Bendito and Molnár 2003). Thalamocortical neurons from the medial thalamic areas express ZIC2 when they are extending their axons to the ipsilateral cortex. Mice mutant for *Zic2* exhibit a strong phenotype of aberrant midline crossing in the thalamus (Escalante et al. 2013).

Therefore, during the development of the vertebrate nervous system, ZIC2 controls axon midline avoidance in at least three different systems, the visual pathway, the ascending dorso-spinal projection, and the thalamocortical tracts (Fig. 11.3).

11.3.3 ZIC2 Is Able to Repress the Axonal Contralateral Program

As abovementioned, the dI1 domain in the developing spinal cord is integrated by ipsilateral and contralateral projecting neurons. Contralateral axons from dI1 neurons cross the floor plate midline, and their somas settle medially at the base of the dorsal horns (Fig. 11.4). Contralateral dl1 neurons express the LIM-HD transcription factor LHX2 (Wilson et al. 2008) which in turn regulates the guidance receptor RIG1, essential for midline crossing (Camurri et al. 2005; Chédotal 2011)

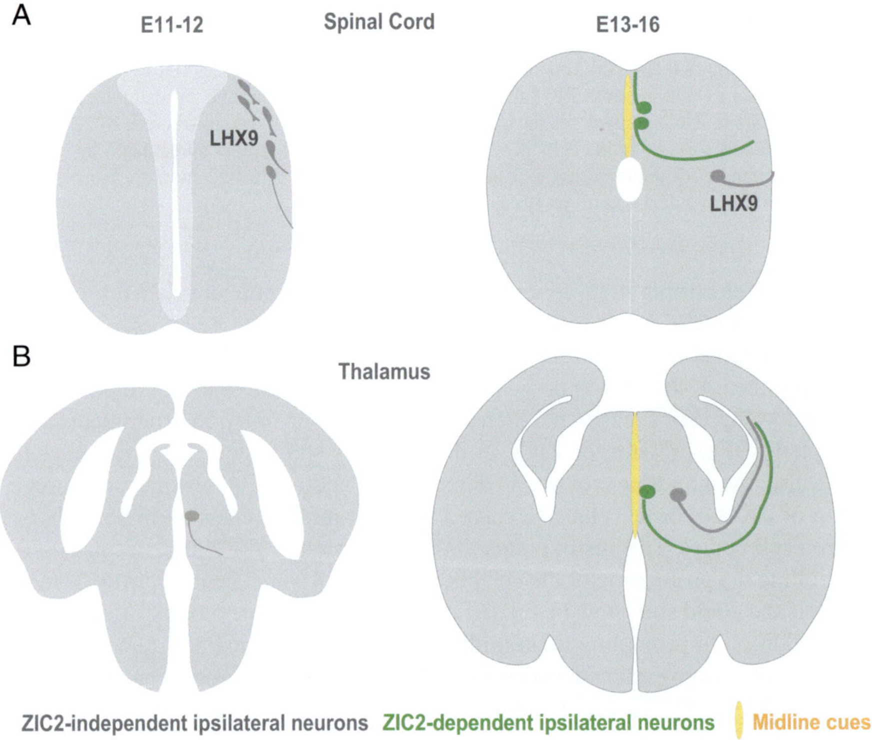

Fig. 11.4 Zic does not specify ipsilaterality per se but induces axon midline avoidance. (**a**) In the developing spinal cord, ipsilateral LHX9-positive interneurons from the dl1 domain become postmitotic around E10 and start to migrate ventrally to then extend their axons laterally. At E13, the first dIL_B late-born neurons are generated in close proximity to the midline. Postmitotic ZIC2 expression peaks from E13 to E16, and during this period, ZIC2 neurons extend their axons ipsilaterally to the DF or to the DLF. In contrast to the early-born dI1 interneurons, whose axons never approach the midline, late-born ZIC2 neurons are exposed to midline cues soon after differentiation. (**b**) In the developing thalamus, axons from E11–E13-born neurons extend ventrally at the time that the thalamic halves are physically separated by the third ventricle. Later, as the thalamic halves fuse and the midline is created, later-born neurons located in medial positions express ZIC2 to respond to midline cues and avoid crossing

(Fig. 11.3). Ectopic expression of ZIC2 in dI1 neurons forces a switch in axonal laterality and downregulation in the expression of LHX2 and RIG1. Ectopic expression of ZIC2 does not affect however the expression of LHX9, another LIM-HD transcription factor expressed in the ipsilateral dI1 interneurons (Ding et al. 2012). As in the spinal cord, ectopic expression of ZIC2 in the retina is sufficient to switch axonal laterality at the midline (García-Frigola et al. 2008). However, in the visual system, because the transcription factors determining midline crossing have not been uncovered, whether ZIC2 represses the expression of contralateral fate determinants is unknown.

11.3.4 ZIC2 Controls Axon Midline Avoidance but Not Ipsilaterality Per Se

One striking question about the role of *Zic* genes in axonal laterality arises from the fact that a number of ipsilateral tracts in the CNS do not express *Zic* genes. This is the case for the different types of ipsilateral interneurons that integrate the dI1-6 domains, ventral spinal neurons or early-born thalamocortical neurons. It has been proposed that similar circumstances may explain the lack of ZIC expression in all of these ipsilateral neurons. For example, ipsilateral dI1 interneurons start axonogenesis early in development, when they are located laterally in the mantle, and instead of being in contact, the two sides of the neural tube are still separated by the ventricular zone. At this time, dI1 neuronal axons extend circumferentially along the basal membrane in an opposite direction to that of the future midline and never encounter midline cues along their trajectory (Wilson et al. 2008; Avraham et al. 2010). Under these circumstances, it seems that the establishment of an ipsilateral projection may rely on fasciculation with the contralateral axons coming from the other side rather than from the activation of ZIC2, as it appears to be the case for ipsilateral LHX9 dI1 neurons (Fig. 11.4a). A similar explanation may apply to the ventral spinal cord neurons. Although a number of ventral interneurons project ipsilaterally (reviewed in Goulding 2009), they do not express ZIC proteins. Again, this may be due to the existence of the ventricle while ventral spinal axons are extending the existence of the ventricle which may be acting as a physical barrier preventing axon midline crossing. In contrast, dIL_B interneurons are generated in a second round of neurogenesis and extend their axons once the two sides of the tube are fused and the midline has formed. To avoid aberrant midline crossing, ipsilateral axons require ZIC2 expression to trigger a repulsive response to the midline. In the ventral cord, a second round of neural differentiation, once the ventricle is gone, never occurs. Therefore, there would be no need for ZIC2 expression at the developing ventral cord.

The scenario observed in the spinal cord neurons respecting Zic2 expression seems to be similar to the situation observed during the development of thalamocortical tracts. When early-born thalamic neurons differentiate and start to extend their axons, the fusion of the two thalamic halves has not yet happened; therefore, axons would never have the opportunity to cross the midline because they cannot transverse the empty space occupied by the third ventricle. Consequently, they do not need to express ZIC2 to avoid midline crossing. Later, when the ventricle shrinks and the two thalamic leaves fuse, new postmitotic neurons extend their axons in close proximity to the recently created midline, and thalamocortical neurons need ZIC2 to prevent axonal crossing (Fig. 11.4).

In summary, ZIC2 does not determine ipsilaterality per se in differentiated neurons but rather regulates axon midline avoidance in those ipsilateral axons that approach the midline at some point of their trajectory during embryonic development.

11.4 Downstream Targets of ZIC Proteins in Neural Circuit Wiring

ZIC proteins are transcription factors. As such, they must regulate the expression of effector molecules associated with axon guidance and cell migration processes. In this section, we review some of the ZIC effectors that have been already defined during axon guidance and axon terminal refinement.

In the retina, ZIC2 expression matches the spatiotemporal expression of the guidance receptor EPHB1, and functional experiments in vivo have demonstrated that ZIC2 regulates EPHB1 expression in ipsilateral RGCs (García-Frigola et al. 2008). EPHRINB2, the ligand for EPHB1, is expressed in glial cells located at the optic chiasm exactly at the time when ipsilateral axons reach the midline (Williams et al. 2003). When EPHB1-expressing RGC axons contact midline glial cells, EPHB1/EPHRINB2 binding mediates a repulsive response that forces RGC axons to turn and continue growing into the same hemisphere (Williams et al. 2003). Although it is clear that ZIC2 regulates the expression of EPHB1 in the ventrotemporal retina, the phenotype of *Zic2*$^{kd/kd}$ mice does not perfectly match that of mice lacking EPHB1. The reduction in the number of ipsilateral axons in *Zic2*$^{kd/kd}$ mice is higher than in *EphB1* mutants (Herrera et al. 2003). In addition, ectopic expression of ZIC2 in contralateral RGCs leads to a stronger phenotype than ectopic EPHB1 expression (Petros et al. 2010). Also, ectopic expression of ZIC2 in an EphB1 null background produces some ectopic ipsilateral axons (García-Frigola et al. 2008). Together these results suggest that ZIC2 mediates axonal laterality by regulating the expression of EPHB1 but also additional molecules.

In contrast to what happens in the visual system, ZIC2 does not regulate the expression of EPHB1 in the ipsilateral dorsal cord neurons, but it does control another member of the same family, EPHA4 (Escalante et al. 2013). In fact, it has been determined via chromatin immunoprecipitation assays that ZIC2 directly binds to the EPHA4 promoter (Escalante et al. 2013), a result that has been subsequently confirmed by genome-wide analysis in a stem cell line (Luo et al. 2015).

It is not surprising that instead of EPHB1, the EPH receptor regulated by ZIC2 in the dorsal cord is EPHA4 because both receptors bind to the same ligand, EPHRINB2. It is likely that regulatory sequences that respond to ZIC2 activity have been evolutionarily selected in the EphB1 and EphA4 locus. In fact, EPHRINBs, the ligands for EPHA4, are highly expressed in the dorsal midline, and EPHA4/EPHRINB signaling mediates a repulsive axon midline response in EPHA4-expressing axons that exit the spinal cord through the DF (Paixão et al. 2013). Interestingly, the subpopulation of ZIC2 spinal neurons that leave the spinal cord through the DLF do not express EPHA4 or EPHB1, indicating that in this subset of ipsilaterally projecting $dL1_B$ neurons, ZIC2 must be regulating the expression of a different set of guidance receptors. In summary, ZIC2 determines axon midline avoidance by regulating the expression of EPH receptors in different contexts, but several evidences indicate that ZIC2 must control additional target molecules in both the visual system and the spinal cord.

In addition, to determine axonal laterality at the midline, ZIC2 also modulates a subsequent step in the wiring of the visual system, the refinement of ipsilateral axon terminals at the visual targets. At birth, when RGC axons reach the visual targets, they initially invade the entire superior colliculus and dorsal lateral geniculate nucleus (Sretavan et al. 1988; McLaughlin et al. 2003), but after a precise refinement process that depends on spontaneous activity in the retina, ipsilateral and contralateral axon terminals project in complementary projection patterns at the target tissues. It is known that the serotonin transporter (SERT) is required for this refinement process because mice lacking SERT expression exhibit defects in the eye-specific refinement of ipsilateral RGC axons (Salichon et al. 2001; Upton et al. 1999). It has been shown that *Zic2* $^{kd/kd}$ mice have strongly reduced SERT expression in ipsilateral RGCs, and manipulation of ZIC2 in the retina leads to defects in axon refinement at the targets, which reveals a role for ZIC2 in a late connectivity step (García-Frigola and Herrera 2010) (Fig. 11.2).

11.5 ZIC Proteins in the Mature Nervous System

As we have seen in previous sections, during embryonic development *Zic* genes are expressed in a variety of transition zones where cells detach from a palisade to join a particular stream or path and start to migrate. This is the case for neural crest cells, hem-derived CR cells, TPS cells, CAS cells, the granule progenitors coming from the rhombic lip, and some ipsilateral RGCs detaching from the CMZ. In the adult brain, some of the few places where *Zic* genes are expressed is the subventricular zone (SVZ) and the rostral migratory stream (RMS) that ends in the olfactory bulb (OB) (Brown and Brown 2009; Merkle et al. 2014). This region may recapitulate some aspects of embryonic ZIC-positive areas because, as in the abovementioned embryonic ZIC-migrating populations, cells in the SVZ detach from a neuroepithelial palisade to repolarize and initiate their migration (Fig. 11.1).

In the SVZ, neuroblasts (also called A cells) originate in clusters under the ventricular epithelium. Each cluster is wrapped by glial-type cells (B cells), which are the principal self-generating cells. The B cells also generate neuroblasts and rapidly dividing C cells, which amplify the production of neuroblasts (Conover and Notti 2008). Neuroblasts migrate out of the SVZ in chains that are mainly aligned in the rostrocaudal direction under the ventricular epithelium at the surface of the corpus striatum. These chains converge at the anterior angle of the ventricle and form the RMS, a compact single pathway that enters the core of the olfactory bulb (Altman 1969; Doetsch and Alvarez-Buylla 1996; Doetsch et al. 1997). Upon reaching the OB, type A cells depart from the tangentially oriented RMS and migrate radially to different layers of the OB. Type A cells move in a stepwise manner, similar to the movement of migrating ZIC-positive neuroblasts in the embryo. They first extend a leading process with a growth cone and then translocate the cell body toward the growth cone tip. This process is repeated, leading to a "saltatory" cellular movement. A dense meshwork of astroglial cells forming long tube-like structures that contain and presumably guide the migrating cells also integrates the RMS (Peretto et al. 1997).

The septum, a structure that is highly positive for *Zic* genes expression, is thought to contribute developmentally to the medial wall of the anterior ventricular SVZ (V-SVZ). $Zic1^{-/-}$ $Zic3^{-/-}$ newborn mice have smaller OBs (Inoue et al. 2007), which may be the consequence of defects in the migration of OB cells during embryogenesis. In the adult, the SVZ lies below the medioventral striatal surface of the lateral ventricle (Quiñones-Hinojosa et al. 2006), and the expression of *Zic* genes is maintained in subpopulations of type A and C cells located in this area (Merkle et al. 2014). Unfortunately, $Zic1^{-/-}$ $Zic3^{-/-}$ and $Zic2^{kd/kd}$ mice are all embryonic lethal. Thus, whether they have defects in RMS migration in adult stages has not yet been resolved.

Despite being a poorly understood process, several studies have identified some of the factors that guide neuroblasts along their SVZ-RMS path. An increasing number of studies implicate EPHB-EPHRINB interactions in the migration of neuroblasts during their trip. First, strong expression of EPHB2, EPHB3, and EPHA4 has been reported in the adult SVZ in mice (Conover et al. 2000). Second, ventricular injection of EPHRINB2/B3 disturbed the formation and migration of neuroblast chains (Conover et al. 2000) (Katakowski et al. 2005; Ricard et al. 2006). Therefore, it is possible that, as it happens during development, ZIC proteins induce the expression of EPH receptors in SVZ type A cells, which will then respond to EPHRINB2, likely expressed in the astroglial cells along the RMS (Wilkinson 2001; Himanen et al. 2004).

The function of ZIC proteins in SVZ and RMS has not being determined, but they have been proposed to be molecular mediators for the specification of medially produced OB interneuron types (Merkle et al. 2014). However, taking into account what we know today about the role of ZIC proteins in neuronal migration, it seems more likely that ZIC proteins instead play a role in the delamination, repolarization, and/or migration of specific types of SVZ neuroblasts.

11.6 Concluding Remarks

More than a decade ago, ZIC2 was first described as a determinant of axon guidance decisions in the developing visual system of mice. Since then, ZIC proteins have been implicated in different neural navigation processes in other contexts, including the guidance of dorsal spinal cord tracts and thalamocortical axons. In addition, these genes have been involved in the migration of several populations of immature neurons. It is known that signaling pathways mediated by EPH-EPHRINB and other guidance molecules have much in common when acting in axon guidance and neuronal migration because they share molecular components and cellular effects that ultimately influence cytoskeleton remodeling (Mellitzer et al. 2000; Zimmer et al. 2007). It is thus likely that ZIC proteins regulate axon guidance and migration using similar mechanisms.

The roles of *Zic* genes in the adult brain are currently unknown. The observations described in the previous section about putative similarities between the adult SVZ and other neuroepithelial regions positive for ZIC expression during development suggest that these proteins may function in delamination/polarization or migration also in the adult brain. Together with the dentate gyrus of the hippocampus, the SVZ is one of the few places in the adult mammalian brain where neurogenesis has been observed (Ming and Song 2011). However, hippocampal neuroblasts do not engage in a long-range trip to migrate further away as they do in the SVZ. Therefore, anatomical and functional analysis of ZIC proteins in the adult dentate gyrus may further help to clarify the role of these proteins in neurogenesis, detachment, cell polarization, and/or migration.

References

Alaynick WA, Jessell TM, Pfaff SL (2011) SnapShot: spinal cord development. Cell 146:178–178.e1. https://doi.org/10.1016/j.cell.2011.06.038

Altman J (1969) Autoradiographic and histological studies of postnatal neurogenesis. IV. Cell proliferation and migration in the anterior forebrain, with special reference to persisting neurogenesis in the olfactory bulb. J Comp Neurol 137:433–457. https://doi.org/10.1002/cne.901370404

Andoniadou CL, Martinez-Barbera JP (2013) Developmental mechanisms directing early anterior forebrain specification in vertebrates. Cell Mol Life Sci 70:3739–3752. https://doi.org/10.1007/s00018-013-1269-5

Aruga J, Yokota N, Hashimoto M et al (1994) A novel zinc finger protein, zic, is involved in neurogenesis, especially in the cell lineage of cerebellar granule cells. J Neurochem 63:1880–1890

Aruga J, Nagai T, Tokuyama T et al (1996a) The mouse zic gene family. Homologues of the Drosophila pair-rule gene odd-paired. J Biol Chem 271:1043–1047

Aruga J, Yozu A, Hayashizaki Y et al (1996b) Identification and characterization of Zic4, a new member of the mouse Zic gene family. Gene 172:291–294

Aruga J, Inoue T, Hoshino J, Mikoshiba K (2002) Zic2 controls cerebellar development in cooperation with Zic1. J Neurosci 22:218–225

Avraham O, Hadas Y, Vald L et al (2009) Transcriptional control of axonal guidance and sorting in dorsal interneurons by the Lim-HD proteins Lhx9 and Lhx1. Neural Dev 4:21. https://doi.org/10.1186/1749-8104-4-21

Avraham O, Hadas Y, Vald L et al (2010) Motor and dorsal root ganglion axons serve as choice points for the ipsilateral turning of dI3 axons. J Neurosci 30:15546–15557. https://doi.org/10.1523/JNEUROSCI.2380-10.2010

Beccari L, Marco-Ferreres R, Bovolenta P (2013) The logic of gene regulatory networks in early vertebrate forebrain patterning. Mech Dev 130:95–111. https://doi.org/10.1016/j.mod.2012.10.004

Bélanger M-C, Robert B, Cayouette M (2016) Msx1-positive progenitors in the retinal ciliary margin give rise to both neural and non-neural progenies in mammals. Dev Cell. https://doi.org/10.1016/j.devcel.2016.11.020

Bielle F, Griveau A, Narboux-Nême N et al (2005) Multiple origins of Cajal-Retzius cells at the borders of the developing pallium. Nat Neurosci 8:1002–1012. https://doi.org/10.1038/nn1511

Borrell V, Marín O (2006) Meninges control tangential migration of hem-derived Cajal-Retzius cells via CXCL12/CXCR4 signaling. Nat Neurosci 9:1284–1293. https://doi.org/10.1038/nn1764

Brown L, Brown S (2009) Zic2 is expressed in pluripotent cells in the blastocyst and adult brain expression overlaps with makers of neurogenesis. Gene Expr Patterns 9:43–49. https://doi.org/10.1016/j.gep.2008.08.002

Camurri L, Mambetisaeva E, Davies D et al (2005) Evidence for the existence of two Robo3 isoforms with divergent biochemical properties. Mol Cell Neurosci 30:485–493. https://doi.org/10.1016/j.mcn.2005.07.014

Carreres MI, Escalante A, Murillo B et al (2011) Transcription factor Foxd1 is required for the specification of the temporal retina in mammals. J Neurosci 31:5673–5681. https://doi.org/10.1523/JNEUROSCI.0394-11.2011

Chédotal A (2011) Further tales of the midline. Curr Opin Neurobiol 21:68–75. https://doi.org/10.1016/j.conb.2010.07.008

Colamarino SA, Tessier-Lavigne M (1995) The role of the floor plate in axon guidance. Annu Rev Neurosci 18:497–529. https://doi.org/10.1146/annurev.ne.18.030195.002433

Conover JC, Notti RQ (2008) The neural stem cell niche. Cell Tissue Res 331:211–224. https://doi.org/10.1007/s00441-007-0503-6

Conover JC, Doetsch F, García-Verdugo JM et al (2000) Disruption of Eph/ephrin signaling affects migration and proliferation in the adult subventricular zone. Nat Neurosci 3:1091–1097. https://doi.org/10.1038/80606

D'Arcangelo G, Miao GG, Chen SC et al (1995) A protein related to extracellular matrix proteins deleted in the mouse mutant reeler. Nature 374:719–723. https://doi.org/10.1038/374719a0

Ding Q, Joshi PS, Xie Z-H et al (2012) BARHL2 transcription factor regulates the ipsilateral/contralateral subtype divergence in postmitotic dI1 neurons of the developing spinal cord. Proc Natl Acad Sci U S A 109:1566–1571. https://doi.org/10.1073/pnas.1112392109

Doetsch F, Alvarez-Buylla A (1996) Network of tangential pathways for neuronal migration in adult mammalian brain. Proc Natl Acad Sci U S A 93:14895–14900

Doetsch F, García-Verdugo JM, Alvarez-Buylla A (1997) Cellular composition and three-dimensional organization of the subventricular germinal zone in the adult mammalian brain. J Neurosci 17:5046–5061

Engle EC (2010) Human genetic disorders of axon guidance. Cold Spring Harb Perspect Biol 2:a001784–a001784. https://doi.org/10.1101/cshperspect.a001784

Escalante A, Murillo B, Morenilla-Palao C et al (2013) Zic2-dependent axon midline avoidance controls the formation of major ipsilateral tracts in the CNS. Neuron 80:1392–1406. https://doi.org/10.1016/j.neuron.2013.10.007

Fischer AJ, Bosse JL, El-Hodiri HM (2014) Reprint of: the ciliary marginal zone (CMZ) in development and regeneration of the vertebrate eye. Exp Eye Res 123:115–120. https://doi.org/10.1016/j.exer.2014.04.019

Frotscher M (1998) Cajal-Retzius cells, Reelin, and the formation of layers. Curr Opin Neurobiol 8:570–575

García-Frigola C, Herrera E (2010) Zic2 regulates the expression of Sert to modulate eye-specific refinement at the visual targets. EMBO J 29:3170–3183. https://doi.org/10.1038/emboj.2010.172

García-Frigola C, Carreres MI, Vegar C et al (2008) Zic2 promotes axonal divergence at the optic chiasm midline by EphB1-dependent and -independent mechanisms. Development 135:1833–1841. https://doi.org/10.1242/dev.020693

Gil-Sanz C, Franco SJ, Martinez-Garay I et al (2013) Cajal-Retzius cells instruct neuronal migration by coincidence signaling between secreted and contact-dependent guidance cues. Neuron 79:461–477. https://doi.org/10.1016/j.neuron.2013.06.040

Goulding M (2009) Circuits controlling vertebrate locomotion: moving in a new direction. Nat Rev Neurosci 10:507–518. https://doi.org/10.1038/nrn2608

Griveau A, Borello U, Causeret F et al (2010) A novel role for Dbx1-derived Cajal-Retzius cells in early regionalization of the cerebral cortical neuroepithelium. PLoS Biol 8:e1000440. https://doi.org/10.1371/journal.pbio.1000440

Gross MK, Dottori M, Goulding M (2002) Lbx1 specifies somatosensory association interneurons in the dorsal spinal cord. Neuron 34:535–549

Hatten ME (1999) Central nervous system neuronal migration. Annu Rev Neurosci 22:511–539. https://doi.org/10.1146/annurev.neuro.22.1.511

Helms AW, Johnson JE (2003) Specification of dorsal spinal cord interneurons. Curr Opin Neurobiol 13:42–49

Herrera E, Brown L, Aruga J et al (2003) Zic2 patterns binocular vision by specifying the uncrossed retinal projection. Cell 114:545–557. https://doi.org/10.1038/nrn1267

Himanen J-P, Chumley MJ, Lackmann M et al (2004) Repelling class discrimination: ephrin-A5 binds to and activates EphB2 receptor signaling. Nat Neurosci 7:501–509. https://doi.org/10.1038/nn1237

Inoue T, Ota M, Ogawa M, Mikoshiba K, Aruga J (2007) Zic1 and Zic3 regulate medial forebrain development through expansion of neuronal progenitors. J Neurosci 27(20):5461–5473. https://doi.org/10.1523/JNEUROSCI.4046-06.2007

Inoue T, Ogawa M, Mikoshiba K, Aruga J (2008) Zic deficiency in the cortical marginal zone and meninges results in cortical lamination defects resembling those in type II lissencephaly. J Neurosci 28:4712–4725. https://doi.org/10.1523/JNEUROSCI.5735-07.2008

Izzi L, Charron F (2011) Midline axon guidance and human genetic disorders. Clin Genet 80:226–234. https://doi.org/10.1111/j.1399-0004.2011.01735.x

Jessell TM (2000) Neuronal specification in the spinal cord: inductive signals and transcriptional codes. Nat Rev Genet 1:20–29. https://doi.org/10.1038/35049541

Karam SD, Burrows RC, Logan C et al (2000) Eph receptors and ephrins in the developing chick cerebellum: relationship to sagittal patterning and granule cell migration. J Neurosci 20:6488–6500

Katakowski M, Zhang Z, deCarvalho AC, Chopp M (2005) EphB2 induces proliferation and promotes a neuronal fate in adult subventricular neural precursor cells. Neurosci Lett 385:204–209. https://doi.org/10.1016/j.neulet.2005.05.060

Liu Y, Golden JA (2007) Patterning the developing diencephalon. Brain Res Rev 53:17–26. https://doi.org/10.1016/j.brainresrev.2006.06.004

López-Bendito G, Molnár Z (2003) Thalamocortical development: how are we going to get there? Nat Rev Neurosci 4:276–289. https://doi.org/10.1038/nrn1075

Luo Z, Gao X, Lin C et al (2015) Zic2 is an enhancer-binding factor required for embryonic stem cell specification. Mol Cell 57:685–694. https://doi.org/10.1016/j.molcel.2015.01.007

Machold R, Fishell G (2005) Math1 is expressed in temporally discrete pools of cerebellar rhombic-lip neural progenitors. Neuron 48:17–24. https://doi.org/10.1016/j.neuron.2005.08.028

Marcucci F, Murcia-Belmonte V, Wang Q et al (2016) The ciliary margin zone of the mammalian retina generates retinal ganglion cells. Cell Rep 17:3153–3164. https://doi.org/10.1016/j.celrep.2016.11.016
Marín O, Rubenstein JLR (2003) Cell migration in the forebrain. Annu Rev Neurosci 26:441–483. https://doi.org/10.1146/annurev.neuro.26.041002.131058
McLaughlin T, Hindges R, O'Leary DDM (2003) Regulation of axial patterning of the retina and its topographic mapping in the brain. Curr Opin Neurobiol 13:57–69
Mellitzer G, Xu Q, Wilkinson DG (2000) Control of cell behaviour by signalling through Eph receptors and ephrins. Curr Opin Neurobiol 10:400–408
Merkle FT, Fuentealba LC, Sanders TA et al (2014) Adult neural stem cells in distinct microdomains generate previously unknown interneuron types. Nat Neurosci 17:207–214. https://doi.org/10.1038/nn.3610
Meyer G, Perez-Garcia CG, Gleeson JG (2002) Selective expression of doublecortin and LIS1 in developing human cortex suggests unique modes of neuronal movement. Cereb Cortex 12:1225–1236
Ming G-L, Song H (2011) Adult neurogenesis in the mammalian brain: significant answers and significant questions. Neuron 70:687–702. https://doi.org/10.1016/j.neuron.2011.05.001
Müller T, Brohmann H, Pierani A et al (2002) The homeodomain factor lbx1 distinguishes two major programs of neuronal differentiation in the dorsal spinal cord. Neuron 34:551–562
Murillo B, Ruiz-Reig N, Herrera M et al (2015) Zic2 controls the migration of specific neuronal populations in the developing forebrain. J Neurosci 35:11266–11280. https://doi.org/10.1523/JNEUROSCI.0779-15.2015
Nagai T, Aruga J, Takada S et al (1997) The expression of the mouse Zic1, Zic2, and Zic3 gene suggests an essential role for Zic genes in body pattern formation. Dev Biol 182:299–313. https://doi.org/10.1006/dbio.1996.8449
Nugent AA, Kolpak AL, Engle EC (2012) Human disorders of axon guidance. Curr Opin Neurobiol 22:837–843. https://doi.org/10.1016/j.conb.2012.02.006
Ogawa M, Miyata T, Nakajima K et al (1995) The reeler gene-associated antigen on Cajal-Retzius neurons is a crucial molecule for laminar organization of cortical neurons. Neuron 14:899–912
Ozawa PMM, Ariza CB, Ishibashi CM et al (2016) Role of CXCL12 and CXCR4 in normal cerebellar development and medulloblastoma. Int J Cancer 138:10–13. https://doi.org/10.1002/ijc.29333
Paixão S, Balijepalli A, Serradj N et al (2013) EphrinB3/EphA4-mediated guidance of ascending and descending spinal tracts. Neuron 80:1407–1420. https://doi.org/10.1016/j.neuron.2013.10.006
Pak W, Hindges R, Lim Y-S et al (2004) Magnitude of binocular vision controlled by islet-2 repression of a genetic program that specifies laterality of retinal axon pathfinding. Cell 119:567–578. https://doi.org/10.1016/j.cell.2004.10.026
Peretto P, Merighi A, Fasolo A, Bonfanti L (1997) Glial tubes in the rostral migratory stream of the adult rat. Brain Res Bull 42:9–21
Petkó M, Antal M (2012) Propriospinal pathways in the dorsal horn (laminae I–IV) of the rat lumbar spinal cord. Brain Res Bull 89:41–49. https://doi.org/10.1016/j.brainresbull.2012.06.005
Petros TJ, Bryson JB, Mason C (2010) Ephrin-B2 elicits differential growth cone collapse and axon retraction in retinal ganglion cells from distinct retinal regions. Dev Neurobiol 70:781–794. https://doi.org/10.1002/dneu.20821
Puelles L, Rubenstein JLR (2003) Forebrain gene expression domains and the evolving prosomeric model. Trends Neurosci 26:469–476. https://doi.org/10.1016/S0166-2236(03)00234-0
Quiñones-Hinojosa A, Sanai N, Soriano-Navarro M et al (2006) Cellular composition and cytoarchitecture of the adult human subventricular zone: a niche of neural stem cells. J Comp Neurol 494:415–434. https://doi.org/10.1002/cne.20798
Remedios R, Huilgol D, Saha B et al (2007) A stream of cells migrating from the caudal telencephalon reveals a link between the amygdala and neocortex. Nat Neurosci 10:1141–1150. https://doi.org/10.1038/nn1955
Ricard J, Salinas J, Garcia L, Liebl DJ (2006) EphrinB3 regulates cell proliferation and survival in adult neurogenesis. Mol Cell Neurosci 31:713–722. https://doi.org/10.1016/j.mcn.2006.01.002

Rice DS, Curran T (2001) Role of the reelin signaling pathway in central nervous system development. Annu Rev Neurosci 24:1005–1039. https://doi.org/10.1146/annurev.neuro.24.1.1005

Rice DS, Williams RW, Goldowitz D (1995) Genetic control of retinal projections in inbred strains of albino mice. J Comp Neurol 354:459–469. https://doi.org/10.1002/cne.903540312

Robertshaw E, Kiecker C (2012) Phylogenetic origins of brain organisers. Scientifica (Cairo) 2012:475017–475014. https://doi.org/10.6064/2012/475017

Salichon N(1), Gaspar P, Upton AL, Picaud S, Hanoun N, Hamon M, De Maeyer E, Murphy DL, Mossner R, Lesch KP, Hen R, Seif I (2001, February 1) Excessive activation of serotonin (5-HT) 1B receptors disrupts the formation of sensory maps in monoamine oxidase a and 5-ht transporter knock-out mice. J Neurosci 21(3):884–896

Santiago AC, Shammah-Lagnado SJ (2004) Efferent connections of the nucleus of the lateral olfactory tract in the rat. J Comp Neurol 471:314–332. https://doi.org/10.1002/cne.20028

Scholpp S, Lumsden A (2010) Building a bridal chamber: development of the thalamus. Trends Neurosci 33:373–380. https://doi.org/10.1016/j.tins.2010.05.003

Seth A, Culverwell J, Walkowicz M et al (2006) Belladonna/(Ihx2) is required for neural patterning and midline axon guidance in the zebrafish forebrain. Development 133:725–735. https://doi.org/10.1242/dev.02244

Shinozaki K, Miyagi T, Yoshida M et al (2002) Absence of Cajal-Retzius cells and subplate neurons associated with defects of tangential cell migration from ganglionic eminence in Emx1/2 double mutant cerebral cortex. Development 129:3479–3492

Sotelo C (2004) Cellular and genetic regulation of the development of the cerebellar system. Prog Neurobiol 72:295–339. https://doi.org/10.1016/j.pneurobio.2004.03.004

Sretavan DW, Shatz CJ, Stryker MP (1988) Modification of retinal ganglion cell axon morphology by prenatal infusion of tetrodotoxin. Nature 336:468–471. https://doi.org/10.1038/336468a0

Subramanian L, Tole S (2009) Mechanisms underlying the specification, positional regulation, and function of the cortical hem. Cereb Cortex 19(Suppl 1):i90–i95. https://doi.org/10.1093/cercor/bhp031

Takiguchi-Hayashi K, Sekiguchi M, Ashigaki S et al (2004) Generation of reelin-positive marginal zone cells from the caudomedial wall of telencephalic vesicles. J Neurosci 24:2286–2295. https://doi.org/10.1523/JNEUROSCI.4671-03.2004

Tissir F, Goffinet AM (2003) Reelin and brain development. Nat Rev Neurosci 4:496–505. https://doi.org/10.1038/nrn1113

Upton AL(1), Salichon N, Lebrand C, Ravary A, Blakely R, Seif I, Gaspar P (1999) Excess of serotonin (5-HT) alters the segregation of ispilateral and contralateral retinal projections in monoamine oxidase A knock-out mice: possible role of 5-HT uptake in retinal ganglion cells during development. J Neurosci 19(16):7007–7024

Vallstedt A, Kullander K (2013) Dorsally derived spinal interneurons in locomotor circuits. Ann N Y Acad Sci 1279:32–42. https://doi.org/10.1111/j.1749-6632.2012.06801.x

Villar-Cerviño V, Molano-Mazón M, Catchpole T et al (2013) Contact repulsion controls the dispersion and final distribution of Cajal-Retzius cells. Neuron 77:457–471. https://doi.org/10.1016/j.neuron.2012.11.023

Wang VY, Rose MF, Zoghbi HY (2005) Math1 expression redefines the rhombic lip derivatives and reveals novel lineages within the brainstem and cerebellum. Neuron 48:31–43. https://doi.org/10.1016/j.neuron.2005.08.024

Wilkinson DG (2001) Multiple roles of EPH receptors and ephrins in neural development. Nat Rev Neurosci 2:155–164. https://doi.org/10.1038/35058515

Williams SE, Mann F, Erskine L et al (2003) Ephrin-B2 and EphB1 mediate retinal axon divergence at the optic chiasm. Neuron 39:919–935. https://doi.org/10.1038/nrn1267

Wilson SW, Houart C (2004) Early steps in the development of the forebrain. Dev Cell 6:167–181

Wilson SI, Shafer B, Lee KJ, Dodd J (2008) A molecular program for contralateral trajectory: Rig-1 control by LIM homeodomain transcription factors. Neuron 59:413–424. https://doi.org/10.1016/j.neuron.2008.07.020

Yoshida M, Assimacopoulos S, Jones KR, Grove EA (2006) Massive loss of Cajal-Retzius cells does not disrupt neocortical layer order. Development 133:537–545. https://doi.org/10.1242/dev.02209

Zimmer G, Kästner B, Weth F, Bolz J (2007) Multiple effects of ephrin-A5 on cortical neurons are mediated by SRC family kinases. J Neurosci 27:5643–5653. https://doi.org/10.1523/JNEUROSCI.0954-07.2007

Zou YR, Kottmann AH, Kuroda M et al (1998) Function of the chemokine receptor CXCR4 in haematopoiesis and in cerebellar development. Nature 393:595–599. https://doi.org/10.1038/31269

Part II
Zic Family in Medicine

Chapter 12
Zic Family Proteins in Emerging Biomedical Studies

Jun Aruga

Abstract Zic family proteins have been investigated in various biomedical studies. Here we summarize the contact points between Zic proteins and recent medical research. The topics cover a wide range, reflecting the pleiotropic roles of these proteins in early embryogenesis and organogenesis. Zic1, Zic2, and Zic3 proteins play important roles in the development of axial and limb bones, and of muscles, among the derivatives of the notochord and somites. Zic1 is involved in bone's response to mechanical stress, and it also serves as a marker specific for brown adipocytes. Zic1, Zic2, Zic3, and Zic5 proteins are required for the development of neural crest derivatives, including the meningeal membrane and facial bones, and deficiency of these proteins causes cortical lamination defects resembling those in type II lissencephaly. In vascular systems, Zic3 is associated not only with normal cardiovascular development, failure of which causes congenital heart anomalies, but also controls maturation of the blood-brain barrier. Zic1 is also expressed in the brain pericytes possessing stem cell properties that control the blood-brain barrier activity and capillary hemodynamic responses. The possible involvement of Zic proteins in neuropsychiatric disorders has been indicated by the analyses of mutant mice behaviors. *Zic1* and *Zic3* mutant mice show hypotonia and decreased locomotor activities. *Zic2* hypomorphic mutant mice exhibit schizophrenia-related behavioral abnormalities such as cognitive dysfunction and impaired sensorimotor gating and social behaviors, and *ZIC2* mutations found in schizophrenia patients included a severely functionally defective one. Based on these facts, the application of Zic protein activities in translational medicine might be considered.

Keywords Somite · Osteocyte · Adipocyte · Neural crest · Meningeal cell · Blood-brain barrier · Behavior

J. Aruga (✉)
Department of Medical Pharmacology, Nagasaki University Graduate School of Biomedical Sciences, Nagasaki, Japan
e-mail: aruga@nagasaki-u.ac.jp

J. Aruga (ed.), *Zic family*, Advances in Experimental Medicine and Biology 1046,
https://doi.org/10.1007/978-981-10-7311-3_12

12.1 Overview of the Role of Zic Proteins in Notochord/ Somite Development

Chordate Zic proteins are essential for the generation of mesoderm and its direct derivatives, including the notochord and somites, as discussed in Chaps. 7, 8, 9, and 10. Here, we briefly review the involvement of mammalian *Zic* genes in the development of somites.

In developing somites, *Zic2* and *Zic3* are expressed in the presomitic mesoderm, and *Zic1* is expressed at a later stage (Fig. 12.1). *Zic2* and *Zic3* synergistically control multiple steps of the mesodermal segmentation process in mammals (Fig. 12.2) (Inoue et al. 2007). In *Zic1* mutants, abnormality was most pronounced in the dorsal components of the sclerotome-derived structures (Aruga et al. 1999). As such, mutant mice with *Zic1*, *Zic2*, and *Zic3* single and *Zic2/3* combined mutations showed a variety of skeletal malformations (Fig. 12.2). These included bridging between adjacent vertebral bodies or arches, or splitting of the vertebral bodies, and impaired midline fusion of vertebral arches (spina bifida occulta). Thus, *Zic1*, *Zic2*, and *Zic3* have a common role in establishing metamerism by maintaining the fidelity of somite development (see Fig. 1.11 for bilaterian metamerism).

Some genetic interactions involving *Zic* genes have been reported. *Zic1* and *Gli3* cooperatively control the patterning of vertebral arches. *Zic1* and *Pax1* also show cooperation, but to a lesser extent (Fig. 12.3a) (Aruga et al. 1999). *Zic1* and *Zic2* control the expression of Myf5, a transcription factor that has a key role in regulating muscle differentiation at the dorsomedial part of the dermomyotome (dorsomedial lip) (Fig. 12.3b). *Zic1/2*, *Gli2*, and *Pax3* synergistically activated the Myf5 epaxial somite enhancer in reporter gene assays (Pan et al. 2011; Himeda et al. 2013).

12.2 Zic1 in Mechanotransduction of Osteocytes

A recent study indicates the role of Zic1 in the mechanotransduction of osteocytes (Kalogeropoulos et al. 2010). The bones of mature vertebrate skeletons constantly remodel in response to the ever-changing mechanical load. Osteocytes transduce the mechanical loading forces into inter- and intracellular biochemical signals. Kalogeropoulos et al. (2010) compared between the transcript profiles of bones exposed to high and low levels of mechanical stress (lumbar spine and iliac crest, respectively) from human subjects and identified ZIC1 as the most significantly upregulated gene in the bone exposed to high mechanical stress. This analysis also identified *PTCH1* and *GLI3* among the upregulated components in the lumber spine.

The same researchers identified Zic1 protein in two murine cell lines, MC3T3-E1 (osteoblast-like cell) and MLO-Y4 (osteocyte-like cell), subjected to mechanical stress (fluid shear stress), to examine the *Zic1* expression levels in them. Interestingly, Zic1 was preferentially localized in the cell nuclei of MLO-Y4 after exposure to mechanical stress, while in the unstressed cells, Zic1 showed uniform distribution

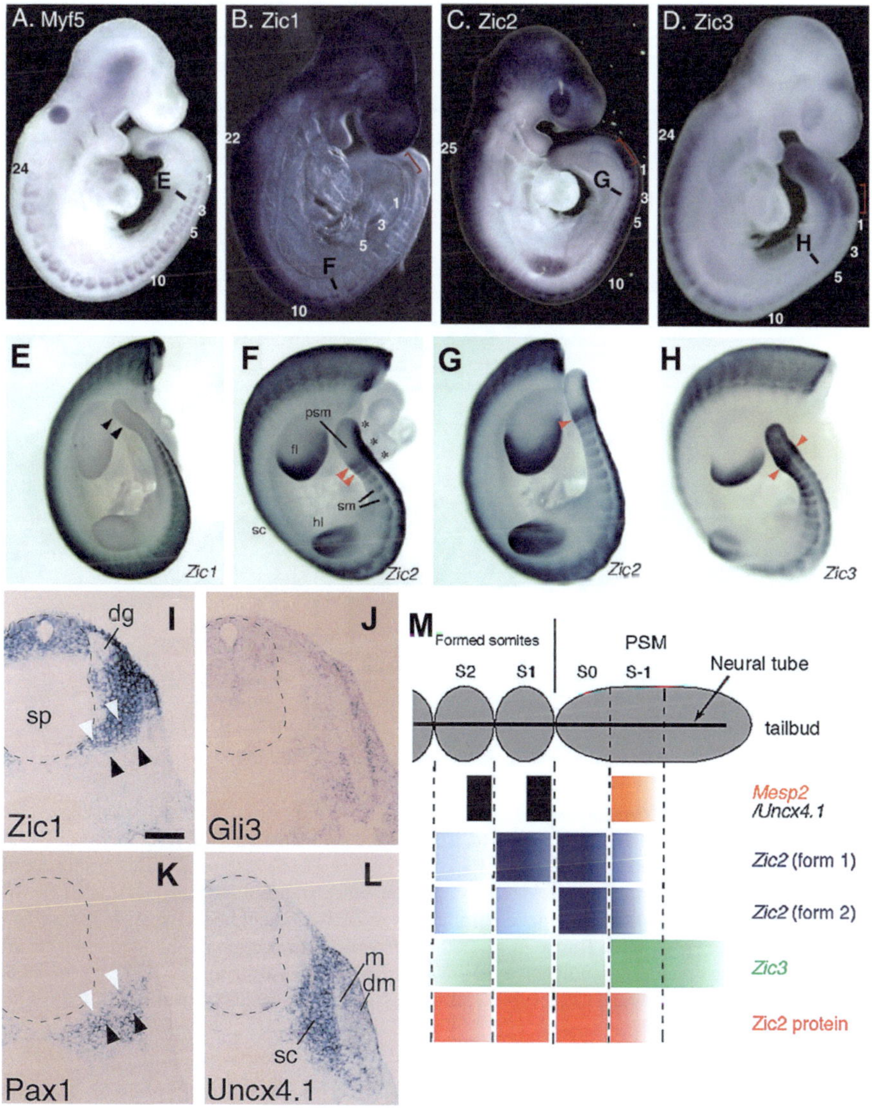

Fig. 12.1 Expression patterns of the indicated genes in somites of mouse embryos. (**a**–**h**) E9.5 embryos (Pan et al. 2011). (**i**–**m**) E10.5 embryos, (**n**) summary of *Zic2* and *Zic3* expression during somitogenesis (Inoue et al. 2007). (**o**–**u**) E10.5 sections (Aruga et al. 1999). *a* anterior compartment of a somite, *p* posterior compartment, *nt* neural tube, *sc* sclerotome, *dm* dermomyotome, *dg* dorsal root ganglion, *psm* presomitic mesoderm ((**a**–**d**) are reprinted from Pan et al. 2011 with permission. (**e**–**h**) and (**m**) are reprinted from Inoue et al. 2007 with permission. (**i**–**l**) are reprinted from Aruga et al. 1999 with permission)

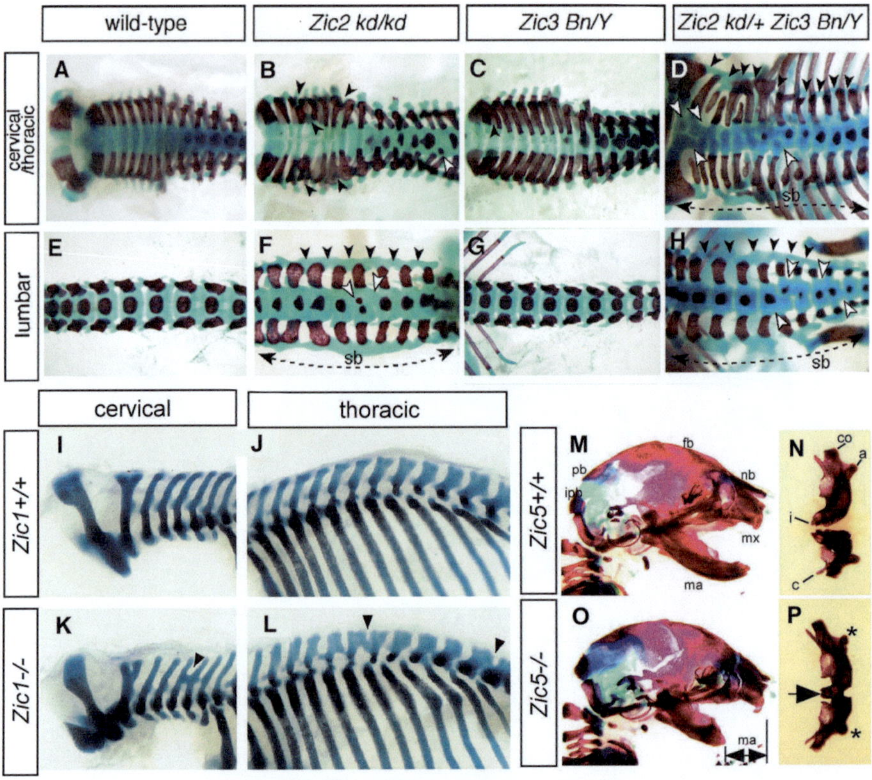

Fig. 12.2 Skeletal phenotype of *Zic* mutant mice. Skeletal staining (red, bones; blue, cartilage) for wild type (**a**, **e**), *Zic2kd/kd* (**b**, **f**), *Zic3 Bn/Y* (**c**, **g**), and *Zic2 kd/+ Zic3 Bn/Y* (**d**, **h**) embryos at E17.5 (Inoue et al. 2007). Skeletal phenotype of E15.5 *Zic11/1* (**i**, **j**), *Zic12/2* (**c**, **d**). The primordia of vertebral arches were abnormally bridged to adjacent ones (indicated by arrowheads in (**k**, **l**)) (Aruga et al. 1999). (**m**–**p**) Craniofacial abnormalities in *Zic5*-deficient mice. (**m**, **o**) wild type and (**n**, **p**) *Zic5−/−* newborn mice (Inoue et al. 2004). Notice the truncation of the mandible. (**o**, **p**) frontal view of the dissected mandible ((**a**–**h**) are reprinted from Inoue et al. 2007 with permission. (**i**–**l**) are Reprinted from Aruga et al. 1999 with permission. (**m**–**p**) are reprinted from Inoue et al. 2004 with permission.

in the cell nuclei and cytoplasm. Furthermore, *Zic1* knockdown in MLO-Y4 was shown to decrease Wnt/β-catenin signaling activation that usually follows mechanical stress. Wnt/β-catenin signaling is required for normal bone homeostasis (Kramer et al. 2010). These facts suggest the involvement of *Zic1* in the bone's response to mechanical stress. Although mechanical stress-induced nuclear import of Zic1 and *Zic1* knockdown-induced β-catenin signal attenuation were observed in MLO-Y4 cells, MC3T3-E1 cells did not show any change. Thus, the findings by Kalogeropoulos et al. (2010) warrant further studies to address the role of *Zic* genes in bone homeostasis in vivo. To our knowledge, bone-related phenotypes of the murine *Zic* mutants are limited to the developmental defects of skeletal patterns or of gross morphology. So far, there are no studies that have focused on the role of *Zic* in the physiological regulation or the fine structure of bones.

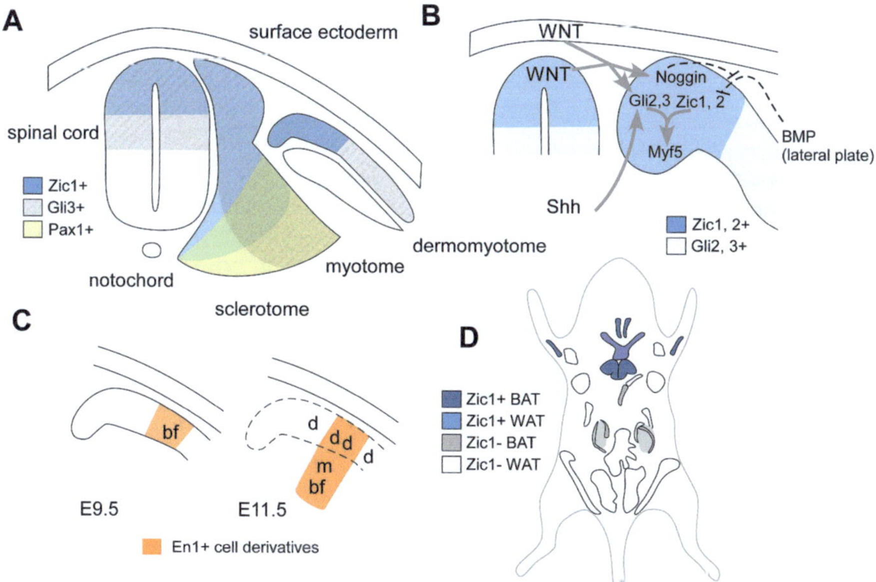

Fig. 12.3 Illustrations for Zic in somites. (**a**) Zic1-Gli3 interaction. A transverse section through E10.5 trunk is schematically drawn. The synergism between Zic1 and Gli3 can be considered based on the Zic1/Gli3 co-expression at dorsolateral sclerotome and the dorsal spinal cord where the interaction may be involved in the segmentation of the dorsal sclerotome and the proper development of the dorsal most vertebral arches. Illustration is based on the results in Aruga et al. (1999). (**b**) A model describing the roles of Shh, Wnt, and BMP signaling and Gli and Zic transcription factors in the activation and patterning of Myf5 for epaxial myogenesis. Illustration is based on the results in Pan et al. (2011). (**c**) Expression of Zic1 in each adipose tissue. Illustration is based on the results in de Jong et al. (2015). (**d**) Development of brown fat tissue from En1+ dermomyotome (orange) (Atit et al. 2006)

12.3 *Zic1* as a Marker for Brown Adipocytes

Recent studies on human adipose tissues have revealed the presence of white and brown adipocytes, both of which are able to store energy in the form of triglycerides (reviewed in Wang and Seale 2016). But brown adipocytes are metabolically more active than white adipocytes and generate heat through the action of mitochondrial uncoupling protein UCP1. The developmental origin and molecular signatures of brown adipocytes are of interest to many researchers, because of brown adipocyte's potential as a therapeutic target to treat obesity and its complications such as type 2 diabetes.

During vertebrate development, the paraxial mesoderm is divided into somites. The medial part of the dermomyotome, a dorsally located somite derivative, is known to generate muscle and dermis. Transcriptome analysis indicated that the immature brown adipocytes (preadipocytes) express a set of genes involved in skeletal muscle development, including basic helix-loop-helix type myogenic regulatory transcription factors, myogenin and myf5 (Timmons et al. 2007). The result suggests a common developmental origin for brown preadipocytes and skeletal muscle cells (Timmons et al. 2007). Further analysis to clarify the differentiation processes of brown and white adipocytes identified Zic1 and Lhx8 as being preferentially expressed in brown adipocytes (Seale et al. 2007; Timmons et al. 2007). It is now widely accepted that the developmental origin of brown adipocytes includes $Myf5^{+}Pax7^{+}En1^{+}$ dermomyotomes (Wang and Seale 2016).

Following the discovery of *Zic1* as a signature of brown adipocytes, several studies used *Zic1* as a molecular marker for brown adipocytes (Jespersen et al. 2013; Petrovic et al. 2010; Rockstroh et al. 2015; Cereijo et al. 2015; Qin et al. 2016; Hafner et al. 2016). These include the study on brown fat cell derivation from induced pluripotent stem (iPS) cells, where human iPS cells showed expression of UCP1 and ZIC1 following treatment with TGFβ signaling inhibitor, ascorbic acid, and EGF (Hafner et al. 2016). These studies have perhaps validated the greater expression of *Zic1* in brown adipocytes than in white adipocytes.

However, whether *Zic1* is a necessary and sufficient marker for brown adipocytes is, in fact, not quite clear. In this regard, an intriguing observation has been reported by De Jong et al. (2015), who examined the expression of previously studied adipocyte subtype markers in an extensive set of adipose tissues (14 classes) sampled from warm (30 °C)- or cold (4 °C)-acclimated mice. In this study, *Zic1* was found to be expressed not only in the brown adipose tissues in the cervical, axillary, and interscapular regions but also in the white adipose tissues in the interscapular region (Fig. 12.3d). Furthermore, the brown adipose tissues in the mediastinal and perirenal regions were shown to not express *Zic1* (Fig. 12.3d). Thus, *Zic1* is expressed in all adipose tissues, whether it is brown or white, located rostrally at the scapula/axilla level, but not in the ones located at the caudal level. *Zic1* expression levels in the adipose tissue types in the warm- and cold-acclimated mice (de Jong et al. 2015) did not vary. These results suggest that the preferential *Zic1* expression in adipose tissues reflect regional properties that are probably established during development.

With respect to paraxial mesoderm, *Zic1* is expressed only in the rostral somites (Fig. 12.1) (Pan et al. 2011; Inoue et al. 2007). In particular, at E9.5 (22 somite stage), *Zic1* expression was detected in 16 rostral somites, in which the most caudal one was located at the forelimb level, nearly matching the caudal boundary of *Zic1* expression in the adipose tissues. Meanwhile, a lineage mapping study has shown that En1-positive cells in the central dermomyotome give rise to brown fat cells at E9.5–E11.5 (Fig. 12.3c) (Atit et al. 2006). It would be tempting to postulate that the brown adipocyte progenitors generated at E9.5 maintain their *Zic1* expression as an inherent signature. However, more comprehensive lineage tracing analysis is needed before this is conclusively proved, because the vasculature also provides a niche supporting both brown and white adipocyte progenitors (Sanchez-Gurmaches and Guertin 2014). Besides its usability as a marker, the role of *Zic1* in adipocyte development has not yet been addressed. Pathophysiological significance of *Zic1* in adipose tissues, if any, also remains to be investigated.

12.4 Overview of the Role of Zic Proteins in Neural Crest Development

Neural crest is a stem/progenitor cell population that differentiates into numerous derivatives, including fibroblasts, chondrocytes, osteocytes, and adipocytes (Hall 1999). Zic genes have been placed in a gene regulatory network controlling neural crest development (Simoes-Costa and Bronner 2015) (Fig. 12.4). In this model, WNT and BMP inhibitors generate balanced WNT and BMP activities at the lateral border. These signals, together with FGFs and Notch signaling, are involved in the specification of neural plate border region. *Zic1*, along with *Msx1*, *Pax3/7*, etc., acts as neural plate border specifier gene. These genes activate another group of genes (*FoxD3*, *Ets1*, and *Snail1/2*) at the time of neural crest specification. *Zic1* and *Pax3* cooperate to directly activate *Snail1* and *Snail2* (also known as Slug) promoters in *Xenopus* (Milet et al. 2013); they also directly regulate *FoxD3* expression in trunk neural crest cells in chick (Simoes-Costa et al. 2012). In mouse, *Zic2* and *Zic5* are required for the development of the cephalic neural crest (Elms et al. 2003; Inoue et al. 2004). These results, together with those reported from the studies on *Xenopus* (Chap. 7) and zebrafish (Chap. 9), corroborate neural crest-inducing activity as one of the shared roles of the vertebrate *Zic* family genes. Furthermore, the necessity of a *Zic* gene was shown in lampreys (jawless fishes, an ancient vertebrate lineage), and neural crest rudiments were detected in the ascidian *Zic-r.b*-positive blastomeres (Abitua et al. 2012) (Chap. 6), suggesting the preexistence of neural crest-like structure-inducing activity of Zic proteins in chordate ancestors. However, the involvement of Zic proteins in the later stages of neural crest development may not be common, and it depends on the type of tissue or cell. For example, in chick, Zic1/2/3 expression is observed in the periotic mesenchyme (Warner et al. 2003; McMahon and Merzdorf 2010; Chervenak et al. 2013), but Zic1 expression was not

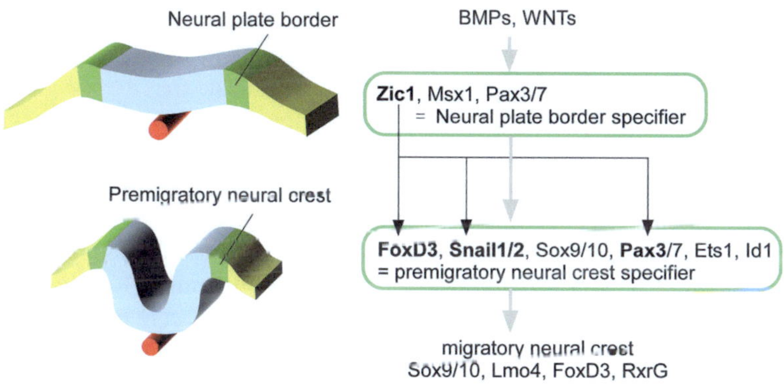

Fig. 12.4 Genes controlling the formation of neural plate border and specification of the neural crest. Zic1 expression occurs in the neural plate border and Zic1 activates the expression of neural crest specifier genes

detected in the migrating neural crest cells in the trunk (Sun Rhodes and Merzdorf 2006). In mice, the expression was also detected in the periotic mesenchyme (Chervenak et al. 2013), and Zic2 hypomorphic mutants (*Zic2kd/kd* and *Zic2ku/ku*) show the impaired inner morphology (Chervenak et al. 2014). However, it is not clear the abnormality reflects the abnormality of otic cyst itself (otocyst derives from rhombomere 5 where Zic2 is expressed) or that of periotic mesenchyme (Chervenak et al. 2014). It would be essential to clarify if Zic2 has any roles in the development of neural crest-derived periotic mesenchyme. Although clinical manifestation of neural crest abnormalities due to Zic alteration is not fully understood, the roles of *Zic* family genes in development of meningeal cells and cephalic mesenchyme forming cranial bones have been studied as below.

12.5 Regulation of Meningeal Cell Development by the *Zic* Family Genes

Meninges (plural of meninx) derive from the neural crest in the head. Its primary function, as has been often described, is to protect brains, but recent studies also indicate their critical roles in the development and maintenance of higher brain functions. Inoue et al. (2008) showed that *Zic1*, *Zic2*, and *Zic3* were expressed in the meningeal cells, and majority of the Cajal-Retzius (CR) cells distributed in the medial and dorsal cortex also expressed Zic proteins in the mid-late embryonic and postnatal cortical marginal zones (Fig. 12.5). During embryonic cortical development, *Zic1/3* double-mutant and hypomorphic *Zic2* mutant mice showed a reduction in the number of CR cells in the rostral cortex, whereas this cell number remained unaffected in the caudal cortex. These mutants also showed mislocalization of the CR cells and cortical lamination defects throughout the brain, resembling the changes noted in type II (cobblestone) lissencephaly. In the *Zic1/3* mutant, reduced proliferation of the meningeal cells was observed before the thinner and disrupted organization of the pial basement membrane (BM) with reduced expression of the BM components and the meningeal cell-derived secretory factor. These defects correlated with the changes in the end feet morphology of the radial glial cells. These findings indicate that the *Zic* genes play critical roles in cortical development by regulating the proliferation of the meningeal cells and the pial BM assembly (Fig. 12.5).

In mature meninges of both human and mouse, *ZIC*-like immunoreactivities were detected in vimentin-expressing arachnoid cells (Fig. 12.5) (Aruga et al. 2010). Since meningiomas are known to arise from arachnoid cells, we examined the expression of human *ZIC* in diverse types of meningiomas. Results indicated that *ZIC1*, *ZIC2*, and *ZIC5* transcript levels in meningiomas were higher than those in the whole brain or in the normal dura mater. The expression levels of *ZIC1*, as per public microarray data, were greater in meningiomas classified as World Health Organization grade II (atypical) than in those classified as grade I (benign). Immunoscreening using anti-ZIC antibodies revealed that 23 out of 23 meningioma cases were ZIC1/2/3/5-immunopositive. By comparison, nuclear staining by the

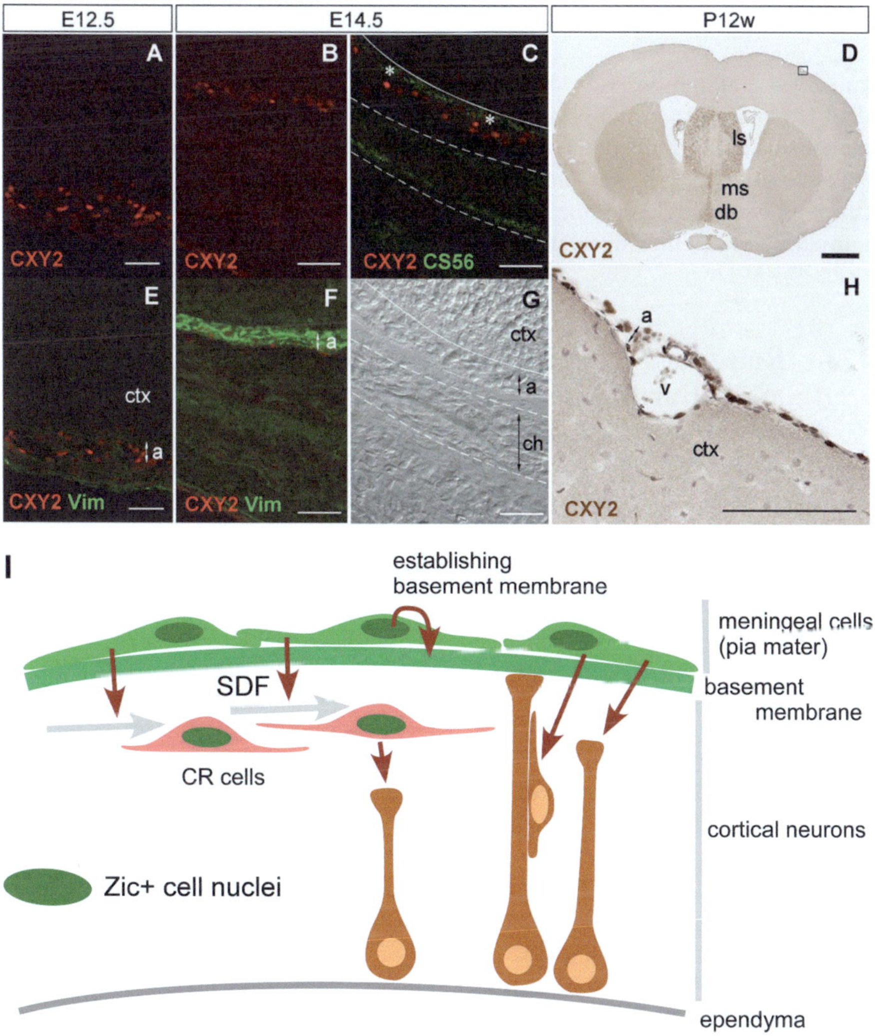

Fig. 12.5 Distribution of Zic proteins in mouse meningeal membrane. Mouse sections were derived from coronal sections through the forebrain at embryonic day 12.5, E12.5 (**a**, **e**), embryonic day 14.5, E14.5 (**b**, **c**, **f**, **g**) or 12 weeks after birth, P12w (**d**, **h**). Images (**a**, **e**), (**b**, **f**), and **c**, **g** indicate the same regions, respectively. Image in (**h**) is a higher magnification of the boxed region in (**d**). Immunostaining with CXY2 (**a**, **b**, **d**, **h**), CXY2/vimentin (**e**, **f**), and CXY2/CS56 (**c**), and differential interference contrast images (**g**) are shown. Zic signals are indicated by red in (**a**–**c**, **e**, **f**) and brown in (**d**, **h**). The color green indicates CS56 (**c**) and vimentin (**c**, **f**) immunoreactivities. In (**c**), the ZIC-like and CS56 immunoreactivities overlapped within the arachnoid layer that expresses vimentin, whereas only CS56-positive signals were detected within the chondrogenic area, between the two broken lines. The white line indicates the interface between meningeal tissue and cerebral cortex. CS56-positive signals were detected in the region that forms the basal lamina (asterisks). In the adult mouse brain, Zic proteins were still produced in the meningeal cells, as well as in the lateral septal nucleus, medial septal nucleus, and diagonal band (**d**). Positive signals were detected in the arachnoid and perivascular cells (**h**). *a* arachnoid layer, *ch* chondrogenic region (prospective temporal bone), *ctx* cerebral cortex, *db* diagonal band, *ls* lateral septal nucleus, *ms* medial septal nucleus, *v* vessel. *Thick scale bar*, 1 mm; *thin scale bar*, 100 μm. (**a**-**h**) are reprinted from Aruga et al. 2010). (**i**) Roles of Zic proteins in the meningeal cells

anti-ZIC4 antibody was not observed in any meningioma case, but was strongly detected in all four medulloblastomas. ZIC-positive meningiomas included meningothelial, fibrous, transitional, and psammomatous histological subtypes. A recent study revealed that specific mutations of RNA polymerase II subunit POLR2A are causative of meningiomas; the mutations cause dysregulations of *ZIC1/4* and *WNT6*, introduced as key meningeal identity genes (Clark et al. 2016).

12.6 Role of Zic Proteins in the Development of the Cardiovascular System

Both human *ZIC3* and mouse *Zic3* are involved in the development of the cardiovascular system, reflecting its role in embryonic left-right patterning by defining node and cilia functionality (Chap. 15), and mouse *Zic2* is also involved in the cardiac situs determination controlling ciliation of node cells during gastrulation (Barratt et al. 2014). On one hand, Zic proteins are expressed in the brain vasculature (Fig. 12.6) (Siegenthaler et al. 2013). In particular, some transcription factors, including Zic3, Foxf2, and Foxq1, are specifically expressed in the endothelial cells of the brain vessels among other organs (Ben-Zvi et al. 2014; Nolan et al. 2013). In functional terms, Hupe et al. (2017) showed that the ectopic expression of Zic3 in human umbilical vein endothelial cells induced the expression of blood-brain barrier differentiation markers, such as SLCO2B1 (anion transporter) and EGFL8 (a secretory protein produced by the brain endothelial cells) (Fig. 12.6). Additionally, some of the blood-brain barrier markers are induced by the cooperative action of Zic3, Foxq1, and Foxf2 (Hupe et al. 2017). These results suggest that Zic3 is involved in the development of blood-brain barrier properties by the endothelial cells of the brain capillaries.

Both Zic3 and Zic1 are known to be expressed in brain vasculature. Siegenthaler et al. (2013) showed that Zic1-like immunoreactivity can be detected in brain pericytes. Brain pericytes are placed externally, adjacent to the endothelial cells, and play a critical role in blood vessel stability and blood-brain barrier maturation (Sweeney et al. 2016). In addition, brain pericytes are derived from the neural crest and possess an ability to differentiate into various cells types, thereby contributing to both neurogenesis and vasculogenesis (Nakagomi et al. 2015). In developmental terms, brain vessel endothelial cells are believed to be derived from the vascular progenitor cells that lie in the perineural vascular plexus (Walchli et al. 2015). Therefore, *Zic* family genes in the neurovascular unit are likely to meet Zic proteins in two types of adjacent cells (endothelial cells and pericytes) which derive from two independent cell lineages.

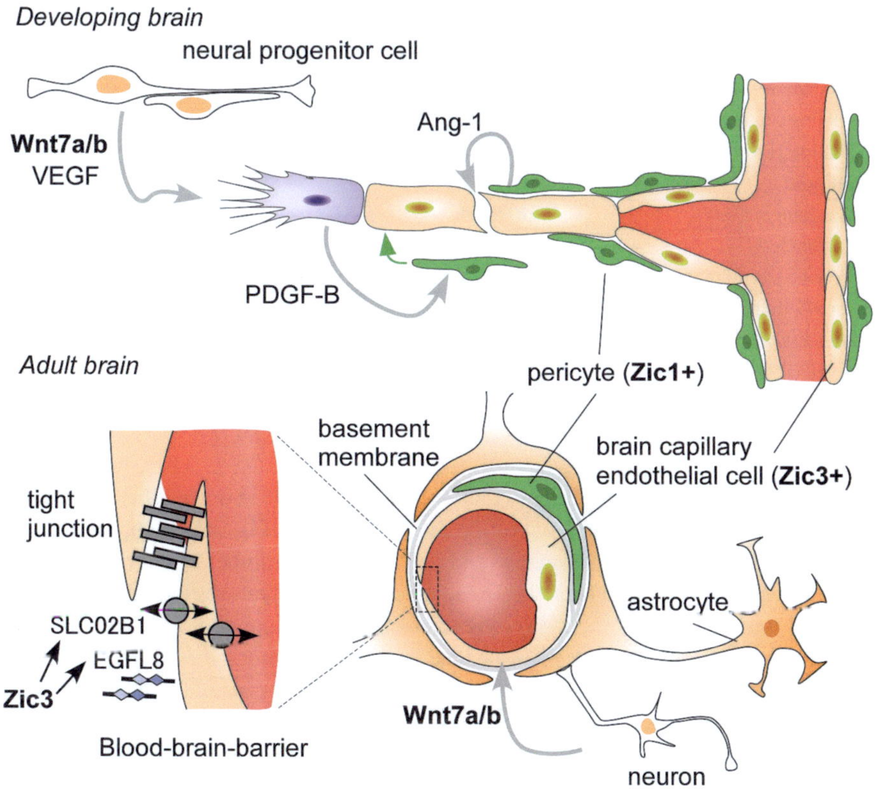

Fig. 12.6 Zic proteins in CNS vasculature

12.7 Behavioral Abnormalities in *Zic* Mutant Mice

Several behavioral abnormalities have been observed in *Zic1*, *Zic2*, and *Zic3* mutant mice (Table 12.1) (Ogura et al. 2001; Aruga et al. 2004; Hatayama et al. 2011). Both *Zic1*+/− and *Zic3*-deficient (Bn/Y) mice displayed reduced horizontal locomotor activities in the open-field test and reduced stay time in the wire-hanging test. Other tests that revealed behavioral abnormalities in the *Zic1* mutant mice were the stationary rod test and the traction test, which assessed behavioral traits similar to those assessed by the wire-hanging test. The observed behavioral deficits are consistent with those reported for cerebellar dysfunction. In contrast, *Zic2+/kd* mutants exhibited a completely different spectrum of behavioral deficits (Table 12.1).

Based on the above result, we extended the analysis of *Zic2+/kd* phenotype (Hatayama et al. 2011). As discussed in Chap. 14, *ZIC2* is a causal gene for holoprosencephaly, and *Zic2kd/kd* mice show embryonic or perinatal lethality with HPE-like symptoms and other anomalies (Nagai et al. 2000). *Zic2+/kd* mice are characterized by a moderate (40%) reduction in *Zic2* expression. As a result, *Zic2+/kd* mice showed increased locomotor activity in novel environments, cognitive and

Table 12.1 Comparison of the behavioral traits of the *Zic* mutants

	Zic1+/–	Zic2+/kd	Zic3 *Bn/Y*
Spontaneous motor activity	Decreased	–	Decreased
Rotating rod test (stationary)	Slightly impaired	–	–
Rotating rod test (rotating)	–	–	–
Footprint test	–	–	–
Wire-hanging test	Impaired	–	Impaired
Traction test	Tend to be reduced	–	–
Acoustic startle response	–	Enhanced	–
Prepulse inhibition	–	Impaired	–
Forced swimming test	–	–	–
Elevated plus maze test	–	–	–

All indicate the results of behavioral tests using mature male mice. (–) indicates no change as compared to the wild type

sensorimotor gating dysfunctions, and social behavioral abnormalities. *Zic2+/kd* mouse brain showed enlargement of the lateral ventricle (Fig. 12.7a, b), thinning of the cerebral cortex and corpus callosum, and decreased number of cholinergic neurons in the basal forebrain (Fig. 12.7c, d). Because these features are reminiscent of schizophrenia, we examined the frequencies of *ZIC2* variant alleles in schizophrenics and non-schizophrenics in the Japanese population. Among three novel missense mutations in *ZIC2* (Fig. 12.7e), *R409P* was only found in patients with schizophrenia, and it was located in a highly conserved position in the zinc finger domain. Mouse *Zic2* with the corresponding mutation showed lowered transcription-activating capacity (Fig. 12.7f) and had impaired target-DNA binding and cofactor binding capacities (Fig. 12.7g).

Basal forebrain cholinergic neurons are thought to be capable of regulating the cortical processing of sensory stimuli within various domains. In addition, recent studies indicate that the cholinergic system modulates cognitive deficits in schizophrenia and that cholinergic transmission is a potential target of therapeutics for the improvement of cognitive functions (refs in Hatayama et al. 2011). Thus, further evaluation of the cholinergic transmission dynamics in *Zic2* mutants would help in understanding better the role of *Zic2* in cognitive functioning. Besides establishing the occurrence of functionally defective *ZIC2* mutation in patients with schizophrenia, our findings raise the possibility that mildly impaired *ZIC2* function does not result in HPE, but in psychiatric illnesses.

Fig 12.7 (continued) box, ZOC domain. (**f**) NIH3T3 cells were transfected with a Zic2-responsive luciferase reporter vector together with a vector expressing wild-type FLAGZic2 (WT) or the FLAG-Zic2-A95T, -R409P, -S444R mutant proteins. (**g**) FLAG-Zic2-WT (wild type) or FLAG-Zic2-R409P coprecipitated with DNA-PK and RHA (see Chap. 18) in an immunoprecipitation assay. R409P mutation decreases the Zic2-DNA-PK interaction. *$P < 0.05$, **$P < 0.01$, ***$P < 0.001$ in *t*-test. Data is presented as means ± SEM (Reprinted from Hatayama et al. 2011)

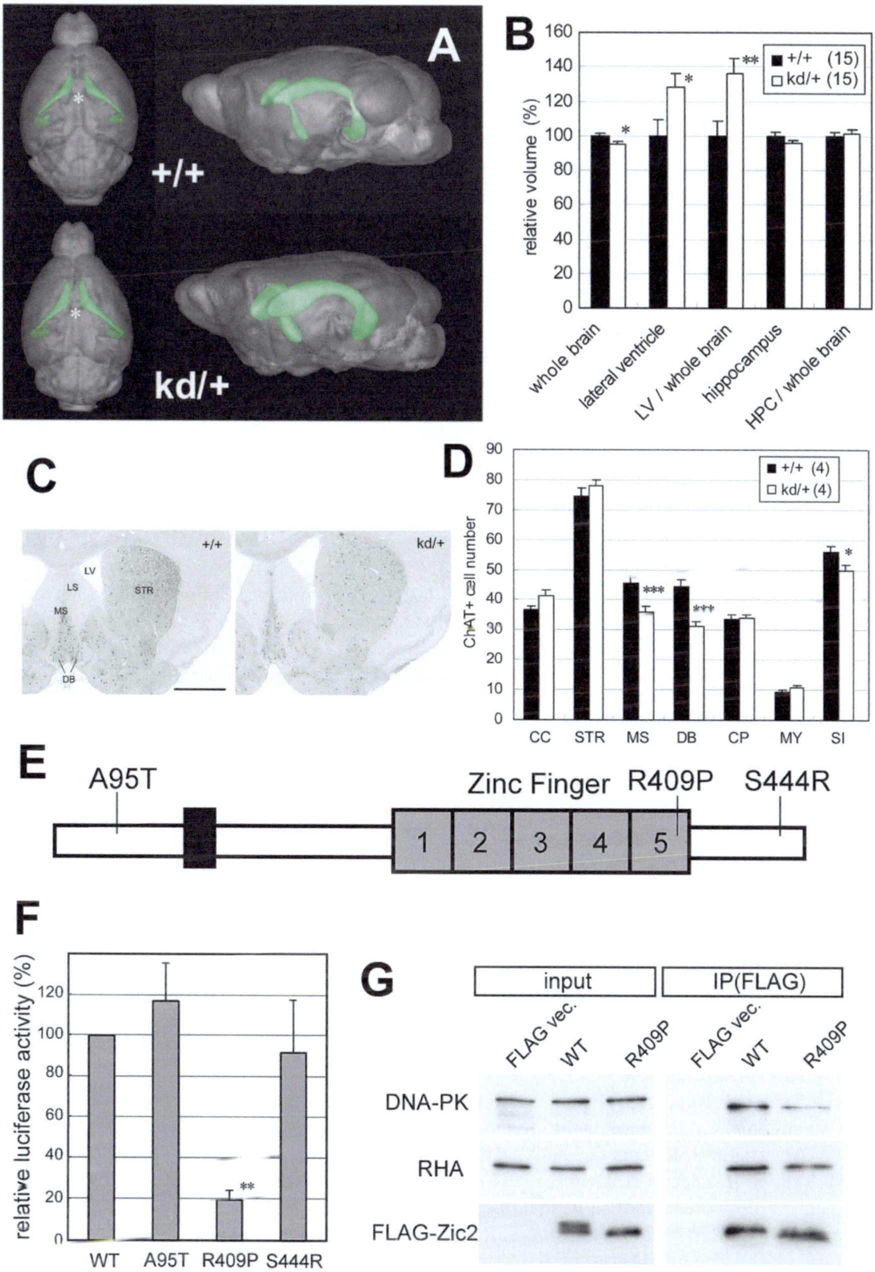

Fig. 12.7 ZIC2 hypomorphic mutation and schizophrenia. (**a**) 3D reconstruction of Zic2+/+ and Zic2kd/+ mice. Green, lateral ventricle, Asterisk septum. (**b**) Volumetric analysis of the entire brain, lateral ventricle (LV), and hippocampus (HPC). WT values = 100%. (**c**) Immunostaining of the brains of Zic2+/+ and Zic2kd/+ mice with anti-ChAT antibody. Scale bar – 1 mm. (**d**) Number of the ChAT+ neurons in the sections. (**e**) Missense mutations identified in ZIC2 protein. Black

References

Abitua PB, Wagner E, Navarrete IA, Levine M (2012) Identification of a rudimentary neural crest in a non-vertebrate chordate. Nature 492(7427):104–107. https://doi.org/10.1038/nature11589

Aruga J, Mizugishi K, Koseki H, Imai K, Balling R, Noda T, Mikoshiba K (1999) Zic1 regulates the patterning of vertebral arches in cooperation with Gli3. Mech Dev 89(1–2):141–150

Aruga J, Nozaki Y, Hatayama M, Odaka YS, Yokota N (2010) Expression of ZIC family genes in meningiomas and other brain tumors. BMC Cancer 10:79. https://doi.org/10.1186/1471-2407-10-79

Aruga J, Ogura H, Shutoh F, Ogawa M, Franke B, Nagao S, Mikoshiba K (2004) Locomotor and oculomotor impairment associated with cerebellar dysgenesis in Zic3-deficient (Bent tail) mutant mice. Eur J Neurosci 20(8):2159–2167. https://doi.org/10.1111/j.1460-9568.2004.03666.x

Atit R, Sgaier SK, Mohamed OA, Taketo MM, Dufort D, Joyner AL, Niswander L, Conlon RA (2006) Beta-catenin activation is necessary and sufficient to specify the dorsal dermal fate in the mouse. Dev Biol 296(1):164–176. doi:S0012-1606(06)00732-9. https://doi.org/10.1016/j.ydbio.2006.04.449

Barratt KS, Glanville-Jones HC, Arkell RM (2014) The Zic2 gene directs the formation and function of node cilia to control cardiac situs. Genesis 52(6):626–635. https://doi.org/10.1002/dvg.22767

Ben-Zvi A, Lacoste B, Kur E, Andreone BJ, Mayshar Y, Yan H, Gu C (2014) Mfsd2a is critical for the formation and function of the blood-brain barrier. Nature 509(7501):507–511. https://doi.org/10.1038/nature13324

Cereijo R, Gallego-Escuredo JM, Moure R, Villarroya J, Domingo JC, Fontdevila J, Martinez E, Gutierrez Mdel M, Mateo MG, Giralt M, Domingo P, Villarroya F (2015) The molecular signature of HIV-1-associated lipomatosis reveals differential involvement of brown and beige/brite adipocyte cell lineages. PLoS One 10(8):e0136571. https://doi.org/10.1371/journal.pone.0136571/PONE-D-15-09971

Chervenak AP, Hakim IS, Barald KF (2013) Spatiotemporal expression of Zic genes during vertebrate inner ear development. Dev Dyn 242(7):897–908. https://doi.org/10.1002/dvdy.23978

Chervenak AP, Bank LM, Thomsen N, Glanville-Jones HC, Jonathan S, Millen KJ, Arkell RM, Barald KF (2014) The role of Zic genes in inner ear development in the mouse: exploring mutant mouse phenotypes. Dev Dyn 243(11):1487–1498. https://doi.org/10.1002/dvdy.24186

Clark VE, Harmanci AS, Bai H, Youngblood MW, Lee TI, Baranoski JF, Ercan-Sencicek AG, Abraham BJ, Weintraub AS, Hnisz D, Simon M, Krischek B, Erson-Omay EZ, Henegariu O, Carrion-Grant G, Mishra-Gorur K, Duran D, Goldmann JE, Schramm J, Goldbrunner R, Piepmeier JM, Vortmeyer AO, Gunel JM, Bilguvar K, Yasuno K, Young RA, Gunel M (2016) Recurrent somatic mutations in POLR2A define a distinct subset of meningiomas. Nat genet 48(10):1253–1259. https://doi.org/10.1038/ng.3651

de Jong JM, Larsson O, Cannon B, Nedergaard J (2015) A stringent validation of mouse adipose tissue identity markers. Am J Physiol Endocrinol Metab 308(12):E1085–E1105. https://doi.org/10.1152/ajpendo.00023.2015ajpendo.00023.2015

Elms P, Siggers P, Napper D, Greenfield A, Arkell R (2003) Zic2 is required for neural crest formation and hindbrain patterning during mouse development. Dev Biol 264(2):391–406

Hafner AL, Contet J, Ravaud C, Yao X, Villageois P, Suknuntha K, Annab K, Peraldi P, Binetruy B, Slukvin II, Ladoux A, Dani C (2016) Brown-like adipose progenitors derived from human induced pluripotent stem cells: identification of critical pathways governing their adipogenic capacity. Sci Rep 6:32490. https://doi.org/10.1038/srep32490/srep32490

Hall BK (1999) The neural crest in development and evolution. Springer, New York

Hatayama M, Ishiguro A, Iwayama Y, Takashima N, Sakoori K, Toyota T, Nozaki Y, Odaka YS, Yamada K, Yoshikawa T, Aruga J (2011) Zic2 hypomorphic mutant mice as a schizophrenia model and ZIC2 mutations identified in schizophrenia patients. Sci Rep 1:16. https://doi.org/10.1038/srep00016

Himeda CL, Barro MV, Emerson CP Jr (2013) Pax3 synergizes with Gli2 and Zic1 in transactivating the Myf5 epaxial somite enhancer. Dev Biol 383(1):7–14. https://doi.org/10.1016/j.ydbio.2013.09.006

Hupe M, Li MX, Kneitz S, Davydova D, Yokota C, Kele-Olovsson J, Hot B, Stenman JM, Gessler M (2017) Gene expression profiles of brain endothelial cells during embryonic development at bulk and single-cell levels. Sci Signal 10(487). https://doi.org/10.1126/scisignal.aag2476

Inoue T, Hatayama M, Tohmonda T, Itohara S, Aruga J, Mikoshiba K (2004) Mouse Zic5 deficiency results in neural tube defects and hypoplasia of cephalic neural crest derivatives. Dev Biol 270(1):146–162. https://doi.org/10.1016/j.ydbio.2004.02.017

Inoue T, Ota M, Mikoshiba K, Aruga J (2007) Zic2 and Zic3 synergistically control neurulation and segmentation of paraxial mesoderm in mouse embryo. Dev Biol 306(2):669–684. https://doi.org/10.1016/j.ydbio.2007.04.003

Inoue T, Ogawa M, Mikoshiba K, Aruga J (2008) Zic deficiency in the cortical marginal zone and meninges results in cortical lamination defects resembling those in type II lissencephaly. J Neurosci 28(18):4712–4725. https://doi.org/10.1523/JNEUROSCI.5735-07.2008

Jespersen NZ, Larsen TJ, Peijs L, Daugaard S, Homoe P, Loft A, de Jong J, Mathur N, Cannon B, Nedergaard J, Pedersen BK, Moller K, Scheele C (2013) A classical brown adipose tissue mRNA signature partly overlaps with brite in the supraclavicular region of adult humans. Cell Metab 17(5):798–805. https://doi.org/10.1016/j.cmet.2013.04.011/S1550-4131(13)00154-X

Kalogeropoulos M, Varanasi SS, Olstad OK, Sanderson P, Gautvik VT, Reppe S, Francis RM, Gautvik KM, Birch MA, Datta HK (2010) Zic1 transcription factor in bone: neural developmental protein regulates mechanotransduction in osteocytes. FASEB J 24(8):2893–2903. https://doi.org/10.1096/fj.09-148908

Kramer I, Halleux C, Keller H, Pegurri M, Gooi JH, Weber PB, Feng JQ, Bonewald LF, Kneissel M (2010) Osteocyte Wnt/beta-catenin signaling is required for normal bone homeostasis. Mol Cell Biol 30(12):3071–3085. https://doi.org/10.1128/MCB.01428-09

McMahon AR, Merzdorf CS (2010) Expression of the zic1, zic2, zic3, and zic4 genes in early chick embryos. BMC Res Notes 3:167. https://doi.org/10.1186/1756-0500-3-167

Milet C, Maczkowiak F, Roche DD, Monsoro-Burq AH (2013) Pax3 and Zic1 drive induction and differentiation of multipotent, migratory, and functional neural crest in Xenopus embryos. Proc Natl Acad Sci U S A 110(14):5528–5533. https://doi.org/10.1073/pnas.1219124110

Nagai T, Aruga J, Minowa O, Sugimoto T, Ohno Y, Noda T, Mikoshiba K (2000) Zic2 regulates the kinetics of neurulation. Proc Natl Acad Sci U S A 97(4):1618–1623

Nakagomi T, Nakano-Doi A, Kawamura M, Matsuyama T (2015) Do vascular pericytes contribute to neurovasculogenesis in the central nervous system as multipotent vascular stem cells? Stem Cells Dev 24(15):1730–1739. https://doi.org/10.1089/scd.2015.0039

Nolan DJ, Ginsberg M, Israely E, Palikuqi B, Poulos MG, James D, Ding BS, Schachterle W, Liu Y, Rosenwaks Z, Butler JM, Xiang J, Rafii A, Shido K, Rabbany SY, Elemento O, Rafii S (2013) Molecular signatures of tissue-specific microvascular endothelial cell heterogeneity in organ maintenance and regeneration. Dev Cell 26(2):204–219. https://doi.org/10.1016/j.devcel.2013.06.017

Ogura H, Aruga J, Mikoshiba K (2001) Behavioral abnormalities of Zic1 and Zic2 mutant mice: implications as models for human neurological disorders. Behav Genet 31(3):317–324

Pan H, Gustafsson MK, Aruga J, Tiedken JJ, Chen JC, Emerson CP Jr (2011) A role for Zic1 and Zic2 in Myf5 regulation and somite myogenesis. Dev Biol 351(1):120–127. https://doi.org/10.1016/j.ydbio.2010.12.037

Petrovic N, Walden TB, Shabalina IG, Timmons JA, Cannon B, Nedergaard J (2010) Chronic peroxisome proliferator-activated receptor gamma (PPARgamma) activation of epididymally derived white adipocyte cultures reveals a population of thermogenically competent, UCP1-containing adipocytes molecularly distinct from classic brown adipocytes. J Biol Chem 285(10):7153–7164. https://doi.org/10.1074/jbc.M109.053942/M109.053942

Qin X, Park HG, Zhang JY, Lawrence P, Liu G, Subramanian N, Kothapalli KS, Brenna JT (2016) Brown but not white adipose cells synthesize omega-3 docosahexaenoic acid in culture. Prostaglandins Leukot Essent Fat Acids 104:19–24. https://doi.org/10.1016/j.plefa.2015.11.001/S0952-3278(15)30046-6

Rockstroh D, Landgraf K, Wagner IV, Gesing J, Tauscher R, Lakowa N, Kiess W, Buhligen U, Wojan M, Till H, Bluher M, Korner A (2015) Direct evidence of brown adipocytes in different fat depots in children. PLoS One 10(2):e0117841. https://doi.org/10.1371/journal.pone.0117841/PONE-D-14-33238

Sanchez-Gurmaches J, Guertin DA (2014) Adipocytes arise from multiple lineages that are heterogeneously and dynamically distributed. Nat Commun 5:4099. https://doi.org/10.1038/ncomms5099

Seale P, Kajimura S, Yang W, Chin S, Rohas LM, Uldry M, Tavernier G, Langin D, Spiegelman BM (2007) Transcriptional control of brown fat determination by PRDM16. Cell Metab 6(1):38–54. doi:S1550-4131(07)00157-X/10.1016/j.cmet.2007.06.001

Siegenthaler JA, Choe Y, Patterson KP, Hsieh I, Li D, Jaminet SC, Daneman R, Kume T, Huang EJ, Pleasure SJ (2013) Foxc1 is required by pericytes during fetal brain angiogenesis. Biol Open 2(7):647–659. https://doi.org/10.1242/bio.20135009

Simoes-Costa M, Bronner ME (2015) Establishing neural crest identity: a gene regulatory recipe. Development 142(2):242–257. https://doi.org/10.1242/dev.105445

Simoes-Costa MS, McKeown SJ, Tan-Cabugao J, Sauka-Spengler T, Bronner ME (2012) Dynamic and differential regulation of stem cell factor FoxD3 in the neural crest is encrypted in the genome. PLoS Genet 8(12):e1003142. https://doi.org/10.1371/journal.pgen.1003142

Sun Rhodes LS, Merzdorf CS (2006) The zic1 gene is expressed in chick somites but not in migratory neural crest. Gene Expr Patterns 6(5):539–545. https://doi.org/10.1016/j.modgep.2005.10.006

Sweeney MD, Ayyadurai S, Zlokovic BV (2016) Pericytes of the neurovascular unit: key functions and signaling pathways. Nat Neurosci 19(6):771–783. https://doi.org/10.1038/nn.4288

Timmons JA, Wennmalm K, Larsson O, Walden TB, Lassmann T, Petrovic N, Hamilton DL, Gimeno RE, Wahlestedt C, Baar K, Nedergaard J, Cannon B (2007) Myogenic gene expression signature establishes that brown and white adipocytes originate from distinct cell lineages. Proc Natl Acad Sci U S A 104(11):4401–4406. doi:0610615104/10.1073/pnas.0610615104

Walchli T, Wacker A, Frei K, Regli L, Schwab ME, Hoerstrup SP, Gerhardt H, Engelhardt B (2015) Wiring the vascular network with neural cues: a CNS perspective. Neuron 87(2):271–296. https://doi.org/10.1016/j.neuron.2015.06.038

Wang W, Seale P (2016) Control of brown and beige fat development. Nat Rev Mol Cell Biol 17(11):691–702. https://doi.org/10.1038/nrm.2016.96

Warner SJ, Hutson MR, Oh SH, Gerlach-Bank LM, Lomax MI, Barald KF (2003) Expression of ZIC genes in the development of the chick inner ear and nervous system. Dev Dyn 226(4):702–712. https://doi.org/10.1002/dvdy.10262

Chapter 13
ZIC1 Function in Normal Cerebellar Development and Human Developmental Pathology

Jun Aruga and Kathleen J. Millen

Abstract *Zic* genes are strongly expressed in the cerebellum. This feature leads to their initial identification and their name "zic," as the abbreviation of "zinc finger protein of the cerebellum." *Zic* gene function in cerebellar development has been investigated mainly in mice. However, association of heterozygous loss of *ZIC1* and *ZIC4* with Dandy-Walker malformation, a structural birth defect of the human cerebellum, highlights the clinical relevance of these studies. Two proposed mechanisms for Zic-mediated cerebellar developmental control have been documented: regulation of neuronal progenitor proliferation-differentiation and the patterning of the cerebellar primordium. Clinical studies have also revealed that *ZIC1* gain of function mutations contribute to coronal craniosynostosis, a rare skull malformation. The molecular pathways contributing to these phenotypes are not fully explored; however, embryonic interactions with sonic hedgehog signaling, retinoic acid signaling, and TGFβ signaling have been described during mouse cerebellar development. Further, Zic1/2 target a multitude of genes associated with cerebellar granule cell maturation during postnatal mouse cerebellar development.

Keywords ZIC1 · Cerebellum · Neural development · Dandy-Walker malformation · Craniosynostosis

J. Aruga (✉)
Department of Medical Pharmacology, Nagasaki University Graduate School of Biomedical Sciences, Nagasaki, Japan
e-mail: aruga@nagasaki-u.ac.jp

K. J. Millen (✉)
Department of Pediatrics, University of Washington School of Medicine, Seattle Children's Hospital Research Institute, Center for Integrative Brain Research, Seattle, WA, USA
e-mail: kathleen.millen@seattlechildrens.org

J. Aruga (ed.), *Zic family*, Advances in Experimental Medicine and Biology 1046,
https://doi.org/10.1007/978-981-10-7311-3_13

13.1 Cerebellar Anatomy

The cerebellum is located on top of the brainstem and has important roles in motor control and cognition (Ghez 1991; Ito 2011; Koziol et al. 2014; Hoche et al. 2016). Without exception, cerebella or cerebellar-like structures exist in most vertebrates in which the multiple Zic genes were reported to date (see Chap. 1). In all vertebrates, the cerebellum has a conserved laminar organization composed of relatively few, morphologically distinct neuronal types communicating in circuit that has been defined for decades. Its near stereotypical structure and simplicity relative to other brain regions has made it an attractive target for study of its underlying molecular developmental mechanisms.

The mature mammalian cerebellum is grossly composed of two bilateral hemispheres flanking a medial vermis. Large lobes, subdivided by a conserved pattern of smaller lobules, are arrayed along the anterior-posterior (rostral-caudal) axis (Fig. 13.1). The cortex of the cerebellum is composed of two primary neuronal types. Tightly packed, small cerebellar granule cells (GCs) constitute the internal granule cell layer (IGL) and represent the largest neuronal population not only of

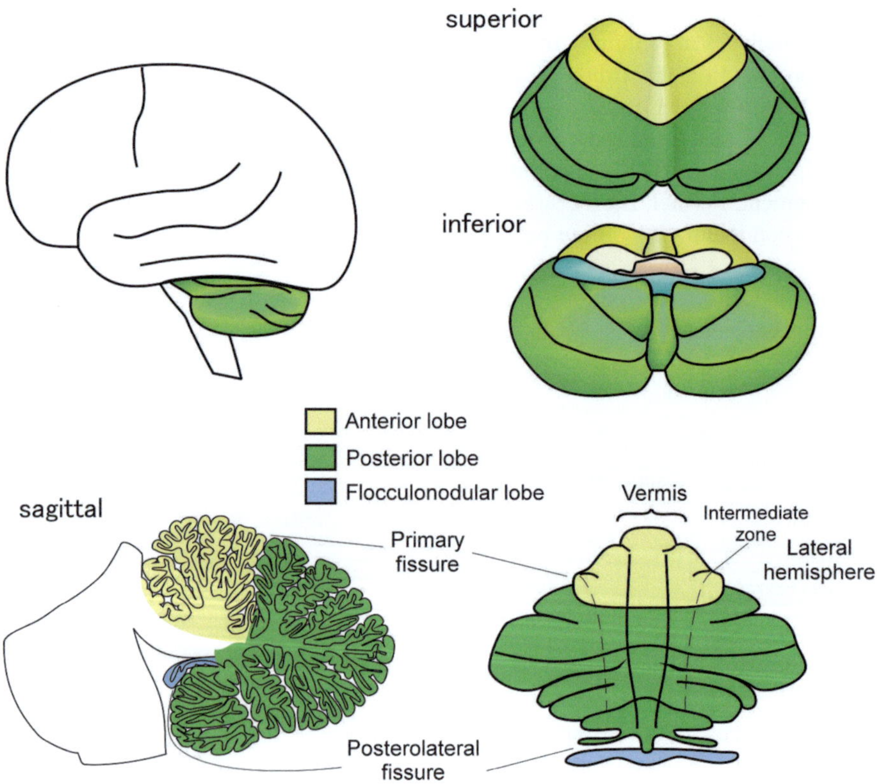

Fig. 13.1 Overview of human cerebellar morphology

the cerebellum but also of the entire brain. In humans approximately 80% of all brain neurons are cerebellar GCs (Azevedo et al. 2009). Purkinje cells (PCs) are large neurons aligned in a monolayer lying directly on top of the IGL. PCs extend flat yet elaborate, tree-like dendrites into the molecular layer in the parasagittal plane. The outermost cortical layer is composed of bifurcate T-shaped GC axons (parallel fibers) which run along the medial lateral axis, each intersecting hundreds to thousands of perpendicular PC dendrites. Several other morphologically and functionally distinct neuronal cell types are found in the cerebellar cortex, including basket and stellate cells in the molecular layer, and Golgi, Lugaro, and unipolar brush cells in the IGL. The neurons of the cerebellar cortex process incoming information from spinal cord- and brain-derived mossy fiber and brain stem-derived climbing fiber afferents. PC conveys the major output inhibitory signals to cerebellar nuclei. Cerebellar nuclei are clusters of neurons within the white matter at the core of the cerebellum and largely are the sole source of efferent output from the cerebellum (White and Sillitoe 2013).

13.2 Cerebellar Development and Zic Gene Expression

The cerebellar primordium or anlage develops from the dorsal, anterior-most hindbrain (Fig. 13.2). The cerebellar territory is positioned along the anterior/posterior axis of the forming neural tube by Fgf and Wnt signals emanating from the isthmic organizer, a transient signaling center located at the mid-hindbrain junction (Leto et al. 2016). The developing cerebellum is also influenced by signaling from the adjacent fourth ventricle roof plate, another transient signaling center, which secretes Bmp, Wnt, and retinoic acid (Chizhikov et al. 2010), and differentiates into the hindbrain choroid plexus epithelium. As neurogenesis progresses, the bilateral primordia fuse on the dorsal midline over the fourth ventricle to establish the medial vermis and lateral cerebellar hemispheres (Louvi et al. 2003).

Cerebellar neurons are generated from two anatomically and molecularly distinct progenitor zones. The cerebellar ventricular zone (VZ) expresses the basic helix-loop-helix (bHLH) factor Ptf1a and gives rise to PCs and all other GABAergic cerebellar neurons, which migrate radially outward into the anlage. The dorsally located rhombic lip (RL) expresses the bHLH factor Atoh1 and gives rise to glutamatergic neurons, most notably cerebellar GC. In contrast to VZ derived PCs, GC progenitors are still mitotically active as they exit the RL. Initially, they migrate tangentially over the anlage to form an external granule layer (EGL). GC progenitors within the EGL are driven to proliferate through reception of a mitotic Sonic hedgehog (Shh) signal received from the underlying differentiating PCs within the anlage (Dahmane and Ruiz i Altaba 1999; De Luca et al. 2016) (Fig. 13.2). Inward radial migration of differentiating EGL cell under the PCs results in the localization of mature GCs within the IGL, of the mature cerebellum.

Zic gene expression profiles in development have been studied in various animal models (Chaps. 2, 3, 4, 5, 6, 7, 8, 9, 10, 11, and 12). Here we will focus on *Zic* gene

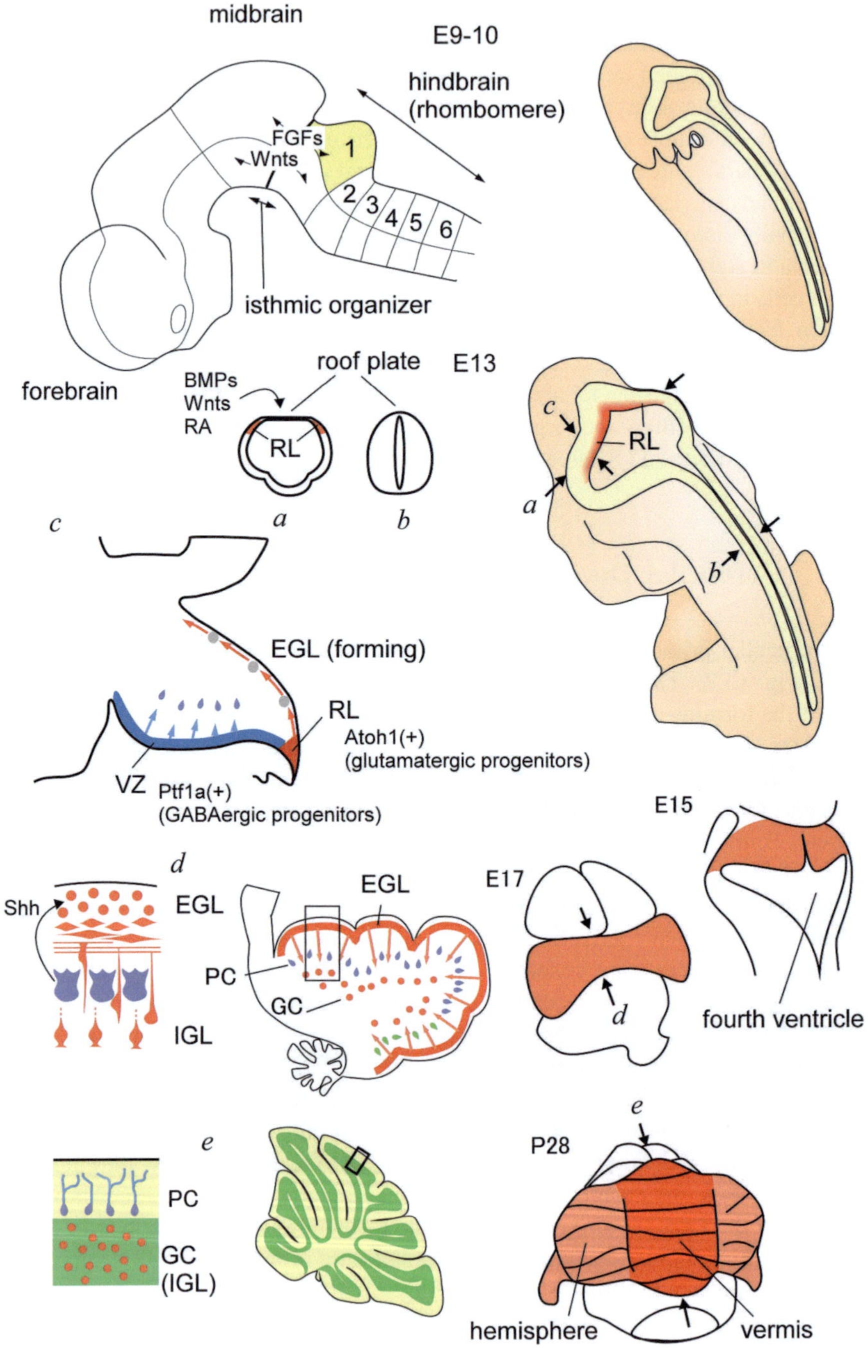

Fig. 13.2 Cerebellar development in mice. *E*, Embryonic day; *P*, postnatal day (indicating the age)

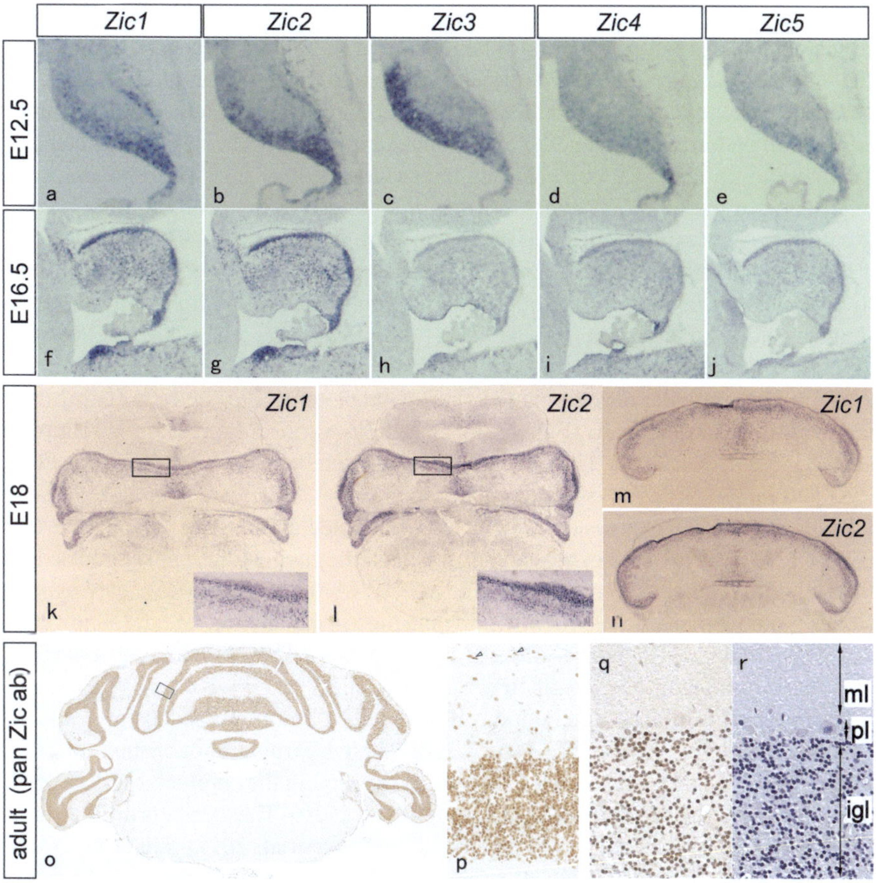

Fig. 13.3 *Zic* expression in the developing and mature cerebellum. (**a–p**) mouse; (**q**, **r**) human. **a–n**, in situ hybridization with indicated gene probes; **o–q**, immunohistochemical staining using anti-pan-Zic antibody recognizing Zic1, Zic2, Zic3, and Zic5 (CXY2 in Aruga et al. 2010); **r**, hematoxylin eosin staining of the section adjacent to (**q**). **a–j**, sagittal sections; **k**, **j**, **o**, **p**, coronal sections; **m**, **n**, horizontal sections (**a–j**, **o**, **p** are prepared from unpublished data T. Inoue, M. Odagawa, and J. Aruga. **k–n** and **q** and **r** are adapted from previous reports Aruga et al. 2002a, 2010 with permission)

expression during mouse cerebellar development (Fig. 13.3). *Zic* gene expression initiates prior to the appearance of the cerebellar primordium and can be traced back to neuroectoderm formation, prior to neural induction, at the earliest point of neural development (See Chap. 11). *Zic* expression in neural plate is enhanced in the midbrain-hindbrain border region in vertebrates including mice (*Zic2*, *Zic3*, and *Zic5*) (Inoue et al. 2004, 2007a; Furushima et al. 2000), chick (*Zic3*) (McMahon and Merzdorf 2010), Xenopus (*Zic3*) (Fujimi et al. 2006), and zebra fish (*zic3* and *zic6*) (Keller and Chitnis 2007).

In mice, *Zic1*, *Zic2*, *Zic3*, and *Zic5* are expressed in overlapping dorsal domains in the cerebellar region by neural tube closure at the hindbrain level at embryonic day (E) 8.5 (Nagai et al. 1997; Furushima et al. 2000; Inoue et al. 2004). As the cerebellar anlage develops into the mature cerebellum, the *Zic* genes show dynamic expression patterns. All *Zic* genes are strongly expressed in the developing cerebellar rhombic lip (Fig. 13.3a–e). Strong expression of *Zic1* and *Zic2* and moderate expression of the other *Zic* genes are then maintained in the proliferating GC progenitors as they exit the rhombic lip to form the EGL (Fig. 13.3f–j) (Aruga et al. 1994, 2002a; Gaston-Massuet et al. 2005; Blank et al. 2011). All *Zic* genes continue to be expressed in GC progenitors as they differentiate and migrate from the EGL into the cerebellar anlage to form the IGL. Although the *Zic* genes have highly overlapping expression patterns in the developing cerebellum, there are some differences in *Zic* cerebellar developmental expression patterns. *Zic1* and *Zic2* expression is predominant in the EGL (Fig. 13.3a, b, g, h, k–n). *Zic3* expression is relatively strong in ventricular zone (Fig. 13.3c, h). In the cerebellar anlage under the EGL, *Zic1* and *Zic2* expression is scattered at E16 (Fig. 13.3f, g) (Blank et al. 2011). By E18, *Zic1* and *Zic2* expression is detected in the medial anlage in developing PCs (Fig. 13.3m, n) with *Zic2* expression also detected in two additional sagittal stripes forming three fan-shaped radiations (Fig. 13.3n) (Aruga et al. 2002a). *Zic4* expression is more restricted to the rhombic lip than the *Zic* genes (Fig. 13.3d) (Blank et al. 2011). Detailed expression profiles for all *Zic* genes with cerebellar cell-type specific markers remain to be conducted.

In the mature mouse cerebellum, all *Zic* genes are expressed (Fig. 13.3o–q) (Aruga et al. 1994, 1996a, b) (*Zic5*, T. Inoue and J. Aruga, unpublished observation). In humans, *ZIC1* and other *ZIC* gene expression in the cerebellar IGL is obvious (Fig. 13.3q, r) (Yokota et al. 1996; Aruga et al. 2010). These results are validated in recent transcriptomics studies (such as Kasukawa et al. 2011; She et al. 2009). [The results can be seen in NCBI's Gene Expression Omnibus database by using keywords ("GDS3917" or "GDS3834") and "zic*"]. Zic1 transcript levels in the cerebellum are consistently the highest among the five Zic family members irrespective of assay methods. Zic1 and Zic2 are commonly used as cerebellar GC markers in assays analyzing the GC migration (Miyata et al. 1996) and inducing cerebellar GCs from ES or iPS cells (Salero and Hatten 2007; Srivastava et al. 2013; Holmes and Heine 2017). However, it should be noted that in addition to strong GC expression, moderate expression is seen in scattered molecular layer cells and cells in the meninges (Fig. 13.3p).

Cerebellar expression of Zic family members has also been described in chick embryos. Lin and Cepko (1998) firstly described the structure and expression of chick *zic1* and *zic3*. Both genes are expressed in the cerebellum, and the migrating GCs are well detected as "granule cell raphes" that correspond to groups of inward migrating GCs in stereotypical series of parasagittal linear arrays. Chick *Zic1* was also used to identify ectopic cerebellum induced by experimental manipulation (Matsumoto et al. 2004).

13.3 Roles of Zic1, Zic2, Zic3, and Zic4 in Murine Cerebellar Development

Zic1 is required for normal cerebellar development, based on analysis of *Zic1−/−* mice. Specifically, the cerebellum of *Zic1−/−* mice is hypoplastic both in the vermis and hemispheres. Reduced granule cell progenitor proliferation in the EGL is almost certainly a major contributor to this hypoplasia. Although cerebellar lamination in *Zic1−/−* is normal, the stereotypical pattern of cerebellar foliation is perturbed (Fig. 13.4) (Aruga et al. 1998; Blank et al. 2011).

Besides Zic1 in the cerebellar primordium, Zic1 in the pontine nucleus is involved in the development of mossy fiber projections into the cerebellum (Dipietrantonio and Dymecki 2009). In utero electroporation-mediated overexpression of Zic1 in the pontine gray nucleus neurons increases the ipsilateral cerebellar projection, representing the neural circuit wiring role of Zic family in the hindbrain (Dipietrantonio and Dymecki 2009).

The cerebellum of Zic2 knockdown homozygotes (*Zic2kd/kd*) showed a medial fusion defect of the cerebellar anlage at E16, with no other obvious histological abnormalities (Aruga et al. 2002a). Neonatal lethality prevented analysis of older *Zic2kd/kd* animals. Although *Zic1+/−* mice do not have a cerebellar foliation phenotype, a role for Zic1-Zic2 interaction was revealed through analysis of *Zic1+/−;Zic2+/kd* trans-heterozygotes. These double heterozygous mutants showed an altered foliation pattern in the vermis and significant size reductions of both the vermis and hemispheres (Fig. 13.4) (Aruga et al. 2002a). The foliation pattern

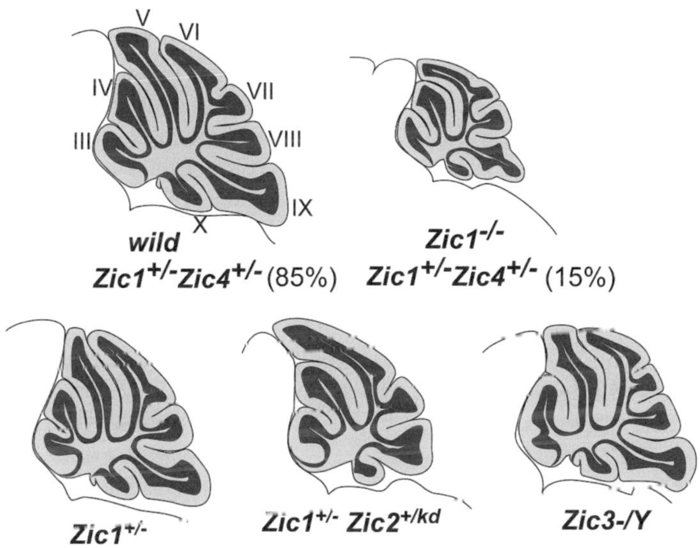

Fig. 13.4 Folial pattern abnormalities found in Zic mutant mice. Illustrations are based on the sagittal sections of cerebellar vermis

abnormality was associated with the reduction of the cerebellar anterior lobe-forming region as indicated by a rostral shift of the cerebellar zonal marker EphA3 at E16. In contrast to the rostral-caudal patterning abnormality, the medial-lateral pattern was not grossly affected. The rostral-caudal patterning abnormality was accompanied by a reduction of mitotic index in the EGL. In terms of proliferation, expression *cyclin D1* (*Ccnd1*) was reduced both in *Zic1−/−* and *Zic1+/−;Zic2+/kd* cerebella, and mitotic inhibitor *p27* (*Cdkn1b*) expression was increased in Zic1−/− cerebella. p27 was shown to inhibit the granule cell precursor proliferation (Miyazawa et al. 2000). In addition, premature neuronal differentiation occurred in the cerebellar nuclei as indicated by enhanced expression of Wnt7a (presynaptic differentiation inducer, secreted by GC), βIII tubulin, and p16 (a mitotic inhibitor) (Aruga et al. 2002a).

In sum, these early studies revealed roles for *Zic1* and *Zic2* in the expansion of the neuronal precursors in cerebellar development. An analogous role of *Zic1* has been described in spinal cord and forebrain development. In the developing spinal cord, *Zic1* is dorsally expressed. Spinal cord dorsal horn cell number is reduced, and premature neuronal differentiation was observed in *Zic1−/−* (Aruga et al. 2002a). This study confirmed the hypothesis developed from mis- and overexpression experiments in chick and mice. *Zic1* misexpression in chick spinal cord resulted in the inhibition of neuronal differentiation irrespective of progenitor location along the dorsoventral axis, and *Zic1* overexpression in mouse dorsal spinal cord resulted in an expansion of *Notch1*- and *Hes1*-expressing region with concurrent maintenance of proliferating neuronal precursors (Aruga et al. 2002b). In *Zic1−/−;Zic3Bn/Y* forebrain, hypoplasia of the hippocampus, septum, and olfactory bulb is observed, and the cell cycle exit rate was increased in the neuronal precursors in septum ventricular layer where *Zic1* and *Zic3* are co-expressed (Inoue et al. 2007b).

Cerebellar hypoplasia and folial mispatterning are also features seen in *Zic3*-deficient (*Bent tail/Y*, *Zic3 Bn/Y*) mice (Aruga et al. 2004). *Bent-tail* is a spontaneous mutation with a deletion including *Zic3*. The *Zic3*-deficient cerebellum was regionally hypoplastic (Fig. 13.4e). Significant mass reduction was observed in the vermis and anterior part of the hemispheres and paraflocculus-flocculus complex. Cerebellar dysgenesis in *Zic3*-deficient mice is proposed to reflect a patterning abnormality at early neurulation stages based on the enhanced *Zic3* expression in midbrain-hindbrain border region before completion of neurulation (Inoue et al. 2004; Aruga et al. 2004).

In contrast to the other *Zic* genes, loss of *Zic4* or *Zic5* function alone does not cause gross cerebellar size or folial patterning phenotypes (Blank et al. 2011; Inoue et al. 2004). An independent line of *Zic4* mutant on the C57/BL6 background also shows no clear difference from wild-type animals in terms of cerebellar morphology (Aruga et al. unpublished observation). However, there is evidence that Zic1 and Zic4 cooperate in both EGL proliferation and folial patterning based on analysis of a *Zic1;Zic4* null mutant allelic series (Blank et al. 2011). Size deficits were shown to be caused by a reduction in EGL proliferation due to a reduced ability of mutant EGL cells to respond to the mitotic Shh signal from the underlying developing PC layer. Causes of the folial patterning defects in *Zic1;Zic4* null and, indeed, all *Zic* mutants remain unknown. However, based on analysis of *Zic1;Zic4* null mutants with reduced Shh function, it is likely that Zic-dependent cerebellar folial patterning in the embryonic anlage is Shh-independent.

13.4 The Role of Zic1 in Hindbrain Development of Nonmammalian Vertebrate Species

Two zebra fish studies show roles for zic genes in hindbrain development. Elsen et al. (2008) carried out *zic1* and *zic4* knockdown analysis using morpholino antisense oligodeoxynucleotides. They found that the zic1/zic4 knockdown resulted in the reduced proliferation of hindbrain progenitor cells, whereas the hindbrain regionalization and neural crest development remain unaffected. A lack of zebra fish cerebellar markers at the time precluded further investigation of cerebellar development in these fish. However, there was impaired expression of fourth ventricle roof plate markers lmx1b.1 and lmx1b.2. By combining the knockdown of lmx1b genes, they found that lmx1b function is required for the roof plate development and morphogenesis of the fourth ventricle. These results indicate an evolutionally conserved mechanism of *Zic* and *Lmx* gene function.

Another critical finding from zebra fish studies was the involvement of retinoic acid signaling as a zic1/zic2 downstream target in the hindbrain development. Specifically, *Aldh1a2* was identified as a zic2 downstream target (Drummond et al. 2013). Zic2 knockdown resulted in the reduction of *aldh1a2* expression, and exogenously supplied retinoic acid rescued the phenotype. Since this study evaluated motor neuron phenotypes, the involvement of retinoic acid signaling downstream of zic genes in the context of cerebellar development is yet to be clarified.

13.5 Identification of Zic1 and Zic2 Downstream Targets in Postnatal Cerebellar GCs by Transcriptomics

Downstream targets are essential components to consider the biological role of Zic proteins as gene expression regulators. While preceding studies revealed several candidate Zic downstream target genes, Frank et al. (2015) carried out a comprehensive analysis of Zic1 and Zic2 downstream targets in cerebellar development. Specifically, they conducted out chromatin immunoprecipitation-sequencing (ChIP-seq) analysis using Zic1 and Zic2 antibodies to clarify the role of chromatin-dependent regulation of enhancer function in later stages of cerebellar GC neuronal maturation.

By comparing the ChIP-seq results of P7 and P60 mouse cerebellum, they identified numerous binding sites that were differentially bound by Zic between these two stages. Both binding-increased sites (n = 15,691) and the binding-decreased sites

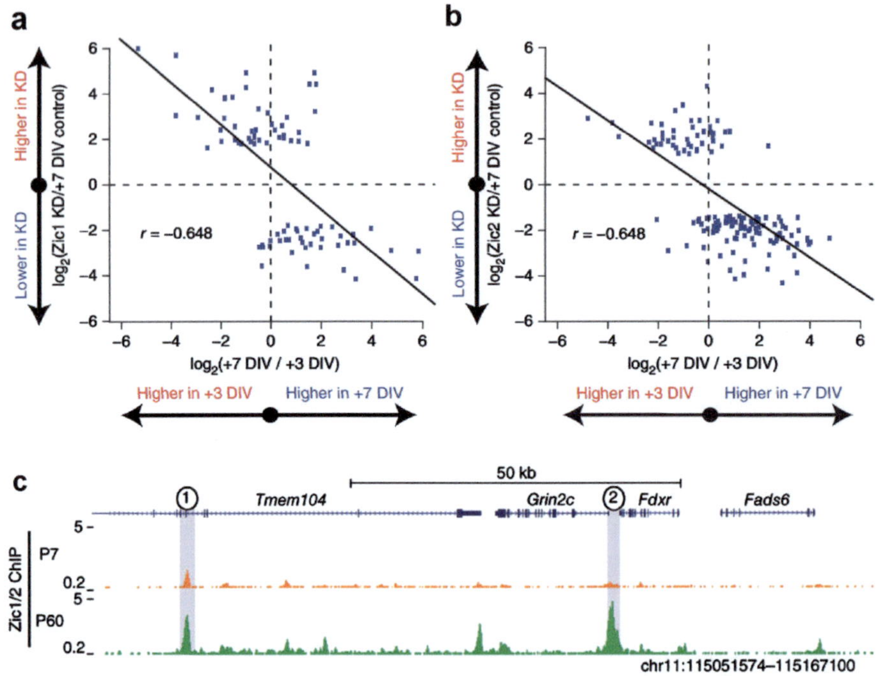

Fig. 13.5 Zic1, Zic2 downstream targets in GC maturation. Correlation between the gene expression level changes between Zic1 knockdown (*KD*) (**a**) or Zic2 knockdown (**b**) and GC maturation in vitro (*+3 DIV*, immature; *+7 DIV*, mature). (**c**) Chromatin immunoprecipitation sequencing experiment results. Zic1 and Zic2 bind enhancer regions 1 and 2, which are activated between postnatal day 7 and 60 of mouse cerebellar GC (Reprinted from Frank et al. 2015 with permission)

(n = 17,452) during P7->P60 development were identified. The P7->P60 binding-increased sites were mostly contained in newly opened chromatin (as determined by DNase I-hypersensitivity), while the decreased sites were in closed chromatin. Increases and decreases in Zic binding were correlated with histone H3 lysine 27 (H3K27) acetylation which is an active chromatin signature. They therefore considered that Zic binding to enhancers is regulated by chromatin dynamics and that contributes to the differential gene transcription during GC differentiation.

Based on these results, Frank et al. (2015) hypothesized that Zic might function to globally coordinate the maturation of gene expression programs in differentiating GCs. Accordingly, Zic binding and developmental upregulation of the nearby gene expression were positively correlated (Fig. 13.5a, b). Then they addressed how the global gene expression was affected by Zic1 or Zic2 knockdown. To this end, they carried out RNA-sequencing (RNA-seq) analysis in cultured GC treated with short hairpin RNA (sh-RNA) targeted to decrease gene expression levels of Zic1 or Zic2. They identified genes with significant expression change by Zic1 knockdown (n = 81) or Zic2 knockdown (n = 147). Interestingly the gene expression changes caused by Zic1 or Zic2 knockdown were negatively correlated with the change of gene expression during in vitro GC differentiation. Namely Zic1 or Zic2 knockdown inhibited the in vitro GC differentiation. Genes affected by the Zic1 or Zic2 knockdown included a NMDA receptor subunit (Grin2c) that is expressed in mature GCs and are involved in mossy fiber-GC excitatory transmission (Kadotani et al. 1996). Grin2c expression is upregulated by Zic1 and Zic2 through their combined actions on a 3′ enhancer (Fig. 13.5c).

In sum, the transcriptomics approach by Frank et al. (2015) corroborates that both Zic1 and Zic2 are critical regulators of gene expression during postnatal cerebellar GC differentiation. The approach was successful in large part because mature GCs represent the most populous cell type in the postnatal cerebellum. Further, granule cell progenitors can be readily purified, cultured, and manipulated from early postnatal mouse cerebellum. Implementation of unbiased genome-wide strategies to find Zic target genes during earlier stages of cerebellar development is complicated by the complex mixture cell types in the embryonic anlage, although emerging single-cell genomic technologies now offer a means to address this heterogeneity.

13.6 Heterozygous Loss of *ZIC1* and *ZIC4* Cause Human Dandy-Walker Malformation

Clinical studies have also shown roles for *ZIC1* and *ZIC4* in human cerebellar development. Dandy-Walker malformation (DWM, MIM%220200) is a rare congenital anomaly of the cerebellum and posterior skull (Fig. 13.6). It is a diagnosis based on brain imaging studies and involves the following constellation of features: (1) a massively enlarged fourth ventricle (often called a DW cyst) in an enlarged posterior fossa (posterior skull), (2) cerebellar vermis hypoplasia preferentially affecting the posterior vermis, and (3) a rotation of the cerebellar vermis away from the brainstem, upward toward an elevated tentorium (the dura between the cerebellum and occipital lobes of the cerebrum). Patients have variable motor and cognitive outcomes, from normal to significantly motor and/or intellectual disabilities. Hydrocephalus and agenesis of the corpus callosum are also variable features of DWM, with incidence not well defined.

In addition to clinical heterogeneity, DWM is genetically heterogeneous. DWM is also a sporadic disorder, very rarely inherited in a familial manner. Several loci have been associated with DWM (Aldinger and Doherty 2016), including interstitial deletions involving long arm of chromosome 3. Grinberg et al. (2004) identified 1.9 Mb DWM critical region based on eight DWM individuals with overlapping interstitial deletions in 3q22.2–3q25.33 (Fig. 13.7). Although this 1.9 Mb region did not contain good candidate genes, the closely linked *ZIC1* and *ZIC4* genes were just 250 kb away from the centromeric end of the critical region. Among the eight individuals in this study, seven individuals were heterozygous for *ZIC1*-*ZIC4* deletion.

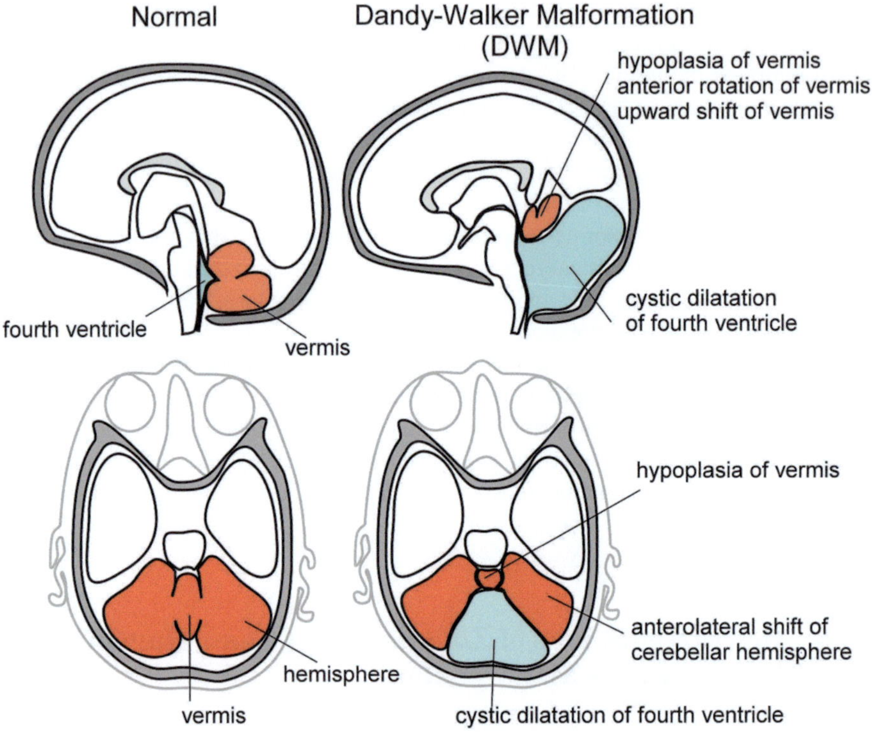

Fig. 13.6 Dandy-Walker malformation. Some diagnostic criteria are illustrated

Notably, the remaining individual without deletion of the *ZIC1* and *ZIC4* genes had reduced expression levels of *ZIC1* and *ZIC4* in lymphoblasts (asterisk in Fig. 13.7).

Following the initial report of involvement of *ZIC1* and *ZIC4* in DWM, additional studies replicated *ZIC1* and *ZIC4* gene deletion in DWM. A case of 3q23-q25.1 interstitial deletion (Tohyama et al. 2011), another case with de novo interstitial deletion of 3q22.3-q25.2 (Lim et al. 2011), and two cases with ca. 20 Mb in 3q22.3-q25.31 or ca. 20 Mb in 3q23-q26.1 (Ferraris et al. 2013) included heterozygous deletion of both *ZIC1* and *ZIC4* genes (Fig. 13.7). However, it is clear that the DWM phenotype is not completely penetrant, as some individuals with *ZIC1-ZIC4* interstitial deletions do not show DWM (Ferraris et al. 2013). Among 19 patients with heterozygous *ZIC1-ZIC4* deletions in 3q22.1-3q26.1, 15 patients had DWM, yet the 4 cases did not. However, among these four patients with *ZIC1-ZIC4* heterozygous deletion, all had cerebellar-related neurological symptoms including gait ataxia, generalized hypotonia, speech impairment, and ataxic movements/tics, in addition to cognitive deficits (Ferraris et al. 2013). Despite these additional patients, deletion of 3q24 *ZIC1* and *ZIC4* genes remains a very rare cause of DWM accounting for less than 1% of all cases. The number of patients is still too limited to fully describe the penetrance or clinical features associated with *ZIC1/4*-related DWM.

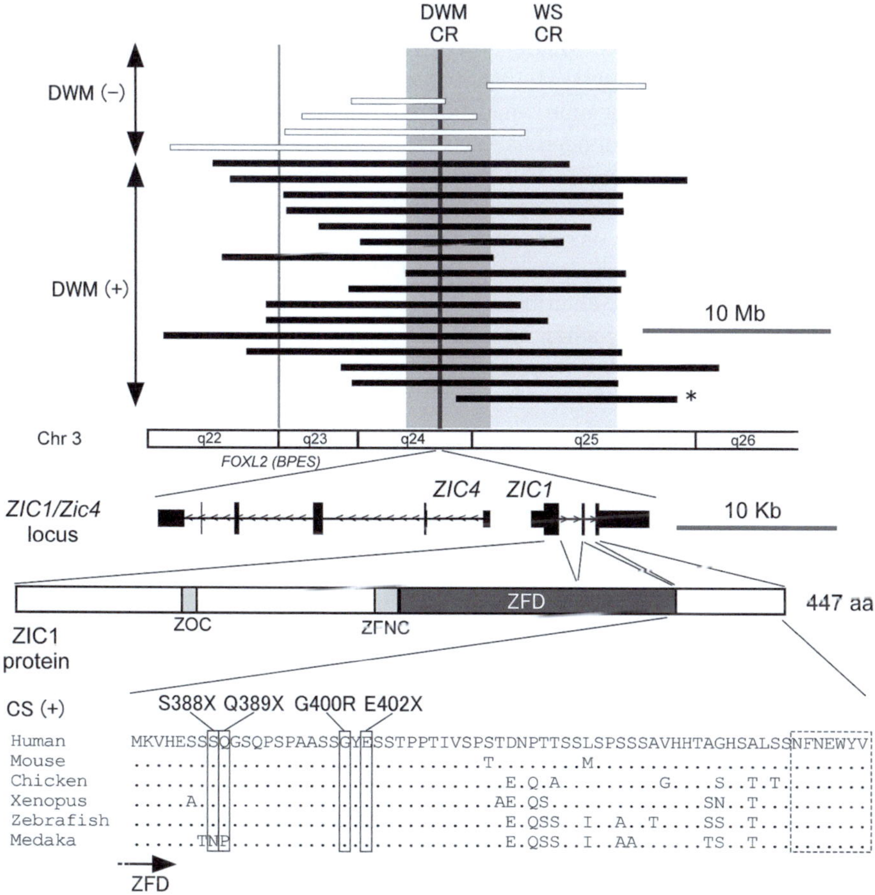

Fig. 13.7 Loss-of-function and gain-of-function mutations in human *ZIC1* associated with Dandy-Walker malformation and craniosynostosis. *Top*, horizontally stacked *open* and *closed bars* indicate the deletions (heterozygous) in the chromosome 3q without or with DWM diagnosis, respectively. *Asterisk* indicates a case without direct deletion of the *ZIC1* and *ZIC4* genes but showing reduced expression levels of *ZIC1* and *ZIC4* expression in lymphoblasts. *Middle*, genomic organization of *ZIC1* and *ZIC4*. *Bottom*, nonsense or missense mutations found in the ZIC1 C-terminal regions of craniosynostosis patients. *Solid-line boxes*, the mutated amino acid residues or their counterparts in the vertebrate Zic1. *Broken-line box*, conserved C terminal residues. *Dots* in the alignment indicate the identical amino acid residues to those of human ZIC1

The human and mouse *ZIC1/4* loss of function phenotypes are difficult to directly correlate. As described above, *Zic* genes have multiple roles in cerebellar development including embryonic cerebellar folial patterning and regulation of cerebellar GC proliferation and differentiation, and it is likely that this is shared across species. Additionally, it is possible that *Zic1* meningeal expression adjacent to the developing cerebellum is important for cerebellar development (Blank et al. 2011). As reviewed in Chap. 12, *Zic* genes play a role in the development of cerebral cortical meningeal membrane, and reduction of SDF1α was shown in *Zic1/Zic3* combined mutants (Inoue et al. 2008). Haldipur et al. (2015, 2017) recently showed that meningeal SDF1α is an essential regulator of multiple aspects of cerebellar development, including cerebellar ventricular zone and rhombic lip neurogenesis, migration of most if not all cerebellar neuronal types, and proliferation of granule cell progenitors. Indeed, we hypothesize that disruption of meningeal signaling, including SDF1α signaling into the developing cerebellum, is central to the pathogenesis of DWM based on our analysis of another rare cause of DWM—heterozygous deletion of *FOXC1* on human chromosome 6p25.3 (Aldinger et al. 2009; Haldipur et al. 2015, 2017). The specific contribution of *Zic* genes in regulating cerebellar meningeal development and signaling into the developing cerebellar development awaits examination of meningeal cell-restricted *Zic* gene knockouts in mice.

13.7 *ZIC2-ZIC5* Mutations in Holoprosencephaly Complicated by DWM

While *ZIC2* is a major causative gene for holoprosencephaly (see Chap. 14), co-occurrence of holoprosencephaly and DWM has been reported for four cases that involve partial deletion of chromosome 13q (*ZIC2* and *ZIC5* are in 13q32.2-32.3) (McCormack et al. 2002; Mimaki et al. 2015). Together with the hypoplastic phenotype of *Zic1+/−;Zic2+/ kd* cerebella (Aruga et al. 2002a), and the reduction of neural crest tissue from the dorsal neural tube in Zic5-deficient mice (Inoue et al. 2004), it is possible that *ZIC2* and *ZIC5* are additional genetic factors contributing to DWM. Although the case numbers are absolutely small, further reports on the coincidence of DWM and holoprosencephaly with 13q partial deletion would attract our attention.

13.8 *ZIC1* Gain-of-Function Mutations in Craniosynostosis

While the loss of function of *ZIC1-ZIC4* genes is causative for DWM, another study revealed that *ZIC1* gain-of-function mutations are involved in yet another congenital anomaly, craniosynostosis. Craniosynostosis is caused by premature fusion of one or more sutures of the skull vault (Fig. 13.8). There are multiple types of craniosynostosis. A subtype involves fusion of the coronal sutures which can be present bilaterally or unilaterally, resulting in brachycephaly or plagiocephaly, respectively. Twigg et al. (2015) carried out whole genome/exome sequencing to identify the genetic etiology of a cohort of coronal synostosis patients and identified heterozygous *ZIC1* nonsense or missense mutations (S388X, Q389X, E402X, and G400R) in the third exon of *ZIC1* in five families with severe coronal synostosis (Fig. 13.7). Because these mutations are clustered in C-terminal region of the protein, the

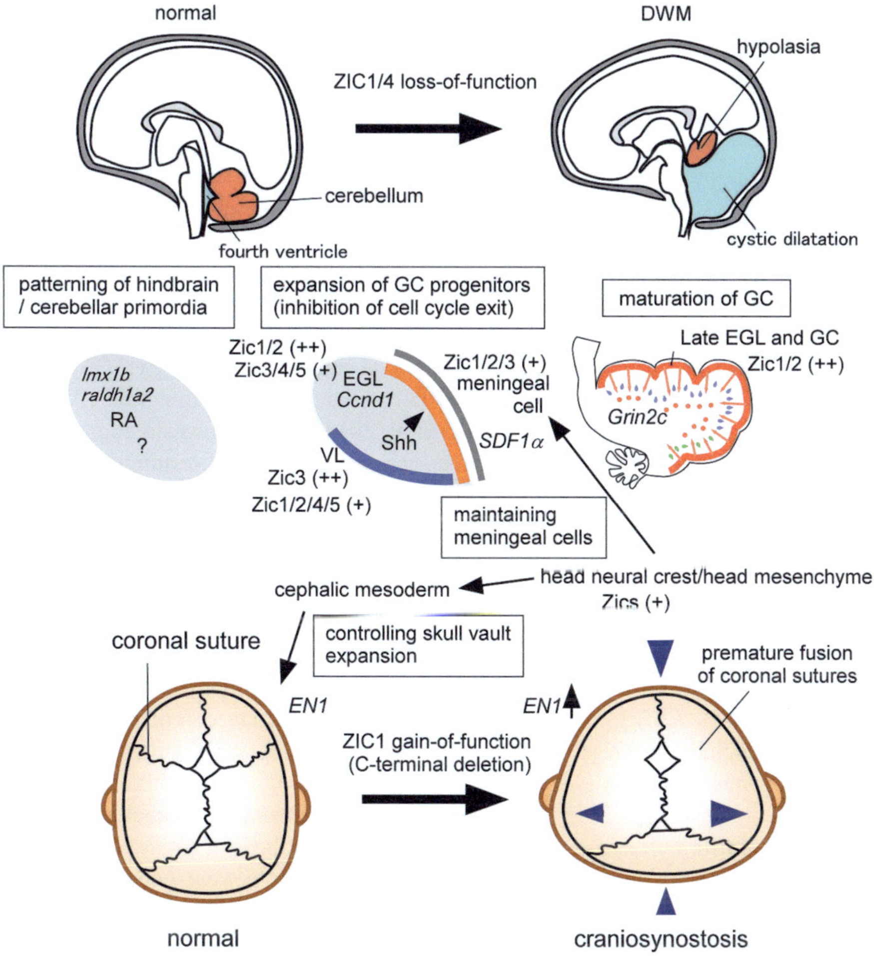

Fig. 13.8 Summary of the known role of Zic family in cerebellar development and the diseases caused by altered ZIC1 function. *Boxes* indicate the major role of Zic1 reviewed in this chapter. *Italicized*, representative positively regulated genes by Zic proteins in each process

mutants retain intact in N-terminal and zinc finger regions. In a previous study, C-terminally deleted Xenopus zic1 caused increased expression of the homeodomain factor *en-2* (Kuo et al. 1998). The *ZIC1 S388X*, *E402X*, and *G400R* variants were subjected to the same assay and also caused increased en-2 expression. The results indicate that the *ZIC1* mutations identified in severe coronal craniosynostosis patients are gain-of-function mutations.

Engrailed1 (*En1*) was shown to be critical for early biogenesis of the mouse coronal suture (Deckelbaum et al. 2012). Twigg et al. (2015) further clarified that mouse *Zic1* expression precedes and overlaps with *En1* expression in a distinct domain in the supraorbital region and cephalic mesoderm between E11.5 and E12.5.

Based on these results, they proposed that the coronal synostosis phenotype associated with the human mutations might be attributable to alteration of *EN1* expression in the supraorbital regulatory center.

In terms of gene regulatory relationships, the induction of *engrailed* expression by ZIC1 orthologues is known in *Drosophila* and *Xenopus*. In addition, a LIM-homeodomain protein, Lmx1b might be also included in this regulatory cascade. In mice Lmx1b is expressed in supraorbital region, and in humans a specific missense mutation in the N-terminal arm of the LMX1B homeodomain is associated with craniosynostosis. Thus the biological role of ZIC1 in the cranial suture development would be interesting from phylogenetic point of view.

This study also highlighted the function of C-terminal domain in ZIC1 (Fig. 13.7) (Twigg et al. 2015). The C-terminus NEWYV is conserved among Zic1, Zic2, and Zic3. In ZIC3 (467 aa), a single case of C-terminal missense mutation (A447G) increased transactivation activity (Cowan et al. 2014). Further investigation of both intramolecular and intermolecular interactions with the ZIC1 C-terminus is needed to fully elucidate the function of this domain.

13.9 Perspectives

As summarized in Fig. 13.8, *Zic1* and other *Zic* genes have crucial roles in cerebellar development. Changes in ZIC1 function lead to human cerebellar and skull pathology, including DWM and craniosynostosis. During development, *Zic1* and other Zic genes are dynamically expressed in the dorsal hindbrain and in the developing cerebellar anlage. Zic gene expression remains strong in cerebellar GC progenitors and differentiated cerebellar GCs. Based on analysis of mouse phenotypes and transcriptome profiling, it is apparent that *Zic1* and other *Zic* genes have multiple roles in regulation of cerebellar progenitor proliferation and differentiation. The molecular pathways upstream and downstream of the *Zic* genes in these contexts are just beginning to be elucidated.

References

Aldinger KA, Doherty D (2016) The genetics of cerebellar malformations. Semin Fetal Neonatal Med 21(5):321–332. https://doi.org/10.1016/j.siny.2016.04.008

Aldinger KA, Lehmann OJ, Hudgins L, Chizhikov VV, Bassuk AG, Ades LC, Krantz ID, Dobyns WB, Millen KJ (2009) FOXC1 is required for normal cerebellar development and is a major contributor to chromosome 6p25.3 Dandy-Walker malformation. Nat Genet 41(9):1037–1042. https://doi.org/10.1038/ng.422

Aruga J, Yokota N, Hashimoto M, Furuichi T, Fukuda M, Mikoshiba K (1994) A novel zinc finger protein, zic, is involved in neurogenesis, especially in the cell lineage of cerebellar granule cells. J Neurochem 63(5):1880–1890

Aruga J, Nagai T, Tokuyama T, Hayashizaki Y, Okazaki Y, Chapman VM, Mikoshiba K (1996a) The mouse zic gene family. Homologues of the Drosophila pair-rule gene odd-paired. J Biol Chem 271(2):1043–1047

Aruga J, Yozu A, Hayashizaki Y, Okazaki Y, Chapman VM, Mikoshiba K (1996b) Identification and characterization of Zic4, a new member of the mouse Zic gene family. Gene 172(2):291–294

Aruga J, Minowa O, Yaginuma H, Kuno J, Nagai T, Noda T, Mikoshiba K (1998) Mouse Zic1 is involved in cerebellar development. J Neurosci 18(1):284–293

Aruga J, Inoue T, Hoshino J, Mikoshiba K (2002a) Zic2 controls cerebellar development in cooperation with Zic1. J Neurosci 22(1):218–225

Aruga J, Tohmonda T, Homma S, Mikoshiba K (2002b) Zic1 promotes the expansion of dorsal neural progenitors in spinal cord by inhibiting neuronal differentiation. Dev Biol 244(2):329–341. https://doi.org/10.1006/dbio.2002.0598

Aruga J, Ogura H, Shutoh F, Ogawa M, Franke B, Nagao S, Mikoshiba K (2004) Locomotor and oculomotor impairment associated with cerebellar dysgenesis in Zic3-deficient (Bent tail) mutant mice. Eur J Neurosci 20(8):2159–2167. https://doi.org/10.1111/j.1460-9568.2004.03666.x

Aruga J, Nozaki Y, Hatayama M, Odaka YS, Yokota N (2010) Expression of ZIC family genes in meningiomas and other brain tumors. BMC Cancer 10:79. https://doi.org/10.1186/1471-2407-10-79

Azevedo FA, Carvalho LR, Grinberg LT, Farfel JM, Ferretti RE, Leite RE, Jacob Filho W, Lent R, Herculano-Houzel S (2009) Equal numbers of neuronal and nonneuronal cells make the human brain an isometrically scaled-up primate brain. J Comp Neurol 513(5):532–541. https://doi.org/10.1002/cne.21974

Blank MC, Grinberg I, Aryee E, Laliberte C, Chizhikov VV, Henkelman RM, Millen KJ (2011) Multiple developmental programs are altered by loss of Zic1 and Zic4 to cause Dandy-Walker malformation cerebellar pathogenesis. Development 138(6):1207–1216. https://doi.org/10.1242/dev.054114

Chizhikov VV, Lindgren AG, Mishima Y, Roberts RW, Aldinger KA, Miesegaes GR, Currle DS, Monuki ES, Millen KJ (2010) Lmx1a regulates fates and location of cells originating from the cerebellar rhombic lip and telencephalic cortical hem. Proc Natl Acad Sci U S A 107(23):10725–10730. https://doi.org/10.1073/pnas.0910786107

Cowan J, Tariq M, Ware SM (2014) Genetic and functional analyses of ZIC3 variants in congenital heart disease. Hum Mutat 35(1):66–75. https://doi.org/10.1002/humu.22457

Dahmane N, Ruiz i Altaba A (1999) Sonic hedgehog regulates the growth and patterning of the cerebellum. Development 126(14):3089–3100

De Luca A, Cerrato V, Fuca E, Parmigiani E, Buffo A, Leto K (2016) Sonic hedgehog patterning during cerebellar development. Cell Mol Life Sci 73(2):291–303. https://doi.org/10.1007/s00018-015-2065-1

Deckelbaum RA, Holmes G, Zhao Z, Tong C, Basilico C, Loomis CA (2012) Regulation of cranial morphogenesis and cell fate at the neural crest-mesoderm boundary by engrailed 1. Development 139(7):1346–1358. https://doi.org/10.1242/dev.076729

Dipietrantonio HJ, Dymecki SM (2009) Zic1 levels regulate mossy fiber neuron position and axon laterality choice in the ventral brain stem. Neuroscience 162(3):560–573. https://doi.org/10.1016/j.neuroscience.2009.02.082

Drummond DL, Cheng CS, Selland LG, Hocking JC, Prichard LB, Waskiewicz AJ (2013) The role of Zic transcription factors in regulating hindbrain retinoic acid signaling. BMC Dev Biol 13:31. https://doi.org/10.1186/1471-213X-13-31

Elsen GE, Choi LY, Millen KJ, Grinblat Y, Prince VE (2008) Zic1 and Zic4 regulate zebrafish roof plate specification and hindbrain ventricle morphogenesis. Dev Biol 314(2):376–392. https://doi.org/10.1016/j.ydbio.2007.12.006

Ferraris A, Bernardini L, Sabolic Avramovska V, Zanni G, Loddo S, Sukarova-Angelovska E, Parisi V, Capalbo A, Tumini S, Travaglini L, Mancini F, Duma F, Barresi S, Novelli A, Mercuri E, Tarani L, Italian CSG, Bertini E, Dallapiccola B, Valente EM (2013) Dandy-Walker malformation and Wisconsin syndrome: novel cases add further insight into the genotype-phenotype correlations of 3q23q25 deletions. Orphanet J Rare Dis 8:75. https://doi.org/10.1186/1750-1172-8-75

Frank CL, Liu F, Wijayatunge R, Song L, Biegler MT, Yang MG, Vockley CM, Safi A, Gersbach CA, Crawford GE, West AE (2015) Regulation of chromatin accessibility and Zic binding at enhancers in the developing cerebellum. Nat Neurosci 18(5):647–656. https://doi.org/10.1038/nn.3995

Fujimi TJ, Mikoshiba K, Aruga J (2006) Xenopus Zic4: conservation and diversification of expression profiles and protein function among the Xenopus Zic family. Dev Dyn 235(12):3379–3386. https://doi.org/10.1002/dvdy.20906

Furushima K, Murata T, Matsuo I, Aizawa S (2000) A new murine zinc finger gene. Opr Mech Dev 98(1–2):161–164

Gaston-Massuet C, Henderson DJ, Greene ND, Copp AJ (2005) Zic4, a zinc-finger transcription factor, is expressed in the developing mouse nervous system. Dev Dyn 233(3):1110–1115. https://doi.org/10.1002/dvdy.20417

Ghez J (1991) The cerebellum. In: Kandel ER, Schwartz JH, Jessell TM (eds) Principles of neural science, 3rd edn. Elsevier, New York, pp 626–646

Grinberg I, Northrup H, Ardinger H, Prasad C, Dobyns WB, Millen KJ (2004) Heterozygous deletion of the linked genes ZIC1 and ZIC4 is involved in Dandy-Walker malformation. Nat Genet 36(10):1053–1055. https://doi.org/10.1038/ng1420

Haldipur P, Sivaprakasam I, Periasamy V, Govindan S, Mani S (2015) Asymmetric cell division of granule neuron progenitors in the external granule layer of the mouse cerebellum. Biol Open 4(7):865–872. https://doi.org/10.1242/bio.009886

Haldipur P, Dang D, Aldinger KA, Janson OK, Guimiot F, Adle-Biasette H, Dobyns WB, Siebert JR, Russo R, Millen KJ (2017) Phenotypic outcomes in mouse and human Foxc1 dependent Dandy-Walker cerebellar malformation suggest shared mechanisms. Elife 6. https://doi.org/10.7554/eLife.20898

Hoche F, Guell X, Sherman JC, Vangel MG, Schmahmann JD (2016) Cerebellar contribution to social cognition. Cerebellum 15(6):732–743. https://doi.org/10.1007/s12311-015-0746-9

Holmes DB, Heine VM (2017) Simplified 3D protocol capable of generating early cortical neuroepithelium. Biol Open 6(3):402–406. https://doi.org/10.1242/bio.021725

Inoue T, Hatayama M, Tohmonda T, Itohara S, Aruga J, Mikoshiba K (2004) Mouse Zic5 deficiency results in neural tube defects and hypoplasia of cephalic neural crest derivatives. Dev Biol 270(1):146–162. https://doi.org/10.1016/j.ydbio.2004.02.017

Inoue T, Ota M, Mikoshiba K, Aruga J (2007a) Zic2 and Zic3 synergistically control neurulation and segmentation of paraxial mesoderm in mouse embryo. Dev Biol 306(2):669–684. https://doi.org/10.1016/j.ydbio.2007.04.003

Inoue T, Ota M, Ogawa M, Mikoshiba K, Aruga J (2007b) Zic1 and Zic3 regulate medial forebrain development through expansion of neuronal progenitors. J Neurosci 27(20):5461–5473. https://doi.org/10.1523/JNEUROSCI.4046-06.2007

Inoue T, Ogawa M, Mikoshiba K, Aruga J (2008) Zic deficiency in the cortical marginal zone and meninges results in cortical lamination defects resembling those in type II lissencephaly. J Neurosci 28(18):4712–4725. https://doi.org/10.1523/JNEUROSCI.5735-07.2008

Ito M (2011) The cerebellum: brain for an implicit self. FT Press Science, New Jersey

Kadotani H, Hirano T, Masugi M, Nakamura K, Nakao K, Katsuki M, Nakanishi S (1996) Motor discoordination results from combined gene disruption of the NMDA receptor NR2A and NR2C subunits, but not from single disruption of the NR2A or NR2C subunit. J Neurosci 16(24):7859–7867

Kasukawa T, Masumoto KH, Nikaido I, Nagano M, Uno KD, Tsujino K, Hanashima C, Shigeyoshi Y, Ueda HR (2011) Quantitative expression profile of distinct functional regions in the adult mouse brain. PLoS One 6(8):e23228. https://doi.org/10.1371/journal.pone.0023228

Keller MJ, Chitnis AB (2007) Insights into the evolutionary history of the vertebrate zic3 locus from a teleost-specific zic6 gene in the zebrafish, Danio rerio. Dev Genes Evol 217(7):541–547. https://doi.org/10.1007/s00427-007-0161-4

Koziol LF, Budding D, Andreasen N, D'Arrigo S, Bulgheroni S, Imamizu H, Ito M, Manto M, Marvel C, Parker K, Pezzulo G, Ramnani N, Riva D, Schmahmann J, Vandervert L, Yamazaki T (2014) Consensus paper: the cerebellum's role in movement and cognition. Cerebellum 13(1):151–177. https://doi.org/10.1007/s12311-013-0511-x

Kuo JS, Patel M, Gamse J, Merzdorf C, Liu X, Apekin V, Sive H (1998) Opl: a zinc finger protein that regulates neural determination and patterning in Xenopus. Development 125(15):2867–2882

Leto K, Arancillo M, Becker EB, Buffo A, Chiang C, Ding B, Dobyns WB, Dusart I, Haldipur P, Hatten ME, Hoshino M, Joyner AL, Kano M, Kilpatrick DL, Koibuchi N, Marino S, Martinez S, Millen KJ, Millner TO, Miyata T, Parmigiani E, Schilling K, Sekerkova G, Sillitoe RV, Sotelo C, Uesaka N, Wefers A, Wingate RJ, Hawkes R (2016) Consensus paper: cerebellar development. Cerebellum 15(6):789–828. https://doi.org/10.1007/s12311-015-0724-2

Lim BC, Park WY, Seo EJ, Kim KJ, Hwang YS, Chae JH (2011) De novo interstitial deletion of 3q22.3-q25.2 encompassing FOXL2, ATR, ZIC1, and ZIC4 in a patient with blepharophimosis/ptosis/epicanthus inversus syndrome, Dandy-Walker malformation, and global developmental delay. J Child Neurol 26(5):615–618. https://doi.org/10.1177/0883073810384996

Lin JC, Cepko CL (1998) Granule cell raphes and parasagittal domains of Purkinje cells: complementary patterns in the developing chick cerebellum. J Neurosci 18(22):9342–9353

Louvi A, Alexandre P, Metin C, Wurst W, Wassef M (2003) The isthmic neuroepithelium is essential for cerebellar midline fusion. Development 130(22):5319–5330. https://doi.org/10.1242/dev.00736

Matsumoto K, Nishihara S, Kamimura M, Shiraishi T, Otoguro T, Uehara M, Maeda Y, Ogura K, Lumsden A, Ogura T (2004) The prepattern transcription factor Irx2, a target of the FGF8/MAP kinase cascade, is involved in cerebellum formation. Nat Neurosci 7(6):605–612. https://doi.org/10.1038/nn1249

McCormack WM Jr, Shen JJ, Curry SM, Berend SA, Kashork C, Pinar H, Potocki L, Bejjani BA (2002) Partial deletions of the long arm of chromosome 13 associated with holoprosencephaly and the Dandy-Walker malformation. Am J Med Genet 112(4):384–389. https://doi.org/10.1002/ajmg.10659

McMahon AR, Merzdorf CS (2010) Expression of the zic1, zic2, zic3, and zic4 genes in early chick embryos. BMC Res Notes 3:167. https://doi.org/10.1186/1756-0500-3-167

Mimaki M, Shiihara T, Watanabe M, Hirakata K, Sakazume S, Ishiguro A, Shimojima K, Yamamoto T, Oka A, Mizuguchi M (2015) Holoprosencephaly with cerebellar vermis hypoplasia in 13q deletion syndrome: critical region for cerebellar dysgenesis within 13q32.2q34. Brain and Development 37(7):714–718. https://doi.org/10.1016/j.braindev.2014.10.009

Miyata T, Nakajima K, Aruga J, Takahashi S, Ikenaka K, Mikoshiba K, Ogawa M (1996) Distribution of a reeler gene-related antigen in the developing cerebellum: an immunohistochemical study with an allogeneic antibody CR-50 on normal and reeler mice. J Comp Neurol 372(2):215–228. https://doi.org/10.1002/(SICI)1096-9861(19960819)372:2<215::AID-CNE5>3.0.CO;2-6

Miyazawa K, Himi T, Garcia V, Yamagishi H, Sato S, Ishizaki Y (2000) A role for p27/Kip1 in the control of cerebellar granule cell precursor proliferation. J Neurosci 20(15):5756–5763

Nagai T, Aruga J, Takada S, Gunther T, Sporle R, Schughart K, Mikoshiba K (1997) The expression of the mouse Zic1, Zic2, and Zic3 gene suggests an essential role for Zic genes in body pattern formation. Dev Biol 182(2):299–313. https://doi.org/10.1006/dbio.1996.8449

Salero E, Hatten ME (2007) Differentiation of ES cells into cerebellar neurons. Proc Natl Acad Sci U S A 104(8):2997–3002. https://doi.org/10.1073/pnas.0610879104

She X, Rohl CA, Castle JC, Kulkarni AV, Johnson JM, Chen R (2009) Definition, conservation and epigenetics of housekeeping and tissue-enriched genes. BMC Genomics 10:269. https://doi.org/10.1186/1471-2164-10-269

Srivastava R, Kumar M, Peineau S, Csaba Z, Mani S, Gressens P, El Ghouzzi V (2013) Conditional induction of Math1 specifies embryonic stem cells to cerebellar granule neuron lineage and promotes differentiation into mature granule neurons. Stem Cells 31(4):652–665. https://doi.org/10.1002/stem.1295

Tohyama J, Kato M, Kawasaki S, Harada N, Kawara H, Matsui T, Akasaka N, Ohashi T, Kobayashi Y, Matsumoto N (2011) Dandy-Walker malformation associated with heterozygous ZIC1 and ZIC4 deletion: report of a new patient. Am J Med Genet A 155A(1):130–133. https://doi.org/10.1002/ajmg.a.33652

Twigg SR, Forecki J, Goos JA, Richardson IC, Hoogeboom AJ, van den Ouweland AM, Swagemakers SM, Lequin MH, Van Antwerp D, SJ MG, Westbury I, Miller KA, Wall SA, WGS C, van der Spek PJ, Mathijssen IM, Pauws E, Merzdorf CS, Wilkie AO (2015) Gain-of-function mutations in ZIC1 are associated with coronal craniosynostosis and learning disability. Am J Hum Genet 97(3):378–388. https://doi.org/10.1016/j.ajhg.2015.07.007

White JJ, Sillitoe RV (2013) Development of the cerebellum: from gene expression patterns to circuit maps. Wiley Interdiscip Rev Dev Biol 2(1):149–164. https://doi.org/10.1002/wdev.65

Yokota N, Aruga J, Takai S, Yamada K, Hamazaki M, Iwase T, Sugimura H, Mikoshiba K (1996) Predominant expression of human zic in cerebellar granule cell lineage and medulloblastoma. Cancer Res 56(2):377–383

Chapter 14
ZIC2 in Holoprosencephaly

Kristen S. Barratt and Ruth M. Arkell

Abstract The *ZIC2* transcription factor is one of the most commonly mutated genes in Holoprosencephaly (HPE) probands. HPE is a severe congenital defect of forebrain development which occurs when the cerebral hemispheres fail to separate during the early stages of organogenesis and is typically associated with mispatterning of the embryonic midline. Recent study of genotype-phenotype correlations in HPE cases has defined distinctive features of *ZIC2*-associated HPE presentation and genetics, revealing that *ZIC2* mutation does not produce the craniofacial abnormalities generally thought to characterise HPE but leads to a range of non-forebrain phenotypes. Furthermore, the studies confirm the extent of *ZIC2* allelic heterogeneity and that pathogenic variants of *ZIC2* are associated with both classic and middle interhemispheric variant (MIHV) HPE which arise from defective ventral and dorsal forebrain patterning, respectively. An allelic series of mouse mutants has helped to delineate the cellular and molecular mechanisms by which one gene leads to defects in these related but distinct embryological processes.

Keywords Zic2 · Holoprosencephaly · Syntelencephaly · Prechordal plate · Telencephalon · Dorsal-ventral pattern · Nodal · Hedgehog · Wnt · BMP

14.1 Holoprosencephaly

The evolution of bilaterians ~600 million years ago paved the way for the establishment of a vertebrate brain in which two hemispheres, divided along the embryonic midline, develop from a single group of cells. The embryological process that leads to hemisphere separation is estimated to completely or partially fail in approximately 1/250 human conceptuses, resulting in the most common structural defect of the human forebrain, Holoprosencephaly (HPE) (Matsunaga and Shiota 1977). The crucial nature of hemisphere separation is evidenced by the positive correlation

K. S. Barratt · R. M. Arkell (✉)
Early Mammalian Development Laboratory, John Curtin School of Medical Research, The Australian National University, Canberra, ACT, Australia
e-mail: ruth.arkell@anu.edu.au

J. Aruga (ed.), *Zic family*, Advances in Experimental Medicine and Biology 1046,
https://doi.org/10.1007/978-981-10-7311-3_14

between the degree of brain malformation and HPE-induced mortality (Solomon et al. 2010b) and the reduction in HPE frequency to 1/10,000 by birth (Orioli and Castilla 2010). The high rate of HPE occurrence suggests that the hemisphere separation process is incredibly fragile, comprised of multiple interconnected steps and vulnerable to interference. Furthermore, the phenotypic heterogeneity characteristic of this condition suggests that interference can come from a variety of sources (i.e. environmental and/or genetic factors) and that multiple factors act cooperatively in at least some HPE cases. The degree of brain separation and whether the failure occurs ventrally or dorsally distinguish two main classes of HPE: classic and middle interhemispheric variant (MIHV).

14.1.1 Classic HPE

In classic HPE, the lack of separation is most severe ventrally, extending to the rostral, dorsal and posterior domains of the forebrain in a graded fashion. This leads to a spectrum of classic HPE, of which alobar HPE (a monoventricle with no hemispheric separation) is the most severe form, followed by semilobar (partial hemispheric separation, resulting in fused left and right frontal and parietal lobes but retaining the posterior portion of the interhemispheric fissure) and lobar HPE (hemispheric and lateral vesicle separation is retained, except for in the rostral and ventral frontal lobes) (reviewed in Marcorelles and Laquerriere 2010). It is estimated that 10–40%, 43–45% and 17–33% of HPE cases are alobar, semilobar and lobar, respectively (Solomon et al. 2010a); however, the number of cases at the severe end of the spectrum is likely underestimated due to early embryonic lethality.

In addition to hemisphere separation, HPE pathogenesis also comprises craniofacial and midline defects as common co-morbidities. The frequent coincidence of, and correlation between, the severity of brain and craniofacial symptoms in HPE is the basis for the notion that ‘the face predicts the brain’, as observed by DeMeyer and colleagues in 1964 (DeMyer et al. 1964). Craniofacial phenotypes accompanying severe HPE often include microcephaly, cyclopia or synophthalmia, and a proboscis. Less-severely affected cases present with microcephaly, hypotelorism, midface hypoplasia, flat nasal bridges, cleft lip and/or palate and/or a single maxillary incisor (Solomon et al. 2010b). Additionally, midline defects such as undivided thalami, absent corpora callosa and absent or hypoplastic olfactory and optic bulbs and tracts occur (Solomon et al. 1993, 2010b). A microform HPE also exists, where subtle facial phenotypes such as hypotelorism, a sharp and narrow nasal bridge and single maxillary incisor are present in the absence of structural brain abnormalities. These cases are often not recognised until a severely affected relative with HPE is identified (Solomon et al. 2010b), and thus the frequency of microform HPE is underestimated.

14.1.2 MIHV HPE

In contrast to classic HPE, MIHV HPE (also known as syntelencephaly) presents with normal separation of the basal forebrain, anterior frontal lobes and occipital regions but a failure to divide the posterior frontal and parietal regions of the cerebral hemispheres along the dorsal midline (Barkovich and Quint 1993; Simon et al. 2002; Lewis et al. 2002). Additional structures, such as the caudate nuclei, thalami and mesencephalon, can also be affected in MIHV cases. Whilst some similarities occur between classic and MIHV HPE (namely non-cleavage of a portion of the cerebral hemispheres), MIHV HPE is rarer and milder than classic HPE (Simon et al. 2002) and likely has a distinct embryological origin (Fernandes et al. 2007).

14.2 The Genetics of Human Holoprosencephaly

When considered as a single disorder, HPE genetics exhibits extreme heterogeneity with multiple classes of causative mutations and numerous modes of heredity. Up to 50% of HPE cases are attributable to chromosomal abnormalities, whilst a further 25% of cases are syndromic and 25% occur in isolation. Whilst some autosomal recessive cases of nonchromosomal, non-syndromic HPE have been reported, the condition is generally considered to be autosomal dominant (Barr and Cohen 2002; Ming et al. 2002; Mercier et al. 2011; Roessler et al. 2012b; Mouden et al. 2016). HPE, however, does not exhibit simple Mendelian inheritance as evidenced by the facts that only ~70% of individuals who carry HPE pathogenic mutations exhibit HPE symptoms (i.e. the condition is incompletely penetrant) (Mercier et al. 2011) and that the same mutation can confer vastly different phenotypes in different carriers as a result of variable expressivity. There is evidence from human and animal studies that both genetic and environmental factors influence the HPE end phenotype (Hong et al. 2012; Kietzman et al. 2014; Mouden et al. 2016); therefore the ultimate consideration when studying HPE is the total activity achieved along the particular embryonic signalling pathways that direct hemisphere separation (Roessler and Muenke 2010). Evidently, some cases of HPE have a digenic basis (Mouden et al. 2016; Dubourg et al. 2016). For the majority of HPE probands, however, the putative second hit factors remain unidentified, and it seems a model of autosomal dominant with modifier effects most aptly describes HPE heritability (Odent et al. 1998; Roessler et al. 2012b).

One clear source of HPE genetic variability is locus heterogeneity (a single disorder is caused by mutations at different chromosomal loci). Fourteen genes have been implicated in non-syndromic classic and microform HPE in humans (*SHH*, *ZIC2*, *TGIF*, *SIX3*, *CDON*, *DISP1*, *DLL*, *FGF8*, *FGFR1*, *FOXH1*, *GAS1*, *PTCH1*, *NODAL*, *TDGF1*) (Roessler et al. 1996; Belloni et al. 1996, Brown et al. 1998, 2001; Gripp et al. 2000; de la Cruz et al. 2002; Ming et al. 2002; Dubourg et al. 2007, 2016; Roessler et al. 2009a, b, c; Lacbawan et al. 2009; Arauz et al. 2010;

Ribeiro et al. 2010; Mercier et al. 2011; Dupé et al. 2011; Bae et al. 2011; Pineda-Alvarez et al. 2012). These genes are classified as either 'large effect' (major) or 'small effect' (minor) HPE genes, according to how often they are mutated in the disorder. In classic HPE, the two genes most commonly mutated are *SHH* (12%) and *ZIC2* (9%), which together account for ~85% of solved probands (Roessler et al. 2009a; Dubourg et al. 2011, 2016). Additional genes have been associated with HPE in the mouse, indicating that they may be minor HPE genes and/or genetic modifiers in humans (reviewed in Schachter and Krauss 2008).

Notably, many genes associated with classic HPE do not cause MIHV HPE. In mice, *Zic2* and *Fgf8*, as well as the BMP ligands, BMP antagonists *Chrd* and *Nog*, BMP receptors *Bmpr1a* and *Bmpr1b* and the transcription factors *Lhx5* and *Rfx4*, have all been associated with MIHV (Nagai et al. 2000; Simon et al. 2002; Anderson et al. 2002; Lewis et al. 2002; Cheng et al. 2006; Storm et al. 2006; Fernandes et al. 2007; Warr et al. 2008; Solomon et al. 2010a; Dubourg et al. 2016). In contrast, relatively few genes (*ZIC2*, *FGF8*, deletion of *EYA4*) have been associated with MIHV HPE in humans (Abe et al. 2009; Solomon et al. 2010a; Dubourg et al. 2016), of which *ZIC2* and *FGF8* are also associated with classic HPE. Animal models of *Zic2* dysfunction suggest that classic and MIHV HPE each have a distinct embryological basis, consistent with their observed differential involvement of the ventral versus dorsal brain regions in human HPE. Thus, some phenotype variance in HPE occurs as a result of the disruption of related but distinct embryonic processes.

Despite recent progress in elucidating the genetic aetiology of human HPE, only 25% of cases of nonchromosomal, non-syndromic HPE have been attributed to pathogenic mutations in known HPE genes (Dubourg et al. 2007; Roessler et al. 2009a). The remaining unsolved (nonchromosomal) cases are thought to be caused by unidentified HPE genes, non-coding region variants that alter the expression of known HPE-associated genes and/or environmental factors. An emerging area of HPE research aims to assess the likely contribution of putative risk factors (such as maternal diabetes, ethyl alcohol, cigarette smoking and retinoic acid) to the aetiology of HPE, either as single factors or in combination with pathogenic genetic lesions. Another current research focus is to catalogue the DNA variants at known HPE causative gene loci in HPE probands and family members using a targeted next-generation sequencing strategy. Studies to date have already revealed that aspects of HPE pathogenesis vary according to the affected genetic locus, and thus, it is already possible to delineate several distinct features of *ZIC2*-associated HPE.

14.3 The Genetics of *ZIC2*-Associated Holoprosencephaly

One unusual aspect of ZIC2 involvement in HPE is that probands with *ZIC2* mutations have been found across the entire HPE phenotypic spectrum. Generally, classic HPE genes are not associated with MIHV. Although the vast majority of *ZIC2* mutations result in classic HPE, a few are associated with MIHV (Table 14.1). A striking demonstration of the variable expressivity of *ZIC2* mutations comes from

Table 14.1 Frequency of known HPE mutations in human *ZIC2* arranged by mutation type and phenotype

	Unknown	Alobar	Semilobar	Lobar	Mic	MIHV
Missense	18 (10.34%)	7 (4.02%)	7 (4.02%)	4 (2.30%)	2 (1.15%)	0 (0.00%)
Nonsense	7 (4.02%)	5 (2.87%)	8 (4.60%)	2 (1.15%)	1 (0.57%)	0 (0.00%)
Frameshift	19 (10.92%)	10 (5.75%)	23 (13.22%)	2 (1.15%)	5 (2.87%)	0 (0.00%)
Insertion	1 (0.57%)	0 (0.00%)	1 (0.57%)	0 (0.00%)	0 (0.00%)	0 (0.00%)
Deletion	2 (1.15%)	1 (0.57%)	1 (0.57%)	1 (0.57%)	0 (0.00%)	1 (0.57%)
Duplication	8 (4.60%)	5 (2.87%)	9 (5.17%)	1 (0.57%)	1 (0.57%)	3 (1.72%)
Splice variant (intron)	6 (3.45%)	2 (1.15%)	3 (1.72%)	0 (0.00%)	0 (0.00%)	1 (0.57%)
SNV (3′UTR)	3 (1.72%)	1 (0.57%)	1 (0.57%)	0 (0.00%)	2 (1.15%)	0 (0.00%)
Total (n = 174)	**64 (36.78%)**	**31 (17.82%)**	**53 (30.46%)**	**10 (5.75%)**	**11 (6.32%)**	**5 (2.87%)**
Total (n = 110)		**31 (28.18%)**	**53 (48.18%)**	**10 (9.09%)**	**11 (10.00%)**	**5 (4.55%)**

Polymorphisms were not included. Identical mutations that occurred in multiple patients were counted as independent instances. Mutation data was collated from Roessler et al. (2009a, 2012a, b), Paulussen et al. (2010), Solomon et al. (2010a), Mercier et al. (2011), Ribeiro et al. (2012), Nakayama et al. (2016), and Dubourg et al. (2016)
Mic Microform, *MIHV* middle interhemispheric variant, *SNV* single nucleotide variant

the report of monozygotic twins with the same de novo mutation at a splice donor site in *ZIC2* (c.1239+1G>C). These twins exhibit different classes of HPE: one twin developed semilobar HPE whilst the other had MIHV (Nakayama et al. 2016). As discussed later, animal models suggest the classic and MIHV forms of HPE have a distinct embryological basis, and it appears that ZIC2 is required for both processes in man and mice. In comparison to HPE as a whole, *ZIC2*-associated HPE manifests as highly penetrant (93%) with relatively few mild phenotype individuals (Solomon et al. 2010a). In fact, it is estimated that 90% of patients with a *ZIC2* mutation exhibit structural brain anomalies (Solomon et al. 2010b). Additionally, *ZIC2* mutation is more frequently associated with severe structural brain anomalies (alobar or semilobar presentations) which account for 75% of the *ZIC2*-associated HPE cases in which phenotype class is recorded (Table 14.1). In contrast, a similar analysis of 92 individuals (probands and family members) with clinically apparent HPE and *SHH* mutations found that 39% exhibited alobar or semilobar brain abnormalities, that 48% presented with no brain abnormalities but had craniofacial characteristics of microform HPE and that none had MIHV (Solomon et al. 2010b). *ZIC2* mutation therefore accounts for the majority of severely affected HPE cases (Solomon et al. 2010b).

Another striking observation based on the molecular subtyping of HPE cases is that *ZIC2* mutation breaks the mantra 'the face predicts the brain', with the craniofacial defects typically associated with classic HPE absent in those patients assessed

(Brown et al. 1998; Solomon et al. 2010a, b). In particular, Solomon et al. (2010a) found no *ZIC2*-associated HPE case presenting with facial findings at the severe end of the spectrum (cyclopia, synophthalmia or a proboscis) or a combination of facial features similar to those caused by mutations in other HPE genes. Instead, a distinct phenotype can be seen in *ZIC2*-associated HPE, consisting of (but not always containing) bitemporal narrowing, upslanted palprebral fissures, a flat nasal bridge, short nose with anteverted nares, a broad and deep philtrum and the appearance of large ears. Whilst facial clefts occurred in both non-*ZIC2* and *ZIC2*-associated HPE, the frequency is reduced by 1/3 in the latter (Solomon et al. 2010a). Despite the mild facial phenotype, these patients often have severe HPE and neurologic impairment (Solomon et al. 2010a, b).

Additionally, other non-forebrain phenotypes are often associated with *ZIC2*-HPE. For example, HPE patients with intragenic *ZIC2* mutations have been found to exhibit neural tube defects (4%), hydrocephalus (12%), skeletal anomalies (14%), cardiac anomalies (9%) and renal anomalies (7%) (Solomon et al. 2010a). Mouse models that recapitulate *ZIC2*-associated HPE also display a subset of these co-morbidities. Both severe and mild ZIC2 loss-of-function lead to incompletely penetrant neural tube defects in murine embryos such as spina bifida, due to a requirement for ZIC2 during neurulation (Nagai et al. 2000; Elms et al. 2003; Ybot-Gonzalez et al. 2007). Severe loss-of-function murine alleles also present with cardiac defects such as randomised heart situs, which is attributed to defective cilia formation and function in the embryonic node (Barratt et al. 2014). This results in errors generating and sensing leftward nodal flow and, thus, aberrant left-right axis formation. Whilst the exact manifestations of non-forebrain anomalies in human patients are not well documented, it is possible that a subset may be due to defects in cilia which are known to be involved in the development and function of each of the affected body systems (reviewed in Fliegauf et al. 2007; Waters and Beales 2011). Thus, *ZIC2* may be an as yet unidentified influence for ciliopathies.

Comparison of relatively large HPE cohorts has also revealed aspects of *ZIC2*-HPE heritability. For example, family analysis has shown that *ZIC2* mutations are largely de novo, with an inheritance rate of only 27–30% (Solomon et al. 2010b; Mercier et al. 2011; Mouden et al. 2016). In contrast, 70% of *SHH* and *SIX3* mutations are inherited (Solomon et al. 2010a). Of the inherited *ZIC2* cases, two-thirds were maternally inherited and one-third paternally inherited. Additionally, families with *ZIC2* mutations in greater than two generations have not been reported. A small subset of *ZIC2*-associated HPE cases, however, appear to occur due to allelic dropout or germline mosaicism, resulting in parents negative for mutations siring multiple affected children (Solomon et al. 2010a, b). The low rate of inheritance in *ZIC2* cohorts may be due to severely affected individuals being unable to reproduce (Solomon et al. 1993) and suggests that mutations in *ZIC2* produce a higher level of lethality in comparison to other HPE genes. Despite this, a recent analysis by Weiss et al. found that individuals with *ZIC2*-associated HPE were more likely to survive to adolescence than those with HPE due to mutations in other implicated genes (Weiss et al. 2017).

Table 14.2 Type and frequency of known HPE mutations in human ZIC2

	Number	Percent of CDS mutations	Percent of non-coding mutations	Percent of all mutations
CDS				
Missense	30	28.57%		25.42%
Nonsense	15	14.29%		12.71%
Frameshift	47	44.76%		39.83%
Duplication	5	4.76%		4.24%
Insertion	3	2.86%		2.54%
Deletion	5	4.76%		4.24%
Total	105			
Non-coding DNA				
Splice variant (intron)	7		53.85%	5.93%
SNV (3′UTR)	6		46.15%	5.08%
Total	13			118

Polymorphisms were not included. Identical mutations that occurred in multiple patients were counted as one instance. Mutation data was collated from Roessler et al. (2009a, 2012a, b), Paulussen et al. (2010), Solomon et al. (2010a), Mercier et al. (2011), Ribeiro et al. (2012), Nakayama et al. (2016), and Dubourg et al. (2016)
SNV Single nucleotide variant, *CDS* coding DNA sequence

It remains unclear from the targeted sequencing analysis to date whether heterozygous *ZIC2* mutation is sufficient to cause HPE. There are two reported cases of *ZIC2* CDS mutations in conjunction with other major HPE genes (*ZIC2/SHH and ZIC2/SIX3*), and mutations in the 3′UTR of *ZIC2* have also been found in conjunction with mutations in other HPE genes (Nanni et al. 1999; Lacbawan et al. 2009; Roessler et al. 2012a). Studies of other HPE-associated genes show that, in mice, exposure to alcohol during gestation in conjunction with a pre-existing mutation increases the frequency and severity of HPE cases (Aoto et al. 2008; Hong et al. 2012; Kietzman et al. 2014). It is therefore possible that at least some cases of HPE arise when a *ZIC2* mutation sensitises the developing embryo to one or multiple teratogens. The involvement of gene x gene and gene x environment interactions may explain why for *ZIC2* (and other HPE associated genes), heterozygous mutations give rise to the HPE phenotype in man but not mouse (Chiang et al. 1996; Nagai et al. 2000; Brown et al. 2005; Schachter and Krauss 2008; Warr et al. 2008; Roessler et al. 2009a; Petryk et al. 2015).

This analysis of HPE cohorts has enabled assessment of the *ZIC2* mutational spectrum, confirming that *ZIC2*-associated HPE exhibits allelic heterogeneity (whereby different mutations in the same gene give rise to the same disease). In the last 12 years, the number of published unique *ZIC2*-associated HPE mutations has grown from 20 (Brown et al. 2005) to 118 at the time of this review (Table 14.2) and will continue to expand as sequencing techniques become more readily available and the unique *ZIC2*-associated HPE phenotype is refined. Amongst the 105 documented HPE cases with mutations in the *ZIC2* coding sequence, the majority are predicted to substantially alter the *ZIC2* transcript (44.76% are frameshift, 14.29%

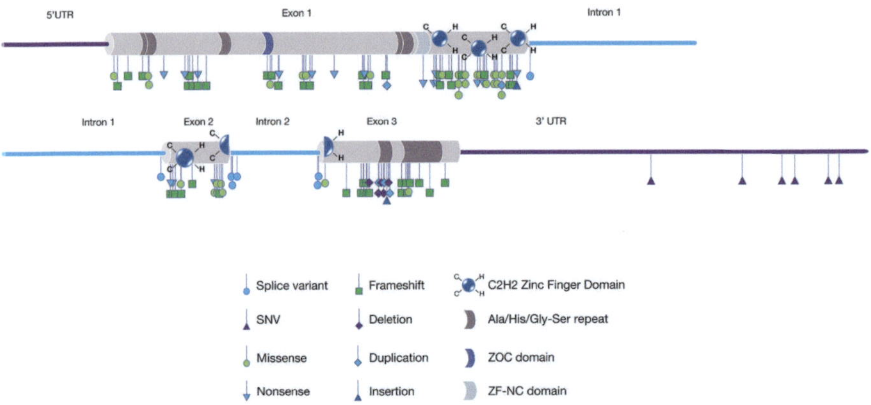

Fig. 14.1 The genomic and protein structure of ZIC2, showing the known human HPE-associated variants. Polymorphisms were not included. Identical mutations that occurred in multiple patients were counted as one instance (Mutation data was collated from Roessler et al. 2009a, 2012a, b; Paulussen et al. 2010; Solomon et al. 2010a; Mercier et al. 2011; Ribeiro et al. 2012; Nakayama et al. 2016; Dubourg et al. 2016). *SNV* Single nucleotide variant, *ZOC* ZIC/odd-paired conserved motif, *ZF-NC* zinc finger N-terminally conserved domain

nonsense, 4.76% duplication, 4.76% deletion and 2.86% insertion) rather than a single amino acid (28.57% missense) (Table 14.2, Fig. 14.1). Similarly, of the 13 mutations in *ZIC2* non-coding DNA, 53.85% are splice variants and predicted to substantially alter the transcript. Intriguingly, the remaining non-coding mutations in *ZIC2* are single nucleotide variants in a conserved region of the 3′UTR (Table 14.1). In 2012 Roessler et al. showed this element is under selective pressure and postulated that the variants contribute to the pathogenicity of HPE by dysregulating *ZIC2* expression during embryo development (Roessler et al. 2012a). Consistent with the mutational landscape of *ZIC2*, analysis by Solomon et al. (2010a) found that 98% of all *ZIC2* mutations were predicted or proven to be loss-of-function.

Analysis of mutation type and location underscores the functional importance of the zinc finger domain (ZFD). For example, 45.71% of all *ZIC2* mutations occur in the ZFD (Table 14.3), and 73% (22/30) of ZFD mutations are missense, demonstrating that single amino acid changes in the ZFD are sufficient to cause disease. Furthermore, no nonsense mutations, nor most frameshift mutations, produce a complete ZFD when translated (Roessler et al. 2009a). In contrast, few missense variants occur outside of the ZFD where 86% of known variants are predicted to drastically alter the *ZIC2* transcript. Another notable enrichment in mutation type is found in the C-terminal alanine repeat (Table 14.3), a region that influences the strength of ZIC2 DNA binding and transcriptional activity. Eight duplication, deletions or insertions have been identified in this small (15 amino acid) region out of 13 variants (62%) of this type across the whole gene. Functional analysis of these variants shows that expansion from 15As to 25As results in near-complete loss of

Table 14.3 Protein location and frequency of known HPE mutations in the human ZIC2 protein

Protein domain	Location (aa)	Missense	Nonsense	Frameshift	Duplication	Insertion	Deletion	Total (n = 105)
Histidine repeat	20–24							**0 (0.00%)**
Alanine repeat	24–34			1				**1 (0.95%)**
Alanine repeat	89–98							**0 (0.00%)**
ZOC	115–126			1				**1 (0.95%)**
Alanine repeat	226–231							**0 (0.00%)**
Histidine repeat	231–240							**0 (0.00%)**
ZF-NC	241–256		1					**1 (0.95%)**
C2H2 zinc finger domain	258–416	22	8	16	1	1		**48 (45.71%)**
Alanine repeat	455–471				3	1	4	**8 (7.62%)**
Serine/glycine repeat	477–516	1		8				**9 (8.57%)**
Non-domain		7	6	21	1	1	1	**37 (35.24%)**
Total (n = 105)		**30 (28.57%)**	**15 (14.29%)**	**47 (44.76%)**	**5 (28.57%)**	**3 (2.86%)**	**5 (4.76%)**	

Polymorphisms and chromosome deletions were not included. Identical mutations that occurred in multiple patients were counted as one instance. Mutation data was collated from Roessler et al. (2009a, 2012b), Paulussen et al. (2010), Solomon et al. (2010a), Mercier et al. (2011), Ribeiro et al. (2012), Nakayama et al. (2016), and Dubourg et al. (2016)

ZOC ZIC/odd-paired conserved motif, *ZF-NC* zinc finger N-terminally conserved domain transactivation, yet reduction to 2As results in both an increase and decrease in transactivation, dependant on the promoter that is used. This indicates that the alanine tract can modulate ZIC2 transactivation, contingent on the DNA sequence being targeted (Brown et al. 2005). Expansion of the *ZIC2* alanine repeat to 25A has been identified in multiple unrelated families and, in some cases, is hypothesised to occur via errors in somatic recombination in the patient's fathers (Brown et al. 2001).

The ongoing efforts to delineate genotype-phenotype correlations in HPE have shed light on the genetic and embryonic origin of HPE and, with respect to aetiology, have led to the following generalisations:

- HPE arises as a consequence of failed dorsal-ventral (D-V) forebrain patterning.
- Classic HPE arises due to failed ventral patterning, whereas MIHV HPE is associated with failed dorsal patterning.
- Classic HPE brain abnormalities can occur with or without associated facial abnormalities.
- For some molecular subclasses of HPE, classic or MIHV HPE brain abnormalities can occur alongside abnormalities not associated with the face and forebrain (e.g. *Zic2*).

To consider how *ZIC2* mutation causes the typical brain abnormalities of HPE requires understanding how D-V forebrain pattern is established. Much of our knowledge regarding mammalian embryonic development and the mechanism of ZIC2 activity during brain development is derived from the analysis of mouse mutant phenotypes, and the next sections will refer to mouse development.

14.4 Ventral Forebrain Patterning During Murine Development

The generation of a correctly patterned embryo is dependent upon inductive interactions between progenitor tissues that direct differentiation, regionalisation and morphogenesis. In the ventral forebrain, patterning is a consequence of inductive interactions between the anterior neurectoderm and the underlying prechordal plate (PrCP) (reviewed in Placzek and Briscoe 2005). Both of these tissues arise during gastrulation (reviewed in Arkell and Tam 2012), prior to which the neurectoderm precursor cells are located within the distal epiblast where they are encased by the distal visceral endoderm cells (Fig. 14.2a). These visceral endoderm cells secrete factors that antagonise TGF-β and WNT signalling activities, protecting the distal ectoderm cells of the epiblast from the signals that would otherwise cause them to differentiate. By the time gastrulation begins, the future neurectoderm cells are found at the anterior of the epiblast where they continue to be protected from differentiation into endoderm or mesoderm by antagonists secreted from the

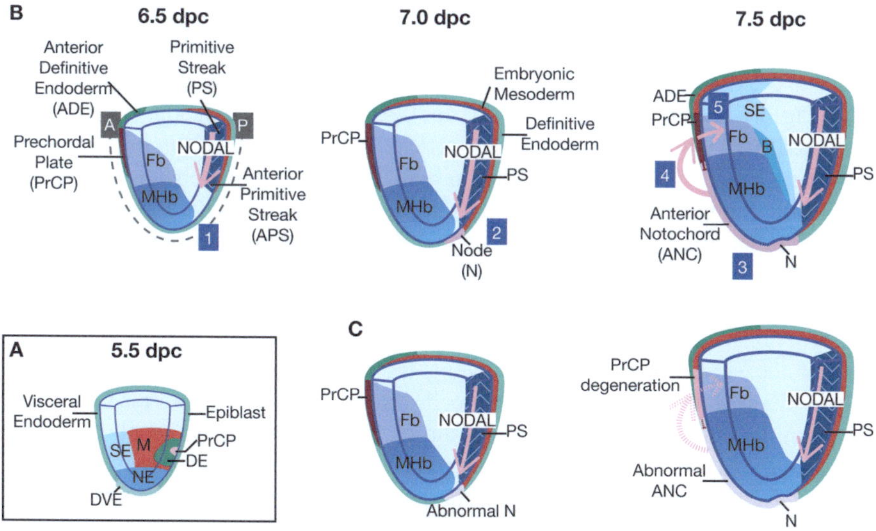

Fig. 14.2 The origin of the prechordal plate and other tissues involved in dorsal-ventral patterning of the murine telencephalon. (**a**) A cutaway diagram of the pre-gastrula mouse embryo (5.5 dpc) with a superimposed cellular fate map. The pre gastrula embryo is bi-laminar, with the inner ectoderm tissue (the epiblast) enveloped by the visceral endoderm. The position of the precursor cells for the ectoderm (neural, NE, and surface, SE), mesoderm (M), definitive endoderm (DE) and anterior prechordal plate (PrCP) is shown. (**b**) Cutaway diagrams of the embryonic portion of the early (6.5 dpc), mid- (7.0 dpc) and late (7.5 dpc) gastrulas. The anterior (A) and posterior (P) of the embryo are marked and the dotted line indicates the A-P axis. Wild-type embryos use five steps to establish the ventral signalling centre in the forebrain neurectoderm. *Step 1*: NODAL signal in the posterior directs the initial differentiation of the PrCP and anterior definitive endoderm cells which transit the anterior primitive streak and migrate to the anterior midline of the embryo. *Step 2*: NODAL signal induces the transition of the anterior primitive streak cells into the overt node. *Step 3*: the anterior notochord cells transit the node and migrate to the anterior midline, coming to lie caudal of the PrCP. *Step 4*: inductive interactions between the PrCP and anterior notochord (negative from PrCP to ANC and positive from ANC to PrCP) stabilise PrCP fate enabling SHH secretion. *Step 5*: SHH signals vertically to the overlying forebrain neurectoderm to establish *Shh* expression in the rostral ventral neural midline and overlay ventral identity information on the neurectoderm which initially is dorsal in character. **c** The alterations in PrCP development in *Zic2*$^{Ku/Ku}$ embryos are shown. At 6.5 dpc *Zic2*$^{Ku/Ku}$ embryos are indistinguishable from wild type, and Step 1 proceeds as normal. At 7.0 dpc, the lack of functional ZIC2 alters the level of perceived NODAL signal such that the overt node is formed (Step 2), but gene expression at the node is abnormal. Step 3 fails in the *Zic2*$^{Ku/Ku}$ mutants and by 7.5 dpc there is no anterior notochord. Consequently, the PrCP degenerates, does not secrete SHH and *Shh* expression and does not initiate in the rostral ventral neural midline, resulting in classic HPE. *B* neural plate border, *DE* definitive endoderm, *dpc* Days post-coitum, *DVE* distal visceral endoderm, *Fb* forebrain neurectoderm, *M* mesoderm, *MHb* mid- and hindbrain neurectoderm, *NE* neurectoderm, *SE* surface ectoderm

enveloping endoderm, which at this point is called the anterior visceral endoderm (AVE). During gastrulation, the anterior neurectoderm population expands anteriorly and proximally and occupies two-thirds of the ectoderm layer by mid-gastrulation (Fig. 14.2b, 7.0 days post-coitum; dpc). Although the descendants of some of these cells will colonise more posterior parts of the brain, the cells in this region are becoming progressively restricted to forebrain fate. Approximately half of the neurectoderm that has been produced at this stage will give rise to the forebrain; therefore perturbations in neural development at this stage disproportionately affect the forebrain.

The tissues that will form the PrCP are initially co-localised with the precursors of the definitive endoderm and prior to gastrulation are found at the future posterior side of the embryo about halfway along the embryonic portion of the epiblast (Fig. 14.2a). During gastrulation, these cells ingress through the anterior segment of the primitive streak and extend along the embryonic midline to reach the entire length of the body axis. The first cells to pass through the primitive streak will come to lie at the embryonic anterior, with the later cells taking up an axial position in accordance with their time of passage through the primitive streak. When the cells first emerge from the primitive streak and migrate along the midline, they do so as a contiguous sheet of cells in which the midline mesoderm is flanked by definitive endoderm. At this stage, this sheet of cells is named the anterior mesendoderm (AME). Later in development, the cells at the midline separate and take up a position amongst the mesoderm tissues. The resulting axial mesoderm structures are named for their position along the body axis: the axial mesoderm that underlies the forebrain is called the PrCP and that which associates with the rest of the brain is the anterior notochord.

Clues to the molecular nature of the ventral forebrain patterning signals are provided by the resulting phenotypes in murine embryos with particular genetic mutations. Classic HPE arises following the loss of morphogenetic signalling activity induced by disruption of the HH ligand SHH or combined mutation of the TGF-β antagonists *Chrd* and *Nog* in murine embryos (Chiang et al. 1996; Anderson et al. 2002). This suggests that the HH pathway must be activated and the TGF-β pathway inhibited in order for ventral forebrain patterning to proceed. In the HH pathway (Fig. 14.3b), the SHH ligand, along with co-receptors CDON, BOC, LRP2 and GAS1, binds to Patched (PTCH1), a transmembrane receptor. This interaction with PTCH1 relieves inhibition of Smoothened (SMO), which then facilitates the production and nuclear transport of full length, activating forms of GLI transcription factors to promote transcription of SHH targets (reviewed in Xavier et al. 2016). As described previously, mutation of not only *SHH* itself but also components of the HH transduction pathway has been found to be mutated in HPE probands, indicating HH signalling is also required for human forebrain ventral patterning. *Shh* transcripts are initially found in the PrCP itself and subsequently in the overlying rostral-ventral neural midline (RVNM). The downstream components and target genes of the SHH transduction pathway are expressed in the RVNM. Mutations in the response components of the SHH pathway produce neurectoderm that is incompetent to respond to the SHH signal, resulting in HPE (Fuccillo et al. 2004; Spoelgen et al. 2005).

In the TGF-β pathway (Fig. 14.3a, c), ligands bind to and activate a type I and type II receptor complex, causing phosphorylation of some members of the SMAD family called the receptor-associated SMADs (R-SMADs). This enables their interaction with the common mediator SMAD (co-SMAD) resulting in nuclear localization, the formation of higher-order transcriptional complexes and regulation of TGF-β target genes. In contrast, BMP ligands signal through different receptors and SMAD molecules than the NODAL and GDF molecules, discussed below. Both gain- and loss-of-function experiments in the mouse indicate that BMP signalling represses *Shh* expression in the RVNM (Anderson et al. 2002). The BMP antagonists *Chrd* and *Nog* are expressed in the node and axial mesoderm derivatives (notochord and PrCP), and in $Chrd^{-/-};Nog^{+/-}$ mutant embryos, BMP antagonism is reduced, resulting in embryos that exhibit cyclopia and HPE in conjunction with loss of *Shh* expression in the PrCP (Anderson et al. 2002). Counterintuitively, *Bmp7* is expressed in the node, notochord and caudal PrCP alongside *Chrd* and *Nog* (Arkell and Beddington 1997; Anderson et al. 2002). Here, BMP7 may (as in the chick) modify the response of ventral midline cells to SHH and induce a rostral identity, instead of a floor-plate identity that would be induced by SHH on its own (Dale et al. 1997). Unlike *Bmp7*, *Bmp2* and *Bmp4* are expressed in the surface ectoderm and, in the case of *Bmp2*, paraxial mesoderm adjacent to the PrCP, but not in the PrCP itself (Anderson et al. 2002). Mutations in the BMP part of the TFG-β pathway have not yet been associated with human HPE, perhaps because mutations that result in elevated signalling are relatively rare compared to those that cause loss of signalling.

The model of ventral forebrain patterning that best fits the experimental data is that SHH, expressed in and secreted from the PrCP, signals vertically to the overlying neurectoderm. One consequence of SHH signal reception and transduction by the neurectoderm is that SHH itself becomes expressed in the RVNM, but only if BMP signalling is low in this tissue (Anderson et al. 2002). SHH expression in the RVNM then leads to neuronal patterning and maintenance of ventral forebrain tissue. The anterior mesendoderm of the late gastrula is also the source of WNT antagonism due to the expression of *Dkk1* under the control of the transcription factor *Otx2*. When *Otx2* is conditionally inactivated in the AME, *Dkk1* expression is not activated in the same tissue, and head truncations characteristic of elevated WNT signalling result (Ip et al. 2014). *Shh* expression is not perturbed by the lack of *Dkk1* expression, and HPE does not arise, clearly indicating that WNT ligands do not direct morphogenetic activity required for forebrain separation. Nonetheless, classic HPE is associated with loss of forebrain tissue (i.e. anterior truncation). It is possible that this is a consequence of loss of midline tissue which, at the anterior, expresses FGF ligands from the anterior neural ridge (ANR) as part of anterior-posterior neural patterning. Depletion of midline FGF will cause posteriorisation and forebrain hypoplasia. Thus, anterior truncation often occurs as a secondary consequence of an aberrant ventral pattern. An anterior truncation phenotype can be seen in isolation from ventral defects, however, and should not of itself be used to infer a role in HPE.

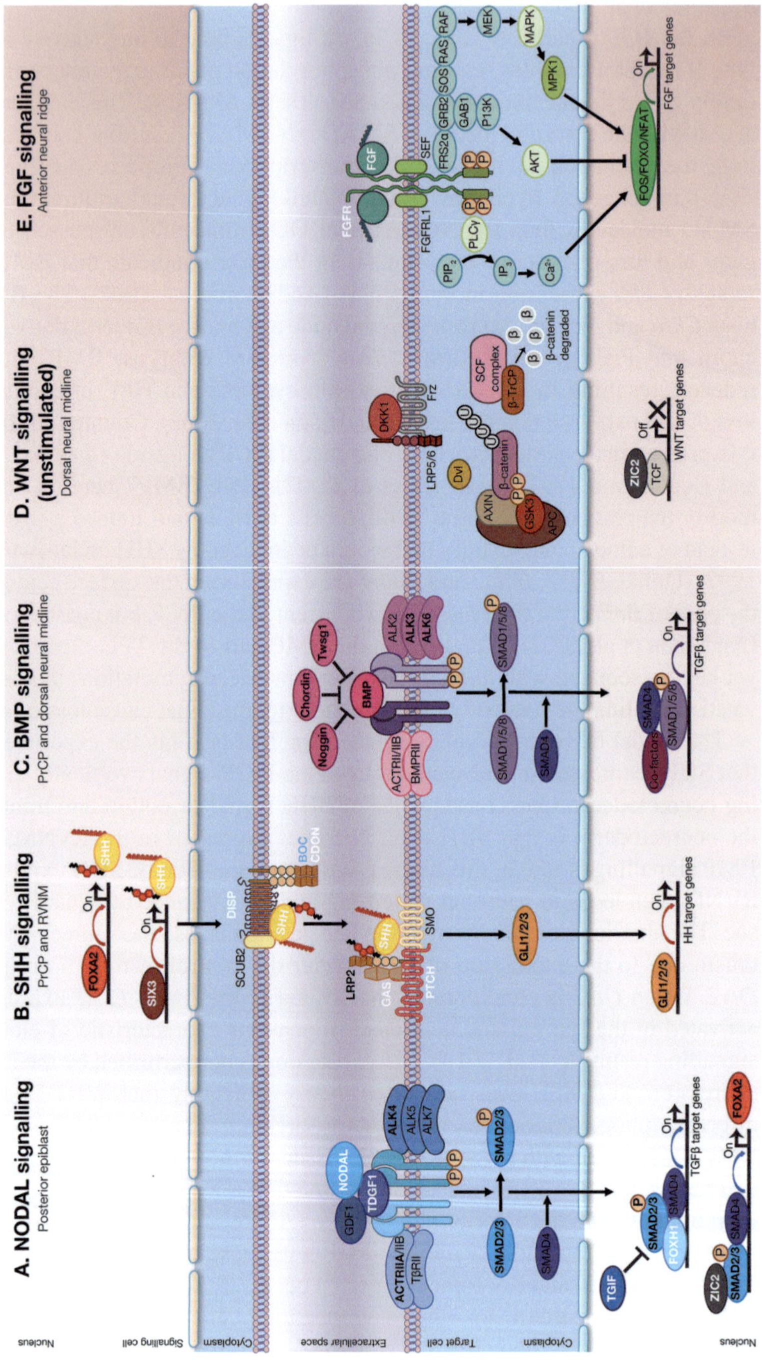

A. NODAL signalling
Posterior epiblast
B. SHH signalling
PrCP and RVNM
C. BMP signalling
PrCP and dorsal neural midline
D. WNT signalling (unstimulated)
Dorsal neural midline
E. FGF signalling
Anterior neural ridge
Nucleus
Signalling cell
Cytoplasm
Extracellular space
Target cell
Cytoplasm
Nucleus
GDF1
NODAL
TDGF1
ACTRIIA/IIB
TβRII
ALK4
ALK5
ALK7
SMAD2/3
SMAD4
TGIF
FOXH1
ZIC2
FOXA2
TGFβ target genes
On
SIX3
SHH
DISP
SCUB2
BOC
CDON
LRP2
CAS
PTCH
SMO
GLI1/2/3
HH target genes
Noggin
Chordin
Twsg1
BMP
ACTRII/IIB
BMPRII
ALK2
ALK3
ALK6
SMAD1/5/8
Co-factors
DKK1
LRP5/6
Frz
Dvl
AXIN
GSK3
APC
β-catenin
SCF complex
β-TrCP
β-catenin degraded
TCF
Off
WNT target genes
FGF
FGFR
FGFRL1
SEF
PIP2
PLCγ
IP3
Ca2+
FRS2α
GRB2
SOS
RAS
RAF
GAB1
PI3K
MEK
MAPK
MPK1
AKT
FOS/FOXO/NFAT
FGF target genes

Fig. 14.3 Signalling pathways involved in dorsal-ventral patterning of the murine telencephalon. (**a**) *Nodal signalling*. Mature nodal ligands complex with the EGF-CFC cofactor TDGF1 (Cripto), type I receptors (ALK4/5/7) and type II receptors (ActRII or ActRIIB). Receptor activation leads to the phosphorylation of the type I receptor by the type II kinase, as well as phosphorylation of SMAD2 or SMAD3, which dimerizes with SMAD4. The SMAD2/3-4 complex translocates to the nucleus and interacts with the transcription factor FOXH1 and promotes transcription of TGFβ target genes or with ZIC2 to promote transcription of FOXA2. (**b**) *SHH signalling*. In the absence of ligand, the transmembrane domain protein Patched1 (PTCH1) inhibits the activity of Smoothened (SMO) in the target cell. In a signalling cell, SHH expression is initiated by the transcription factors FOXA2 or SIX3 binding to enhancers. In the extracellular space, the binding of secreted SHH to PTCH releases SMO, allowing for regulation of HH target genes by the transcription factors GLI1, GLI2 and GLI3. (**c**) *BMP signalling*. The secreted BMP ligand binds to type I (ALK2/3/6) and type II receptors (ACTRII and ACTIIB). This results in the phosphorylation of DNA binding proteins SMAD1/5/8, which complex with SMAD4. The SMAD1/5/8-4 complex translocates to the nucleus and, via interactions with cofactors, regulates transcription of TGFβ target genes; they regulate transcription of BMP target genes. In the extracellular space, the secreted antagonists Chordin (CHRD) and Noggin (NOG) directly interact with BMP ligands to prevent the activation of downstream effectors, whilst twisted gastrulation (TWSG1) can act as both an antagonist and agonist of BMP signalling in a cell-specific manner. (**d**) *WNT signalling*. In the absence of ligand/presence of inhibitors such as DKK1, β-catenin is phosphorylated (P) by the kinase activity of the destruction complex (consisting of Axin, APC, and GSK3), polyubiquitylated (U) by the SCF (SKP1, Cullin, F-box)/BTrCP complex and degraded by the proteasome. ZIC2 acts as a transcriptional corepressor, complexing with TCF proteins to prevent transcription of WNT target genes. (**e**) *FGF signalling*. The secreted FGF ligand binds to tyrosine kinase receptors (FGFR), activating multiple signalling pathways such as the RAS/MAPK, PLC-γ, PI3K and STAT. These pathways culminate in the promotion or repression of FGF target genes by the transcription factor FOS and the transcription factor families NFAT and FOXO. White bold, genes implicated in classic and MIHV HPE in humans; black bold, genes implicated in classic and MIHV HPE in mice; blue bold, genes identified as modifiers in classic HPE in humans. *PrCP* Prechordal plate, *RVNM* rostral-ventral neural midline

14.5 PrCP Morphogenesis During Murine Development

As described above, cells in the PrCP synthesise and secrete SHH, the critical ligand for ventral neural patterning. It follows that classic HPE can also arise if the cells of the PrCP are not formed or do not function properly. The PrCP arises due to a series of inductive interactions between cells at the anterior primitive streak (APS) and then amongst the axial tissues generated by movement of cells through the APS (Fig. 14.2b). As described above, the PrCP is formed from cells that pass through the APS early in gastrulation (Kinder et al. 2001). Some cells that transit the APS and migrate anteriorly adopt a paraxial fate (i.e. either side of the midline) and form the anterior definitive endoderm (ADE). By mid-gastrulation, the APS cells have been organised into a structure called the node. The cells of the epiblast which transit through the node of the mid-gastrula differentiate into AME and migrate to the anterior midline, taking up a position adjacent to, but posterior of, the cells of the PrCP. These cells form the anterior notochord (ANC) (Kinder et al. 2001). Inductive interactions between these anterior tissues (ADE, PrCP and ANC) influence not only forebrain patterning (Hallonet et al. 2002) but also the fate and survival of the PrCP. If the PrCP is removed, the ANC reconstitutes new PrCP tissue. Conversely, removal of the ANC results in failed PrCP development, indicating that the ANC promotes survival of the PrCP (Camus et al. 2000).

The molecular nature of the signals required for PrCP development can also be inferred by the phenotype of mouse mutants. An extensive series of murine alleles in *Nodal* itself, or components of the NODAL signal transduction pathway (Fig. 14.3a), demonstrate that NODAL signalling during gastrulation is required for steps 1 and 2 of PrCP development (Fig. 14.2b). Even a small decrease in NODAL signalling activity prevents differentiation of the ADE and PrCP precursors via transit through the APS (Norris et al. 2002; Vincent et al. 2003) and manifests as moderate anterior truncation. Further loss also interferes with node induction by the APS and results in severe anterior truncation, absent AME and somite fusion across the midline (Episkopou et al. 2001; Chu et al. 2004; Dunn et al. 2004; Liu et al. 2004). These mutations affect processes beyond ventral neural patterning and obscure the role for NODAL in preventing HPE. However, embryos that are heterozygous null for *Nodal* and null for a closely related TGF-β signalling molecule (*Gdf1*) develop HPE which arises due to abberant anterior notochord and PrCP development (Andersson et al. 2006) clearly indicating the role for the non-BMP part of the TGF-β pathway in PrCP development. The identity of the signals between the anterior tissues (ADE, PrCP and ANC) that stabilise PrCP fate is unknown, but SHH is a candidate for the survival signal sent from the ANC to the PrCP as indicated by chimeric experiments between $Shh^{-/-}$ and wild-type cells (Aoto et al. 2009).

14.6 *Zic2* Mutation and Ventral Forebrain Patterning

Severe loss-of-function alleles of murine *Zic2* result in classic HPE, suggesting an involvement in ventral neural patterning. Gene expression studies rule out the possibility that the ZIC2 transcription factor directly regulates *Shh* expression in either the PrCP or RVNM since neither tissue is a site of *Zic2* expression at the appropriate stage (Nagai et al. 1997; Elms et al. 2004). Similarly, the lack of RVNM expression at the time at which ventral pattern is imposed implies that *Zic2* is not part of the transcriptional response to HH signalling in the murine forebrain. Nonetheless, the high similarity between ZIC and GLI zinc finger domains and the finding that ZIC and GLI proteins can physically interact via their ZFDs (Kinzler and Vogelstein 1990; Pavletich and Pabo 1993; Mizugishi et al. 2001) suggested ZIC2 could act downstream of SHH signalling in the forebrain (Roessler and Muenke 2001). This hypothesis was directly tested by cross of the *Kumba* (*Ku*) allele of *Zic2* ($Zic2^{Ku}$ MGI:106679) with the *Shh* null allele (Shh^{tm1Chg} MGI: 1857796) (Warr et al. 2008). The *Ku* mutant carries a missense mutation in the fourth zinc finger that abolishes the DNA binding and transcriptional activation ability of ZIC2 (Elms et al. 2003; Brown et al. 2005). When intercrossed, it was observed that neither gene was sensitive to a decreased dose of the other, and double homozygous embryos exhibited a novel phenotype demonstrating that ZIC2 does not act downstream of SHH in murine forebrain development (Warr et al. 2008). Moreover, the same study showed that a phenotype was present in $Zic2^{Ku/Ku}$ embryos before the stage at which *Shh* expression is first detected and that germline loss of all HH signalling (via *Smo* deletion) does not reproduce the early aspects of the $Zic2^{Ku/Ku}$ phenotype (Warr et al. 2008).

Instead, it appears that ZIC2 intersects the NODAL signalling pathway at mid-gastrulation. *Nodal* loss-of-function is lethal at gastrulation, and compound heterozygous embryos for both ZIC2 and NODAL do not survive to the forebrain stage of development. Sequentially decreasing the dose of NODAL activity on the $Zic2^{Ku/Ku}$ background shifts the *Nodal* phenotype towards the more severe end of the spectrum (increased frequency and severity of anterior truncation) (Houtmeyers et al. 2016). Evidently, in the absence of ZIC2 function, the embryos perceive a lower dose of NODAL signalling, suggesting that *Zic2* normally promotes NODAL signalling at the APS (Fig. 14.2b). This is supported by the analysis of $Zic2^{Ku/Ku}$ embryos which show that in the absence of *Zic2* function, the derivatives of the APS (i.e. the ADE and PrCP cells) are specified and migrate to the embryonic anterior to take up their normal position (Warr et al. 2008) and the node is induced (Elms et al. 2003). However, gene expression at the newly induced node is highly aberrant; the expression of every node specific gene so far examined at mid-gastrulation in $Zic2^{Ku/Ku}$ embryos is depleted (Warr et al. 2008; Barratt et al. 2014). Cell death and proliferation of the ANC cells that emerge from the mid-gastrula node are unaltered, but the transcripts of genes that mark the emerging ANC are depleted, suggesting that this tissue is not specified. Despite the earlier evidence of PrCP formation, by late gastrulation, the expression of markers characteristic of the PrCP is absent in *Ku* embryos. Consequently, *Shh* expression in the PrCP is not activated, and the expression of *Shh* and SHH target genes in the RVNM is not initiated (Warr et al. 2008).

In addition to the failure of ANC development, the mid-gastrula expression of genes required for node cell cilia formation and function is also perturbed and is the likely cause of the cardiac situs abnormalities observed in *Zic2*$^{Ku/Ku}$ mouse embryos (Barratt et al. 2014).

The analysis of the *Zic2*$^{Ku/Ku}$ phenotype suggests that PrCP development fails at stage 3 and that the earliest identified molecular and functional abnormalities are at the mid-gastrula node (Fig. 14.2b, c). This is therefore considered the stage and site of primary *Zic2* function (Warr et al. 2008). This functional analysis is consistent with the node of the mid-gastrula embryo (the structure that produces the ANC) being the only unique site of *Zic2* gene expression at this stage of development compared to other ZIC family members. Other closely related Zic genes (*Zic3* and *Zic5*) are co-expressed with *Zic2* in all other areas of the gastrula at this stage and likely compensate for ZIC2 loss-of-function in these cells (Furushima et al. 2000; Elms et al. 2004). The precise molecular role of ZIC2 at the mid-gastrula node remains unclear. The level of *Nodal* transcript is unaltered in *Zic2*$^{Ku/Ku}$ embryos, indicating that ZIC2 does not promote NODAL activity by directly controlling *Nodal* expression but instead acts downstream of the NODAL signal (Houtmeyers et al. 2016). Another hypothesis is that ZIC2 directly regulates expression of the *Foxa2* transcription factor. *Foxa2*, a NODAL target gene (Hoodless et al. 2001) expressed in the APS, node and AME (Sasaki and Hogan 1993; Monaghan et al. 1993; Ang et al. 1993; Ruiz i Altaba et al. 1993; Ang and Rossant 1994; Dufort et al. 1998), is known to control *Shh* expression (Jeong and Epstein 2003). In turn, SHH can induce *Foxa2* expression (Echelard et al. 1993). A scenario in which, during normal development, ZIC2 controls *Foxa2* expression in the node and its derivative ANC cells to initiate the *Foxa2*/*Shh* autoinduction loop and eventually provide the SHH-based survival signal to stabilise PrCP cell development is consistent with the phenotype analysis of the *Zic2*$^{Ku/Ku}$ embryos.

When overexpressed in mammalian cell lines, ZIC2 is able to physically interact with both SMAD2 and SMAD3 (the receptor-activated proteins that control transcription in a NODAL-dependent manner) (Fig. 14.3a). When bound to SMAD proteins, ZIC2 opposes SMAD activity (it dampens SMAD-dependent transcription or overcomes SMAD-dependent repression). In cultured human cells, ZIC2 can act in concert with SMAD3 to promote *FOXA2* expression, but the ZIC2 protein encoded by the *Ku* allele of ZIC2 is unable to do so, despite still physically interacting with SMAD (Houtmeyers et al. 2016). Overall, the cell-based data, in combination with the genetic evidence that ZIC2 is required to promote NODAL signalling, supports a model in which expression of node-specific enhancers is initially repressed and subsequently converted to expression activation in the presence of SMAD/ZIC2 complexes. The proposed molecular interactions between ZIC2 and SMAD molecules, and between this complex and SMAD DNA binding elements, are yet to be demonstrated in vitro.

14.7 Dorsal Forebrain Patterning During Murine Development

Initial dorsal forebrain patterning is a consequence of inductive interactions between the anterior neurectoderm and the adjacent surface ectoderm (Liem et al. 1995; Furuta et al. 1997). Both of these tissues arise during gastrulation (for review see Arkell and Tam 2012) with the origin of the anterior neurectoderm already described above. Prior to gastrulation, precursor cells for the surface ectoderm are found at the future anterior side of the embryo, about halfway down the embryonic portion of the epiblast (Fig. 14.2a). Like the prospective neurectoderm, these cells do not pass through the primitive streak during gastrulation but differentiate in situ into surface ectoderm. During gastrulation, the cells are arranged in an anterior-posterior order but in a more lateral position than the neurectoderm cells. The progenitors of another nonneural ectoderm derivative, the neural crest cells (which gives rise to the ecto-mesenchyme and cranial ganglia in the head), are juxtaposed between the neurectoderm and surface ectoderm cells at a region known as the neural plate border (Fig. 14.2b, 7.5 dpc). The entire ectoderm arises as a contiguous sheet which, in a process called neurulation, folds to form the neural tube and overlying surface ectoderm. Consequently, cells from the medial neural plate take up a ventral position, and cells from the lateral neural plate adopt a dorsal position. Once the process of neurulation is complete, the forebrain neurectoderm at the dorsal midline continues its morphogenesis and, by the combined strategies of low mitosis and high apoptosis, undergoes thinning and invagination to divide the cerebrum into two hemispheres (Fig. 14.4a) (Furuta et al. 1997; Groves and LaBonne 2014). In a secondary phase of dorsal patterning, inductive interactions between the invaginated cells instruct the differentiation of specialised midline dorsal structures. These include the choroid plexus which will secrete cerebrospinal fluid and the cortical hem which instructs adjacent neurectoderm to differentiate into the hippocampus (Hébert et al. 2002; Groves and LaBonne 2014).

Much of our knowledge regarding dorsal patterning of the neurectoderm comes mainly from studies of the spinal cord (reviewed in Le Dréau and Martí 2012), but aspects of forebrain neurectoderm patterning involve distinct mechanisms. For example, in the forebrain, dorsal patterning occurs in close proximity to anterior neural ridge (ANR), a source of morphogenetic signals which sits at the rostral junction of the neural and surface ectoderm (Crossley and Martin 1995; Shimamura and Rubenstein 1997). The establishment of forebrain dorsal-ventral pattern therefore intersects and interacts with the anterior-posterior patterning system. Additionally, the forebrain neurectoderm arises earliest of all the neurectoderm subdivisions, and dorsally restricted gene expression patterns (such as that of *Pax3* [Goulding et al. 1991]) are evident from the time of somite formation. Despite this, the first overt signs of dorsal differentiation in the forebrain (i.e. roof-plate thinning) occur in embryos with approximately 20 somites. In contrast, dorsal development is evident in the hindbrain 24 h earlier when the neural crest first emerges from the dorsal neurectoderm of the five-somite embryo (Serbedzija et al. 1990). The influences that promote dorsal differentiation in the hindbrain region must operate

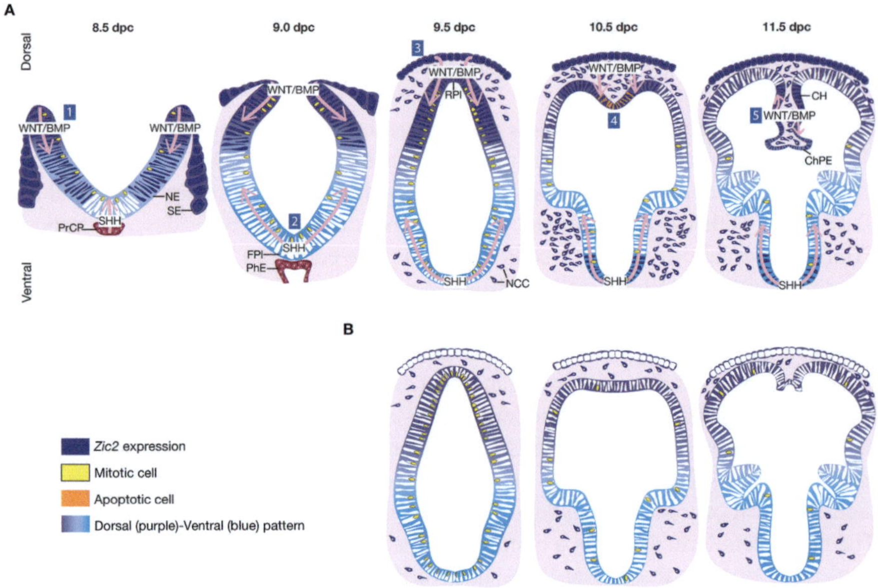

Fig. 14.4 Proposed model of dorsal telencephalic midline development and *Zic2*-associated MIHV HPE. (**a**) Diagrams of transverse sections of the telencephalon at 8.5–11.5 dpc. *Step 1*: due to the influence of WNT and BMP signals from the border region, the forebrain neurectoderm is initially dorsal in character, as shown by *Zic2* expression throughout most of the neurectoderm. SHH signals ventrally from the PrCP to the overlying neurectoderm. *Step 2*: *Shh* expression is established in the rostral-ventral neural midline, which overlays ventral identity information onto the neurectoderm. WNT and BMP signalling and *Zic2* expression become restricted to the dorsal half of the neurectoderm. *Step 3*: once neurulation is complete and the surface ectoderm overlies the neurectoderm, neural crest cells migrate in from the hindbrain and surround the dorsal neurectoderm. The dorsal-most neurectoderm cells exit the cell cycle. *Step 4*: the neural crest cells expand under the influence of WNT signalling, and the dorsal neurectoderm cells undergo apoptosis, leading to thinning and invagination of the roof-plate. *Zic2* expression initiates in the ventral midline. *Step 5*: BMP and WNT signalling in the invaginated tissues induce the cortical hem, choroid plexus epithelium and the hippocampus. (**b**) In *Zic2*$^{kd/kd}$ embryos at 9.0–9.5 dpc, WNT expression in the surface ectoderm is delayed. Consequently, dorsal neurectoderm cells do not exit the cell cycle, and neural crest infiltration is reduced. At 10.5 dpc the dorsal cells do not undergo apoptosis and invagination of the roof-plate does not occur. Though there are few neural crest cells, WNT expression is initiated. By 11.5 dpc, dorsal midline structures are absent or hypoplastic, resulting in MIHV. *CH* cortical hem, *ChPE* choroid plexus epithelium *FPl* floor-plate, *NCC* neural crest cells, *NE* neurectoderm, *PhE* pharangeal endoderm, *PrCP* Prechordal plate, *RPl* roof-plate, *SE* surface ectoderm

in concert with, or very soon after, the signals late in gastrulation that induce the anterior epiblast to adopt a neural fate. Presumably dorsal-ventral patterning of the forebrain also begins during gastrulation, and as in other A–P regions, the early forebrain neurectoderm has a dorsal character which subsequently undergoes modification to direct the secondary differentiation events of midline development. Given the timing of dorsal differentiation, it is possible that the neural crest-derived mesenchyme overlying the dorsal forebrain provides instructive signals to drive invagination of the telencephalic midline (Choe et al. 2014), similar to the role of retinoic acid in the chick (Gupta and Sen 2016). There remains much to learn about the timing and source of the various morphogenetic signals that control dorsal forebrain fate, and a proposed model of this multistep process is shown in Fig. 14.4a.

The identity of the molecular signals that instruct dorsal patterning has been elusive, perhaps because of redundancy or the iterative use of the same signalling pathway during dorsal patterning and midline structure differentiation (Fig. 14.4). Members of the BMP and WNT signalling molecule families are expressed in a spatial-temporal manner consistent with a role in dorsal neural patterning. Five BMP ligands (BMP2, BMP4–7) are expressed at the future dorsal midline in the forebrain neurectoderm and surface ectoderm. Embryos null for either *Bmp4* or *Bmp2* die before neurulation is complete (Winnier et al. 1995; Zhang and Bradley 1996), whilst *Bmp4* conditional mutants develop a phenotypically normal telencephalon (Hébert et al. 2003). Similarly, no neural phenotype is seen in *Bmp5*, *6* and *7* mutants (Kingsley et al. 1992; Dudley et al. 1995; Luo et al. 1995; Solloway et al. 1998). A role in the relatively late events of dorsal forebrain patterning is, however, revealed when the BMP-specific receptors are mutated in a time- and tissue-dependent manner. Animals that are constitutive null for *Bmpr1b* and in which *Bmpr1a* is deleted in the telencephalon at ~9.0 dpc do not show the characteristic thinning of the roof-plate at 10.5 dpc that is required for dorsal hemisphere separation. Subsequently, they exhibit MIHV HPE and loss of all dorsal midline cell types (i.e. the choroid plexus and cortical hem fail to form) despite maintenance of *Zic2* expression. The specification of ventral and cortical cell types, however, remains unaffected (Fernandes et al. 2007). Once the midline cells invaginate, BMP expression is initiated in the choroid plexus epithelium anlagen (Currle et al. 2005), and overexpression of a constitutively active BMPR1a transforms cortical precursors into choroid plexus cells (Panchision et al. 2001), suggesting BMP signalling induces choroid plexus cell fate. When the roof-plate is ablated prior to differentiation of the choriod plexus and cortical hem, these structures fail to form. The expression of some dorsal midline genes (but not *Zic2*) can be rescued in tissue explants from roof-plate-ablated embryos, however, by culture in BMP4 (Cheng et al. 2006). These experiments, therefore, establish a primary role for BMP signalling in the prevention of MIHV HPE.

Similarly, *Wnt1* and *Wnt3a* are expressed in the future dorsal neurectoderm along the length of the axis prior to neural tube closure and in the roof-plate following closure (Parr et al. 1993, Megason and McMahon 2002). In the canonical WNT signalling pathway (Fig. 14.3d), binding of WNT ligand to a cognate receptor complex stimulates a cascade of cytoplasmic events culminating in β-catenin nuclear entry. Nuclear β-catenin associates with transcription factors of the TCF/LEF family and converts target gene repression to activation (reviewed in Arkell et al. 2013). WNTs were primarily considered mitogenic signals for the neurectoderm until elevated WNT signalling was shown to alter progenitor gene expression along the dorsal-ventral axis, promoting the production of dorsal progenitors and suppressing ventral progenitors. WNT signalling acts via TCF/LEF-dependent evolutionarily conserved enhancers to establish the dorsal domain of *Gli3* expression (Yu et al.

2008). In turn, GLI3, acting as a transcriptional repressor, inhibits the ventrally produced SHH signal (Alvarez-Medina et al. 2007). This interplay of SHH and WNT signalling may also be relevant for telencephalon patterning as evidenced by the effect of differential regulation of GLI3 upon the ventral and dorsal telencephalic neuronal subtypes generated from human embryonic stem cells (Li et al. 2009). A further role for WNT signalling at later stages of roof-plate development is revealed by conditional deletion of β-catenin in the neural crest cells abutting the telencephalic neurectoderm which causes a failure of neural crest cell expansion and of telencephalic midline invagination (Choe et al. 2014).

In the forebrain, the dorsal-ventral morphogenetic signals of SHH, BMP and WNT intersect those provided by the ANR, a source of FGF ligands (FGF8, FGF15/19, FGF17 and FGF18). When FGF ligands bind to tyrosine kinase receptors (FGFR), multiple signalling pathways are activated such as the RAS/MAPK, PLC-γ, PI3K and STAT pathways (Dailey et al. 2005) (Fig. 14.3e). When FGF signalling in the telencephalon is attenuated or abolished via mutation of *Fgf8*, the FGF receptor *Fgfr1* alone or other receptor combinations, mice exhibit telencephalic hypoplasia. In embryos lacking *Fgfr1* and *Fgfr2* in the telencephalon, the decreased size of the forebrain is attributed to reduced cell proliferation in the ventral midline along with increased apoptosis in the dorsal midline (Storm et al. 2006; Gutin et al. 2006). Defects in these animals resemble those defects associated with the ventral forms of HPE. As reviewed by Hoch et al. (2009), there is a complex series of interdependencies between the forebrain signalling centres. For example, the ventral SHH signal maintains rostral midline FGF ligand expression (Ohkubo et al. 2002; Hayhurst et al. 2008), and simultaneously, the dorsal neurectoderm expression of the repressor form of GLI3 (GLI3R) represses FGF ligand dorsally (Theil et al. 1999). Similarly, the rostral source of FGF works via the *Foxg1* transcription factor (Gutin et al. 2006; Hébert and Fishell 2008) to inhibit BMP signalling, thus maintaining rostral proliferation and preventing premature differentiation of neuronal progenitor cells (Shimamura and Rubenstein 1997; Dou et al. 1999, 2000; Hanashima et al. 2004). Human genetics indicates that anterior FGF signalling impacts dorsal forebrain patterning (Dubourg et al. 2016), but despite evidence that FGF signalling can control *Zic2* expression in the 9.5 dpc telencephalon (Okada et al. 2008; Hayhurst et al. 2008), it remains unclear whether FGF control of *Zic2* expression plays a role in *Zic2*-associated MIHV.

14.8 *Zic2* Mutation and Dorsal Forebrain Patterning

Partial loss-of-function alleles of murine *Zic2* result in MIHV HPE, suggesting an involvement for ZIC2 protein in dorsal neural patterning and differentiation (Nagai et al. 2000). Late in gastrulation, as the neural plate is forming, *Zic2* expression recedes from the posterior embryo proper and becomes restricted to the anterior neurectoderm. By this time, *Zic3* and *Zic5* expression is also restricted to this region. The expression of each of these genes then subsides in the medial neural plate, becoming progressively confined to the lateral (future dorsal) neurectoderm and the flanking surface ectoderm (Elms et al. 2004; Houtmeyers et al. 2013). After the neural tube closes, high levels of *Zic2* expression can be detected along the entire anterior-posterior extent of the neural tube including the dorsal telencephalon (roof-plate and hippocampal primordium) (Cheng et al. 2006; Okada et al. 2008) with extensive overlap in this expression domain of all Zic family members. *Zic2* is therefore expressed in a manner consistent with a role in dorsal patterning, and indeed, loss-of-function mutations in *Zic2* lead to defects in neural crest development, a dorsal cell type of the hindbrain and spinal cord (Nagai et al. 2000; Elms et al. 2003).

Embryos homozygous for a hypomorphic allele of *Zic2* ($Zic2^{tm1Jaru}$ MGI:2156825; aka *kd*) in which reduced amounts of the normal transcript are produced (Nagai et al. 2000) lack a telencephalic roof-plate at mid-gestation. After neural tube closure in wild-type embryos, the dorsal midline of the telencephalon immediately becomes devoid of mitotic cells, and within 24 h, apoptosis in this tissue is noticeably higher than in the surrounding tissues (Fig. 14.4a). $Zic2^{kd/kd}$ embryos exhibit neither of these features, and consequently, roof-plate thinning and invagination does not occur. Subsequently, the structures that should be derived from the dorsal midline are either severely hypoplastic or absent (Fig. 14.4b). At this stage of development, *Wnt3a* should be expressed in the dorsal midline of the forebrain and along the spinal cord until the position of the forelimb bud. In $Zic2^{kd/kd}$ embryos, this expression is delayed such that it has only just been initiated in the dorsal forebrain (Nagai et al. 2000). This work firmly connects ZIC2 function to the MIHV form of HPE, but many questions remain unanswered regarding the role of ZIC2 in roof-plate formation. For example, it is not known precisely when the process of dorsal patterning and roof-plate induction fails in $Zic2^{kd/kd}$ embryos. This may reflect a role for ZIC2 in earlier dorsal patterning events rather than roof-plate induction, per se. It is also not clear whether the expression of *Zic2* in the neurectoderm, flanking surface ectoderm or both tissues is required for roof-plate formation. Alternatively, it is possible that the documented role of ZIC2 in neural crest cell development is important since in $Zic2^{kd/kd}$ embryos there is an evident lack of neural crest cells in the intrahemispheric mesenchyme. Furthermore, ZIC2 is known to be able to physically interact with transcriptional mediators of the WNT, TGF-β and SHH pathways (Koyabu et al. 2001; Pourebrahim et al. 2011; Fujimi et al. 2012; Houtmeyers et al. 2016), and whether it does so during dorsal patterning is unknown. The experiments on $Zic2^{kd/kd}$ embryos highlight that roof-plate formation and dorsal midline development are particularly sensitive to loss of ZIC2 levels, with a reduction of *Zic2* expression to ~20% sufficient to generate MIHV HPE in mice (Nagai et al. 2000).

14.9 Conclusion

Current knowledge indicates that ZIC2 plays two distinct roles in forebrain development. At mid-gastrulation, ZIC2 functions at the node to shape the NODAL gradient. Severe *Zic2* loss-of-function mutations lead to a transient failure in the production of the ANC, and consequently, PrCP fate is not stabilised. As such, the SHH signal is not sent from the PrCP, and the ventral forebrain signalling centre is not established, resulting in gross perturbations in forebrain D-V pattern and classic HPE. This requirement appears well buffered, as mild loss-of-function is compatible with correct function of the ventral forebrain signalling centre in man and mouse. A second requirement for ZIC2 function occurs in the dorsal neurectoderm of the developing telencephalon where it acts to promote formation of the dorsal signalling centre responsible for roof-plate and choroid plexus development. It is clear that a failure of ZIC2 function in the developing dorsal forebrain results in MIHV, but just how ZIC2 intersects the interconnected forebrain signalling network is unknown. This requirement appears more sensitive to ZIC2 levels than ventral forebrain patterning, as hypomorphic mutations have been associated with MIHV in both man and mouse (Nagai et al. 2000; Brown et al. 2001, 2005; Solomon et al. 2010a). The precise molecular role of ZIC2 in both classic and MIHV HPE, however, remains a subject of ongoing investigation. It will be particularly interesting to determine whether it is ZIC transcription factor or cofactor activity that is essential for the relevant steps in forebrain patterning and tissue differentiation.

References

Abe Y, Oka A, Mizuguchi M et al (2009) EYA4, deleted in a case with middle interhemispheric variant of holoprosencephaly, interacts with SIX3 both physically and functionally. Hum Mutat 30:946–955. https://doi.org/10.1002/humu.21094

Alvarez-Medina R, Cayuso J, Okubo T et al (2007) Wnt canonical pathway restricts graded Shh/Gli patterning activity through the regulation of Gli3 expression. Development 135:237–247. https://doi.org/10.1242/dev.012054

Anderson RM, Lawrence AR, Stottmann RW et al (2002) Chordin and noggin promote organizing centers of forebrain development in the mouse. Development 129:4975–4987

Andersson O, Reissmann E, Jörnvall H, Ibáñez CF (2006) Synergistic interaction between Gdf1 and Nodal during anterior axis development. Dev Biol 293:370–381. https://doi.org/10.1016/j.ydbio.2006.02.002

Ang SL, Rossant J (1994) HNF-3 beta is essential for node and notochord formation in mouse development. Cell 78:561–574

Ang SL, Wierda A, Wong D et al (1993) The formation and maintenance of the definitive endoderm lineage in the mouse: involvement of HNF3/forkhead proteins. Development 119:1301–1315

Aoto K, Shikata Y, Higashiyama D, Shiota K, Motoyama J (2008) Fetal ethanol exposure activates protein kinase a and impairsShh expression in prechordal mesendoderm cells in the pathogenesis of holoprosencephaly. Birth defects research part A: clinical and molecular. Teratology 82(4):224–231

Aoto K, Shikata Y, Imai H et al (2009) Mouse Shh is required for prechordal plate maintenance during brain and craniofacial morphogenesis. Dev Biol 327:106–120. https://doi.org/10.1016/j.ydbio.2008.11.022

Arauz RF, Solomon BD, Pineda-Alvarez DE et al (2010) A hypomorphic allele in the FGF8 gene contributes to holoprosencephaly and is allelic to gonadotropin-releasing hormone deficiency in humans. Mol Syndromol 1:59–66. https://doi.org/10.1159/000302285

Arkell R, Beddington RS (1997) BMP-7 influences pattern and growth of the developing hindbrain of mouse embryos. Development 124(1):1–12

Arkell RM, Tam PPL (2012) Initiating head development in mouse embryos: integrating signalling and transcriptional activity. Open Biol 2:120030–120030. https://doi.org/10.1098/rsob.120030

Arkell RM, Fossat N, Tam PPL (2013) Wnt signalling in mouse gastrulation and anterior development: new players in the pathway and signal output. Curr Opin Genet Dev 23:454–460. https://doi.org/10.1016/j.gde.2013.03.001

Bae G-U, Domené S, Roessler E et al (2011) Mutations in CDON, encoding a hedgehog receptor, result in holoprosencephaly and defective interactions with other hedgehog receptors. Am J Hum Genet 89:231–240. https://doi.org/10.1016/j.ajhg.2011.07.001

Barkovich AJ, Quint DJ (1993) Middle interhemispheric fusion: an unusual variant of holoprosencephaly. Am J Neuroradiol 14:431–440

Barr M, Cohen MM (2002) Autosomal recessive alobar holoprosencephaly with essentially normal faces. Am J Med Genet 112:28–30. https://doi.org/10.1002/ajmg.10587

Barratt KS, Glanville-Jones HC, Arkell RM (2014) The Zic2 gene directs the formation and function of node cilia to control cardiac situs. Genesis 52:626–635. https://doi.org/10.1002/dvg.22767

Belloni E, Muenke M, Roessler E, Traverse G, Siegel-Bartelt J, Frumkin A, Mitchell HF, Donis-Keller H, Helms C, Hing AV, Heng HHQ, Koop B, Martindale D, Rommens JM, Tsui L-C, Scherer SW (1996) Identification of Sonic hedgehog as a candidate gene responsible for holoprosencephaly. Nat Genet 14(3):353–356

Brown S, Warburton D, Brown LY et al (1998) Holoprosencephaly due to mutations in ZIC2, a homologue of Drosophila odd-paired. Nat Genet 20:180–183. https://doi.org/10.1038/2484

Brown LY, Odent S, David V et al (2001) Holoprosencephaly due to mutations in ZIC2: alanine tract expansion mutations may be caused by parental somatic recombination. Hum Mol Genet 10:791–796

Brown LY, Paraso M, Arkell RM, Brown S (2005) In vitro analysis of partial loss-of-function ZIC2 mutations in holoprosencephaly: alanine tract expansion modulates DNA binding and transactivation. Hum Mol Genet 14:411–420. https://doi.org/10.1093/hmg/ddi037

Camus A, Davidson BP, Billiards S et al (2000) The morphogenetic role of midline mesendoderm and ectoderm in the development of the forebrain and the midbrain of the mouse embryo. Development 127:1799–1813

Cheng X, Hsu C, Currle DS et al (2006) Central roles of the roof plate in telencephalic development and holoprosencephaly. J Neurosci 26:7640–7649. https://doi.org/10.1523/JNEUROSCI.0714-06.2006

Chiang C, Litingtung Y, Lee E et al (1996) Cyclopia and defective axial patterning in mice lacking sonic hedgehog gene function. Nature 383:407–413

Choe Y, Zarbalis KS, Pleasure SJ et al (2014) Neural crest-derived mesenchymal cells require Wnt signaling for their development and drive invagination of the telencephalic midline. PLoS One 9:e86025. https://doi.org/10.1371/journal.pone.0086025

Chu GC, Dunn NR, Anderson DC et al (2004) Differential requirements for Smad4 in TGFbeta-dependent patterning of the early mouse embryo. Development 131:3501–3512. https://doi.org/10.1242/dev.01248

Crossley PH, Martin GR (1995) The mouse Fgf8 gene encodes a family of polypeptides and is expressed in regions that direct outgrowth and patterning in the developing embryo. Development 121:439–451

Currle DS, Cheng X, Hsu C-M, Monuki ES (2005) Direct and indirect roles of CNS dorsal midline cells in choroid plexus epithelia formation. Development 132:3549–3559. https://doi.org/10.1242/dev.01915

Dailey L, Ambrosetti D, Mansukhani A, Basilico C (2005) Mechanisms underlying differential responses to FGF signaling. Cytokine Growth Factor Rev 16:233–247. https://doi.org/10.1016/j.cytogfr.2005.01.007

Dale JK, Vesque C, Lints TJ et al (1997) Cooperation of BMP7 and SHH in the induction of forebrain ventral midline cells by prechordal mesoderm. Cell 90:257–269

de la Cruz JM, Bamford RN, Burdine RD et al (2002) A loss-of-function mutation in the CFC domain of TDGF1 is associated with human forebrain defects. Hum Genet 110:422–428. https://doi.org/10.1007/s00439-002-0709-3

DeMyer W, Zeman W, Palmer CG (1964) The face predicts the brain: diagnostic significance of median facial anomalies for holoprosencephaly (arhinencephaly). Pediatrics 34:256–263

Dou CL, Li S, Lai E (1999) Dual role of brain factor-1 in regulating growth and patterning of the cerebral hemispheres. Cereb Cortex 9:543–550

Dou C, Lee J, Liu B et al (2000) BF-1 interferes with transforming growth factor beta signaling by associating with Smad partners. Mol Cell Biol 20:6201–6211

Dubourg C, Bendavid C, Pasquier L et al (2007) Holoprosencephaly. Orphanet J Rare Dis 2:8. https://doi.org/10.1186/1750-1172-2-8

Dubourg C, David V, Gropman A et al (2011) Clinical utility gene card for: holoprosencephaly. Eur J Hum Genet 19:preceeding 118–120. https://doi.org/10.1038/ejhg.2010.110

Dubourg C, Carré W, Hamdi-Rozé H et al (2016) Mutational spectrum in holoprosencephaly shows that FGF is a new major signaling pathway. Hum Mutat 37:1329–1339

Dudley AT, Lyons KM, Robertson EJ (1995) A requirement for bone morphogenetic protein-7 during development of the mammalian kidney and eye. Genes Dev 9:2795–2807. https://doi.org/10.1101/GAD.9.22.2795

Dufort D, Schwartz L, Harpal K, Rossant J (1998) The transcription factor HNF3beta is required in visceral endoderm for normal primitive streak morphogenesis. Development 125:3015–3025. doi: 7588062

Dunn N, Vincent S, Oxburgh L et al (2004) Combinatorial activities of Smad2 and Smad3 regulate mesoderm formation and patterning in the mouse embryo. Development 131:1717–1728. https://doi.org/10.1242/dev.01072

Dupé V, Rochard L, Mercier S et al (2011) NOTCH, a new signaling pathway implicated in holoprosencephaly. Hum Mol Genet 20:1122–1131. https://doi.org/10.1093/hmg/ddq556

Echelard Y, Epstein DJ, St-Jacques B et al (1993) Sonic hedgehog, a member of a family of putative signaling molecules, is implicated in the regulation of CNS polarity. Cell 75:1417–1430

Elms P, Siggers P, Napper D et al (2003) Zic2 is required for neural crest formation and hindbrain patterning during mouse development. Dev Biol 264:391–406. https://doi.org/10.1016/j.ydbio.2003.09.005

Elms P, Scurry A, Davies J et al (2004) Overlapping and distinct expression domains of Zic2 and Zic3 during mouse gastrulation. Gene Expr Patterns 4:505–511. https://doi.org/10.1016/j.modgep.2004.03.003

Episkopou V, Arkell R, Timmons PM et al (2001) Induction of the mammalian node requires Arkadia function in the extraembryonic lineages. Nature 410:825–830. https://doi.org/10.1038/35071095

Fernandes M, Gutin G, Alcorn H et al (2007) Mutations in the BMP pathway in mice support the existence of two molecular classes of holoprosencephaly. Development 134:3789–3794. https://doi.org/10.1242/dev.004325

Fliegauf M, Benzing T, Omran H (2007) When cilia go bad: cilia defects and ciliopathies. Nat Rev Mol Cell Biol 8:880–893. https://doi.org/10.1038/nrm2278

Fuccillo M, Rallu M, McMahon AP, Fishell G (2004) Temporal requirement for hedgehog signaling in ventral telencephalic patterning. Development 131(20):5031–5040

Fujimi TJ, Hatayama M, Aruga J (2012) Xenopus Zic3 controls notochord and organizer development through suppression of the Wnt/β-catenin signaling pathway. Dev Biol 361:220–231. https://doi.org/10.1016/j.ydbio.2011.10.026

Furushima K, Murata T, Matsuo I, Aizawa S (2000) A new murine zinc finger gene, Opr. Mech Dev 98:161–164. https://doi.org/10.1016/S0925-4773(00)00456-1

Furuta Y, Piston DW, Hogan BL (1997) Bone morphogenetic proteins (BMPs) as regulators of dorsal forebrain development. Development 124:2203–2212

Goulding MD, Chalepakis G, Deutsch U et al (1991) Pax-3, a novel murine DNA binding protein expressed during early neurogenesis. EMBO J 10:1135–1147

Gripp KW, Wotton D, Edwards MC et al (2000) Mutations in TGIF cause holoprosencephaly and link NODAL signalling to human neural axis determination. Nat Genet 25:205–208. https://doi.org/10.1038/76074

Groves AK, LaBonne C (2014) Setting appropriate boundaries: fate, patterning and competence at the neural plate border. Dev Biol 389:2–12. https://doi.org/10.1016/j.ydbio.2013.11.027

Gupta S, Sen J (2016) Roof plate mediated morphogenesis of the forebrain: new players join the game. Dev Biol 413:145–152. https://doi.org/10.1016/j.ydbio.2016.03.019

Gutin G, Fernandes M, Palazzolo L et al (2006) FGF signalling generates ventral telencephalic cells independently of SHH. Development 133:2937–2946. https://doi.org/10.1242/dev.02465

Hallonet M, Kaestner KH, Martin-Parras L et al (2002) Maintenance of the specification of the anterior definitive endoderm and forebrain depends on the axial mesendoderm: a study using HNF3β/Foxa2 conditional mutants. Dev Biol 243:20–33. https://doi.org/10.1006/dbio.2001.0536

Hanashima C, Li SC, Shen L et al (2004) Foxg1 suppresses early cortical cell fate. Science 303:56–59. https://doi.org/10.1126/science.1090674

Hayhurst M, Gore BB, Tessier-Lavigne M, McConnell SK (2008) Ongoing sonic hedgehog signaling is required for dorsal midline formation in the developing forebrain. Dev Neurobiol 68:83–100. https://doi.org/10.1002/dneu.20576

Hébert JM, Mishina Y, McConnell SK (2002) BMP signaling is required locally to pattern the dorsal telencephalic midline. Neuron. https://doi.org/10.1016/S0896-6273(02)00900-5

Hébert JM, Fishell G (2008) The genetics of early telencephalon patterning: some assembly required. Nat Rev Neurosci 9(9):678–685

Hébert JM, Hayhurst M, Marks ME et al (2003) BMP ligands act redundantly to pattern the dorsal telencephalic midline. Genesis 35:214–219. https://doi.org/10.1002/gene.10183

Hoch RV, Rubenstein JLR, Pleasure S (2009) Genes and signaling events that establish regional patterning of the mammalian forebrain. Semin Cell Dev Biol 20:378–386

Hong M, Krauss RS, Hannenhalli S et al (2012) Cdon mutation and fetal ethanol exposure synergize to produce midline signaling defects and holoprosencephaly spectrum disorders in mice. PLoS Genet 8:e1002999. https://doi.org/10.1371/journal.pgen.1002999

Hoodless PA, Pye M, Chazaud C et al (2001) FoxH1 (Fast) functions to specify the anterior primitive streak in the mouse. Genes Dev 15:1257–1271. https://doi.org/10.1101/gad.881501

Houtmeyers R, Souopgui J, Tejpar S, Arkell RM (2013) The ZIC gene family encodes multi-functional proteins essential for patterning and morphogenesis. Cell Mol Life Sci 70:3791–3811. https://doi.org/10.1007/s00018-013-1285-5

Houtmeyers R, Tchouate Gainkam O, Glanville-Jones HC et al (2016) Zic2 mutation causes Holoprosencephaly via disruption of NODAL signalling. Hum Mol Genet 61:ddw235. https://doi.org/10.1093/hmg/ddw235

Ip CK, Fossat N, Jones V et al (2014) Head formation: OTX2 regulates Dkk1 and Lhx1 activity in the anterior mesendoderm. Development 141:3859–3867. https://doi.org/10.1242/dev.114900

Jeong Y, Epstein DJ (2003) Distinct regulators of Shh transcription in the floor plate and notochord indicate separate origins for these tissues in the mouse node. Development 130:3891–3902. https://doi.org/10.1242/dev.00590

Kietzman HW, Everson JL, Sulik KK, Lipinski RJ (2014) The teratogenic effects of prenatal ethanol exposure are exacerbated by sonic hedgehog or Gli2 haploinsufficiency in the mouse. PLoS One 9:e89448. https://doi.org/10.1371/journal.pone.0089448

Kinder SJ, Tsang TE, Wakamiya M et al (2001) The organizer of the mouse gastrula is composed of a dynamic population of progenitor cells for the axial mesoderm. Development 128:3623–3634
Kingsley DM, Bland AE, Grubber JM et al (1992) The mouse short ear skeletal morphogenesis locus is associated with defects in a bone morphogenetic member of the TGF beta superfamily. Cell 71:399–410
Kinzler KW, Vogelstein B (1990) The GLI gene encodes a nuclear protein which binds specific sequences in the human genome. Mol Cell Biol 10:634–642
Koyabu Y, Nakata K, Mizugishi K et al (2001) Physical and functional interactions between Zic and Gli proteins. J Biol Chem 276:6889–6892. https://doi.org/10.1074/jbc.C000773200
Lacbawan F, Solomon BD, Roessler E et al (2009) Clinical spectrum of SIX3-associated mutations in holoprosencephaly: correlation between genotype, phenotype and function. J Med Genet 46:389–398. https://doi.org/10.1136/jmg.2008.063818
Le Dréau G, Martí E (2012) Dorsal-ventral patterning of the neural tube: a tale of three signals. Dev Neurobiol 72:1471–1481. https://doi.org/10.1002/dneu.22015
Lewis AJ, Simon EM, Barkovich AJ et al (2002) Middle interhemispheric variant of holoprosencephaly: a distinct cliniconeuroradiologic subtype. Neurology 59:1860–1865
Li X-J, Zhang X, Johnson MA et al (2009) Coordination of sonic hedgehog and Wnt signaling determines ventral and dorsal telencephalic neuron types from human embryonic stem cells. Development 136:4055–4063. https://doi.org/10.1242/dev.036624
Liem KF, Tremml G, Roelink H, Jessell TM (1995) Dorsal differentiation of neural plate cells induced by BMP-mediated signals from epidermal ectoderm. Cell 82:969–979
Liu Y, Festing M, Thompson JC et al (2004) Smad2 and Smad3 coordinately regulate craniofacial and endodermal development. Dev Biol 270:411–426. https://doi.org/10.1016/j.ydbio.2004.03.017
Luo G, Hofmann C, Bronckers AL et al (1995) BMP-7 is an inducer of nephrogenesis, and is also required for eye development and skeletal patterning. Genes Dev 9:2808–2820
Marcorelles P, Laquerriere A (2010) Neuropathology of holoprosencephaly. Am J Med Genet C Semin Med Genet 154C:109–119. https://doi.org/10.1002/ajmg.c.30249
Matsunaga E, Shiota K (1977) Holoprosencephaly in human embryos: epidemiologic studies of 150 cases. Teratology 16:261–272. https://doi.org/10.1002/tera.1420160304
Megason SG, McMahon AP (2002) A mitogen gradient of dorsal midline Wnts organizes growth in the CNS. Development 129:2087–2098
Mercier S, Dubourg C, Garcelon N et al (2011) New findings for phenotype-genotype correlations in a large European series of holoprosencephaly cases. J Med Genet 48:752–760. https://doi.org/10.1136/jmedgenet-2011-100339
Ming JE, Kaupas ME, Roessler E et al (2002) Mutations in PATCHED-1, the receptor for SONIC HEDGEHOG, are associated with holoprosencephaly. Hum Genet 110:297–301. https://doi.org/10.1007/s00439-002-0695-5
Mizugishi K, Aruga J, Nakata K, Mikoshiba K (2001) Molecular properties of Zic proteins as transcriptional regulators and their relationship to GLI proteins. J Biol Chem 276:2180–2188. https://doi.org/10.1074/jbc.M004430200
Monaghan AP, Kaestner KH, Grau E, Schütz G (1993) Postimplantation expression patterns indicate a role for the mouse forkhead/HNF-3 alpha, beta and gamma genes in determination of the definitive endoderm, chordamesoderm and neuroectoderm. Development 119:567–578
Mouden C, Dubourg C, Carré W et al (2016) Complex mode of inheritance in holoprosencephaly revealed by whole exome sequencing. Clin Genet. https://doi.org/10.1111/cge.12722
Nagai T, Aruga J, Takada S et al (1997) The expression of the mouse Zic1, Zic2, and Zic3 gene suggests an essential role for Zic genes in body pattern formation. Dev Biol 182:299–313. https://doi.org/10.1006/dbio.1996.8449
Nagai T, Aruga J, Minowa O et al (2000) Zic2 regulates the kinetics of neurulation. Proc Natl Acad Sci U S A 97:1618–1623. https://doi.org/10.1073/pnas.97.4.1618
Nakayama J, Kinugasa H, Ohto T et al (2016) Monozygotic twins with de novo ZIC2 gene mutations discordant for the type of holoprosencephaly. Neurology 86:1456–1458. https://doi.org/10.1212/WNL.0000000000002567

Nanni L, Ming JE, Bocian M et al (1999) The mutational spectrum of the sonic hedgehog gene in holoprosencephaly: SHH mutations cause a significant proportion of autosomal dominant holoprosencephaly. Hum Mol Genet 8:2479–2488. https://doi.org/10.1093/hmg/8.13.2479

Norris DP, Brennan J, Bikoff EK, Robertson EJ (2002) The Foxh1-dependent autoregulatory enhancer controls the level of Nodal signals in the mouse embryo. Development 129:3455–3468

Odent S, Le Marec B, Munnich A et al (1998) Segregation analysis in nonsyndromic holoprosencephaly. Am J Med Genet 77:139–143

Ohkubo Y, Chiang C, Rubenstein JLR (2002) Coordinate regulation and synergistic actions of BMP4, SHH and FGF8 in the rostral prosencephalon regulate morphogenesis of the telencephalic and optic vesicles. Neuroscience 111:1–17

Okada T, Okumura Y, Motoyama J, Ogawa M (2008) FGF8 signaling patterns the telencephalic midline by regulating putative key factors of midline development. Dev Biol 320:92–101. https://doi.org/10.1016/j.ydbio.2008.04.034

Orioli IM, Castilla EE (2010) Epidemiology of holoprosencephaly: prevalence and risk factors. Am J Med Genet C Semin Med Genet 154:13–21

Panchision DM, Pickel JM, Studer L et al (2001) Sequential actions of BMP receptors control neural precursor cell production and fate. Genes Dev 15:2094–2110. https://doi.org/10.1101/gad.894701

Parr BA, Shea MJ, Vassileva G, McMahon AP (1993) Mouse Wnt genes exhibit discrete domains of expression in the early embryonic CNS and limb buds. Development 119:247–261

Paulussen ADC, Schrander-Stumpel CT, Tserpelis DCJ et al (2010) The unfolding clinical spectrum of holoprosencephaly due to mutations in SHH, ZIC2, SIX3 and TGIF genes. Eur J Hum Genet 18:999–1005. https://doi.org/10.1038/ejhg.2010.70

Pavletich NP, Pabo CO (1993) Crystal structure of a five-finger GLI-DNA complex: new perspectives on zinc fingers. Science 261:1701–1707

Petryk A, Graf D, Marcucio R (2015) Holoprosencephaly: signaling interactions between the brain and the face, the environment and the genes, and the phenotypic variability in animal models and humans. Wiley Interdiscip Rev Dev Biol 4:17–32

Pineda-Alvarez DE, Roessler E, Hu P et al (2012) Missense substitutions in the GAS1 protein present in holoprosencephaly patients reduce the affinity for its ligand, SHH. Hum Genet 131:301–310. https://doi.org/10.1007/s00439-011-1078-6

Placzek M, Briscoe J (2005) The floor plate: multiple cells, multiple signals. Nat Rev Neurosci 6:230–240. https://doi.org/10.1038/nrn1628

Pourebrahim R, Houtmeyers R, Ghogomu S et al (2011) Transcription factor Zic2 inhibits Wnt/Beta-catenin protein signaling. J Biol Chem 286:37732–37740. https://doi.org/10.1074/jbc.M111.242826

Ribeiro LA, Quiezi RG, Nascimento A et al (2010) Holoprosencephaly and holoprosencephaly-like phenotype and GAS1 DNA sequence changes: report of four Brazilian patients. Am J Med Genet A 152A:1688–1694. https://doi.org/10.1002/ajmg.a.33466

Ribeiro LA, Roessler E, Hu P et al (2012) Comparison of mutation findings in ZIC2 between microform and classical holoprosencephaly in a Brazilian cohort. Birth Defects Res A Clin Mol Teratol 94:912–917. https://doi.org/10.1002/bdra.23047

Roessler E, Muenke M (2001) Midline and laterality defects: left and right meet in the middle. BioEssays 23:888–900. https://doi.org/10.1002/bies.1130

Roessler E, Muenke M (2010) The molecular genetics of holoprosencephaly. Am J Med Genet C Semin Med Genet 154:52–61

Roessler E, Belloni E, Gaudenz K, Jay P, Berta P, Scherer SW, Tsui L-C, Muenke M (1996) Mutations in the human Sonic Hedgehog gene cause holoprosencephaly. Nat Genet 14(3):357–360

Roessler E, Lacbawan F, Dubourg C et al (2009a) The full spectrum of holoprosencephaly-associated mutations within the ZIC2 gene in humans predicts loss-of-function as the predominant disease mechanism. Hum Mutat. https://doi.org/10.1002/humu.20982.The

Roessler E, Ma Y, Ouspenskaia MV et al (2009b) Truncating loss-of-function mutations of DISP1 contribute to holoprosencephaly-like microform features in humans. Hum Genet 125:393–400. https://doi.org/10.1007/s00439-009-0628-7

Roessler E, Pei W, Ouspenskaia MV et al (2009c) Cumulative ligand activity of NODAL mutations and modifiers are linked to human heart defects and holoprosencephaly. Mol Genet Metab 98:225–234. https://doi.org/10.1016/j.ymgme.2009.05.005

Roessler E, Hu P, Hong SK et al (2012a) Unique alterations of an ultraconserved non-coding element in the 3′UTR of ZIC2 in holoprosencephaly. PLoS One 7:3–8. https://doi.org/10.1371/journal.pone.0039026

Roessler E, Vélez JI, Zhou N, Muenke M (2012b) Utilizing prospective sequence analysis of SHH, ZIC2, SIX3 and TGIF in holoprosencephaly probands to describe the parameters limiting the observed frequency of mutant gene x gene interactions. Mol Genet Metab 105:658–664. https://doi.org/10.1016/j.ymgme.2012.01.005

Ruiz i Altaba A, Prezioso VR, Darnell JE, Jessell TM (1993) Sequential expression of HNF-3 beta and HNF-3 alpha by embryonic organizing centers: the dorsal lip/node, notochord and floor plate. Mech Dev 44:91–108

Sasaki H, Hogan BL (1993) Differential expression of multiple fork head related genes during gastrulation and axial pattern formation in the mouse embryo. Development 118:47–59

Schachter K A, Krauss RS (2008) Murine models of holoprosencephaly, 1st edn. Elsevier

Serbedzija GN, Fraser SE, Bronner-Fraser M (1990) Pathways of trunk neural crest cell migration in the mouse embryo as revealed by vital dye labelling. Development 108:605–612

Shimamura K, Rubenstein JL (1997) Inductive interactions direct early regionalization of the mouse forebrain. Development 124:2709–2718

Simon EM, Hevner RF, Pinter JD et al (2002) The middle interhemispheric variant of holoprosencephaly. Am J Neuroradiol 23:151–155

Solloway MJ, Dudley AT, Bikoff EK et al (1998) Mice lacking Bmp6 function. Dev Genet 22:321–339. https://doi.org/10.1002/(SICI)1520-6408(1998)22:4<321::AID-DVG3>3.0.CO;2-8

Solomon BD, Gropman A, Muenke M (1993) Holoprosencephaly overview. University of Washington, Seattle

Solomon BD, Lacbawan F, Mercier S et al (2010a) Mutations in ZIC2 in human holoprosencephaly: description of a novel ZIC2 specific phenotype and comprehensive analysis of 157 individuals. J Med Genet 47:513–524. https://doi.org/10.1136/jmg.2009.073049

Solomon BD, Mercier S, Vélez JI et al (2010b) Analysis of genotype-phenotype correlations in human holoprosencephaly. Am J Med Genet C Semin Med Genet 154:133–141

Spoelgen R, Hammes A, Anzenberger U et al (2005) LRP2/megalin is required for patterning of the ventral telencephalon. Development 132:405–414. https://doi.org/10.1242/dev.01580

Storm EE, Garel S, Borello U et al (2006) Dose-dependent functions of Fgf8 in regulating telencephalic patterning centers. Development 133:1831–1844. https://doi.org/10.1242/dev.02324

Theil T, Alvarez-Bolado G, Walter A, Rüther U (1999) Gli3 is required for Emx gene expression during dorsal telencephalon development. Development 126:3561–3571. https://doi.org/10.1038/ng1196-353

Vincent SD, Dunn NR, Hayashi S et al (2003) Cell fate decisions within the mouse organizer are governed by graded Nodal signals. Genes Dev 17:1646–1662. https://doi.org/10.1101/gad.1100503

Warr N, Powles-Glover N, Chappell A et al (2008) Zic2-associated holoprosencephaly is caused by a transient defect in the organizer region during gastrulation. Hum Mol Genet 17:2986–2996. https://doi.org/10.1093/hmg/ddn197

Waters AM, Beales PL (2011) Ciliopathies: an expanding disease spectrum. Pediatr Nephrol 26:1039–1056. https://doi.org/10.1007/s00467-010-1731-7

Weiss K, Kruszka P, Guillen Sacoto MJ et al (2017) In-depth investigations of adolescents and adults with holoprosencephaly identify unique characteristics. Genet Med. https://doi.org/10.1038/gim.2017.68

Winnier G, Blessing M, Labosky PA, Hogan BL (1995) Bone morphogenetic protein-4 is required for mesoderm formation and patterning in the mouse. Genes Dev 9:2105–2116

Xavier GM, Seppala M, Barrell W et al (2016) Hedgehog receptor function during craniofacial development. Dev Biol 415:198–215

Ybot-Gonzalez P, Gaston-Massuet C, Girdler G et al (2007) Neural plate morphogenesis during mouse neurulation is regulated by antagonism of Bmp signalling. Development 134:3203–3211. https://doi.org/10.1242/dev.008177

Yu W, McDonnell K, Taketo MM, Bai CB (2008) Wnt signaling determines ventral spinal cord cell fates in a time-dependent manner

Zhang H, Bradley A (1996) Mice deficient for BMP2 are nonviable and have defects in amnion/chorion and cardiac development. Development 122:2977–2986

Chapter 15
ZIC3 in Heterotaxy

Helen M. Bellchambers and Stephanie M. Ware

Abstract Mutation of ZIC3 causes X-linked heterotaxy, a syndrome in which the laterality of internal organs is disrupted. Analysis of model organisms and gene expression during early development suggests ZIC3-related heterotaxy occurs due to defects at the earliest stage of left-right axis formation. Although there are data to support abnormalities of the node and cilia as underlying causes, it is unclear at the molecular level why loss of ZIC3 function causes such these defects. ZIC3 has putative roles in a number of developmental signalling pathways that have distinct roles in establishing the left-right axis. This complicates the understanding of the mechanistic basis of Zic3 in early development and left-right patterning. Here we summarise our current understanding of ZIC3 function and describe the potential role ZIC3 plays in important signalling pathways and their links to heterotaxy.

Keywords Left-right patterning · Gastrulation · Mutation · Planar cell polarity · Node · Cilia

15.1 Introduction

ZIC3 was the first ZIC family member identified as a cause of a human developmental disorder. This finding was facilitated by the fact that ZIC3 is located on the X chromosome and therefore causes disease in an X-linked inheritance pattern in which males are affected. Since its initial identification as the cause of X-linked heterotaxy 20 years ago, there has been significant progress in understanding its role in human congenital anomalies, importance in left-right (LR) patterning and interaction with multiple developmental signalling pathways. In this chapter, we review the current understanding of ZIC3 function and gaps that remain to be investigated.

H. M. Bellchambers
Department of Pediatrics, Indiana University School of Medicine, Indianapolis, IN, USA

S. M. Ware (✉)
Department of Pediatrics, Indiana University School of Medicine, Indianapolis, IN, USA

Department of Medical and Molecular Genetics, Indiana University School of Medicine, Indianapolis, IN, USA
e-mail: stware@iu.edu

J. Aruga (ed.), *Zic family*, Advances in Experimental Medicine and Biology 1046,
https://doi.org/10.1007/978-981-10-7311-3_15

15.2 Heterotaxy

The internal organs show asymmetry when comparing left and right sides of the body. Such asymmetry is identified in organ placement, with organs such as the heart, stomach and spleen found on the left of the body, whereas the liver and gallbladder are located on the right. Insults to early development alter such asymmetry causing a spectrum of disorders from isolated organ abnormalities to complete reversal (termed situs inversus). These abnormalities include loss of normally asymmetric structures, improper symmetry or lateralisation resulting in organ isomerism and/or failure to regress symmetric embryonic structures (e.g. persistent left superior vena cava). Heterotaxy, or situs ambiguus, is a syndrome consisting of multiple such congenital anomalies without complete mirror image reversal.

'Classic' heterotaxy is defined as situs abnormalities in at least two organs or tissues. Historically, it has been further subdivided into a variety of subgroups based on spleen status (polysplenia or asplenia) or type of symmetry (right isomerism, left isomerism). However, the range of anatomic defects that occur in heterotaxy often defies such simplistic classification. For example, the heart defects identified in heterotaxy are tremendously variable ranging from simple septal defects (atrial septal defect, ventricular septal defect), atrioventricular canal defects, conotruncal anomalies (e.g. double outlet left ventricle, d-transposition of the great arteries), ventricular inversion and l-transposition of the great arteries; isomerism of the atria are sometimes present (right atrial isomerism, left atrial isomerism), with an effect on the conduction system; and anomalies of the vessels are also common (e.g. right aortic arch, interrupted inferior vena cava, anomalous pulmonary venous return). Heterotaxy has a population incidence of approximately 1/10,000, with a greater incidence in Asian (Kim et al. 2008) and African American (Correa-Villasenor et al. 1991) races, and a slightly higher prevalence in females (0.9 per 10,000 births) than males (0.7 per 10,000 births) (Lin et al. 2014). It accounts for 3% of all congenital heart defects (CHDs), and of all cardiovascular malformations, it is the most highly heritable (Oyen et al. 2009). An estimated 10% of heterotaxy patients have a close family history of CHDs. Although the genetic cause for heterotaxy in a family is not often identified, it is known from studying X-linked heterotaxy, caused by mutations in ZIC3, that the same mutation that causes heterotaxy in one patient can cause isolated CHD in another.

At the chromosome level, causes of heterotaxy include aneuploidy (trisomy 13 and trisomy 18), small chromosomal rearrangements (inversions, unbalanced translocations) and copy number variants (CNVs: submicroscopic deletions or duplications) with CNVs estimated to explain 20–30% of heterotaxy cases (Fakhro et al. 2011; Rigler et al. 2015). The inheritance patterns of heterotaxy include sporadic, autosomal recessive, autosomal dominant and X-linked. Patients may have two causative loci (Bamford et al. 2000) or exhibit oligogenic inheritance which may partially explain the phenotypic diversity and variable penetrance within families. Additional investigation is required to better understand complex inheritance. Most pathogenic single-gene mutations are dominantly inherited with reduced penetrance, but 6.3%–12.1% of patients with primary ciliary dyskinesia have heterotaxy (Kennedy et al. 2007; Shapiro et al. 2014) which is primarily inherited in an autosomal recessive manner. Genes known to cause heterotaxy include *Nodal*, *ZIC3* (3% of sporadic cases; >75% of familial cases), *CFC1*, *CRELD1*, *FOXH1*, *SESN1*, *LEFTYA*, *GDF1*, *ACVR2B* and *NKX2.5* (Cowan et al. 2014; Sutherland and Ware 2009). Of these genes only *ZIC3* has been assessed in large enough cohorts to give reliable estimates of the rate of mutation in heterotaxy patients; similar estimates in the other genes remain to be determined.

15.3 Establishment of LR Asymmetry

LR asymmetry initiates at a structure termed the embryonic node in mouse with similar structures found in zebrafish (Kupffer's vesicle), *Xenopus* (gastrocoel roof plate), chick (Hensen's node) and rabbit (posterior notochordal plate) (Blum et al. 2014). Formation of these structures requires several transcription factors (including *Brachyury*, *Noto*, *Foxa2* and *Zic3*) and several major signalling pathways. The node forms from epiblast cells in the anterior primitive streak at the midline of the embryo (Hamada and Tam 2014). The node cells are organised into columnar epithelium, which, in the case of mice, forms a fluid-filled cavity covered by the parietal yolk sac. These cells can be divided into two groups: the 200–300 pit cells in the central region of the node and the crown cells around the periphery of the node (Lee and Anderson 2008). Both groups of cells are monociliated, but only the cilia of the pit cells are motile and rotate clockwise when viewed from the 'apical aspect'. The motility or immobility of the cilia in these regions is reflective of the type of 9 + 0 cilia present; the motile central cilia express the dynein encoded by *Dnah11*, whereas cilia at the periphery of the node are thought to be mechanosensory due to expression of the calcium channel *Pkd2* and *Pkd1l1* (Field et al. 2011; McGrath et al. 2003; Yoshiba et al. 2012).

Establishment of LR asymmetry involves integration of the two established embryonic axes (anterior/posterior and dorsal/ventral) to create a third. The node is positioned on the ventral surface of the embryo, with the cilia positioned towards the posterior side of the cells (Babu and Roy 2013). Because the cells are dome-shaped, the posterior localisation causes the cilia to tilt and thus the cilia rotary motion to vary in distance from the cell surface, the end result of which is a leftward flow of fluid across the node (Hamada and Tam 2014). The leftward flow has been shown to be critical for establishment of LR asymmetry as loss of the flow through loss of cilia motility or abnormal cilia structure results in randomisation of LR asymmetry (Babu and Roy 2013), whereas artificial reversal of cilia fluid flow results in activation of left signals on the right of the embryo (Nonaka et al. 2002). There are two main theories as to how the leftward flow creates the asymmetry. The first is that the leftward flow generates a gradient of left determinant particles or morphogens such as signalling ligands (Okada et al. 2005; Tanaka et al. 2005). The second theory is immotile cilia of the crown cells sense the flow and cause an influx in Ca^{2+} ions through the calcium channel Pkd2 (McGrath et al. 2003). The influx results in the activation of *Nodal* in the perinodal crown cells on the left.

However the flow functions to initiate LR asymmetry, the end result is to lower levels of Cerl2 (also called Dand5) on the left side of the node through increasing degradation of *Cerl2* mRNA (Nakamura et al. 2012). Since Cerl2 is a Nodal antagonist, this Cerl2 asymmetry results in an asymmetry in Nodal activity, with higher levels of Nodal activity in the left crown cells than in those right of the node. The increased expression of *Nodal* in the left crown cells is transient as Cerl2 protein is translocated between cells. Thus, the Cerl2 protein that continues to be produced in the crown cells to the right of the node will gradually move to those cells lacking *Cerl2* mRNA. However, the temporary change in Cerl2 levels is critical as loss/lack of *Cerl2* in mice results in bilateral or right-sided expression of *Nodal* in lateral plate mesoderm (LPM) (Marques et al. 2004). The 'leftness' signal created by *Nodal* asymmetry is transferred/transmitted from the crown cells on the left node to activate *Nodal* in the left LPM. How the message is transferred is not completely understood, but it may be transmitted via crown cell secretion of Nodal-Gdf1 heterodimer. Nodal positively regulates itself in the LPM to enhance the signal but also activates feedback inhibitors Lefty1/2 in the midline and left LPM to antagonise Nodal leading to restriction of the extent/duration of Nodal signalling. Lefty also acts to prevent transfer of left-sided signal across the midline to the right side. Expression of *Nodal* in the left LPM also upregulates the homeobox transcription factor *Pitx2*. The expression of *Pitx2* is more stable than *Nodal* and persists after loss of *Nodal* signal in the LPM, thereby acting as the major left determinator.

Situs inversus usually results from defects in 'signalling at the node', whereas heterotaxy can also be caused by early defects in node formation or later defects in transfer of signal to LPM and from the LPM to organ primordia. The latter two steps can also cause isolated CHD.

15.4 ZIC3

ZIC3 is part of the ZIC family and GLI superfamily which are defined by the presence of five highly conserved tandem C_2H_2 zinc fingers (Fig. 15.1b). ZIC proteins (named ZIC1–5) are distinguished from other superfamily members by an atypical first finger which contains 6–38 amino acids between the two cysteine residues rather than the normal 2–4. The atypical finger is highly conserved between ZIC1-3 and more divergent for ZIC4 and ZIC5. Functions of the zinc finger domain include interaction with DNA (Badis et al. 2009; Lim et al. 2010), interaction with proteins (Koyabu et al. 2001; Pourebrahim et al. 2011) and localisation to the nucleus, as both nuclear localisation signals and a nuclear export signal are located within this domain (Bedard et al. 2007; Hatayama et al. 2008). Other domains conserved within the family include the ZF-NC (zinc finger N-terminally conserved) domain, a small (14–21 amino acids) region of unknown function highly conserved in vertebrate and arthropod ZICs except for zebrafish Zic6 (Aruga et al. 2006; Keller and Chitnis

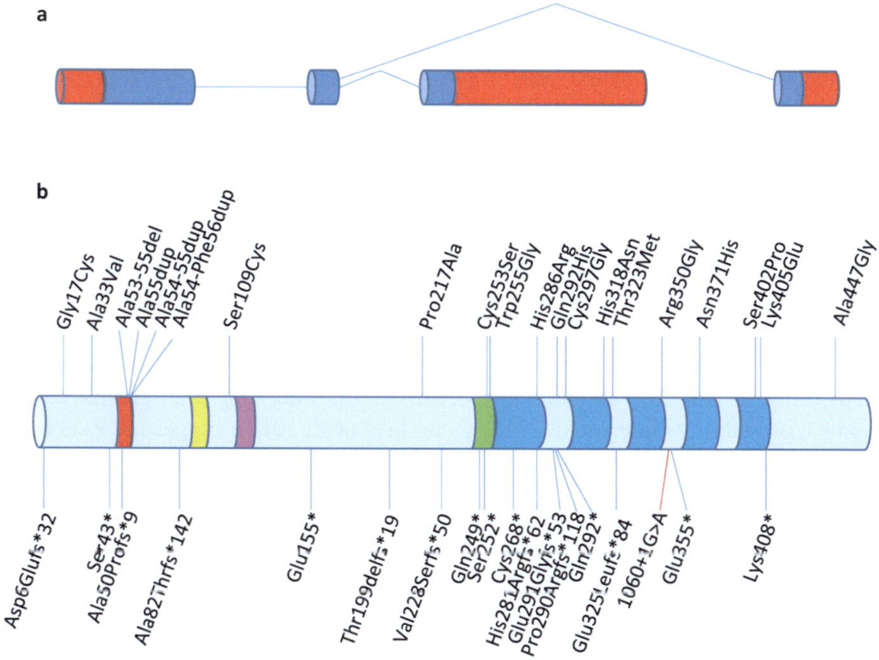

Fig. 15.1 The genomic arrangement and protein domains of ZIC3. (**a**) *ZIC3* consists of four exons, shown here in blue, with alternative splicing of exons 3 and 4 creating two isoforms. Untranslated regions are shown in red. (**b**) *ZIC3* mutations. Missense mutations, in-frame expansions and in-frame deletions that have been identified in heterotaxy patients are shown above the protein, whereas nonsense and frame shift mutations are shown below. Red, polyalanine tract; yellow, polyhistidine tract; pink, ZOC (Zic-odd paired conserved) domain; green, ZF-NC (zinc finger N-terminally conserved); blue, zinc fingers

2007). ZIC3 and most other ZIC proteins also contain the ZOC (Zic-odd-paired conserved) domain which is a small conserved region (9–10 amino acids) found in a subset of the ZICs (ZIC1-3 in mammals) (Aruga et al. 2006). This region has been suggested to be necessary for certain protein-protein interactions, in particular with MDFI (also called I-mfa) (Mizugishi et al. 2004). ZIC3 also contains several low complexity domains, including polyalanine and polyhistidine tracts, and has an alternatively spliced transcript (Fig. 15.1a) (Bedard et al. 2011).

Expression of *Zic3* during early development shows overlap with other family members' expression patterns but also demonstrates unique expression domains. *Zic3* expression is initially found in the embryonic and extraembryonic ectoderm prior to gastrulation (Elms et al. 2004). Expression is lost from the anterior portion of the ectoderm during gastrulation but is expanded to include some of the embryonic mesoderm. Expression of *Zic3* overlaps with *Zic2* and *Zic5* at this stage but has unique expression in the anterior of the gastrula in the prechordal plate and anterior definite endoderm as well as in the embryonic node at 7.75 dpc (Sutherland et al. 2013). By the early head fold stage, *Zic3* ectoderm expression becomes restricted to the neuroectoderm, and as neurulation progresses, *Zic3* neuroectoderm expression becomes restricted to the dorsal regions that represent the future site of neural crest production and dorsal neural production. *Zic3* is also expressed in the presomitic population of the lateral mesoderm and at later stages of development is found in the dorsal spinal cord, the dorsal cranial neural tube, the eye and the limb buds.

Of note, despite the cardiac defects seen in *Zic3* null mice, analysis by in situ hybridisation has shown a distinct lack of *Zic3* expression in the heart at any point in development (Nagai et al. 1997). In one publication, quantitative RT-PCR on cardiac tissue from E10.5 suggested low levels of *Zic3* transcripts were present in the heart (Zhu et al. 2007a), but attempts to confirm this expression using a highly sensitive *Zic3*-LacZ-BAC reporter line have also indicated a lack of *Zic3* in the heart (Sutherland et al. 2013). Furthermore, tissue-specific deletion of *Zic3* in cardiac cells using several different cardiac-specific Cre lines does not alter cardiac development or viability (Jiang et al. 2013; Sutherland et al. 2013), suggesting Zic3 does not have a role in these tissues.

15.5 *Zic3* Mutant Phenotype

Several null *Zic3* mouse lines have been generated, through interstitial deletion, targeted insertion or spontaneous mutations during an ethylnitrosourea (ENU) mutagenesis screen (Ahmed et al. 2013; Bogani et al. 2004; Haaning et al. 2013; Jiang et al. 2013; Purandare et al. 2002; Sutherland et al. 2013). The phenotypes of these *Zic3* null mice indicate the role/function of Zic3 begins early in development, with a portion of null embryos failing to begin gastrulation. Of those that do gastrulate, another 40% show defects during gastrulation including excess mesoderm formation and a subset show axis duplication. Shortly after gastrulation, mouse embryos develop further phenotypes including caudal truncation, heart abnormalities, defects in somitogenesis and delayed growth. Zic3 continues to have roles throughout development; thus later stage *Zic3* nulls show a variety of phenotypes including neural tube defects (both exencephaly and spina bifida), craniofacial defects, omphalocele, skeletal defects, limb defects, defects in folia and olfactory bulb patterning and microphthalmia.

Aside from these defects, the *Zic3* null mice show a number of phenotypes consistent with heterotaxy. The heart defects seen in *Zic3* null mice are varied but include dextro-transposition of the great arteries, interrupted aortic arch, right aortic arch, atrial septal defect, ventricular septal defect and abnormal systemic venous connections (Purandare et al. 2002). Consistent with classic heterotaxy, mice with heart anomalies also have additional anomalies in the LR patterning of other organs including lung isomerisms, right-sided stomach with reversed liver and hypoplastic spleen (Purandare et al. 2002). Quantification of the rate of these heart defects at d10.5–d12.5 has found that roughly 47% of *Zic3* null mice have defects in heart looping that can further be divided into sinistral (leftward) looping (16.7%), ventral (forward) looping (23%) and no looping (6.7%) (Ware et al. 2006a). The *Zic3* mutant mice also show defects in expression of laterality genes during the establishment of the LR axis. Normally *Nodal* expression begins in the crown cells at the 0–2 somite stage and is maintained past the 6 somite stage. *Zic3* null and hypomorphic mice retain the initial expression of *Nodal*, but the expression is not maintained and is lost by the 4–6 somite stage in a subset of embryos. At the later somite stages, when *Nodal* and *Pitx2* are normally expressed in the left LPM, mutants show a variable expression of these genes which can be in the right LPM, left LPM, bilateral or absent (Purandare et al. 2002; Sutherland et al. 2013). The role of Zic3 is conserved between species as loss of *Zic3* in *Xenopus* and zebrafish embryos causes similar phenotypes including delayed gastrulation and defects in left-right patterning including abnormal heart and gut looping (Cast et al. 2012).

Notably, gain of Zic3 function can also result in similar phenotypes in a number of species. Analysis of a mouse line showing increased expression (but not ectopic expression) of *Zic3* due to insertion of a Neo cassette found 6.9% showed dextrocardia at E14.5–15.5 and 7.9% at 4-weeks old, and there were also individual cases of left-sided gall bladder and asplenia (Zhu et al. 2007b). Overexpression of *Xenopus Zic3* mRNA in the right side of *Xenopus* embryos causes defects in the laterality of the heart and gut and induces expression of major left determinants *Pitx2* and *Xnr1* (Kitaguchi et al. 2000). By injecting *Xenopus* embryos with *Zic3* tagged with part of the human glucocorticoid receptor (*Zic3*-GR), which is only active in the presence of dexamethasone, the effect of *Zic3* overexpression on laterality was found to be restricted to the last blastula and early gastrula stage, with later induction of Zic3 activity not able to induce expression of *Pitx2* (Kitaguchi et al. 2000). Similarly, overexpression of human *ZIC3* in zebrafish by mRNA injection results in an altered position of the heart tube in 40% of embryos (Paulussen et al. 2016). In addition it has been found that in at least one mouse line, a truncated allele of *Zic3* escapes nonsense-mediated decay (Ahmed et al. 2013). The stability of such transcripts raises the possibility that Zic3 truncation may produce dominant-negative proteins which potentially interfere with other members of the Zic family, although further analysis of the mouse line has suggested the mutation produces a loss-of-function allele.

Though *Zic3* has a wide expression pattern, the temporal and tissue-specific requirement of *Zic3* expression for the development of heterotaxy has been delineated by use of two conditional mouse lines, both of which delete the first exon of *Zic3* in the presence of recombinase enzyme Cre to abolish its function (Jiang et al. 2013; Sutherland et al. 2013). Knockout of *Zic3* in the epiblast using a *Sox2-Cre* line results in the same phenotypes as *Zic3* nulls including gastrulation, neural tube, exencephaly and laterality defects (Jiang et al. 2013; Sutherland et al. 2013), although lethality was not as high as in *Zic3* nulls despite all adults displaying kinked tails consistent with loss of *Zic3* (Jiang et al. 2013). The tissue for which Zic3 activity is required for normal establishment of LR asymmetry can be further delimited to migrating streak and mesodermal lineage derivatives as knockout of *Zic3* within mesendodermal cells using *T-Cre* results in laterality defects including looping defects of the heart at 9.5 dpc (Jiang et al. 2013; Sutherland et al. 2013) as well abnormal expression of *Pitx2* and *Lefty* and dysmorphic nodes (Sutherland et al. 2013). Notably, knockout of *Zic3* in cardiac progenitors/tissue using five different Cre lines (*Nkx2.5*, *Mef2c*, *Wnt1*, β*MyHC* and *Mesp1*) did not reduce viability or cause laterality defects, indicating the heart malformations produced by loss of *Zic3* are not a result of loss of *Zic3* in cardiomyocytes, cardiac neural crest or the anterior heart field (Jiang et al. 2013; Sutherland et al. 2013). Loss of *Zic3* in the node through use of a *Foxj1-Cre* or a nodal-dependent enhancer (*NDE*)-*Cre* line also did not cause laterality defect (Sutherland et al. 2013), suggesting that Zic3 expression is required prior to node formation to ensure normal cardiac development.

15.6 Interactions of ZIC3 in Developmental Signalling Pathways in Heterotaxy

The ZICs are multifunctional proteins which are able to influence several of the major signalling pathways including hedgehog, Wnt, planar cell polarity (PCP) and transforming growth factor β signalling. Herein we summarise the potential role of ZIC3 in these pathways for the initiation of LR asymmetry.

15.7 Hedgehog Signalling

Mammals have three paralogous hedgehog genes: Sonic hedgehog (*Shh*) which is the most broadly expressed hedgehog gene and is involved in several stages of development including patterning the somites, neural tube and limb as well as in development of the tooth, lung and gut; Indian hedgehog (*Ihh*) which is involved in bone differentiation and development of the mammalian gut (Ramalho-Santos et al. 2000; St-Jacques et al. 1999); and Desert hedgehog (*Dhh*) which is involved in gonad differentiation and development of the perineural sheath (Bitgood et al. 1996; Parmantier et al. 1999). In the absence of these three hedgehog ligands (Fig. 15.2a), the 12-span transmembrane receptors Patched-1 (Ptch1) and Patched-2 (Ptch2) accumulate in and around the cilium to repress activity of 7-span transmembrane protein Smoothened (Smo) which thus inhibits downstream hedgehog signalling (Rohatgi et al. 2007). The repression of Smo occurs through Ptch controlling the localisation, stability and phosphorylation of Smo. For example, in the repressed

Fig. 15.2 (continued) (**b**) In response to binding of Shh ligands, Ptch exits the cilium which causes derepression of Smo. Sufu and Gli still interact, but active Smo inhibits Pka and promotes transport of the Sufu/Gli complex to tip of the cilia where the proteins disassociate. The full-length activator form of Gli then moves into the nucleus to activate hedgehog target genes. Zic3 directly interacts with Gli proteins and potentially increases Gli localisation to the nucleus but has also been found to influence the ratio of activator and repressor forms of Gli

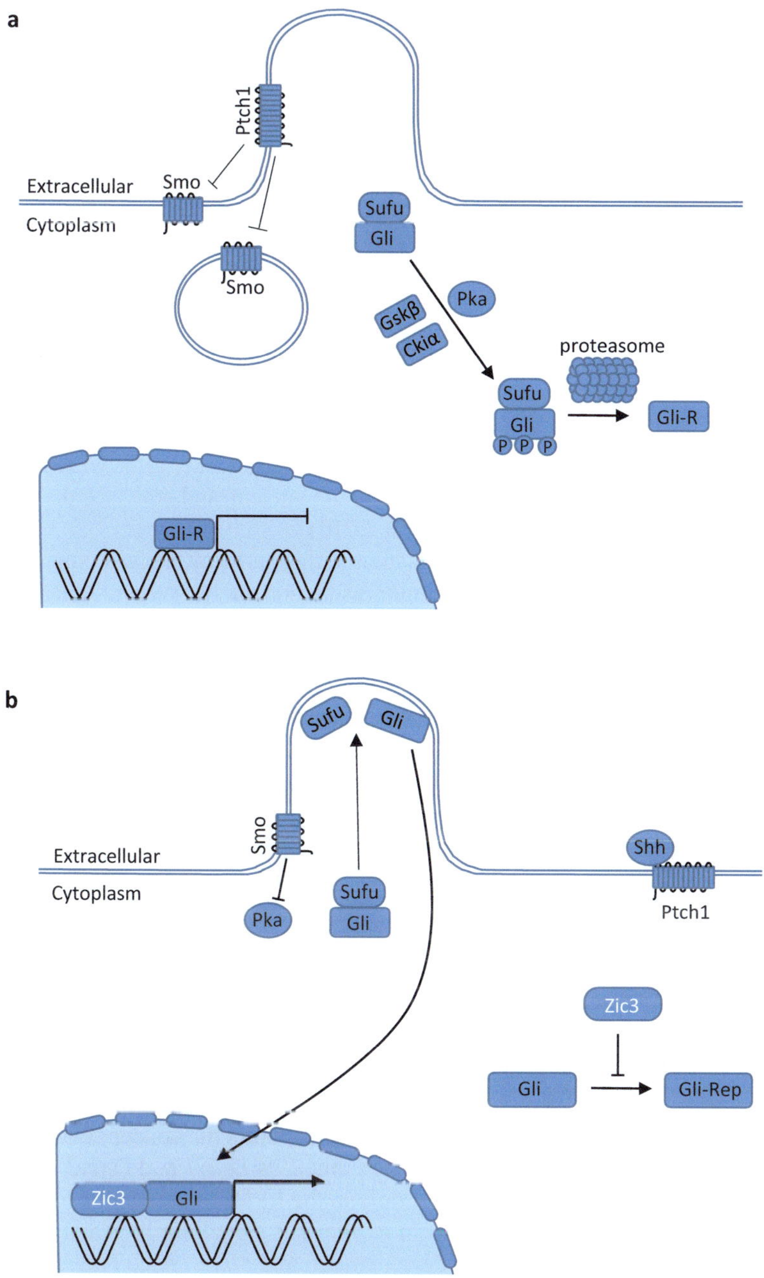

Fig. 15.2 Hedgehog signalling. (**a**) In the absence of hedgehog ligands, Ptch1 accumulates in and around the cilia where it inhibits Smo, partly by preventing it from entering the cilia. When Smo is repressed, Sufu binds Gli, thus preventing Gli from entering the nucleus. The Sufu-Gli interaction also causes phosphorylation of Gli by the kinases Pka, Gskβ and Ckiα which in turn causes partial degradation by the proteasome resulting in the truncated Gli-repressor forms.

state, Smo is localised mainly to plasma and vesicle membranes (Milenkovic et al. 2009) as Ptch prevents localisation to cilia. When Smo is repressed, suppressor of fused (Sufu) negatively regulates the pathway by binding the Gli family transcription factors and thereby retaining Gli proteins in the cytoplasm (Dunaeva et al. 2003; Humke et al. 2010; Kogerman et al. 1999). The sequestering of Gli proteins both prevents their nuclear transcriptional activation of target genes and causes Gli2 and Gli3 to be processed by the proteasome to produce repressor forms of these Gli proteins (Humke et al. 2010; Wang et al. 2000). There is also evidence that Sufu/Gli form a repressor complex in the nucleus to suppress Gli-induced gene expression. In contrast, when hedgehog signalling is active (Fig. 15.2b), the secreted hedgehog ligand binds Ptch1, causing Ptch to exit the cilium leading to derepression of Smo. The exact mechanism of Smo derepression is not clear, but involves trafficking of Smo to the cilia and phosphorylation of the cytoplasmic tail by Ckiα and Grk2 (Chen et al. 2011). Active Smo inhibits Pka and promotes transport of the Gli/Sufu complex to the tips of cilia where the proteins disassociate, after which the full-length active form of the Gli proteins enters the nucleus and activates downstream hedgehog target genes (Tukachinsky et al. 2010).

Given the central role of cilia localisation in this pathway, normal cilia assembly and function are critical for activation/transduction of hedgehog signalling. For example, cilia intraflagellar transport (IFT) controls the bidirectional transport of protein complexes between the base and tip of cilia and is critical for the assembly and function of cilia. Several IFT proteins, including Ift172, Ift88, Kif3a and Dync2h1, have been shown in mouse genetic studies to be required for mouse hedgehog signalling and have been placed in the middle of the pathway downstream of Smo and upstream the Gli proteins (Huangfu and Anderson 2005; Huangfu et al. 2003; May et al. 2005). Disruption of basal body proteins also causes defects in cilia formation including abnormal or absent cilia. Mice with such basal body disruptions show both cilia defects and phenotypes consistent with loss of hedgehog, such as polydactyly (Bangs and Anderson 2017; Vierkotten et al. 2007; Weatherbee et al. 2009).

Several lines of evidence demonstrate hedgehog signalling has a role in the establishment of LR asymmetry. In particular, *Smo*$^{-/-}$ mice show several phenotypes consistent with loss of LR asymmetry including failure of embryonic turning and loss of heart looping (Zhang et al. 2001). Although the failure of such *Smo*$^{-/-}$ embryos to survive past 9.5 dpc precludes analysis of later loss of LR asymmetry phenotypes, analysis of molecular markers also supports the concept that the establishment of the LR axis is disrupted in these mice, as *Lefty1* and *Lefty2* expression is lost, whereas *Nodal* and *Pitx2* is lost only in the LPM. Interestingly, loss of individual hedgehog ligands does not seem to alter LR signalling in exactly the same way, as loss of Ihh does not affect development, and *Shh*$^{-/-}$ mice only show some phenotypes indicating initiation of the asymmetry has been affected, such as bilaterally unilobed lungs and abnormal cardiac looping. In addition, a small proportion of mice (10%) show reversal of cardiac looping (Meyers and Martin 1999; Tsukui et al. 1999), but do not show the same differences as Smo mutants at the molecular level. In particular, *Lefty2*, *Nodal* and *Pitx2* expression is not lost, but instead the three genes are ectopically expressed in the right LPM. Instead, it takes loss of both *Ihh* and *Shh* to produce embryos indistinguishable from the *Smo* mutants. As both genes are weakly expressed in the node, this suggests redundancy in hedgehog signalling in regulating formation of the asymmetry (Zhang et al. 2001). Restoration of hedgehog signalling in the LPM of *Smo*$^{-/-}$ mutants by the use of a LPM enhancer-driven Smo transgenic line is sufficient to re-establish expression of *Nodal*, *Lefty* and *Pitx2* and normal heart looping (Tsiairis and McMahon 2009). However *Lefty* expression at the midline is still absent, suggesting hedgehog may have multiple roles in regulating asymmetry formation.

There is evidence that loss of Zic family members can repress hedgehog signalling in multiple species. For example, loss of zic1 in zebrafish via MO injection produces reduced expression of *ptc1*, *shh* and the zebrafish hedgehog homolog tiggywinkle (*twhh*) (Maurus and Harris 2009), loss of Zic1 in mice causes reduced expression of *Ptch1*, *Gli1* and Shh downstream gene *Mycn* in the cerebellar vermis (Blank et al. 2011) and knockdown of *ZIC2* in HeLa cells using RNAi reduces expression of several hedgehog signalling downstream targets including *GLI1*, *PTCH1* and *CYCLIN* (Chan et al. 2011). The regulation of hedgehog most likely occurs through interaction of ZIC with the GLI proteins. ZIC1, ZIC2 and ZIC3 have all been found to interact with multiple GLI proteins via cell-based immunoprecipitation assays (Koyabu et al. 2001; Zhu et al. 2008). These ZICs also show synergistic activation with the GLI family of various luciferase reporters in cell-based mechanisms (Koyabu et al. 2001; Pan et al. 2011; Zhu et al. 2008). There are several potential mechanisms by which ZICs influence GLI activity, all of which involve interaction of ZIC and GLI proteins through their respective zinc finger domains. Through this interaction the ZICs have been found to increase the localisation of GLIs to the nucleus (Koyabu et al. 2001), which should enhance GLI activity. ZIC3 has also been shown to alter the ratio of the repressor and activator forms of GLI3. In vivo, the loss of Zic3 alters the ratio of Gli3 activator: repressor in their shared limb bud expression domains, leading to the ability of *Zic3* null mice to rescue the polydactyly phenotype in *Gli3*$^{-/-}$ mice (Quinn et al. 2012).

Given these influences on hedgehog signalling, it could be predicted that *Zic3* null mice would have downregulated hedgehog signalling and thus would be similar to *Smo* mutants at the molecular level, with loss of *Pitx2* and *Nodal* in the LPM. As *Zic3* null mice retain expression of left markers such as *Pitx2* and *Nodal* in the LPM, it is unlikely that hedgehog signalling is completely lost in such mutants.

15.8 Canonical Wnt Signalling

There are several signalling pathways regulated by secreted Wnt ligands. Canonical Wnt signalling differs from the other pathways in that it is regulated by controlling the amount of β-catenin that reaches the nucleus. β-Catenin protein is constantly produced by cells and has important roles outside of the canonical pathway in cell-cell adhesion, with total loss of β-catenin resulting in adhesion defects and embryonic disintegration (Hierholzer and Kemler 2010). In the absence of Wnt ligands (Fig. 15.3a), the pathway is in the *off* state and the amount of β-catenin is strongly depleted by activity of the β-catenin cytoplasmic degradation complex. The degradation complex includes scaffolding proteins Axin and adenomatous polyposis coli (Apc) which promote the phosphorylation of β-catenin through action of casein kinase 1 (Ck1) and glycogen synthase kinase (Gsk3). The phosphorylation of β-catenin targets it for ubiquitin-dependent degradation via ubiquitin ligase β-Trcp.

The pathway is switched into the *on* state (Fig. 15.3b) by the extracellular presence of 1 of at least 19 secreted Wnt ligands (other Wnt ligands activate other pathways). The Wnt ligands interact with Frizzled (Fz) and Lrp5/6 co-receptors on the surface of cells to form a receptor complex that results in phosphorylation of the intracellular portion of Lrp5/6. The phosphorylated intracellular domain of Lrp5/6 acts as a docking site for Axin causing Axin to be relocated to the cell membrane in a process dependent on the Dishevelled (Dvl) proteins. Due to the loss of Axin, β-catenin is no longer degraded and thus accumulates in the cytoplasm and translocates into the nucleus. Once in the nucleus, β-catenin activates transcription of target genes through interaction with the T-cell factor (Tcf)/lymphoid enhancer-binding factor 1 (Lef1) family of transcription factors. In the absence of Wnt ligands, Tcf/Lef transcription factors inhibit the transcription of the same target genes through interaction with Tle/Groucho family of corepressors. Nuclear β-catenin displaces corepressors and recruits co-activators including Bcl9, histone acetylation Cbp and p300 to interact with Tcf/Lef to activate target genes.

Multiple Wnt ligands may act during the initial establishment of LR asymmetry in functions that are not completely redundant. For example, *Wnt3a* is symmetrically expressed in the dorsal posterior node, and its loss produces various LR defects including delayed bilateral expression of *Nodal*, *Lefty2* and *Pitx2* in the LPM, as well as randomisation of heart looping, randomisation of embryonic turning, abnormal stomach positioning and abnormal lung lobation (Nakaya et al. 2005). Wnt3a protein, on the other hand, shows an asymmetric expression in the node crown cells, where it is thought to promote degradation of Cerl2 (Kitajima et al. 2013; Nakamura et al. 2012). Initially *Cerl2* and *Wnt3a* are both expressed symmetrically in the crown cells, but both inhibit expression of the other gene: Wnt3a increases degradation of *Cerl2* mRNA due to a Tcf-binding site in the 3′ untranslated region (UTR), whereas Cerl2 likely causes degradation of Wnt protein (Nakamura et al. 2012). The repression of both genes is balanced so that neither is completely lost. Once degradation of

Fig. 15.3 (continued) via the ligase β-Trcp and subsequent degradation. Due to the constant degradation of β-catenin, it fails to enter the nucleus, and the transcription factors *TCF* and *LEF* inhibit transcription of Wnt targets due to interaction with the TLE/Groucho family of corepressors. Zic3 also interacts with *TCF/LEF* to promote repression of Wnt targets without directly interacting with DNA. (**b**) Wnt ligands interact with the extracellular portion of the Fzd receptor and LRP5/6 co-receptor to cause the phosphorylation of the intracellular portion of LRP5/6, thus producing an Axin docking site. Axin and the β-catenin complex are relocated to the cell membrane in a process dependent on the Dvl proteins. β-Catenin is thus no longer degraded and accumulates in the nucleus where it interacts with *TCF/LEF* to displace Groucho and recruit co-activators

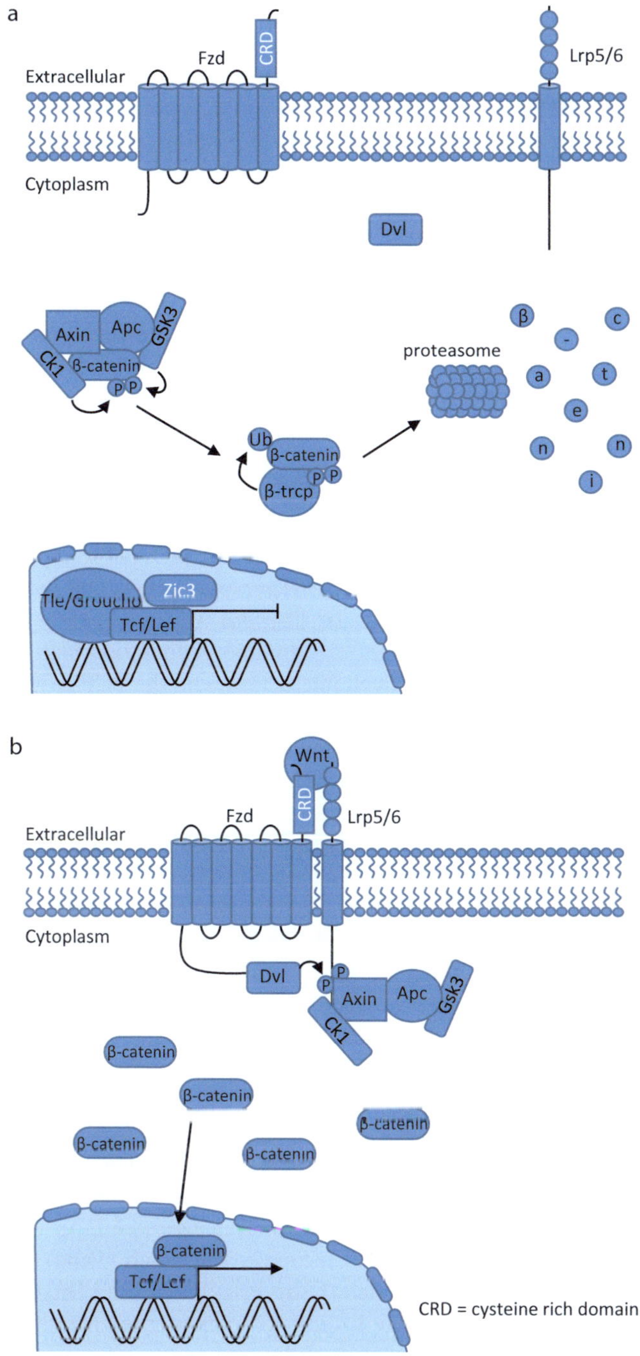

Fig. 15.3 Canonical Wnt Signalling. (**a**) In the absence of Wnt ligands, β-catenin is bound by the β-catenin destruction complex consisting of Axin and APC which promotes phosphorylation of β-catenin by the kinases *CK1* and *Gsk3*. The phosphorylation of β-catenin targets it for ubiquitination

Cerl2 is triggered by nodal flow, the initial loss of Cerl2 in the left crown cells allows accumulation of Wnt3a which in turn increases degradation of *Cerl2* to maintain the reduction in Cerl2 in the left crown cells after flow is lost (Kitajima et al. 2013; Nakamura et al. 2012). It should be noted that so far, loss of Wnt signalling has not been found to affect cilia structure or basal body position in mice (Hashimoto et al. 2010; Nakamura et al. 2012; Nakaya et al. 2005), though in other species such as zebrafish, Wnt signalling regulates ciliogenesis in the equivalent of the node (Caron et al. 2012; Lin and Xu 2009; Zhang et al. 2012; Zhu et al. 2015).

The first suggestion that the Zic family might influence Wnt signalling came from the observation that *Zic3* null mouse embryos showed posterior axis duplications (Ware et al. 2006b), a phenotype that had previously been associated with increased or ectopic Wnt signalling (Popperl et al. 1997; Zeng et al. 1997). How the ZIC family influenced the pathway was not initially clear since some findings, such as the loss of *Wnt3a* expression in *Zic3* null mice (Ware et al. 2006b), suggest the ZICs act upstream of Wnt, whereas other evidence, such as the finding that Wnt pathway activity is necessary for ZIC expression at the neural plate border and tectum in zebrafish (Garnett et al. 2012; Nyholm et al. 2007), placed the ZICs downstream of Wnt. Recently two papers confirmed ZICs act to repress canonical Wnt signalling using cell-based TOPFLASH reporter assays and by showing that Zics could repress Wnt in vivo to rescue *Xenopus* axis duplication (Fujimi et al. 2012; Pourebrahim et al. 2011). Using co-immunoprecipitation both groups found that ZICs influence the pathway though interaction with the TCF/LEF transcription factors. The ZIC/TCF complex is able to interact with DNA, though ZIC DNA-binding ability is not required for such interaction suggesting a cofactor role for the ZICs in inhibiting Wnt signalling (Pourebrahim et al. 2011).

Given that the ZICs inhibit Wnt signalling, it might be expected that loss of *Zic3* in mice would upregulate the pathway and thus produce bilateral activation of *Nodal* in the crown cells. Such activation has not been reported. Given the complex interactions with this pathway, is it possible ZICs regulate Wnt in cell type-specific manners and thus produce a different effect depending on the developmental stage or tissue type.

15.9 PCP Signalling

PCP regulates the alignment of cell polarity across a tissue plane through interaction of proteins at the membranes of adjacent cells. The pathway is considered part of non-canonical Wnt signalling and is best studied in *Drosophila* in which the core pathway proteins include Frizzled (Fz) receptor, Van Gogh (Vang) and Flamingo (Fmi) at the membrane and Dishevelled (Dsh), Prickle (Pk) and Diego (Dgo) in the cytoplasm. These proteins form two distinct complexes at the membrane on either side of a cell: Fz, Dsh and Dgo at one side and Vang and Pk at the opposite side, with atypical transmembrane cadherin Fmi found on both sides (Axelrod 2001; Bastock et al. 2003; Feiguin et al. 2001; Shimada et al. 2001; Strutt 2001). The complexes encourage/enforce the separation of the two complexes by inhibiting one another intracellularly. For example, the Vang/Pk complex inhibits formation of Fz/Dsh by Pk directly binding Dsh to both affect Dsh stability and prevent Dsh from localising to the membrane (Tree et al. 2002). Dgo antagonises the inhibition of Dsh by competing with Pk for the same binding site as Dsh and thus stabilises the complex (Jenny et al. 2005). Outside of the cells, the Fz-Fmi complex interacts with Vang-Fmi complex on a neighbouring cell, forming homodimers between cells. As the extracellular portions of Fz and Vang can be deleted without altering the localisation of these proteins, it is thought that the Fmi forms a bridge between proteins (Chen et al. 2008). The interaction serves to stabilise the complexes and enforce the orientation of PCP signalling across neighbouring cells and thus across tissue planes.

The PCP pathway is largely conserved in mammals (Fig. 15.4), but several vertebrate-specific PCP factors have been identified, such as the Ror2 and Ryk co-receptors. These proteins interact with Fzd (the mammalian homolog of Fz) and Vangl2 to form a receptor complex for the Wnt ligand Wnt5a to trigger phosphorylation of Dvl (the mammalian homolog of Dsh) and Vangl2. In zebrafish the phosphorylation of Vangl2 correlates with increased Vangl2 activity (Gao et al. 2011), thus Wnt5a may act as an instructional signal regulating the initial asymmetry formation of the PCP complexes. Once the asymmetry is established, PCP membrane interactions have various tissue-specific downstream consequences, through manipulation of two distinct pathways. The first pathway is mediated by the Rho subfamily of small GTPases and the second pathway by Rac and Cdc42 proteins. Both pathways activate Jnk to regulate nuclear responses, though the Rho subfamily also acts independently of the nucleus to trigger cytoskeletal rearrangements.

One of the roles of PCP signalling is positioning the basal body and thus the cilia within the cell. The major PCP signalling components are asymmetrically localised within node cells, with Dvl2 and Dvl3 localised to the posterior of node cells (Hashimoto et al. 2010), whereas Prickle2 and Vangl1 are localised to the anterior (Antic et al. 2010). This asymmetry is thought to be caused by a gradient of PCP activity across the node produced by the combination of expression of Wnt5a/Wnt5b ligands posterior of the node and the PCP inhibitors Sfrp1/Sfrp2/Sfrp5 anterior of the node (Minegishi et al. 2017). The node cilia are initially positioned towards the centre of cells, but PCP signalling gradually causes a posterior shift of basal body and cilia of the node cells (Hashimoto et al. 2010), which is needed to produce the leftward flow that activates downstream LR signalling events. Disrupting either of these complexes prevents the posterior shift of cilia, causing loss of flow. For example, in mice without either Vangl1 and/or Vangl2, the cilia remain centrally localised and develop various LR asymmetry defects including failure of heart looping and embryonic turning as well as lung isomerism and randomised *Nodal* and *Pitx2* expression in the LPM (Antic et al. 2010; Song et al. 2010). Loss of multiple *Dvl* genes or inhibition of the downstream PCP gene *Rac1* also causes a similarly impaired basal body shift (Hashimoto et al. 2010). It should be noted that scanning electron microscopy has indicated the cilia are structurally normal in all of these mutants (Antic et al. 2010; Hashimoto et al. 2010; Song et al. 2010). PCP signalling may also be involved earlier as embryos that lack *Rac1* in the epiblast show defects

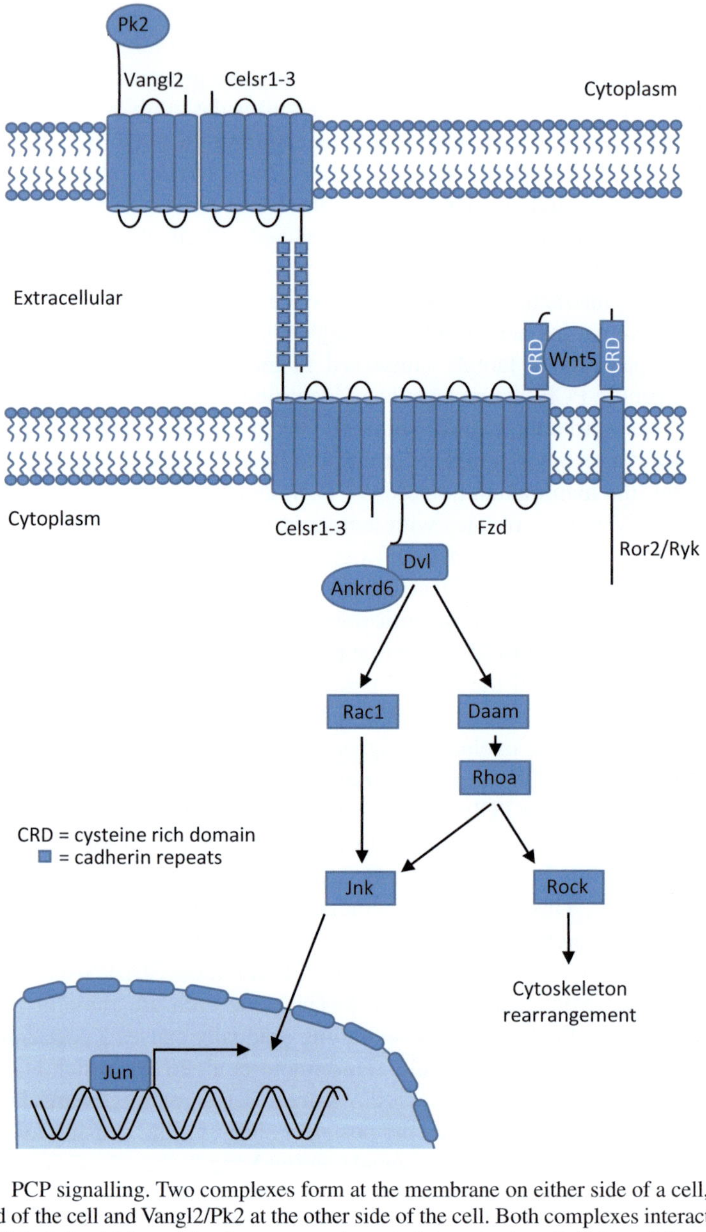

Fig. 15.4 PCP signalling. Two complexes form at the membrane on either side of a cell, Fzd/Dvl at one end of the cell and Vangl2/Pk2 at the other side of the cell. Both complexes interact with the cadherin Celsr1-3, which enables the two complexes to interact extracellularly and form homodimers between cells. Mammalian PCP signalling also involves interaction of the Fzd/Dvl complex with the co-receptors Ror2/Ryk which enables Wnt5a to potentially act as an instructional signal. In response to Wnt ligand Dvl activates Rac1 and Daam to activate transcription of target genes via JNK/Jun and cytoskeletal rearrangement via Rock

in the formation of the node (Migeotte et al. 2011). In these mutants, the pit cells of the node are present in several distinct clusters though the cilia are structurally normal.

There is some evidence that Zic3 regulates PCP signalling. For example, Zic3 null mice show tail kinks, a phenotype that is seen in Prickle1 hypomorphs (Liu et al. 2014) and in loss-of-function PCP alleles of LRP6 (Allache et al. 2014). *zic3* zebrafish and *Xenopus* morphants also show defects in convergence extension morphogenesis, a process regulated by PCP signalling (Cast et al. 2012). Several PCP-related genes have also been identified as putative Zic3 targets in zebrafish by CHIP-seq (Winata et al. 2013). Additionally the nodes of *Zic3* null mice show node abnormalities (Fig. 15.5) (Sutherland et al. 2013) similar to those seen in the PCP effector gene *Rac1* mutants (Migeotte et al. 2011).

15.10 TGFβ (Nodal) Signalling

Nodal is part of the Tgfβ superfamily of ligands which exist as homodimers or in rare cases heterodimers to bind Tgf beta type II receptors. Upon ligand binding the type II receptor (Fig. 15.6), a heterotetrameric complex is formed with type I receptor (Alk1-7), causing the type II receptor to phosphorylate the type I receptor. These initial steps of activation are similar for the other Tgfβ superfamily of ligands, including BMPs (Bmp2, Bmp4 and Bmp7), Tgf-β and Activins, but Nodal signalling is unique in that it also requires the presence of EGF-CFC co-receptor (Cripto and Cryptic in mammals). Thus Nodal signalling has the potential to be inhibited without affecting the other TGFβ signalling through targeting the co-receptor. Formation of the ligand receptor complex triggers the type I receptor to phosphorylate the C-terminal serine residue of cytoplasmic receptor SMADs (R-SMADs). The R-Smad targeted for phosphorylation varies depending on which ligand activated the pathway, with Smad1, Smad5 and Smad8 phosphorylated in the case of BMP signalling and Smad2 and Smad3 phosphorylated in the case of Tgf-β, Activin and Nodal signalling. Phosphorylated R-Smad forms homomeric and heteromeric complexes with Smad4 (also called co-Smad) and accumulates in the nucleus to regulate target genes. Smads alone activate transcription weakly and require additional tissue-specific transcription factor or cofactors. In particular Smad2 does not directly bind DNA.

Nodal signalling is critical for the initiation of LR asymmetry both due to its activity in the crown cells of the node and the LPM. In addition, Nodal is required for formation of the primitive streak and thus the node. In *Zic3* null mice, *Nodal* expression is initiated in the crown cells but not maintained, and *Nodal* LPM expression is randomised, suggesting Zic3 regulates this pathway (Purandare et al. 2002). Partial loss of *Nodal* also increases the severity of *Zic3* phenotypes, which also suggests the genes act within the same pathway (Ware et al. 2006a). Zic3 is able to influence the pathway via a Nodal enhancer to activate expression in mice and *Xenopus* (Ware et al. 2006a), placing Zic3 upstream Nodal signalling. However,

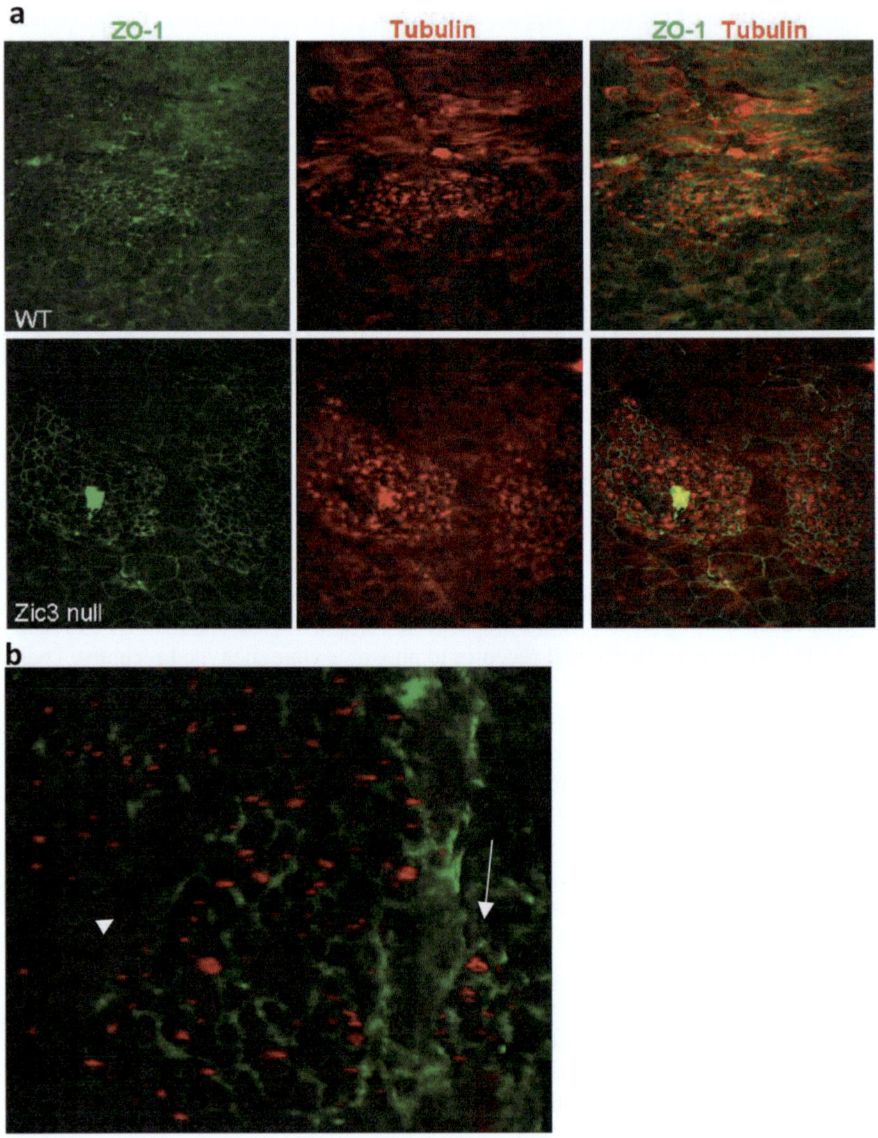

Fig. 15.5 Immunofluorescent image of a Zic3 mutant mouse node. (**a**) ZO1 immunohistochemistry (*green*) marks the cell borders, whereas acetylated tubulin (*red*) marks the cilia. (**b**) A higher magnification view of part of a Zic3 mutant node in which phalloidin staining (*green*) marks the cell borders whereas gamma tubulin (*red*) marks the basal bodies at the base of cilia. Node cells normally cluster in a tear drop shape, but the Zic3 mutants often show ectopic foci of node cells or lack of continuity of the node ultrastructure. *Arrow* indicates a second smaller group of ciliated node cells physically separated from the other node cells. *Arrowhead* indicates a region of cells in the middle of the node which are larger and have no cilia and thus are likely endoderm cells

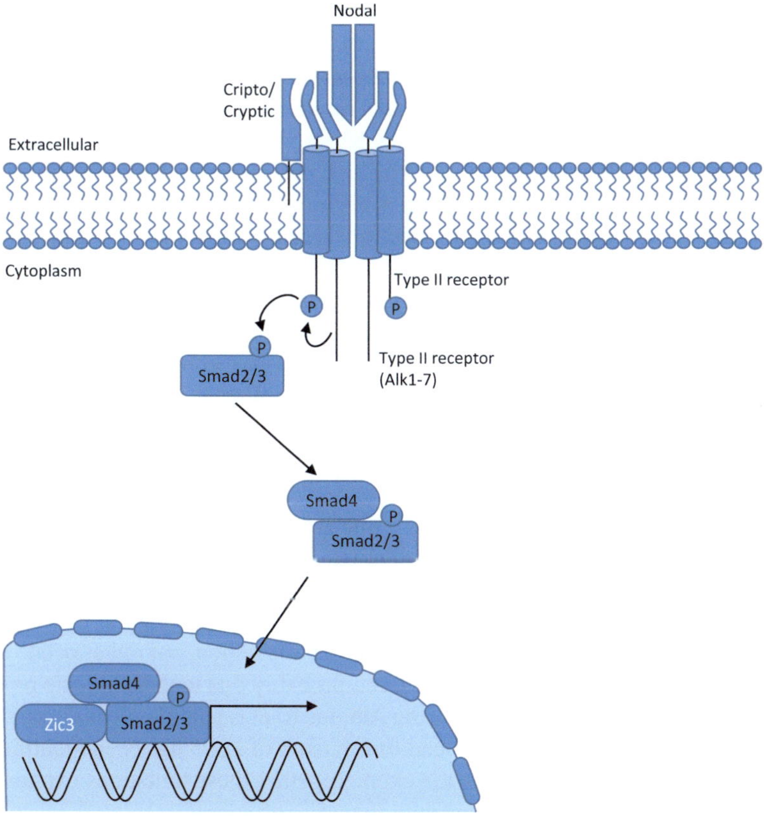

Fig. 15.6 TGFβ signalling. TGFβ ligands interact with a heterotetrameric complex consisting of type I and type II receptors. In the case of Nodal, ligand binding also involves the co-receptor Cripto/Cryptic. The presence of Nodal causes the type II receptor to phosphorylate the type I receptor which in turn causes phosphorylation of cytoplasmic receptors Smad2 and Smad3. Phosphorylated Smad2/3 complexes with Smad4 and accumulates in the nucleus to activate target genes. Activation also requires tissue-specific transcription factors and cofactors which potentially include Zic3

recently ZIC2 was shown to interact with both SMAD2 and SMAD3 (Houtmeyers et al. 2016). If the interaction is conserved within the ZIC family, ZIC3 may influence Nodal signalling at multiple points in the pathway.

15.11 Lessons from Human ZIC3 Mutations

Loss of ZIC3 was originally identified as a potential cause of heterotaxy through mapping of a submicroscopic deletion of the X chromosome in a case of X-linked familial heterotaxy (Casey et al. 1993; Ferrero et al. 1997; Gebbia et al. 1997).

Other cases of ZIC3 deletion have since been found in heterotaxy patients, including patients in which the entire ZIC3 sequence has been lost (Chung et al. 2011) and in one sporadic case a male heterotaxy patient was found to have a deletion of only the third exon of *ZIC3* (Cowan et al. 2016). There have also been at least two instances of translocations involving the X chromosome resulting in heterotaxy (Fritz et al. 2005; Tzschach et al. 2006). In both cases, the female patients were found to have undergone X-inactivation of their normal X chromosome, suggesting that the translocation altered the transcriptional regulation of the *ZIC3* gene and thus resulted in loss-of-function.

To date 38 unique variants altering the sequence of *ZIC3* have been identified in heterotaxy patients (Fig. 15.1b) (Chhin et al. 2007; Cowan et al. 2014; D'Alessandro et al. 2013a b; De Luca et al. 2010; El Malti et al. 2016; Gebbia et al. 1997; Ma et al. 2012; Megarbane et al. 2000; Paulussen et al. 2016; Ware et al. 2004; Wessels et al. 2010). The variants affecting the *ZIC3* sequence include 16 missense mutations of which 13 alter amino acids conserved in all of human, mouse, *Xenopus* and zebrafish. Other variants affecting *ZIC3* include one splice site, nine frame shift and eight nonsense mutations, as well as one in-frame deletion and three in-frame duplications. In total, 20/38 (52.6%) of the mutations are in the ZFD and 76.3% in total disrupt this region. As this region is involved in DNA binding, it is expected that such disruption of the ZFD would prevent transcriptional regulation by ZIC3. The in-frame deletions and duplications do not affect this region, but rather occur within or immediately adjacent to the Alanine tract, probably due to the triplicate repeat in this region. Whether the expansion of the Alanine tract is pathogenic is currently not clear. Although an alternatively spliced fourth exon of *ZIC3* has been identified, no mutations have been found within the exon. Together these mutations explain 75% of familial X-linked heterotaxy but less than 5% of sporadic cases (Cowan et al. 2014; Ware et al. 2004).

There are several pieces of evidence that suggest ZIC3 is intolerant to genetic variation and thus that it is important for human development. In particular ZIC3 is very highly conserved across organisms, particularly between zinc fingers 2–5. In addition, there are few polymorphisms in *ZIC3* in different races/ethnicities in humans, and the introns are much more highly conserved than typically seen.

From studying familial heterotaxy, it appears that the effect of ZIC3 mutation is variable. ZIC3 mutations appear to be highly penetrant, but at least one male with a ZIC3 mutation is known to be unaffected. Phenotypic variability is common within families with ZIC3 mutations. There is often a range of heart defects identified and other situs anomalies can be variable. Both asplenia and polysplenia have been identified as a result of ZIC3 mutations, as well as left isomerism and right isomerism (Chhin et al. 2007; Cowan et al. 2014, 2016; D'Alessandro et al. 2013a; El Malti et al. 2016; Ma et al. 2012; Paulussen et al. 2016; Ware et al. 2004). This suggests that mutations in ZIC3 result in randomisation of LR patterning. In general, male hemizygous patients show more severe phenotypes than female heterozygous patients carrying the same mutation. Nevertheless, female carriers are often found to have phenotypes, albeit at lower rates than males. The phenotypes in these female patients in some cases are attributed to skewed X-inactivation (Chhin et al. 2007).

Animal models suggest a role for ZIC3 in gastrulation, neural tube closure and other congenital anomalies. Large cohorts of patients with neural tube defects have not been screened for ZIC3 mutations, but a smaller screen of 352 patients failed to identify any causative variants in ZIC3 (Klootwijk et al. 2004). Expansion of the Alanine tract in ZIC3 has been reported in association with VACTERL (Wessels et al. 2010), a multiple congenital anomaly association likely resulting from multiple distinct causes that affect gastrulation. VACTERL syndrome is defined as the co-occurrence of three or more vertebral anomalies (V), anal atresia (A), cardiovascular malformations (C), tracheoesophageal (T), renal anomalies (R) and limb anomalies (L). It remains to be determined whether ZIC3 mutations contribute significantly to isolated CHD or other isolated birth defects.

15.12 Conclusion

ZIC3 is a critical regulator of early development, particularly in the establishment of the LR axis. Research over the past 20 years has shown ZIC3 to be a multifunctional protein with significant roles in several of the major signalling pathways, but it remains to be definitively determined which roles lead to heterotaxy when ZIC3 function is compromised.

References

Ahmed JN et al (2013) A murine Zic3 transcript with a premature termination codon evades nonsense-mediated decay during axis formation. Dis Model Mech 6:755–767. https://doi.org/10.1242/dmm.011668

Allache R et al (2014) Novel mutations in Lrp6 orthologs in mouse and human neural tube defects affect a highly dosage-sensitive Wnt non-canonical planar cell polarity pathway. Hum Mol Genet 23:1687–1699. https://doi.org/10.1093/hmg/ddt558

Antic D, Stubbs JL, Suyama K, Kintner C, Scott MP, Axelrod JD (2010) Planar cell polarity enables posterior localization of nodal cilia and left-right axis determination during mouse and Xenopus embryogenesis. PLoS One 5:e8999. https://doi.org/10.1371/journal.pone.0008999

Aruga J et al (2006) A wide-range phylogenetic analysis of Zic proteins: implications for correlations between protein structure conservation and body plan complexity. Genomics 87:783–792. https://doi.org/10.1016/j.ygeno.2006.02.011

Axelrod JD (2001) Unipolar membrane association of dishevelled mediates frizzled planar cell polarity signaling. Genes Dev 15:1182–1187. https://doi.org/10.1101/gad.890501

Babu D, Roy S (2013) Left-right asymmetry: cilia stir up new surprises in the node. Open Biol 3:130052. https://doi.org/10.1098/rsob.130052

Badis G et al (2009) Diversity and complexity in DNA recognition by transcription factors. Science (New York, NY) 324:1720–1723. https://doi.org/10.1126/science.1162327

Bamford RN et al (2000) Loss of function mutations in the EGF-CFC gene CFC1 are associated with human left-right laterality defects. Nat Genet 26:365–369. https://doi.org/10.1038/81695

Bangs F, Anderson KV (2017) Primary Cilia and Mammalian Hedgehog Signaling. Cold Spring Harb Perspect Biol 9. https://doi.org/10.1101/cshperspect.a028175

Bastock R, Strutt H, Strutt D (2003) Strabismus is asymmetrically localised and binds to prickle and dishevelled during drosophila planar polarity patterning. Development (Cambridge, UK) 130:3007–3014

Bedard JE, Purnell JD, Ware SM (2007) Nuclear import and export signals are essential for proper cellular trafficking and function of ZIC3. Hum Mol Genet 16:187–198. https://doi.org/10.1093/hmg/ddl461

Bedard JE, Haaning AM, Ware SM (2011) Identification of a novel ZIC3 isoform and mutation screening in patients with heterotaxy and congenital heart disease. PLoS One 6:e23755. https://doi.org/10.1371/journal.pone.0023755

Bitgood MJ, Shen L, McMahon AP (1996) Sertoli cell signaling by desert hedgehog regulates the male germline. Curr Biol 6:298–304

Blank MC, Grinberg I, Aryee E, Laliberte C, Chizhikov VV, Henkelman RM, Millen KJ (2011) Multiple developmental programs are altered by loss of Zic1 and Zic4 to cause Dandy-Walker malformation cerebellar pathogenesis. Development (Cambridge, UK) 138:1207–1216. https://doi.org/10.1242/dev.054114

Blum M, Feistel K, Thumberger T, Schweickert A (2014) The evolution and conservation of left-right patterning mechanisms. Development (Cambridge, UK) 141:1603–1613. https://doi.org/10.1242/dev.100560

Bogani D et al (2004) New semidominant mutations that affect mouse development. Genesis (New York, NY: 2000) 40:109–117. https://doi.org/10.1002/gene.20071

Caron A, Xu X, Lin X (2012) Wnt/beta-catenin signaling directly regulates Foxj1 expression and ciliogenesis in zebrafish Kupffer's vesicle. Development (Cambridge, UK) 139:514–524. https://doi.org/10.1242/dev.071746

Casey B, Devoto M, Jones KL, Ballabio A (1993) Mapping a gene for familial situs abnormalities to human chromosome Xq24-q27.1. Nat Genet 5:403–407. https://doi.org/10.1038/ng1293-403

Cast AE, Gao C, Amack JD, Ware SM (2012) An essential and highly conserved role for Zic3 in left-right patterning, gastrulation and convergent extension morphogenesis. Dev Biol 364:22–31. https://doi.org/10.1016/j.ydbio.2012.01.011

Chan DW, Liu VW, Leung LY, Yao KM, Chan KK, Cheung AN, Ngan HY (2011) Zic2 synergistically enhances hedgehog signalling through nuclear retention of Gli1 in cervical cancer cells. J Pathol 225:525–534. https://doi.org/10.1002/path.2901

Chen WS et al (2008) Asymmetric homotypic interactions of the atypical cadherin flamingo mediate intercellular polarity signaling. Cell 133:1093–1105. https://doi.org/10.1016/j.cell.2008.04.048

Chen Y, Sasai N, Ma G, Yue T, Jia J, Briscoe J, Jiang J (2011) Sonic hedgehog dependent phosphorylation by CK1alpha and GRK2 is required for ciliary accumulation and activation of smoothened. PLoS Biol 9:e1001083. https://doi.org/10.1371/journal.pbio.1001083

Chhin B et al (2007) Elucidation of penetrance variability of a ZIC3 mutation in a family with complex heart defects and functional analysis of ZIC3 mutations in the first zinc finger domain. Hum Mutat 28:563–570. https://doi.org/10.1002/humu.20480

Chung B, Shaffer LG, Keating S, Johnson J, Casey B, Chitayat D (2011) From VACTERL-H to heterotaxy: variable expressivity of ZIC3-related disorders. Am J Med Genet A 155a:1123–1128. https://doi.org/10.1002/ajmg.a.33859

Correa-Villasenor A, McCarter R, Downing J, Ferencz C (1991) White-black differences in cardiovascular malformations in infancy and socioeconomic factors. The Baltimore-Washington Infant Study Group. Am J Epidemiol 134:393–402

Cowan J, Tariq M, Ware SM (2014) Genetic and functional analyses of ZIC3 variants in congenital heart disease. Hum Mutat 35:66–75. https://doi.org/10.1002/humu.22457

Cowan JR et al (2016) Copy number variation as a genetic basis for heterotaxy and heterotaxy-spectrum congenital heart defects philosophical transactions of the Royal Society of London series B. Biol Sci 371. https://doi.org/10.1098/rstb.2015.0406

D'Alessandro LC, Casey B, Siu VM (2013a) Situs inversus totalis and a novel ZIC3 mutation in a family with X-linked heterotaxy. Congenit Heart Dis 8:E36–E40. https://doi.org/10.1111/j.1747-0803.2011.00602.x

D'Alessandro LC, Latney BC, Paluru PC, Goldmuntz E (2013b) The phenotypic spectrum of ZIC3 mutations includes isolated d-transposition of the great arteries and double outlet right ventricle. Am J Med Genet A 161a:792–802. https://doi.org/10.1002/ajmg.a.35849
De Luca A et al (2010) Familial transposition of the great arteries caused by multiple mutations in laterality genes. Heart (British Cardiac Society) 96:673–677. https://doi.org/10.1136/hrt.2009.181685
Dunaeva M, Michelson P, Kogerman P, Toftgard R (2003) Characterization of the physical interaction of Gli proteins with SUFU proteins. J Biol Chem 278:5116–5122. https://doi.org/10.1074/jbc.M209492200
El Malti R et al (2016) A systematic variant screening in familial cases of congenital heart defects demonstrates the usefulness of molecular genetics in this field. Eur J Hum Genet: EJHG 24:228–236. https://doi.org/10.1038/ejhg.2015.105
Elms P, Scurry A, Davies J, Willoughby C, Hacker T, Bogani D, Arkell R (2004) Overlapping and distinct expression domains of Zic2 and Zic3 during mouse gastrulation. Gene Expr Patterns: GEP 4:505–511. https://doi.org/10.1016/j.modgep.2004.03.003
Fakhro KA et al (2011) Rare copy number variations in congenital heart disease patients identify unique genes in left-right patterning. Proc Natl Acad Sci U S A 108:2915–2920. https://doi.org/10.1073/pnas.1019645108
Feiguin F, Hannus M, Mlodzik M, Eaton S (2001) The ankyrin repeat protein Diego mediates frizzled-dependent planar polarization. Dev Cell 1:93–101
Ferrero GB et al (1997) A submicroscopic deletion in Xq26 associated with familial situs ambiguus. Am J Hum Genet 61:395–401. https://doi.org/10.1086/514857
Field S et al (2011) Pkd1l1 establishes left-right asymmetry and physically interacts with Pkd2. Development (Cambridge, UK) 138:1131–1142. https://doi.org/10.1242/dev.058149
Fritz B et al (2005) Situs ambiguus in a female fetus with balanced (X;21) translocation – evidence for functional nullisomy of the ZIC3 gene? Eur J Hum enet: EJHG 13:34–40. https://doi.org/10.1038/sj.ejhg.5201213
Fujimi TJ, Hatayama M, Aruga J (2012) Xenopus Zic3 controls notochord and organizer development through suppression of the Wnt/beta-catenin signaling pathway. Dev Biol 361:220–231. https://doi.org/10.1016/j.ydbio.2011.10.026
Gao B et al (2011) Wnt signaling gradients establish planar cell polarity by inducing Vangl2 phosphorylation through Ror2. Dev Cell 20:163–176. https://doi.org/10.1016/j.devcel.2011.01.001
Garnett AT, Square TA, Medeiros DM (2012) BMP, Wnt and FGF signals are integrated through evolutionarily conserved enhancers to achieve robust expression of Pax3 and Zic genes at the zebrafish neural plate border. Development (Cambridge, UK) 139:4220–4231. https://doi.org/10.1242/dev.081497
Gebbia M et al (1997) X-linked situs abnormalities result from mutations in ZIC3. Nat Genet 17:305–308. https://doi.org/10.1038/ng1197-305
Haaning AM, Quinn ME, Ware SM (2013) Heterotaxy-spectrum heart defects in Zic3 hypomorphic mice. Pediatr Res 74:494–502. https://doi.org/10.1038/pr.2013.147
Hamada H, Tam PP (2014) Mechanisms of left-right asymmetry and patterning: driver, mediator and responder. F1000prime Rep 6:110. 10.12703/p6-110
Hashimoto M et al (2010) Planar polarization of node cells determines the rotational axis of node cilia. Nat Cell Biol 12:170–176. https://doi.org/10.1038/ncb2020
Hatayama M et al (2008) Functional and structural basis of the nuclear localization signal in the ZIC3 zinc finger domain. Hum Mol Genet 17:3459–3473. https://doi.org/10.1093/hmg/ddn239
Hierholzer A, Kemler R (2010) Beta-catenin-mediated signaling and cell adhesion in postgastrulation mouse embryos. Dev Dyn: Off Publ Am Assoc Anatomists 239:191–199. https://doi.org/10.1002/dvdy.22075
Houtmeyers R et al (2016) Zic2 mutation causes holoprosencephaly via disruption of NODAL signalling. Hum Mol Genet 25:3946–3959. https://doi.org/10.1093/hmg/ddw235
Huangfu D, Anderson KV (2005) Cilia and hedgehog responsiveness in the mouse. Proc Natl Acad Sci U S A 102:11325–11330. https://doi.org/10.1073/pnas.0505328102

Huangfu D, Liu A, Rakeman AS, Murcia NS, Niswander L, Anderson KV (2003) Hedgehog signalling in the mouse requires intraflagellar transport proteins. Nature 426:83–87. https://doi.org/10.1038/nature02061

Humke EW, Dorn KV, Milenkovic L, Scott MP, Rohatgi R (2010) The output of hedgehog signaling is controlled by the dynamic association between suppressor of fused and the Gli proteins. Genes Dev 24:670–682. https://doi.org/10.1101/gad.1902910

Jenny A, Reynolds-Kenneally J, Das G, Burnett M, Mlodzik M (2005) Diego and Prickle regulate frizzled planar cell polarity signalling by competing for dishevelled binding. Nat Cell Biol 7:691–697. https://doi.org/10.1038/ncb1271

Jiang Z, Zhu L, Hu L, Slesnick TC, Pautler RG, Justice MJ, Belmont JW (2013) Zic3 is required in the extra-cardiac perinodal region of the lateral plate mesoderm for left-right patterning and heart development. Hum Mol Genet 22:879–889. https://doi.org/10.1093/hmg/dds494

Keller MJ, Chitnis AB (2007) Insights into the evolutionary history of the vertebrate zic3 locus from a teleost-specific zic6 gene in the zebrafish, Danio rerio. Dev Genes Evol 217:541–547. https://doi.org/10.1007/s00427-007-0161-4

Kennedy MP et al (2007) Congenital heart disease and other heterotaxic defects in a large cohort of patients with primary ciliary dyskinesia. Circulation 115:2814–2821. https://doi.org/10.1161/circulationaha.106.649038

Kim SJ, Kim WH, Lim HG, Lee JY (2008) Outcome of 200 patients after an extracardiac Fontan procedure. J Thorac Cardiovasc Surg 136:108–116. https://doi.org/10.1016/j.jtcvs.2007.12.032

Kitaguchi T, Nagai T, Nakata K, Aruga J, Mikoshiba K (2000) Zic3 is involved in the left-right specification of the Xenopus embryo. Development (Cambridge, UK) 127:4787–4795

Kitajima K, Oki S, Ohkawa Y, Sumi T, Meno C (2013) Wnt signaling regulates left-right axis formation in the node of mouse embryos. Dev Biol 380:222–232. https://doi.org/10.1016/j.ydbio.2013.05.011

Klootwijk R et al (2004) Genetic variants in ZIC1, ZIC2, and ZIC3 are not major risk factors for neural tube defects in humans. Am J Med Genet A 124a:40–47. https://doi.org/10.1002/ajmg.a.20402

Kogerman P et al (1999) Mammalian suppressor-of-fused modulates nuclear-cytoplasmic shuttling of Gli-1. Nat Cell Biol 1:312–319. https://doi.org/10.1038/13031

Koyabu Y, Nakata K, Mizugishi K, Aruga J, Mikoshiba K (2001) Physical and functional interactions between Zic and Gli proteins. J Biol Chem 276:6889–6892. https://doi.org/10.1074/jbc.C000773200

Lee JD, Anderson KV (2008) Morphogenesis of the node and notochord: the cellular basis for the establishment and maintenance of left-right asymmetry in the mouse. Dev Dyn: Off Publ Am Assoc Anatomists 237:3464–3476. https://doi.org/10.1002/dvdy.21598

Lim LS, Hong FH, Kunarso G, Stanton LW (2010) The pluripotency regulator Zic3 is a direct activator of the Nanog promoter in ESCs. Stem Cells (Dayton, Ohio) 28:1961–1969. https://doi.org/10.1002/stem.527

Lin X, Xu X (2009) Distinct functions of Wnt/beta-catenin signaling in KV development and cardiac asymmetry. Development (Cambridge, UK) 136:207–217. https://doi.org/10.1242/dev.029561

Lin AE et al (2014) Laterality defects in the national birth defects prevention study (1998-2007): birth prevalence and descriptive epidemiology. Am J Med Genet A 164a:2581–2591. https://doi.org/10.1002/ajmg.a.36695

Liu C, Lin C, Gao C, May-Simera H, Swaroop A, Li T (2014) Null and hypomorph Prickle1 alleles in mice phenocopy human Robinow syndrome and disrupt signaling downstream of Wnt5a. Biol Open 3:861–870. https://doi.org/10.1242/bio.20148375

Ma L, Selamet Tierney ES, Lee T, Lanzano P, Chung WK (2012) Mutations in ZIC3 and ACVR2B are a common cause of heterotaxy and associated cardiovascular anomalies. Cardiol Young 22:194–201. https://doi.org/10.1017/s1047951111001181

Marques S, Borges AC, Silva AC, Freitas S, Cordenonsi M, Belo JA (2004) The activity of the nodal antagonist Cerl-2 in the mouse node is required for correct L/R body axis. Genes Dev 18:2342–2347. https://doi.org/10.1101/gad.306504

Maurus D, Harris WA (2009) Zic-associated holoprosencephaly: zebrafish Zic1 controls midline formation and forebrain patterning by regulating nodal, hedgehog, and retinoic acid signaling. Genes Dev 23:1461–1473. https://doi.org/10.1101/gad.517009

May SR et al (2005) Loss of the retrograde motor for IFT disrupts localization of Smo to cilia and prevents the expression of both activator and repressor functions of Gli. Dev Biol 287:378–389. https://doi.org/10.1016/j.ydbio.2005.08.050

McGrath J, Somlo S, Makova S, Tian X, Brueckner M (2003) Two populations of node monocilia initiate left-right asymmetry in the mouse. Cell 114:61–73

Megarbane A et al (2000) X-linked transposition of the great arteries and incomplete penetrance among males with a nonsense mutation in ZIC3. Eur J Hum Genet: EJHG 8:704–708. https://doi.org/10.1038/sj.ejhg.5200526

Meyers EN, Martin GR (1999) Differences in left-right axis pathways in mouse and chick: functions of FGF8 and SHH. Science (New York, NY) 285:403–406

Migeotte I, Grego-Bessa J, Anderson KV (2011) Rac1 mediates morphogenetic responses to intercellular signals in the gastrulating mouse embryo. Development (Cambridge, UK) 138:3011–3020. https://doi.org/10.1242/dev.059766

Milenkovic L, Scott MP, Rohatgi R (2009) Lateral transport of smoothened from the plasma membrane to the membrane of the cilium. J Cell Biol 187:365–374. https://doi.org/10.1083/jcb.200907126

Minegishi K et al (2017) A Wnt5 activity asymmetry and intercellular signaling via PCP proteins polarize node cells for left-right symmetry breaking. Dev Cell 40:439–452.e434. https://doi.org/10.1016/j.devcel.2017.02.010

Mizugishi K, Hatayama M, Tohmonda T, Ogawa M, Inoue T, Mikoshiba K, Aruga J (2004) Myogenic repressor I-mfa interferes with the function of Zic family proteins. Biochem Biophys Res Commun 320:233–240. https://doi.org/10.1016/j.bbrc.2004.05.158

Nagai T, Aruga J, Takada S, Gunther T, Sporle R, Schughart K, Mikoshiba K (1997) The expression of the mouse Zic1, Zic2, and Zic3 gene suggests an essential role for Zic genes in body pattern formation. Dev Biol 182:299–313. https://doi.org/10.1006/dbio.1996.8449

Nakamura T et al (2012) Fluid flow and interlinked feedback loops establish left-right asymmetric decay of Cerl2 mRNA. Nat Commun 3:1322. https://doi.org/10.1038/ncomms2319

Nakaya MA, Biris K, Tsukiyama T, Jaime S, Rawls JA, Yamaguchi TP (2005) Wnt3a links left-right determination with segmentation and anteroposterior axis elongation. Development (Cambridge, UK) 132:5425–5436. https://doi.org/10.1242/dev.02149

Nonaka S, Shiratori H, Saijoh Y, Hamada H (2002) Determination of left-right patterning of the mouse embryo by artificial nodal flow. Nature 418:96–99. https://doi.org/10.1038/nature00849

Nyholm MK, Wu SF, Dorsky RI, Grinblat Y (2007) The zebrafish zic2a-zic5 gene pair acts downstream of canonical Wnt signaling to control cell proliferation in the developing tectum. Development (Cambridge, UK) 134:735–746. https://doi.org/10.1242/dev.02756

Okada Y, Takeda S, Tanaka Y, Izpisua Belmonte JC, Hirokawa N (2005) Mechanism of nodal flow: a conserved symmetry breaking event in left-right axis determination. Cell 121:633–644. https://doi.org/10.1016/j.cell.2005.04.008

Oyen N, Poulsen G, Boyd HA, Wohlfahrt J, Jensen PK, Melbye M (2009) Recurrence of congenital heart defects in families. Circulation 120:295–301. https://doi.org/10.1161/circulationaha.109.857987

Pan H, Gustafsson MK, Aruga J, Tiedken JJ, Chen JC, Emerson CP Jr (2011) A role for Zic1 and Zic2 in Myf5 regulation and somite myogenesis. Dev Biol 351:120–127. https://doi.org/10.1016/j.ydbio.2010.12.037

Parmantier E et al (1999) Schwann cell-derived desert hedgehog controls the development of peripheral nerve sheaths. Neuron 23:713–724

Paulussen AD et al (2016) Rare novel variants in the ZIC3 gene cause X-linked heterotaxy. Eur J Hum Genet: EJHG 24:1783–1791. https://doi.org/10.1038/ejhg.2016.91

Popperl H, Schmidt C, Wilson V, Hume CR, Dodd J, Krumlauf R, Beddington RS (1997) Misexpression of Cwnt8C in the mouse induces an ectopic embryonic axis and causes a truncation of the anterior neuroectoderm. Development (Cambridge, UK) 124:2997–3005

Pourebrahim R et al (2011) Transcription factor Zic2 inhibits Wnt/beta-catenin protein signaling. J Biol Chem 286:37732–37740. https://doi.org/10.1074/jbc.M111.242826
Purandare SM et al (2002) A complex syndrome of left-right axis, central nervous system and axial skeleton defects in Zic3 mutant mice. Development (Cambridge, UK) 129:2293–2302
Quinn ME, Haaning A, Ware SM (2012) Preaxial polydactyly caused by Gli3 haploinsufficiency is rescued by Zic3 loss of function in mice. Hum Mol Genet 21:1888–1896. https://doi.org/10.1093/hmg/dds002
Ramalho-Santos M, Melton DA, McMahon AP (2000) Hedgehog signals regulate multiple aspects of gastrointestinal development. Development (Cambridge, UK) 127:2763–2772
Rigler SL et al (2015) Novel copy-number variants in a population-based investigation of classic heterotaxy genetics in medicine: official journal of the American College of Medical. Genetics 17:348–357. https://doi.org/10.1038/gim.2014.112
Rohatgi R, Milenkovic L, Scott MP (2007) Patched1 regulates hedgehog signaling at the primary cilium. Science (New York, NY) 317:372–376. https://doi.org/10.1126/science.1139740
Shapiro AJ et al (2014) Laterality defects other than situs inversus totalis in primary ciliary dyskinesia: insights into situs ambiguus and heterotaxy. Chest 146:1176–1186. https://doi.org/10.1378/chest.13-1704
Shimada Y, Usui T, Yanagawa S, Takeichi M, Uemura T (2001) Asymmetric colocalization of flamingo, a seven-pass transmembrane cadherin, and dishevelled in planar cell polarization. Curr Biol 11:859–863
Song H, Hu J, Chen W, Elliott G, Andre P, Gao B, Yang Y (2010) Planar cell polarity breaks bilateral symmetry by controlling ciliary positioning. Nature 466:378–382. https://doi.org/10.1038/nature09129
St-Jacques B, Hammerschmidt M, McMahon AP (1999) Indian hedgehog signaling regulates proliferation and differentiation of chondrocytes and is essential for bone formation. Genes Dev 13:2072–2086
Strutt DI (2001) Asymmetric localization of frizzled and the establishment of cell polarity in the drosophila wing. Mol Cell 7:367–375
Sutherland MJ, Ware SM (2009) Disorders of left-right asymmetry: heterotaxy and situs inversus. Am J Med Genet C: Semin Med Genet 151c:307–317. https://doi.org/10.1002/ajmg.c.30228
Sutherland MJ, Wang S, Quinn ME, Haaning A, Ware SM (2013) Zic3 is required in the migrating primitive streak for node morphogenesis and left-right patterning. Hum Mol Genet 22:1913–1923. https://doi.org/10.1093/hmg/ddt001
Tanaka Y, Okada Y, Hirokawa N (2005) FGF-induced vesicular release of sonic hedgehog and retinoic acid in leftward nodal flow is critical for left-right determination. Nature 435:172–177. https://doi.org/10.1038/nature03494
Tree DR, Shulman JM, Rousset R, Scott MP, Gubb D, Axelrod JD (2002) Prickle mediates feedback amplification to generate asymmetric planar cell polarity signaling. Cell 109:371–381
Tsiairis CD, McMahon AP (2009) An Hh-dependent pathway in lateral plate mesoderm enables the generation of left/right asymmetry. Curr Biol 19:1912–1917. https://doi.org/10.1016/j.cub.2009.09.057
Tsukui T et al (1999) Multiple left-right asymmetry defects in Shh(-/-) mutant mice unveil a convergence of the shh and retinoic acid pathways in the control of Lefty-1. Proc Natl Acad Sci U S A 96:11376–11381
Tukachinsky H, Lopez LV, Salic A (2010) A mechanism for vertebrate hedgehog signaling: recruitment to cilia and dissociation of SuFu-Gli protein complexes. J Cell Biol 191:415–428. https://doi.org/10.1083/jcb.201004108
Tzschach A et al (2006) Heterotaxy and cardiac defect in a girl with chromosome translocation t(X;1)(q26;p13.1) and involvement of ZIC3. Eur J Hum Genet: EJHG 14:1317–1320. https://doi.org/10.1038/sj.ejhg.5201707
Vierkotten J, Dildrop R, Peters T, Wang B, Ruther U (2007) Ftm is a novel basal body protein of cilia involved in Shh signalling. Development (Cambridge, UK) 134:2569–2577. https://doi.org/10.1242/dev.003715

Wang B, Fallon JF, Beachy PA (2000) Hedgehog-regulated processing of Gli3 produces an anterior/posterior repressor gradient in the developing vertebrate limb. Cell 100:423–434
Ware SM et al (2004) Identification and functional analysis of ZIC3 mutations in heterotaxy and related congenital heart defects. Am J Hum Genet 74:93–105. https://doi.org/10.1086/380998
Ware SM, Harutyunyan KG, Belmont JW (2006a) Heart defects in X-linked heterotaxy: evidence for a genetic interaction of Zic3 with the nodal signaling pathway. Dev Dyn: Off Publ Am Assoc Anatomists 235:1631–1637. https://doi.org/10.1002/dvdy.20719
Ware SM, Harutyunyan KG, Belmont JW (2006b) Zic3 is critical for early embryonic patterning during gastrulation. Dev Dyn: Off Publ Am Assoc Anatomists 235:776–785. https://doi.org/10.1002/dvdy.20668
Weatherbee SD, Niswander LA, Anderson KV (2009) A mouse model for Meckel syndrome reveals Mks1 is required for ciliogenesis and hedgehog signaling. Hum Mol Genet 18:4565–4575. https://doi.org/10.1093/hmg/ddp422
Wessels MW et al (2010) Polyalanine expansion in the ZIC3 gene leading to X-linked heterotaxy with VACTERL association: a new polyalanine disorder? J Med Genet 47:351–355. https://doi.org/10.1136/jmg.2008.060913
Winata CL et al (2013) Genome wide analysis reveals Zic3 interaction with distal regulatory elements of stage specific developmental genes in zebrafish. PLoS Genet 9:e1003852. https://doi.org/10.1371/journal.pgen.1003852
Yoshiba S et al (2012) Cilia at the node of mouse embryos sense fluid flow for left-right determination via Pkd2. Science (New York, NY) 338:226–231. https://doi.org/10.1126/science.1222538
Zeng L et al (1997) The mouse fused locus encodes Axin, an inhibitor of the Wnt signaling pathway that regulates embryonic axis formation. Cell 90:181–192
Zhang XM, Ramalho-Santos M, McMahon AP (2001) Smoothened mutants reveal redundant roles for Shh and Ihh signaling including regulation of L/R symmetry by the mouse node. Cell 106:781–792
Zhang M, Zhang J, Lin SC, Meng A (2012) Beta-catenin 1 and beta-catenin 2 play similar and distinct roles in left-right asymmetric development of zebrafish embryos. Development (Cambridge, UK) 139:2009–2019. https://doi.org/10.1242/dev.074435
Zhu L, Harutyunyan KG, Peng JL, Wang J, Schwartz RJ, Belmont JW (2007a) Identification of a novel role of ZIC3 in regulating cardiac development. Hum Mol Genet 16:1649–1660. https://doi.org/10.1093/hmg/ddm106
Zhu L, Peng JL, Harutyunyan KG, Garcia MD, Justice MJ, Belmont JW (2007b) Craniofacial, skeletal, and cardiac defects associated with altered embryonic murine Zic3 expression following targeted insertion of a PGK-NEO cassette. Front Biosci 12:1680–1690
Zhu L, Zhou G, Poole S, Belmont JW (2008) Characterization of the interactions of human ZIC3 mutants with GLI3. Hum Mutat 29:99–105. https://doi.org/10.1002/humu.20606
Zhu P, Xu X, Lin X (2015) Both ciliary and non-ciliary functions of Foxj1a confer Wnt/beta-catenin signaling in zebrafish left-right patterning. Biol Open 4:1376–1386. https://doi.org/10.1242/bio.012088

Chapter 16
Deregulation of ZIC Family Members in Oncogenesis

Rob Houtmeyers, Jabob Souopgui, and Sabine Tejpar

Abstract In the last decade, the amount of investigations on the involvement of ZIC genes in the cancer field have exponentially expanded. In most cancer types, promoter methylation leads to silenced ZIC family members, but specific subsets of patients clearly show increased expression of one or head-to-head located ZIC genes in the respective tumor tissue. It is unclear at this stage how these transcription factors contribute to tumorigenesis, but the potential implications in pathways that are most frequently mutated in cancer such as the canonical Wnt, TGF-beta, and STAT-3 pathway are evident. By exploring well-established developmental models, researchers were able to position ZIC genes not only as classical transcription factors but also as cofactors of chromatin remodeling complexes that are crucial for maintenance of the cell but also during differentiation and maturation of ZIC-expressing tissues in vivo. The translation of this obtained evidence to the cancer field will be challenging but will indisputably lead to a better understanding how the factors can contribute to the tumor development in the given subsets of patients.

Keywords Methylation · Chromatin remodeling · Tumor suppressor · Canonical Wnt · Nodal · Progenitor cells · Differentiation · ZIC · Cancer

16.1 Reported ZIC Gene Alterations in Cancer

The first discovery of ZIC genes in the cerebellum (Aruga et al. 1994) has led to numerous investigations on their function as neural transcription factors. These studies frequently used early developmental model systems that relate to the

R. Houtmeyers (✉) · S. Tejpar (✉)
Department of Oncology, Katholieke Universiteit Leuven, Leuven, Belgium
e-mail: rob.houtmeyers@kuleuven.be; sabine.tejpar@uzleuven.be

J. Souopgui
Department of Oncology, Katholieke Universiteit Leuven, Leuven, Belgium

Department of Embryology and Biotechnology, Université Libre de Bruxelles, Gosselies, Belgium

J. Aruga (ed.), *Zic family*, Advances in Experimental Medicine and Biology 1046,
https://doi.org/10.1007/978-981-10-7311-3_16

Table 16.1 *ZIC* gene deregulation in cancer

Cancer type	ZIC gene	Effect	Study
Desmoid	*ZIC1, 4*	Demethylation	Pourebrahim et al. (2007)
Gastric	*ZIC1*	Methylation	Wang et al. (2009)
Liposarcoma	*ZIC1*	Increase expression	Brill et al. (2010)
Colorectal	*ZIC1*	Methylation	Gan et al. (2011)
Epithelial ovarian	*ZIC2*	Increase expression	Marchini et al. (2012)
Thyroid	*ZIC1*	Methylation	Rodriguez-Rodero et al. (2013)
Malignant pleural mesothelioma	*ZIC1*	Methylation	Cheng et al. (2013)
Hepatocellular carcinoma	*ZIC1*	Methylation	Wang et al. (2014)
Bladder	*ZIC1*	Methylation	Kitchen et al. (2015)
Breast	*ZIC1*	Methylation	Gan et al. (2011)
Endometrial (positive node)	*ZIC2*	Increase expression	Bidus et al. (2006)
Oral squamous cell carcinoma	*ZIC2*	Increase expression	Sakuma et al. (2010)
Medulloblastoma	*ZIC1–5*	Expression	Aruga et al. (2010)
Meningioma	*ZIC1, 2, 5*	Expression	Aruga et al. (2010)
Cervical	*ZIC2*	Increase expression	Chan et al. (2011)
Bladder	*ZIC4*	Methylation	Kandimalla et al. (2012)
Ovarian	*ZIC1, 4*	Methylation	Huang et al. (2013)
Head and neck	*ZIC1, 4*	Methylation	Guerrero-Preston et al. (2014)
Non-small cell lung cancer	*Zic5*	Increase expression	Sun et al. (2016)
Melanoma	*Zic5*	Increase expression	Satow et al. (2017)

respective phenotypes when they are mutated. However, it lasted until 2006 before the first article came out on any of the ZIC gene members and their putative role in cancer (Bidus et al. 2006). During the last decade, more groups have contributed to address the implication of this gene family in tumorigenesis, but most of the investigations remain descriptive. Table 16.1 summarizes the alterations in ZIC gene expression in tumors compared to normal tissue.

In most of the cancer types or at least in large subpopulations, the expression of ZIC genes seems to be reduced through a mechanism that is called hypermethylation (Table 16.1). It is a part of epigenetic silencing that results in methylation of CpG islands, which are frequently located at promotor regions or DNA sequences that are necessary for gene transcription. This can prevent the transcriptional machinery from binding to the regulatory sequences and causes a blockade for gene expression. This is one way in which tumor cells can silence detrimental genes, the so-called tumor suppressor genes (Esteller 2002). Thanks to the rapid development of sequencing platforms and the publicly available databases, it is now possible to obtain genome-wide methylation patterns that can be linked to RNA expression and the respective protein level. Altogether, this allows the researchers to better understand the role of epigenetics on target gene regulation during tumor development. The mechanism by which critical transcriptional elements get targeted and recognized by epigenetic machinery however remains unknown and particularly challenging to investigate.

Important to note is that most of the projects shown in Table 16.1 were performed on sections of the tumor and, in case of methylation studies, are the result of the average increase of the epigenetic modifications. These studies do not incorporate heterogeneity of the tumor and cannot exclude that small subsets of patients have ZIC-expressing cells. In fact, it would be interesting to at least subdivide the results within the same cancer as this has been reported by Bidus et al. (2006). This study observed increased expression of ZIC2 only in the positive node endometrial tumors which could indicate an important necessity in a subgroup of this cancer. How these subpopulations particularly can benefit from increased ZIC expression will be discussed in the chapter on "Transcriptional Regulation by ZIC Genes."

Although a wide variety of cancer types are found that express ZIC genes, most tumors only express one family member alone or together with the head-to-head located gene. This might indicate that the specificity within the family members as discussed in Houtmeyers et al. (2013) can also be used as a benefit to the tumor. It is not clear at this point if the co-expression of the head-to-head located gene can also contribute to tumor development or that this event is only due to the opened chromatin at the shared promoter site.

16.2 ZIC Genes Can Affect Cancer-Related Pathways

16.2.1 Canonical Wnt Signaling

This signaling cascade has been well established in developmental biology and is shown to be critical for stemness. It is tightly associated with colon cancer but is indisputably linked to many other cancer types as well.

Abnormal activation of chromosome 5q was found to be correlated with the carcinogenic process of both the sporadic and hereditary familial adenomatous polyposis (FAP) tumors (Fearon and Vogelstein 1990). The adenomatous polyposis coli (APC) gene that is encoded at 5q is a crucial component of the protein complex that destroys the key Wnt activator, beta-catenin. More recent studies now indicate that approximately 90% of sporadic colon cancers harbor mutations that activate the canonical Wnt pathway, most of them by destroying the APC function (Najdi et al. 2011). A similar approach was taken by activating the cascade by proviral insertion of int-1 (Wnt1) in mice, one of the pathway ligands. It showed to be a crucial step to induce mammary adenocarcinomas and indicated that ligand activation and associated hyperplasia were initiating a multistep carcinogenic program (Tsukamoto et al. 1988). The canonical Wnt pathway is therefore well studied in oncogenic models, and developing specific inhibitors became a hot research topic. A summary of Wnt targeting strategies was reviewed by Novellasdemunt et al. (2015).

Another type of tumor in which high rates of canonical Wnt mutations were found is desmoid tumors. These tumors are fibromatous lesions that occur sporadically but also in FAP patients. Because of the association with hereditary CRC tumors, Tejpar and collaborators investigated if the molecular events that drive desmoid tumors would be identical to the ones that drive colorectal tumorigenesis (Tejpar et al. 2005). They could show that indeed FAP-associated desmoid tumors are caused by germline APC mutations and somatic inactivation of the other APC allele. The sporadic desmoids are found to have mostly oncogenic mutations in beta-catenin, mutations that corresponded to the ones that were found in CRC.

Two genes that are highly expressed, both at mRNA and protein level, in desmoid tumors versus fascia are ZIC1 and ZIC4. Although their role in tumor development is not elaborated, a hypothesis that ZIC genes could function within the canonical Wnt pathway was formulated and addressed experimentally. These studies were mainly carried out in early embryonic development, using *Xenopus laevis*. This animal model was already frequently used to study the function of ZICs in a more neurological context but also allowed these researchers to investigate the function of each ZIC gene in tissue that was not affected by tumorigenic mutations. It was first described that indeed ZIC2 could act as a potent inhibitor of canonical Wnt signaling both in vitro and in vivo by interacting with the transcription factor TCF4 (Pourebrahim et al. 2011). This was later confirmed by an independent group that showed that this inhibitory role is conserved within all ZIC family members (Fujimi et al. 2012).

These data now could help to understand why in some tumors ZIC gene promoters are methylated that causes transcriptional silencing since it could be beneficial for tumors to prevent potent Wnt inhibitors from signaling. However, it cannot explain why in specific subsets of cancers there is an increase of ZIC expression like in the desmoid tumors. This could indicate that each ZIC gene also has specific functions that are not conserved in the family or that it has potential functions independent of canonical Wnt signaling. Yet, this remains under investigation.

16.2.2 TGF-Beta Signaling

One approach to investigate other potential functions of ZIC genes is to study the phenotypic defects in animal models in which genetic modifications were introduced in one specific ZIC member. A comprehensive review on these models but also on features of the ZIC family that could explain the phenotypic and functional differences within the family was published by Houtmeyers et al. (2013).

Mutations in ZIC2 lead to the holoprosencephaly (HPE) phenotype that occurs both in humans and mice. The researchers traced back the original defect in these mice to defective organizer development, a tissue that is strongly dependent on both canonical Wnt and Nodal signaling. Since the described function of ZIC2 in canonical Wnt signaling alone cannot explain the phenotype, a potential role for ZIC2 in Nodal signaling was postulated (Warr et al. 2008).

Indeed, ZIC2 could interact with Nodal transcription factors SMAD2 and SMAD3 and can cause transcriptional alterations on Nodal-dependent targets and readout. This was evident during early development on organizer-specific targets such as FOXA2 and goosecoid but could also be reproduced in other models indicating that this could be a function that is conserved within different species (Houtmeyers et al. 2016).

Further research on the given tissue also indicated that the levels of activity of at least these two pathways are critical for target gene transcription. It is at this point difficult to understand exactly how ZIC genes could function in this signal integration, but it would indisputably lead to a better understanding of how ZICs can function in a cancer setting since abnormal Nodal and TGF-beta signaling in colon cancer is linked to poor prognosis and can result in metastatic features of the tumor (unpublished data). Since this aberrant signaling of this pathway is attributed to only a subset of CRC tumors, it would correspond better to the expression pattern of ZIC genes in tumors as summarized in Table 16.1.

16.2.3 Other Signaling Pathways

The identification of altered ZIC gene expression in a range of cancer types (Table 16.1) led to the investigation of ZIC gene function in a tumor setting. In particular, this was done for ZIC1 in gastric and thyroid cancer (Wang et al. 2009; Zhong et al. 2012).

Since in both cases ZIC1 expression is silenced by promoter methylation, their attempts were mainly focused on inducing its expression in a carcinogenic background. Although inducing gene expression might force the cells toward nonphysiological responses, both independent experiments resulted in a blockade of PI3K and MAPK signaling. Overexpression of ZIC1 contributed to inhibition of cell proliferation, cell migration, and cell cycle arrest. Taken into account all results, this would support a role for ZIC1 as a potential tumor suppressor which is silenced in both cancer types (Qiang et al. 2014; Zhong et al. 2012).

A third pathway that was inactivated upon ZIC2 overexpression in gastric cancer was the Sonic Hedgehog pathway. This pathway is frequently linked to developmental processes in which ZIC genes also play critical roles such as neural tube development. Hedgehog signals emanating from the ventral side of the neural tube are, namely, critical to create a gradient that leads to correct patterning of the neural tube. Mutations in this pathway are also linked to the HPE phenotype that can be found in ZIC2 mutant patients and mice (Houtmeyers et al. 2013).

ZIC genes have the potency to interact with GLI factors that can be seen as the transcriptional regulators of the Hedgehog pathway. Like ZIC genes, these are zinc finger-containing proteins which cannot only interact with ZICs, but because of their structural similarities, it was even proposed that their function could even be taken over by ZIC family members, although these hypotheses need further evidence before it can be generally accepted (Zhong et al. 2012; Koyabu et al. 2001).

Recent studies reported Zic5 as an oncogene in non-small cell lung cancer (Sun et al. 2016) and melanoma (Satow et al. 2017). Indeed, Zic5 was strongly expressed at the transcriptional and translational levels in these tumors and correlated with tumor aggressiveness. Knocking down Zic5 significantly inhibits cell growth and prevents metastasis formation. Inhibition of cell growth caused cell cycle arrest in the G2 phase. Zic5 expression controls cell cycle by favoring the expression of CDK1/cyclin B1 complex. Moreover Zic5 induced cell motility and invasion by reducing either CDH1 expression or activating focal adhesion kinase (FAK) (Satow et al. 2017; Sun et al. 2016). The latter group showed that Zic5 positively regulate factor platelet-derived growth factor D (PDGFD), a focal adhesion-related gene. They also reported a positive feedback loop between Zic5 expression and FAK and STAT3 activation using melanoma cancer cell lines.

FAK/STAT3 activation has been associated with melanoma malignancy and drug resistance (Hirata et al. 2015). Melanoma recurrence generally occurs as a result of acquired resistance to BRAF mutants (a mutation that lead to constitutive activation of MERK-ERK signaling to promote cell growth and survival). Zic5 or PDGFD suppression in vemurafenib-resistant cell lines significantly reduced cell proliferation and ERK phosphorylation and induced apoptosis suggesting that targeting Zic5 may provide effective therapy to BRAF inhibitor-resistant tumors. This nice story is cell line dependent, and given the diversity of cell in a tumor type, there is no guaranty that similar results will be obtained in other cell line of the same tumor type.

16.3 Transcriptional Regulation by ZIC Genes

Two independent groups (Luo et al. 2015; Frank et al. 2015) have suggested that Zic2 plays a key role in chromatin accessibility and regulation of coordinated transcription programs involved in differentiation. These studies demonstrated that Zic2 is enriched at the enhancers of both active and poised genes in ESCs and the developing cerebellum and identified Zic2 as a cofactor of the Mbd3-NuRD complex, co-occupying enhancer regions genome-wide and functioning together in regulating the ESC chromatin state and gene expression. Zic2 was also described as one of the earliest markers to be upregulated in a murine stem cell model which undergoes neuroectodermal differentiation, a model which we will use for our mechanistic studies in vitro (Liu et al. 2015). Zic2 is one of the earliest markers to be upregulated, so it is interesting for us to hypothesize a role for Zic2 in these chromatin alterations driving the differentiation of murine ES cells, as well as studying the regulation of Zic2 itself.

Not many functional studies happened on neuroectodermal differentiation, but previous works identified ZIC2 as a potent transcriptional regulator of both canonical Wnt (Pourebrahim et al. 2011) and TGF-beta/Nodal signaling during development (Houtmeyers et al. 2016). The same group showed the in vivo relevance of these findings: In *Xenopus* they demonstrated complete disruption of organizer formation, a process known to be dependent on coordinated levels of canonical Wnt and TGF-beta/Nodal signaling, upon disruption of zic2 function. These functions may be by direct promoter regulation by ZIC2 as a DNA binding protein, but equally interesting ZIC2 protein interactions leading to regulation of chromatin access and initiation of transcription or repression exist.

16.4 Conclusive Remarks

Results from the investigations on ZIC gene expression and function in the context of oncogenesis remain in general descriptive, complex, and sometime controversial. Using cell lines as model systems ZIC gene overexpression seems to provide evidence on their potential role in tumor suppression as revealed by studies on Zic1. This is supported by the observation that Zic1 expression is silenced due to promoter methylation in some cancer types. However, such a claim is intriguing, knowing that Zic1 is overexpressed in numerous cancers including desmoids and liposarcoma tumors. In line with this, Zic2 is overexpressed in oral squamous cell carcinoma, ovarian, and endometrial cancers, while studies in cell lines revealed antagonistic effects on TGF-B and Wnt signaling pathways. In contrast Zic5 overexpression in non-small cell lung cancer and melanoma seems to correlate with function studies showing this gene as oncogene in cell line gain- and loss-of-function investigation in vemurafenib-resistant cell lines. Altogether, understanding of the molecular basis of ZIC function as well as the mechanisms and interplays of this gene family in the context of developmental biology is a milestone to address their function in the context of cancer biology. Of course, there is no doubt that ZIC genes could turn out to be key players in tumorigenesis and progression with potential roles in diagnosis, prognosis, and therapy.

References

Aruga J, Yokota N, Hashimoto M, Furuichi T, Fukuda M, Mikoshiba K (1994) A novel zinc finger protein, zic, is involved in neurogenesis, especially in the cell lineage of cerebellar granule cells. J Neurochem 63(5):1880–1890

Aruga J, Nozaki Y, Hatayama M, Odaka YS, Yokota N (2010) Expression of ZIC family genes in meningiomas and other brain tumors. BMC Cancer 10:79. https://doi.org/10.1186/1471-2407-10-79

Bidus MA, Risinger JI, Chandramouli GV, Dainty LA, Litzi TJ, Berchuck A, Barrett JC, Maxwell GL (2006) Prediction of lymph node metastasis in patients with endometrioid endometrial cancer using expression microarray. Clin Cancer Res Off J Am Assoc Cancer Res 12(1):83–88. https://doi.org/10.1158/1078-0432.CCR-05-0835

Brill E, Gobble R, Angeles C, Lagos Quintana M, Crago A, Laxa B, Decarolis P, Zhang L, Antonescu C, Socci ND, Taylor BS, Sander C, Koff A, Singer S (2010) ZIC1 overexpression is oncogenic in liposarcoma. Cancer Res 70(17):6891–6901. https://doi.org/10.1158/0008-5472.CAN-10-0745

Chan DW, Liu VW, Leung LY, Yao KM, Chan KK, Cheung AN, Ngan HY (2011) Zic2 synergistically enhances Hedgehog signalling through nuclear retention of Gli1 in cervical cancer cells. J Pathol 225(4):525–534. https://doi.org/10.1002/path.2901

Cheng YY, Kirschner MB, Cheng NC, Gattani S, Klebe S, Edelman JJ, Vallely MP, McCaughan BC, Jin HC, van Zandwijk N, Reid G (2013) ZIC1 is silenced and has tumor suppressor function in malignant pleural mesothelioma. J Thorac Oncol Off Publ Int Assoc Study Lung Cancer 8(10):1317–1328. https://doi.org/10.1097/JTO.0b013e3182a0840a

Esteller M (2002) CpG island hypermethylation and tumor suppressor genes: a booming present, a brighter future. Oncogene 21(35):5427–5440. https://doi.org/10.1038/sj.onc.1205600

Fearon ER, Vogelstein B (1990) A genetic model for colorectal tumorigenesis. Cell 61(5):759–767

Frank CL, Liu F, Wijayatunge R, Song L, Biegler MT, Yang MG, Vockley CM, Safi A, Gersbach CA, Crawford GE, West AE (2015) Regulation of chromatin accessibility and Zic binding at enhancers in the developing cerebellum. Nat Neurosci 18(5):647–656. https://doi.org/10.1038/nn.3995

Fujimi TJ, Hatayama M, Aruga J (2012) Xenopus Zic3 controls notochord and organizer development through suppression of the Wnt/beta-catenin signaling pathway. Dev Biol 361(2):220–231. https://doi.org/10.1016/j.ydbio.2011.10.026

Gan L, Chen S, Zhong J, Wang X, Lam EK, Liu X, Zhang J, Zhou T, Yu J, Si J, Wang L, Jin H (2011) ZIC1 is downregulated through promoter hypermethylation, and functions as a tumor suppressor gene in colorectal cancer. PLoS One 6(2):e16916. https://doi.org/10.1371/journal.pone.0016916

Guerrero-Preston R, Michailidi C, Marchionni L, Pickering CR, Frederick MJ, Myers JN, Yegnasubramanian S, Hadar T, Noordhuis MG, Zizkova V, Fertig E, Agrawal N, Westra W, Koch W, Califano J, Velculescu VE, Sidransky D (2014) Key tumor suppressor genes inactivated by "greater promoter" methylation and somatic mutations in head and neck cancer. Epigenetics 9(7):1031–1046. https://doi.org/10.4161/epi.29025

Hirata E, Girotti MR, Viros A, Hooper S, Spencer-Dene B, Matsuda M, Larkin J, Marais R, Sahai E (2015) Intravital imaging reveals how BRAF inhibition generates drug-tolerant microenvironments with high integrin β1/FAK signaling. Cancer Cell 27(4):574–588. https://doi.org/10.1016/j.ccell.2015.03.008

Houtmeyers R, Souopgui J, Tejpar S, Arkell R (2013) The ZIC gene family encodes multi-functional proteins essential for patterning and morphogenesis. Cell Mol Life Sci 70(20):3791–3811. https://doi.org/10.1007/s00018-013-1285-5

Houtmeyers R, Tchouate Gainkam O, Glanville-Jones HA, Van den Bosch B, Chappell A, Barratt KS, Souopgui J, Tejpar S, Arkell RM (2016) Zic2 mutation causes holoprosencephaly via disruption of NODAL signalling. Hum Mol Genet 25(18):3946–3959. https://doi.org/10.1093/hmg/ddw235

Huang RL, Gu F, Kirma NB, Ruan J, Chen CL, Wang HC, Liao YP, Chang CC, Yu MH, Pilrose JM, Thompson IM, Huang HC, Huang TH, Lai HC, Nephew KP (2013) Comprehensive methylome analysis of ovarian tumors reveals hedgehog signaling pathway regulators as prognostic DNA methylation biomarkers. Epigenetics 8(6):624–634. https://doi.org/10.4161/epi.24816

Kandimalla R, van Tilborg AA, Kompier LC, Stumpel DJ, Stam RW, Bangma CH, Zwarthoff EC (2012) Genome-wide analysis of CpG island methylation in bladder cancer identified TBX2, TBX3, GATA2, and ZIC4 as pTa-specific prognostic markers. Eur Urol 61(6):1245–1256. https://doi.org/10.1016/j.eururo.2012.01.011

Kitchen MO, Bryan RT, Haworth KE, Emes RD, Luscombe C, Gommersall L, Cheng KK, Zeegers MP, James ND, Devall AJ, Fryer AA, Farrell WE (2015) Methylation of HOXA9 and ISL1 predicts patient outcome in high-grade non-invasive bladder cancer. PLoS One 10(9):e0137003. https://doi.org/10.1371/journal.pone.0137003

Koyabu Y, Nakata K, Mizugishi K, Aruga J, Mikoshiba K (2001) Physical and functional interactions between Zic and Gli proteins. J Biol Chem 276(10):6889–6892. https://doi.org/10.1074/jbc.C000773200

Liu P, Dou X, Liu C, Wang L, Xing C, Peng G, Chen J, Yu F, Qiao Y, Song L, Wu Y, Yue C, Li J, Han JD, Tang K, Jing N (2015) Histone deacetylation promotes mouse neural induction by restricting Nodal-dependent mesendoderm fate. Nat Commun 6:6830. https://doi.org/10.1038/ncomms7830

Luo Z, Gao X, Lin C, Smith ER, Marshall SA, Swanson SK, Florens L, Washburn MP, Shilatifard A (2015) Zic2 is an enhancer-binding factor required for embryonic stem cell specification. Mol Cell 57(4):685–694. https://doi.org/10.1016/j.molcel.2015.01.007

Marchini S, Poynor E, Barakat RR, Clivio L, Cinquini M, Fruscio R, Porcu L, Bussani C, D'Incalci M, Erba E, Romano M, Cattoretti G, Katsaros D, Koff A, Luzzatto L (2012) The zinc finger gene ZIC2 has features of an oncogene and its overexpression correlates strongly with the clinical course of epithelial ovarian cancer. Clin Cancer Res 18(16):4313–4324. https://doi.org/10.1158/1078-0432.CCR-12-0037

Najdi R, Holcombe RF, Waterman ML (2011) Wnt signaling and colon carcinogenesis: beyond APC. J Carcinog 10:5. https://doi.org/10.4103/1477-3163.78111

Novellasdemunt L, Antas P, Li VS (2015) Targeting Wnt signaling in colorectal cancer. A review in the theme: cell signaling: proteins, pathways and mechanisms. Am J Physiol Cell Physiol 309(8):C511–C521. https://doi.org/10.1152/ajpcell.00117.2015

Pourebrahim R, Van Dam K, Bauters M, De Wever I, Sciot R, Cassiman JJ, Tejpar S (2007) ZIC1 gene expression is controlled by DNA and histone methylation in mesenchymal proliferations. FEBS Lett 581(26):5122–5126. https://doi.org/10.1016/j.febslet.2007.09.061

Pourebrahim R, Houtmeyers R, Ghogomu S, Janssens S, Thelie A, Tran HT, Langenberg T, Vleminckx K, Bellefroid E, Cassiman JJ, Tejpar S (2011) Transcription factor Zic2 inhibits Wnt/beta-catenin protein signaling. J Biol Chem 286(43):37732–37740. https://doi.org/10.1074/jbc.M111.242826

Qiang W, Zhao Y, Yang Q, Liu W, Guan H, Lv S, Ji M, Shi B, Hou P (2014) ZIC1 is a putative tumor suppressor in thyroid cancer by modulating major signaling pathways and transcription factor FOXO3a. J Clin Endocrinol Metab 99(7):E1163–E1172. https://doi.org/10.1210/jc.2013-3729

Rodriguez-Rodero S, Fernandez AF, Fernandez-Morera JL, Castro-Santos P, Bayon GF, Ferrero C, Urdinguio RG, Gonzalez-Marquez R, Suarez C, Fernandez-Vega I, Fresno Forcelledo MF, Martinez-Camblor P, Mancikova V, Castelblanco E, Perez M, Marron PI, Mendiola M, Hardisson D, Santisteban P, Riesco-Eizaguirre G, Matias-Guiu X, Carnero A, Robledo M, Delgado-Alvarez E, Menendez-Torre E, Fraga MF (2013) DNA methylation signatures identify biologically distinct thyroid cancer subtypes. J Clin Endocrinol Metab 98(7):2811–2821. https://doi.org/10.1210/jc.2012-3566

Sakuma K, Kasamatsu A, Yamatoji M, Yamano Y, Fushimi K, Iyoda M, Ogoshi K, Shinozuka K, Ogawara K, Shiiba M, Tanzawa H, Uzawa K (2010) Expression status of Zic family member 2 as a prognostic marker for oral squamous cell carcinoma. J Cancer Res Clin Oncol 136(4):553–559. https://doi.org/10.1007/s00432-009-0689-y

Satow R, Nakamura T, Kato C, Endo M, Tamura M, Batori R, Tomura S, Murayama Y, Fukami K (2017) ZIC5 drives melanoma aggressiveness by PDGFD-Mediated activation of FAK and STAT3. Cancer Res 77(2):366–377. https://doi.org/10.1158/0008-5472.CAN-16-0991

Sun Q, Shi R, Wang X, Li D, Wu H, Ren B (2016) Overexpression of ZIC5 promotes proliferation in non-small cell lung cancer. Biochem Biophys Res Commun 479(3):502–509. https://doi.org/10.1016/j.bbrc.2016.09.098.

Tejpar S, Michils G, Denys H, Van Dam K, Nik SA, Jadidizadeh A, Cassiman JJ (2005) Analysis of Wnt/Beta catenin signalling in desmoid tumors. Acta Gastro-Enterol Belg 68(1):5–9

Tsukamoto AS, Grosschedl R, Guzman RC, Parslow T, Varmus HE (1988) Expression of the int-1 gene in transgenic mice is associated with mammary gland hyperplasia and adenocarcinomas in male and female mice. Cell 55(4):619–625

Wang LJ, Jin HC, Wang X, Lam EK, Zhang JB, Liu X, Chan FK, Si JM, Sung JJ (2009) ZIC1 is downregulated through promoter hypermethylation in gastric cancer. Biochem Biophys Res Commun 379(4):959–963. https://doi.org/10.1016/j.bbrc.2008.12.180

Wang YY, Jiang JX, Ma H, Han J, Sun ZY, Liu ZM, Xu ZG (2014) Role of ZIC1 methylation in hepatocellular carcinoma and its clinical significance. Tumour Biol 35(8):7429–7433. https://doi.org/10.1007/s13277-014-1971-4

Warr N, Powles-Glover N, Chappell A, Robson J, Norris D, Arkell RM (2008) Zic2-associated holoprosencephaly is caused by a transient defect in the organizer region during gastrulation. Hum Mol Genet 17(19):2986–2996. https://doi.org/10.1093/hmg/ddn197

Zhong J, Chen S, Xue M, Du Q, Cai J, Jin H, Si J, Wang L (2012) ZIC1 modulates cell-cycle distributions and cell migration through regulation of sonic hedgehog, PI(3)K and MAPK signaling pathways in gastric cancer. BMC Cancer 12:290. https://doi.org/10.1186/1471-2407-12-290

Chapter 17
Roles of ZIC2 in Regulation of Pluripotent Stem Cells

Hisato Kondoh

Abstract Pioneered by the classical mouse embryonic stem cells (ESCs), various stem cell lines representing the peri- and postimplantation stages of embryogenesis have been established. To gain insight into the gene regulatory network operating in these cells, we first investigated epiblast stem cells (EpiSCs), performing ChIP-seq analysis for five major transcription factors (TFs) involved in epiblast regulation. The analysis indicated that SOX2-POU5F1 TF pairs highlighted in mouse ESCs are not the major players in other stem cells. The major acting transcription factors shift from SOX2/POU5F1 in mouse ESCs to ZIC2/OTX2 in EpiSCs, and this shift is primed in ESCs by binding of ZIC2 at relevant genomic positions that later function as enhancers.

Keywords embryonic stem cells (ESCs) · epiblast stem cells (EpiSCs) · ChIP-seq · SOX2/POU5F1 · ZIC2/OTX2

17.1 Introduction

The ZIC family of transcription factors (TFs) was discovered as being involved in neural development at various developmental stages (Aruga et al. 1994; Aruga 2004). Participation of ZIC TFs in earlier developmental stages, i.e., before and during gastrulation, was suggested by their expression patterns, wherein expression of ZIC2 and ZIC3 was prominent (Elms et al. 2004; Inoue et al. 2007; Brown and Brown 2009). Their knockout or hypomorph phenotypes also indicated their involvement in gastrulation or neurulation stages of development (Nagai et al. 2000; Ware et al. 2006; Inoue et al. 2007; Warr et al. 2008), but their phenotypes were variable among embryos possibly reflecting functional redundancy among members of this family of TFs during earlier developmental stages. The experimental evidence of the involvement of ZIC2 and related TFs, particularly ZIC3, in early stages of embryogenesis was first provided by investigation of mouse embryonic stem cells (ESCs) (Lim et al. 2007). Mouse ESCs express both ZIC2 and ZIC3,

H. Kondoh (✉)
Faculty of Life Sciences, Kyoto Sangyo University, Kyoto, Japan
e-mail: kondohh@cc.kyoto-su.ac.jp

J. Aruga (ed.), *Zic family*, Advances in Experimental Medicine and Biology 1046,
https://doi.org/10.1007/978-981-10-7311-3_17

When ZIC3 was knocked down using specific shRNAs, the ESCs were destabilized and promoted to develop into endodermal tissues; the knockdown effects were enhanced by the combined knockdown of ZIC2 and ZIC3 (Lim et al. 2007).

However, a clear view of how ZIC TFs participate in regulation of pluripotent stem cells in a developmental stage-dependent fashion was obtained only recently when ChIP-seq studies on genome-wide ZIC2-binding regions were performed, as will be detailed in this chapter.

17.2 Relationships Among Stem Cells with Pluripotency

Mouse ESCs are the first achieved pluripotent stem cells derived from the inner cell mass (ICM) of preimplantation blastocyst (Evans and Kaufman 1981; Martin 1981) that were established and maintained in the presence of the cytokine leukemia inhibitory factor (LIF). Original ESC cultures used feeder cells; however, at present, various feeder-free culture recipes for ESC cultures are available, which invariantly employ LIF to activate the gp130 cellular receptor and downstream TF STAT3. The pluripotency of mouse ESCs was demonstrated by their production of teratomas upon injection into appropriate host mice (Robertson 1987) and by formation of chimeras either by injection into blastocoel or by aggregation with eight-cell stage embryos (Nagy et al. 2003). Indeed, establishment of mouse ESCs was envisioned by the chimera-forming ability of some mouse teratocarcinoma cell lines (Robertson 1987). However, it should be noted that the LIF-dependent ESC culture does not recapitulate in vivo condition of ICM in the blastocyst, which is LIF-independent (Chen et al. 2000). The LIF-dependent signaling systems inhibit the ICM-derived cells to develop further into the epiblast and other tissues of the egg cylinder stage of mouse embryogenesis.

However, ICMs derived from human blastocysts did not give rise to such cell lines. Human ICMs placed under culture conditions containing activin and bFGF in the culture medium in fibronectin-coated dishes yielded slow-growing cell lines that formed flattened cell clusters in culture, which were called human ESCs. The secretory factor nodal is expressed in the epiblastic component of embryos after implantation. Activin used to derive and maintain human ESCs should be regarded as in vitro replacement of nodal because it acts on the same nodal receptor, which consists of ActRI-ActRII complex, although activin can function in the absence of the co-receptor Crypto required for nodal signaling (Shen 2007). Therefore, it is regarded as a physiological ligand, in contrast to LIF in mouse ESCs. Although both mouse and human ESCs bear the same naming of ESCs and express SOX2, POU5F1(OCT3/4), and NANOG, which are regarded as pluripotency TFs, they are distinct in the extracellular factor requirements, cell morphology, and growth properties. In addition, mouse ESCs express *Fgf4*; in contrast, human ESCs express *Fgf5*. The expression of *Fgf5* rather than *Fgf4* by human ESCs suggested that although these cells started as *Fgf4*-expressing preimplantation ICM, they became

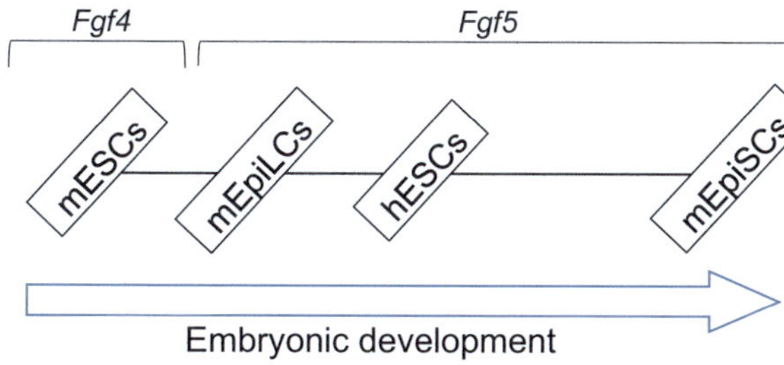

Fig. 17.1 Positioning of various pluripotent stem cells along the developmental stages. hESCs are placed between mESC and mEpiSCs according to the overlap frequency of SOX2 ChIP-seq peaks, whereas mEpiLCs are placed between mESCs and hESCs according to the frequency of POU5F1 ChIP-seq peak overlaps (Data from Matsuda et al. 2017). Expression of *Fgf4* and *Fgf5* by these stem cells is also indicated

stabilized as postimplantation epiblast expressing *Fgf5*. Human ESCs demonstrated pluripotency by their production of teratocarcinomas in immunodeficient host mice.

Following this notion, Brons et al. (2007) and Tesar et al. (2007) successfully isolated cell lines by directly seeding postimplantation epiblast of mouse egg cylinder stage embryos that resembled human ESCs in colony morphology and expression of *Fgf5* but grew faster. The cell lines retained various features of the epiblast (Brons et al. 2007; Tesar et al. 2007; Iwafuchi-Doi et al. 2012; Sumi et al. 2013) and were called epiblast stem cells (EpiSCs). Mouse EpiSCs injected into mouse blastocoel did not participate in embryogenesis (Brons et al. 2007; Tesar et al. 2007) but did participate in embryogenesis when injected into the space between the epiblast and the underlying visceral endoderm of egg cylinder stage epiblast (Sumi et al. 2013). Mouse EpiSCs also formed teratomas in immunodeficient host mice (Brons et al. 2007; Tesar et al. 2007).

When EpiSC cell lines were first established from mouse epiblast, their similarity to human ESCs was emphasized; however, an important difference was also noted. Mouse EpiSCs readily developed into neural tissues when activin was removed from the culture (Brons et al. 2007; Tesar et al. 2007; Iwafuchi-Doi et al. 2012), whereas human ESCs required a day of Fgf-free culture period to develop into neural tissues in the absence of activin (Vallier et al. 2009), indicating that mouse EpiSCs are in a more advanced developmental stage than human ESCs and are ready to develop into somatic cells.

Another developmental stage to be placed between mouse ESCs and mouse EpiSCs, designated as epiblast-like cells (EpiLCs), has also been described. When mouse ESCs are placed in culture conditions for EpiSCs, the cells assume the morphology reminiscent of EpiSCs and start to express *Fgf5* but die shortly (Hayashi et al. 2011). Therefore, EpiLCs do not represent a stable cell state but rather a transient intermediate state. Figure 17.1 schematically illustrates an order of the stem cells along the developmental stages.

17.3 Major TFs Expressed in the Epiblast Including ZIC2

Several TFs are known to be expressed throughout the epiblast of egg cylinder stage embryos before E6.5 but shift their expression domain anteriorly [ZIC2, OTX2, SOX2, and POU3F1(OCT6)] or posteriorly [POU5F1(OCT3/4)] once gastrulation starts after E6.5 (Fig. 17.2). ZIC3 is also expressed in the epiblast without showing much change in the anteroposterior distribution of the expression domains (Elms et al. 2004; Inoue et al. 2007). Although ZIC3 likely possesses regulatory functions analogous to ZIC2, considering their possessing similar protein sequences and domain organizations, ZIC2 appears to dominate over ZIC3 in the regulation of epiblast, judging from immunostaining of cells with antibodies that detect both ZIC2 and ZIC3 (Inoue et al. 2007), a higher transactivation potential of ZIC2 (Iwafuchi-Doi et al. 2012) (Fig. 17.6b), and unique enhancer-priming activity in ESCs to be discussed below.

We performed systematic ChIP-seq analysis of ZIC2 and four other TFs in EpiSCs using biotinylated TFs (Matsuda et al. 2017), with 60,000–120,000 annotated ChIP-seq peaks for each TF. In the ZIC2 ChIP-seq peak regions, the ZIC2-binding motifs, such as CTGTG, CCAGG, and CGA[G/T]TGGA, were significantly enriched. Because TFs usually function in cooperation with other TFs, we compared the overlap frequencies of TF ChIP-seq peaks and found a high-frequency peak overlap between ZIC2 and OTX2. The analysis also revealed the second overlapping TF groups of SOX2 and POU factors (Table 17.1). We also analyzed association of histone H3K4me1/K27ac modification signatures for active enhancers and H3K4me1/K27me3 modifications for poised enhancers (Ong and Corces 2012; Calo and Wysocka 2013) based on data published for the same EpiSCs (Factor et al. 2014). As shown in Table 17.2, ZIC2 ChIP-seq peaks were most frequently associated with the histone modification signature for both active and poised enhancers.

Figure 17.3 shows two examples of association of ZIC2 ChIP-seq peaks with histone modification signatures. One is the genomic region surrounding the *Fgf5* gene that is strongly expressed in EpiSCs, where ZIC2 together with OTX2 bind to five putative enhancer regions (boxed), which are overlapped with H3K4me1 and H3K27ac ChIP-seq peaks. The other is the genomic region surrounding the *Pax6* gene that is not expressed in EpiSCs but is quickly activated when the EpiSCs develop into neural tissues (Iwafuchi-Doi et al. 2012). The *Pax6* locus has a region with many sites for ZIC2 binding, which is heavily covered by H3K4me1 and K27me3 ChIP-seq peaks, indicating that the expression of *Pax6* is in a "poised" state in EpiSCs. All these observations revealed strong association of ZIC2 binding with local histone H3K27 modifications, suggesting an important function of ZIC2 in EpiSCs to specify locations of histone modifications in a context-dependent fashion.

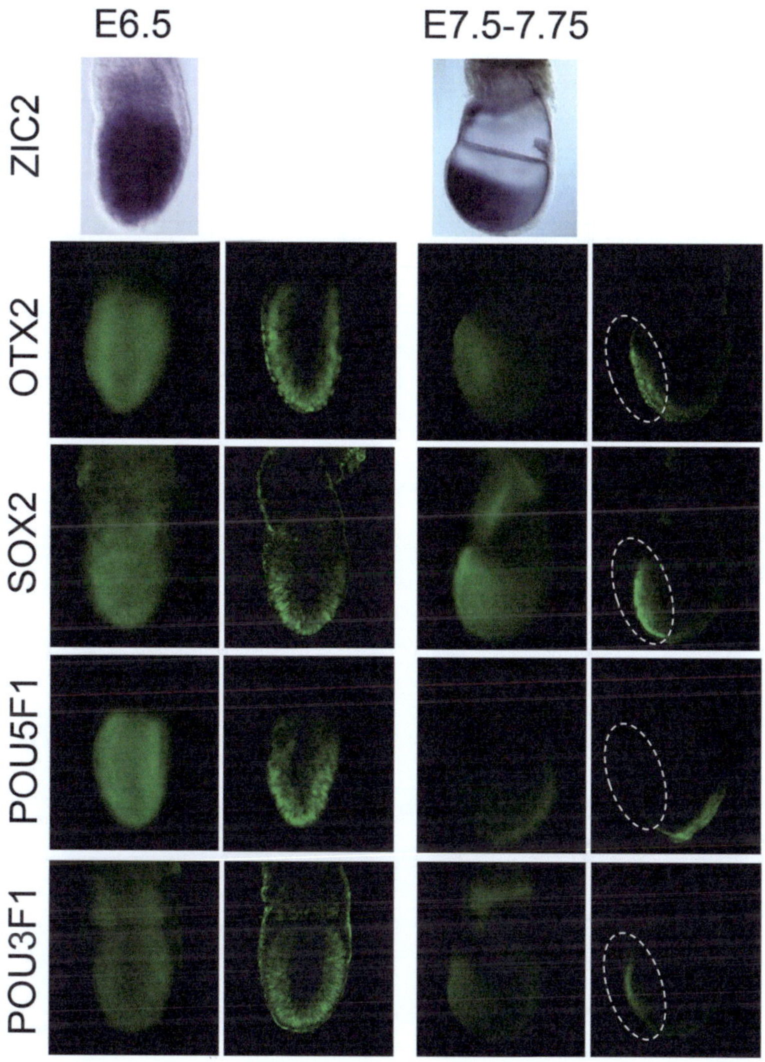

Fig. 17.2 Expression patterns of major TFs at E6.5 and E7.5–7.75 in mouse embryos. These TFs are expressed uniformly in the epiblast at E6.5 but are expressed in the anterior-dominant fashion (or posterior-dominant fashion for POU5F1) at E7.5–7.75 in mouse embryos. ZIC2 expression is shown by whole-mount in situ hybridization (Data from Fig. 1EK of Elms et al. 2004), Elsevier, reproduction granted by the publisher), whereas expression of other TFs is shown by immunostaining in lateral views (left) and optical sections (right) (Data from Fig. 2AB of Iwafuchi-Doi et al. 2012)

Table 17.1 Frequency of the overlap of ChIP-seq peaks

TFs	# ChIP-seq peaks	% peaks of the left column overlapping with peaks of:				
		ZIC2	OTX2	SOX2	POU5F1	POU3F1
ZIC2	122,053		15.9	2.8	2.8	1.2
OTX2	103,281	18.7		7.8	5	7.3
SOX2	63,751	5.3	12.6		8.1	18.4
POU5F1	62,457	5.6	8.2	8.3		53.4
POU3F1	112,753	1.3	6.7	10.4	29.6	

Overlap frequencies higher than 10% and lower than 2% are highlighted in orange and gray, respectively. High-frequency peak overlaps in ZIC2-OTX2 and SOX2-POU TF groups are highlighted by thick-lined boxes

Data from Fig. 4A of Matsuda et al. (2017)

Table 17.2 Overlaps of TF ChIP-seq peaks with histone H3 modification signatures for active or poised-state enhancers

		Active enhancers		Poised enhancers	
		K4me1∩K27ac		K4me1∩K27me3	
TFs	#TF peaks	(19,902)		(4261)	
ZIC2	122,053	15,872	13.0%	3093	2.5%
OTX2	103,281	9960	9.6%	870	0.8%
SOX2	63,751	2985	4.7%	146	0.2%
POU5F1	62,457	1369	2.2%	78	0.1%
POU3F1	112,753	963	0.9%	58	0.1%
ZIC2∩OTX2	19,349	6410	33.1%	637	3.3%

Data from Fig. 4C of Matsuda et al. (2017)

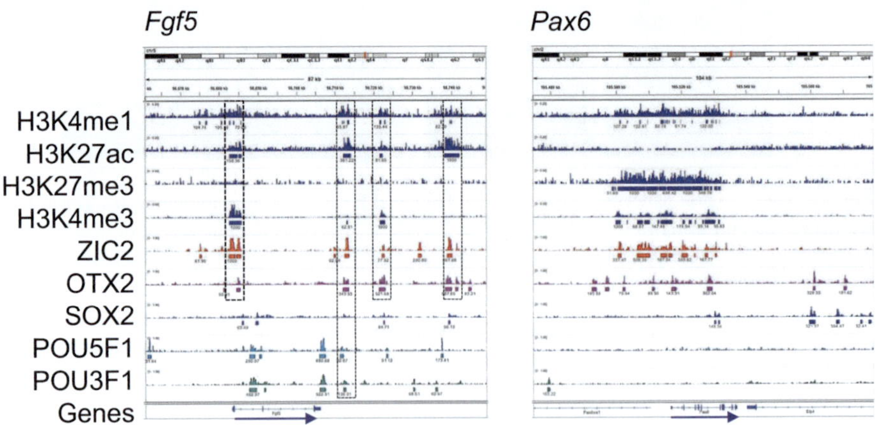

Fig. 17.3 Distribution of ChIP-seq peaks in genomic regions surrounding *Fgf5* and *Pax6* genes. ChIP-seq peaks for TF binding and modified histone H3s surrounding the genomic regions for *Fgf5* (mm9 chr5:98,661,300-98,748,500) and *Pax6* (mm9 chr2:105,475,120-105,578,450) in EpiSCs, representing active and poised-state genes, respectively (Data from Fig. 4D of Matsuda et al. 2017)

Among TF ChIP-seq peaks, the ZIC2 and OTX2 peaks overlapped at the highest frequency (~30%), and these overlapping ChIP-seq peaks were very strongly (~30%) associated with H3K4me1/K27ac modifications (Table 17.2). Many TF-encoding genes that characterize pluripotent EpiSCs, e.g., *Mycn*, *Nanog*, *Otx2*, *Pou5f1*, *Sall4*, and *Sox2*, are within 5 kb of the ZIC2-OTX2 overlapping peaks marked by histone H3K4me1/K27ac modifications and are strongly downregulated following differentiation into nonneural somatic tissues (Matsuda et al. 2017), confirming that the ZIC2-OTX2 TF pairs play major roles in the regulation of EpiSCs.

17.4 SOX2-POU5F1 Pair in ESCs and ZIC2-OTX2 Pair in EpiSCs Regulate Analogous Gene Sets

As shown in Table 17.1, SOX2, POU5F1, and POU3F1 form the second interacting TF group. It was rather surprising that the frequency of overlap between the ChIP-seq SOX2 and POU5F1 peaks was only ~8% in contrast to ~80% of ChIP-seq peak overlap in ESCs (Chen et al. 2008; Lodato et al. 2013), indicating that the functioning of SOX2 POU5F1 heterodimer prevails only in ESCs despite the fact that both ESCs and EpiSCs express high levels of SOX2 and POU5F1 (Fig. 17.4a). However, it has previously been shown that in human ESCs, the overlap frequency of SOX2 and POU5F1 ChIP-seq peaks was much lower (~20%) than in mouse ESCs, placing human ESCs in between mouse ESCs and EpiSCs in the cell state (Fig. 17.1). Therefore, the regulatory function of SOX2-POU5F TF pair, which was prominent in mouse ESCs, appears to be lost stepwisely as development proceeded. Indeed, the SOX2- and POU5F1-binding genomic regions were individually divergent between ESCs and EpiSCs, as shown in Fig. 17.4b.

It was reported that genomic distribution of H3K4me1/K27 ac-marked putative enhancer regions are markedly different between ESCs and EpiSCs (Factor et al. 2014), which is consistent with the divergent binding regions for SOX2 and POU5F1 in these cells. However, the same authors also found that the majority of genes expressed in ESCs are shared by EpiSCs (Factor et al. 2014) with some exceptions such as *Fgf4* and *Fgf5*, which are specifically expressed in mouse ESCs and EpiSCs, respectively. Considering these observations, one possible model was that similar gene sets are regulated by different enhancer sets in mouse ESCs and EpiSCs: those largely dependent on the SOX2 and POU5F1 binding in ESCs vs. those largely dependent on ZIC2 and OTX2 binding in EpiSCs. This model was supported by the observations mentioned below. When SOX2-POU5F1-cobound genomic regions marked by H3K4me1/K27 ac histone modifications in mouse ESCs were compared with ZIC2-OTX2-cobound genomic regions with the same histone marking in EpiSCs, their overlap was limited to a small fraction (~8%), indicating that the major acting enhancers in ESCs and EpiSCs are mostly located at distinct positions (Fig. 17.4c). However, when the genes proximal to these overlapping ChIP-seq peaks with histone modification signatures for active enhancers

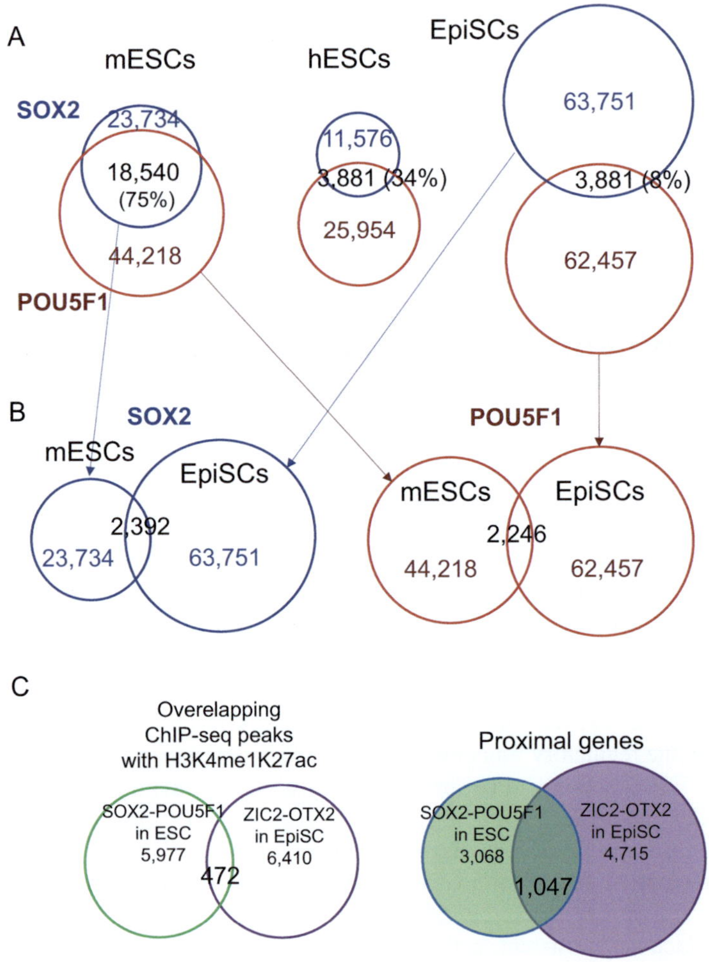

Fig. 17.4 Differences in SOX2 and POU5F1 ChIP-seq peak positions and their overlaps between mouse ESCs and EpiSCs in relation to ZIC2/OTX2 ChIP-seq peaks in EpiSCs. (**a**) Variations in the overlap of SOX2 and POU5F1 ChIP-seq peaks between ESCs and EpiSCs. (**b**) Low frequencies (<10%) of the overlap of SOX2 ChIP-seq peaks in mouse ESCs and EpiSCs and that of POU5F1 peaks in these cells. (**c**) Intersection of SOX2/POU5F1 ChIP-seq peak overlaps marked by H3K4me1/K27ac histone modifications in ESCs and for ZIC2/OTX2 ChIP-seq peak overlaps marked by the same histone modifications EpiSCs (left). Intersection of genes proximal (<50 kb for the former and <5 kb for the latter) to respective ChIP-seq peaks. Different proximity to the ChIP-seq peaks were chosen because SOX2- and POU5F1-dependent enhancers usually act from more distant positions than those dependent on ZIC2 or OTX2 (Data from Fig. 6 of Matsuda et al. 2017)

were compared (3068 genes proximal to the peaks in ESCs and 4715 genes proximal to the peaks in EpiSC), as many as 34% (1047) of the genes were identical (Fig. 17.4c). By contrast, the expected overlapping gene number of the same number of randomly selected RefSeq genes was 398. These observations indicated that SOX2-POU5F1-dependent enhancers in ESCs and ZIC2-OTX2-dependent enhancers in EpiSCs could activate similar sets of genes in these different pluripotent cells.

17.5 ZIC2-OTX2-Binding Regions in EpiSCs Are Primed, at Least in Part, at the ESC State

An immediate question that follows the above observations is how the major acting TFs are shifted from SOX2 and POU5F1 in the ESCs to ZIC2 and OTX2 in the EpiSCs. Recently published ChIP-seq data for ZIC2 binding in mouse ESCs gave 12,828 assignable peaks (Luo et al. 2015), 80% (10,241) of which overlapped with those in EpiSCs (Fig. 17.5), indicating that a majority of ZIC2-binding genomic regions in mouse ESCs are conserved in EpiSCs, and likely contribute to the enhancer activity in cooperation with OTX2 in EpiSCs. Such priming of future enhancer regions by binding of TFs has been discussed in the context of "pioneer factors" if the priming depends on heterologous TFs than the actuating TFs (Zaret and Mango 2016) or "seed enhancers" if the initial enhancer activities are minute (Factor et al. 2014). It is likely that the ZIC2-dependent enhancers that are primed at the ESC stage are activated in cooperation with OTX2 once SOX2-POU5F1 TF pair becomes disengaged from the major regulatory functions observed in ESCs. In contrast to the case of ZIC2 binding, the OTX2-binding regions are much more diversified among mouse ESCs, EpiLCs, and EpiSCs (Fig. 17.5a), which is in support of the model that ZIC2 is the major enhancer-priming TF. Indeed, Luo et al. (2015) showed that ZIC2 knockdown in ESCs using specific shRNA, but not that of ZIC3, strongly affected somatic cell development in the ESC-derived embryoid bodies.

Factor et al. (2014) identified 606 pluripotency-specific genes that are expressed in both mouse ESCs and EpiSCs but downregulated in various somatic tissues. Figure 17.5b represents an example of these genes, with *Sall4* concerned with the ChIP-seq peak distribution for ZIC2, OTX2, SOX2, and POU5F1 in ESCs and EpiSCs. OTX2 and POU5F1 data in EpiLCs, which are taken from a paper reporting that these TF pair play regulatory functions in EpiLCs, are also included (Yang et al. 2014); however, data for ZIC2 and SOX2 in EpiLCs are required to evaluate the actual roles for OTX2 and POU5F1 in EpiLCs.

When a ZIC2 peak is found at a genomic position of EpiSCs proximal to a pluripotency-specific gene, a corresponding ZIC2 peak is also found in the ESC genome (indicated by boxes in dash lines in Fig. 17.5b). The same ZIC2-bound genomic regions are often bound by SOX2, POU5F1, and even OTX2 in ESCs. In EpiLC, the same region is also bound by OTX2 and POU5F1, and in EpiSCs, the ZIC2 peaks are often, but not always, associated with the OTX2 peaks. Even with the *Fgf5* gene, which is not expressed in ESCs, the five putative enhancer regions bound by ZIC2 and OTX2 in EpiSCs (Figs. 17.3 and 17.5b) are also bound by ZIC2 in ESCs, providing a strong support for the priming function of ZIC2 binding in ESCs.

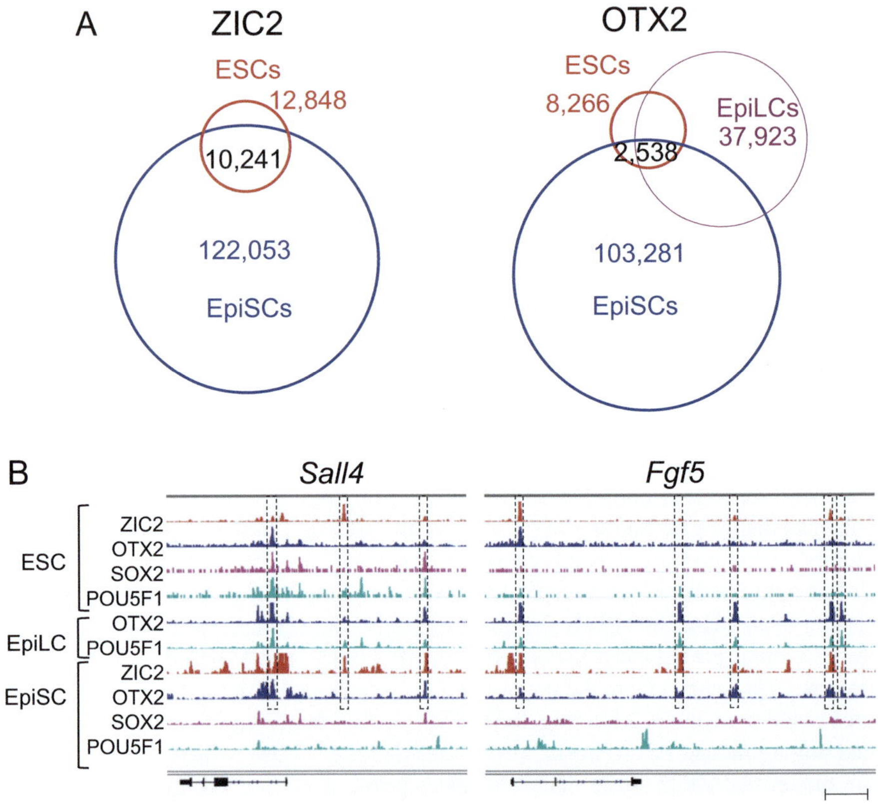

Fig. 17.5 Replacing the regulatory functions of SOX2-POU5F1 TF pair in ESCs by those of ZIC2-OTX2 pair in EpiSCs. (**a**) ZIC2-binding genomic regions in mouse ESCs are mostly conserved in EpiSCs (left), whereas OTX2-binding regions are more variable among mouse ESCs, EpiLCs, and EpiSCs (right). (**b**) Genomic regions surrounding the *Sall4* (pluripotency-related) and *Fgf5* (EpiSC-specific) genes. Dash-lined boxes indicate the genomic regions where ZIC2 binding is indicated by ChIP-seq peaks in both ESCs and EpiSCs. These regions are also bound by other TFs in ESCs, EpiLCs, or EpiSCs but little by SOX2 or POU5F1 in EpiSCs (Data from Fig. 7 of Matsuda et al. 2017)

17.6 Regulation by ZIC2 in the Epiblast and During the Derivation of Somatic Cells

We analyzed the impacts of knockdown or overexpression of six TFs, namely, ZIC2, OTX2, SOX2, POU5F1, POU3F1, and NANOG, on the expression of various TF and signaling factor genes in EpiSCs, where many genes for various somatic lineages were also expressed, albeit at low levels, in addition to the genes that characterize the epiblast. The results shown in Fig. 17.6a indicated that integration of individual action of these TFs results in activation of genes that are epiblast characteristic and/or involved in neural development (*Fgf5*, *Eomes*, *Nanog*, *Sox2*, and *Nkx1.2*) and inhibition of mesoderm- and endoderm-characteristic genes, *Brachyury* and *Sox17*, respectively. Activation of *Fgf5* by ZIC2 was consistent with the assignment of gene-proximal ZIC2-binding region as putative enhancers of respective genes (Fig. 17.3). *Sox2* expression in the epiblast depends on the upstream N2 enhancer, which is activated by the combined actions of ZIC2, OTX2, and a POU factor (Iwafuchi-Doi et al. 2012) (Fig. 17.6b), consistent with the moderate activation of *Sox2* by overexpression of ZIC2 (Fig. 17.6a).

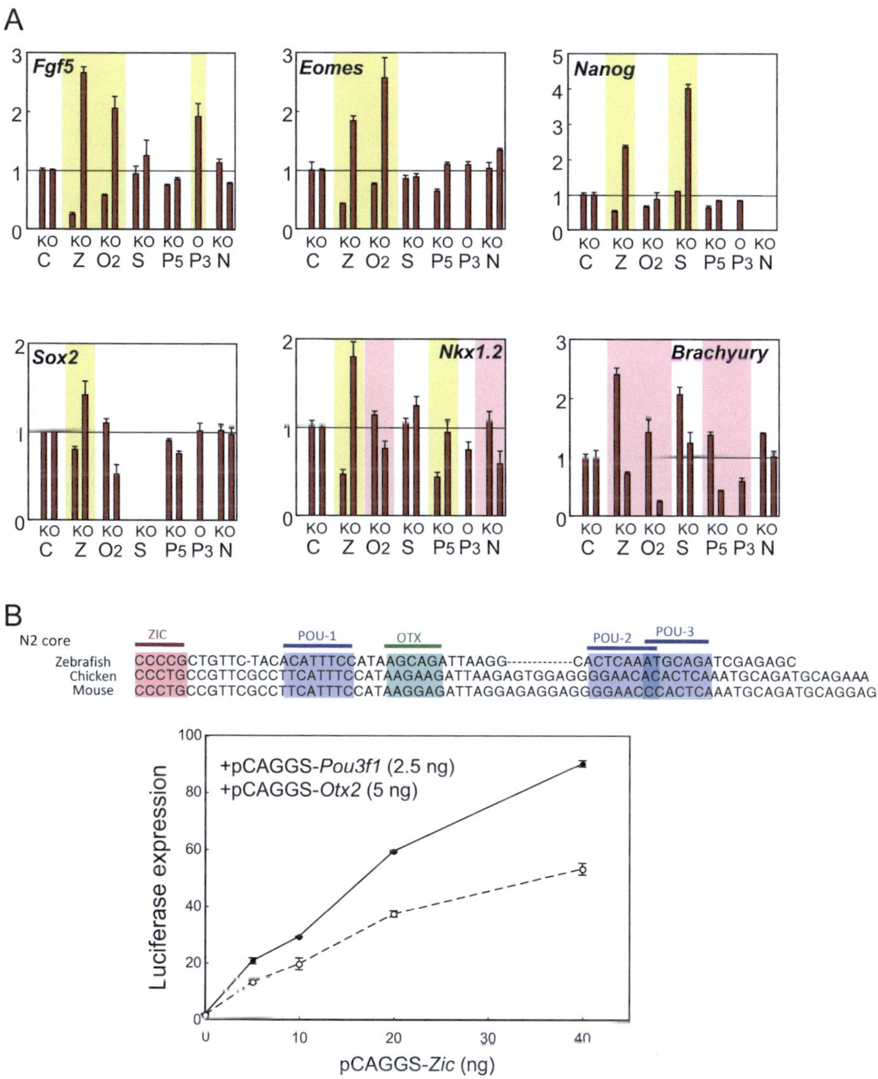

Fig. 17.6 Gene regulatory networks in epiblast stem cells. (**a**) Impacts of knockdown (K) or overexpression (O) of ZIC2/3 (Z), OTX2 (O2), SOX2 (S), POU5F1 (P5), POU3F1 (P3), and NANOG (N) on the indicated genes, as determined by qRT-PCR. When knockdown of a TF downregulates and/or overexpression of a TF activates the gene, the TF is assessed to activate the gene and highlighted in yellow. When opposite effects were observed, the TF is assessed to repress the gene and highlighted in pink. (**b**) The core sequence of the N2 enhancer of *Sox2* (top) and activation by ZIC2 (solid line) or ZIC3 (broken line) of the minimal 73-bp enhancer dimer in 10T1/2 fibroblasts together with POU3F1 and OTX2 (bottom) (Data adopted from Fig. 3 and Fig. S9 of Iwafuchi-Doi et al. 2012)

17.7 Conclusions

The regulatory functions of ZIC2 in the earlier developmental stages and pluripotent stem cells have only recently gained a spotlight. The ChIP-seq and functional analyses of ZIC2 in mouse ESCs and EpiSCs have revealed new aspects of pluripotent stem cell regulation. SOX2, POU5F1, OTX2, and ZIC2 change their regulatory roles among the stem cells, where ZIC2 function in priming future acting enhancers appears to be particularly important in deriving somatic cells from the stem cells. In addition, the possible link between ZIC2 binding to a genomic region and histone H3 modifications for active or poised enhancers may provide a paradigm for the study of the causal link between TF binding and epigenetic histone modifications.

Acknowledgments This study was supported by JSPS Kakenhi Grant P26251024 and JP17H03680.

References

Aruga J (2004) The role of Zic genes in neural development. Mol Cell Neurosci 26:205–221

Aruga J, Yokota N, Hashimoto M, Furuichi T, Fukuda M, Mikoshiba K (1994) A novel zinc finger protein, zic, is involved in neurogenesis, especially in the cell lineage of cerebellar granule cells. J Neurochem 63:1880–1890

Brons IG, Smithers LE, Trotter MW, Rugg-Gunn P, Sun B, Chuva de Sousa Lopes SM, Howlett SK, Clarkson A, Ahrlund-Richter L, Pedersen RA, Vallier L (2007) Derivation of pluripotent epiblast stem cells from mammalian embryos. Nature 448:191–195

Brown L, Brown S (2009) Zic2 is expressed in pluripotent cells in the blastocyst and adult brain expression overlaps with makers of neurogenesis. Gene Expr Patterns 9:43–49

Calo E, Wysocka J (2013) Modification of enhancer chromatin: what, how, and why? Mol Cell 49:825–837

Chen JR, Cheng JG, Shatzer T, Sewell L, Hernandez L, Stewart CL (2000) Leukemia inhibitory factor can substitute for nidatory estrogen and is essential to inducing a receptive uterus for implantation but is not essential for subsequent embryogenesis. Endocrinology 141:4365–4372

Chen X, Xu H, Yuan P et al (2008) Integration of external signaling pathways with the core transcriptional network in embryonic stem cells. Cell 133:1106–1117

Elms P, Scurry A, Davies J, Willoughby C, Hacker T, Bogani D, Arkell R (2004) Overlapping and distinct expression domains of Zic2 and Zic3 during mouse gastrulation. Gene Expr Patterns 4:505–511

Evans MJ, Kaufman MH (1981) Establishment in culture of pluripotential cells from mouse embryos. Nature 292:154–156

Factor DC, Corradin O, Zentner GE, Saiakhova A, Song L, Chenoweth JG, McKay RD, Crawford GE, Scacheri PC, Tesar PJ (2014) Epigenomic comparison reveals activation of "seed" enhancers during transition from naive to primed pluripotency. Cell Stem Cell 14:854–863

Hayashi K, Ohta H, Kurimoto K, Aramaki S, Saitou M (2011) Reconstitution of the mouse germ cell specification pathway in culture by pluripotent stem cells. Cell 146:519–532

Inoue T, Ota M, Mikoshiba K, Aruga J (2007) Zic2 and Zic3 synergistically control neurulation and segmentation of paraxial mesoderm in mouse embryo. Dev Biol 306:669–684

Iwafuchi-Doi M, Matsuda K, Murakami K, Niwa H, Tesar PJ, Aruga J, Matsuo I, Kondoh H (2012) Transcriptional regulatory networks in epiblast cells and during anterior neural plate development as modeled in epiblast stem cells. Development 139:3926–3937

Lim LS, Loh YH, Zhang W, Li Y, Chen X, Wang Y, Bakre M, Ng HH, Stanton LW (2007) Zic3 is required for maintenance of pluripotency in embryonic stem cells. Mol Biol Cell 18:1348–1358

Lodato MA, Ng CW, Wamstad JA, Cheng AW, Thai KK, Fraenkel E, Jaenisch R, Boyer LA (2013) SOX2 co-occupies distal enhancer elements with distinct POU factors in ESCs and NPCs to specify cell state. PLoS Genet 9:e1003288

Luo Z, Gao X, Lin C, Smith ER, Marshall SA, Swanson SK, Florens L, Washburn MP, Shilatifard A (2015) Zic2 is an enhancer-binding factor required for embryonic stem cell specification. Mol Cell 57:685–694

Martin GR (1981) Isolation of a pluripotent cell line from early mouse embryos cultured in medium conditioned by teratocarcinoma stem cells. Proc Natl Acad Sci U S A 78:7634–7638

Matsuda K, Mikami T, Oki S, Iida H, Andrabi M, Boss J, Yamaguchi K, Shigenobu S, Kondoh H (2017) ChIP-seq analysis of genomic binding regions of five major transcription factors highlights a central role for ZIC2 in the mouse epiblast stem cell gene regulatory network. Development 144:1948–1958

Nagai T, Aruga J, Minowa O, Sugimoto T, Ohno Y, Noda T, Mikoshiba K (2000) Zic2 regulates the kinetics of neurulation. Proc Natl Acad Sci U S A 97:1618–1623

Nagy J, Gertsenstein M, Vintersten K, Behringer R (2003) Manipulating the mouse embryo: a lobaoratory manual. Cold Spring Harbor Laboratory Press, Cold Spring Harbor

Ong CT, Corces VG (2012) Enhancers: emerging roles in cell fate specification. EMBO Rep 13:423–430

Robertson EJ (1987) Teratocarcinomas and embryonic stem cells: a practical approach. IRL Press, Oxford

Shen MM (2007) Nodal signaling: developmental roles and regulation. Development 134:1023–1034

Sumi T, Oki S, Kitajima K, Meno C (2013) Epiblast ground state is controlled by canonical Wnt/beta-catenin signaling in the postimplantation mouse embryo and epiblast stem cells. PLoS One 8:e63378

Tesar PJ, Chenoweth JG, Brook FA, Davies TJ, Evans EP, Mack DL, Gardner RL, McKay RD (2007) New cell lines from mouse epiblast share defining features with human embryonic stem cells. Nature 448:196–199

Vallier L, Mendjan S, Brown S, Chng Z, Teo A, Smithers LE, Trotter MW, Cho CH, Martinez A, Rugg-Gunn P, Brons G, Pedersen RA (2009) Activin/Nodal signalling maintains pluripotency by controlling Nanog expression. Development 136:1339–1349

Ware SM, Harutyunyan KG, Belmont JW (2006) Zic3 is critical for early embryonic patterning during gastrulation. Dev Dyn 235:776–785

Warr N, Powles-Glover N, Chappell A, Robson J, Norris D, Arkell RM (2008) Zic2-associated holoprosencephaly is caused by a transient defect in the organizer region during gastrulation. Hum Mol Genet 17:2986–2996

Yang SH, Kalkan T, Morissroe C, Marks H, Stunnenberg H, Smith A, Sharrocks AD (2014) Otx2 and Oct4 drive early enhancer activation during embryonic stem cell transition from naive pluripotency. Cell Rep 7:1968–1981

Zaret KS, Mango SE (2016) Pioneer transcription factors, chromatin dynamics, and cell fate control. Curr Opin Genet Dev 37:76–81

Chapter 18
Role of Zic Family Proteins in Transcriptional Regulation and Chromatin Remodeling

Minoru Hatayama and Jun Aruga

Abstract Proper functions of Zic proteins are essential for animals in health and disease. Here, we summarize our current understanding of the molecular properties and functions of the Zic family across animal species and paralog subtypes. Zics are basic proteins with some posttranslational modifications and can move to the cell nucleus via importin- and CRM1-based nucleocytoplasmic shuttling mechanisms. Degradation is mediated by the ubiquitin proteasome system. Many Zic proteins are capable of binding to two types of target DNA sequences (CTGCTG-core-type and GC-stretch-type). Recent chromatin immunoprecipitation assays showed that CTGCTG-core-type target sequences are enriched in enhancers. Nonetheless, the DNA binding is not always required for transcriptional regulation by Zic proteins. On the other hand, Zic proteins bind many proteins including transcription factors (Gli1–3, Tcf1 or Tcf4, Smad2 or Smad3, Oct4, Pax3, Cdx, and SRF), chromatin-remodeling factors (NuRD and NURF), and other nuclear enzymes (DNA-PK, PARP1, and RNA helicase A). Zic family–mediated gene expression control involves both their actions near the transcription start site and those affecting the global gene expression via binding to enhancers. Although Zic proteins perform essential functions in transcriptional regulation of *Oct4* and *Nanog* expression via their promoters, recent genome-wide analyses of the Zic-binding sites and their downstream targets indicate that Zic proteins are associated with distant regulatory elements and are the critical enhancer-priming nuclear regulators in organismal development. Chromatin-remodeling complexes such as NuRD and NURF that interact with Zic proteins have been shown to participate in Zic-mediated enhancer regulation.

Keywords Protein structure · Protein modification · Protein trafficking · Protein-protein interaction · Prtein-DNA interaction · Gene expression regulation · Epigenetic regulation

---

M. Hatayama (✉) · J. Aruga (✉)
Department of Medical Pharmacology, Nagasaki University Graduate School of Biomedical Sciences, Nagasaki, Japan
e-mail: hatayama@nagasaki-u.ac.jp; aruga@nagasaki-u.ac.jp

J. Aruga (ed.), *Zic family*, Advances in Experimental Medicine and Biology 1046,
https://doi.org/10.1007/978-981-10-7311-3_18

18.1 Molecular Properties of Zic Family Proteins

18.1.1 Biophysical and Structural Properties of Zic Family Proteins

Zic proteins are middle-sized basic proteins (e.g., human ZIC1–5, M_w 37–68 kDa, pI 8.7–9.6). Although varying among species, molecular weight of most of metazoan Zic proteins is within 30–75 kDa, most frequently ~50 kDa. Zic proteins share five C2H2-type zinc finger (ZF) domains (ZFDs) and additional partially conserved domains (ZOC, ZFNC; Chap. 1).

Zic family proteins contain five tandemly repeated Cys2His2 (C2H2)-type ZFDs similar to those of proteins Gli, Glis, and Nkl (Glis2) (Chap. 1 and Fig. 1.2). In one study, three-dimensional (3D) structure of human GLI1 ZFD complexed with its high-affinity DNA-binding site was reported (Pavletich and Pabo 1993). Nonetheless, there are clear functional differences between Zic and Gli ZFDs in terms of a nuclear localization signal (NLS; Chap. 1) and target sequences (see below). We determined the 3D structure of human ZIC3 ZFD ZF1–ZF4 in solution (Hatayama et al. 2008). As a result, ZF2–ZF4 showed typical C2H2 ZF structure in which two cysteines are located in an antiparallel β-sheet region and two histidines are in an α-helical region, and together they are tetrahedrally coordinated to a zinc ion, as in GLI1 ZFs (Fig. 18.1a). A global view of the structure (Fig. 18.1b) indicates that ZF1 also conform to the typical C2H2 ZF structure. In contrast, ZF1 and ZF2 are folded into a single structure (heterodimer), whereas ZF3 and ZF4 exist as independent units. The unification is mediated by hydrophobic interactions between groups of hydrophobic residues in ZF1 and ZF2, which conform to the evolutionarily conserved tandem CWCH2 sequence motif (Chap. 1) (Hatayama and Aruga 2010). In the ZF2+ZF3 region, there are two clusters of NLS-forming basic residues (Fig. 18.2a, b), giving rise to a bipartite-cluster-type NLS (Hatayama et al. 2008). Because GLI1 ZF1 and ZF2 also form a unified structure (Pavletich and Pabo 1993), the overall structure of the ZFD is thought to be similar between ZIC3 ZFD and GLI1 ZFD. In case of the GLI1 ZFD, ZF2–ZF5 bind in the major groove and wrap around the DNA (Fig. 18.2c), and this DNA-binding model is widely known for the C2H2-type ZFD (Wolfe et al. 2000). ZIC3 ZFD is predicted to interact with target DNA in a grossly similar fashion because it also shows partial affinity for the GLI-binding site (Sect. 18.2.1). However, it is possible that Zic ZF2 and ZF3 differently interact with DNA from those of Gli proteins, considering the difference in the binding motifs between Zic and Gli proteins (Sect. 18.2.1).

Alterations in Zic, Gli, and Glis ZFD structure have also been analyzed by circular dichroism spectroscopy. The α-helical part is extremely thermally stable, reflecting the compact globular structure, but decreases upon removal of the zinc ion (Sakai-Kato et al. 2008). After incubation with oligo DNA containing the Gli-binding sequence, the α-helix contents of Zic, Gli, and Glis proteins all increase and are reported to be the highest in Gli proteins among the three families and to correlate well with the affinity of each ZF for the oligo DNA (correlation coefficient

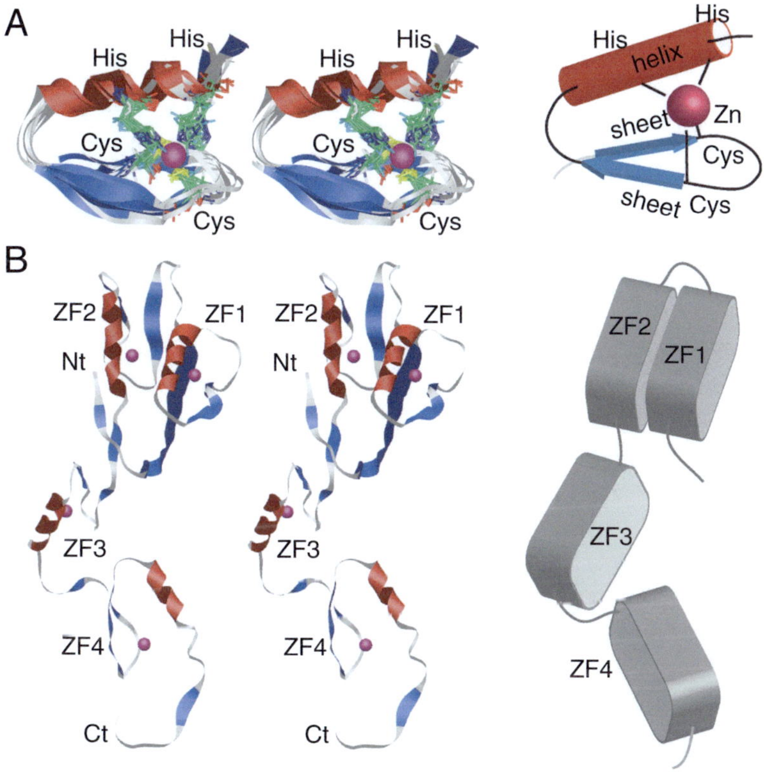

Fig. 18.1 Structure of the Zic zinc finger (ZF) domain (ZFD). (**a**) Conserved structure of ZIC3 and GLI1 ZFs. ZIC3 ZF2–ZF4 and GLI1 ZF1–ZF5 are superimposed (*left*, stereoview). *Ribbons*, polypeptide backbones; *sticks*, side chains of two cysteines and two histidines; *spheres*, zinc ions. *Right*, a schematic drawing. Two β-sheets and one α-helix (ββα) form a basic globular structure of C2H2 ZF. (**b**) ZIC3 ZF1–ZF4 (Protein Data Bank [PDB] ID 2PRC). *Left*, stereoview; *right*, schematic drawing. ZFs are linked via a short linker and ZF1 and ZF2 form a single structure

0.85) (Sakai-Kato et al. 2008). Among the five members of the mouse Zic family, the effect of lowering of pH is the smallest in Zic3, but all manifest the same tendency for decreased α-helix content (Sakai-Kato et al. 2009). Heterotaxy-derived ZIC3 ZF1 mutations (see Chap. 15) increase random coil content (C253S and H286R) or cause only a subtle change, in case of the absence of a tryptophan residue (W255G), thus indicating the differential effects on ZFD conformation among the three mutations (Hatayama et al. 2008).

At present, there are no reports dealing with biophysical and structural properties of Zic proteins besides the ZFD. Future studies are expected to uncover the properties of the other domains including ZOC, ZFNC, and carboxy termini of vertebrate Zic1–3 or the relations among them.

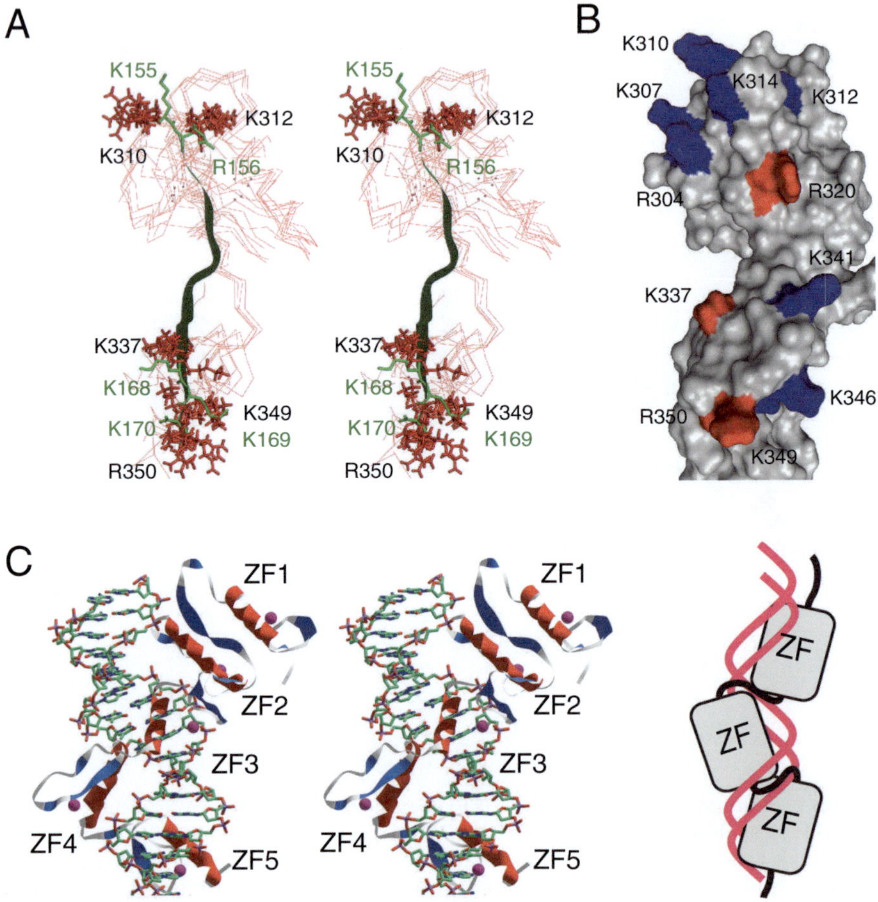

Fig. 18.2 Structure of the Zic NLS and GLI–DNA complex. (**a**) ZIC3 NLS (*red*) superimposed on a typical bipartite NLS from nucleoplasmin (*green*) (PDB ID: 1EJY) (stereoview). (**b**) A surface model of ZIC3 NLS. The core residue of NLS is indicated in *red*, and other NLS-forming residues are highlighted in *blue*. (**c**) Structure of the GLI–DNA complex (PDB ID 2GLI) (stereoview)

18.1.2 Zic Protein Isoforms

Aside from the Zic protein isoforms represented by paralogs (e.g., mammalian Zic1–5), there are isoforms reflecting transcript variation. According to some studies, vertebrate Zic3 genes are composed of three exons (Chap. 1). On the basis of a sequence comparison among mammalian species, Bedard et al. (2011) identified the fourth exon located 5 kb downstream of the conventional *Zic3* gene. The fourth exon is used as an alternative to the third exon, and the resultant transcript is predicted to encode a Zic3 isoform (Zic3-B) in which a part of ZF5 in the C terminus is replaced with a new amino acid (aa) sequence together with a change in the ZF5 C2H2 motif within C2HC. Zic3-B shows increased nuclear accumulation, but the transcription-activating ability is significantly lower in comparison with the conventional Zic3 protein. The Zic3-B transcript has been detected in mouse embryos and in the adult brain by RT-PCR assays. Further research is expected to clarify the significance of Zic3-B in terms of ontogeny and phylogeny.

In the current database of human and mouse protein-coding regions (CCDS Rel 21), we also identified four ZIC4 isoforms (aa sequence lengths 334, 384, 372, and 128). They share a common aa sequence from ZF4 to the C terminus. The 128-aa isoform is generated by alternative splicing with donor usage of the A-intron (Chap. 1), containing only ZF4 and ZF5 in its ZFD. The other three isoforms are generated by alternative splicing of the N-terminal-region-containing exons, comprising intact ZF1–ZF5 ZFD. Possible differences in their molecular and functional properties have not been described. Nevertheless, accumulation of full-length mRNA sequence data in databases should help to understand the significance of isoforms derived from transcript variants.

18.1.3 Posttranslational Modification of Zic Proteins

According to some studies, phosphorylation occurs at position S200 of mouse Zic2 (S199 in human ZIC2), and this aa position is conserved in vertebrates (Ishiguro and Aruga 2008; Olsen et al. 2010). The phosphorylation is mediated by the DNA-dependent protein kinase catalytic subunit (DNA-PKcs), and this phosphorylation is necessary for the cofactor RHA's (RNA helicase) binding to Zic2 (Ishiguro et al. 2007). Zic2 can be ubiquitinated by E3 ubiquitin ligase Rines (Ogawa et al. 2008), and small ubiquitin-like modifier (SUMO) can be covalently attached to ZIC3 K248 (Chen et al. 2013). The biological significance will be reviewed later.

In addition, there are several modification sites in the human and mouse ZIC proteins (Fig. 18.3), and these sites have been mostly identified in proteomic studies (Christensen et al. 2010; Dephoure et al. 2008; Grimsrud et al. 2012; Hendriks et al. 2014; Huttlin et al. 2010; Kettenbach et al. 2011; Olsen et al. 2010; Rigbolt et al. 2011; Zhao et al. 2010). Some of them are found among conserved residues of paralogs or between human and mouse orthologs (Fig. 18.3). In addition to the two abovementioned modifications (ZIC2 phospho-S199 and ZIC3 SUMO-K248), the following modifications are noteworthy. First, phosphorylation of threonine residues in the highly conserved C2H2 ZF linker sequence has been found in ZIC2 (T328, T358) and ZIC5 (T486). This site can be summarized as HT*GEKP, where H indicates the last histidine in the C2H2 motif and T* denotes the phosphothreonine. These modifications may represent the cell cycle dependent inactivation of C2H2 ZF proteins (Rizkallah et al. 2015; Suzuki et al. 2015). In various studies, the corresponding threonine of many C2H2 ZF proteins has been shown to be phosphorylated, and this phosphorylation impairs the ZFD DNA interaction (Dovat

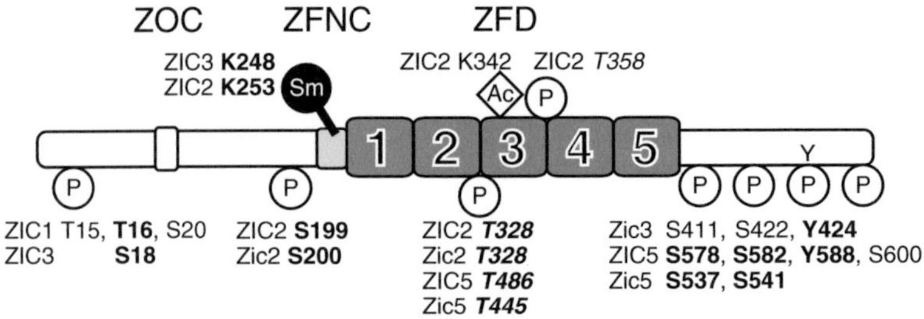

Fig. 18.3 Posttranslational modification of Zic proteins. Generalized structures of human ZIC and mouse Zic proteins are presented, and conserved domains and posttranslational modifications supported by experimental evidence are indicated. *Ac* acetylation, *P* phosphorylation, *Sm* SUMOylation, *Y* position of tyrosine phosphorylation. *Bold*, a conserved aa residue among the paralogs or orthologs; *italic*, conserved in the C2H2 ZF unit

et al. 2002; Jantz and Berg 2004) and takes place during mitotic prophase, metaphase, and anaphase (Rizkallah et al. 2011). The responsible kinases are proposed to be CDK1 (Suzuki et al. 2015) and its target TOPK/PBK (Rizkallah et al. 2015). These findings suggest that the interaction of Zic proteins with DNA is also inhibited in a cell cycle-dependent manner by the threonine phosphorylation in a ZF linker. Second, acetylation can occur at ZIC2 K342, which is equivalent to NLS-forming ZIC3 K337 in ZF3. In some studies, acetylation of the corresponding sites in another ZF protein was found to impair DNA binding (Song et al. 2003), and acetylation of lysine in NLS was shown to impair the nuclear localization of some transcription factors (Sadoul et al. 2011). Third, there are conserved phosphorylation sites both in N- and C-terminal regions. Phosphorylation sites in the ZIC3 C-terminal region have also been identified in our analyses (Hatayama et al., unpublished results). Some are conserved among the paralogs (ZIC1 T16 and ZIC3 S18, ZIC3 Y424 and ZIC5 Y588), indicating the presence of an evolutionarily conserved mechanism of phosphorylation.

Proteolytic processing is one of several modifications of protein function. In the Gli2 or Gli3 protein, cleavage is important for functional regulation (Hatayama and Aruga 2012). Nonetheless, processing of Zic proteins has not been reported to date.

18.2 Molecular Interactions Involving Zic Proteins

18.2.1 DNA-Binding Properties of Zic Family Proteins

DNA–Zic protein interaction has been investigated by means of synthetic DNA or genome-derived DNA sequences. One of the synthetic DNA approaches is called SELEX (systematic evolution of ligands by exponential enrichment) or CASTing (cyclic amplification and selection of targets); this classical method has enabled researchers to identify the sequences that bind to Zic proteins or a purified ZFD

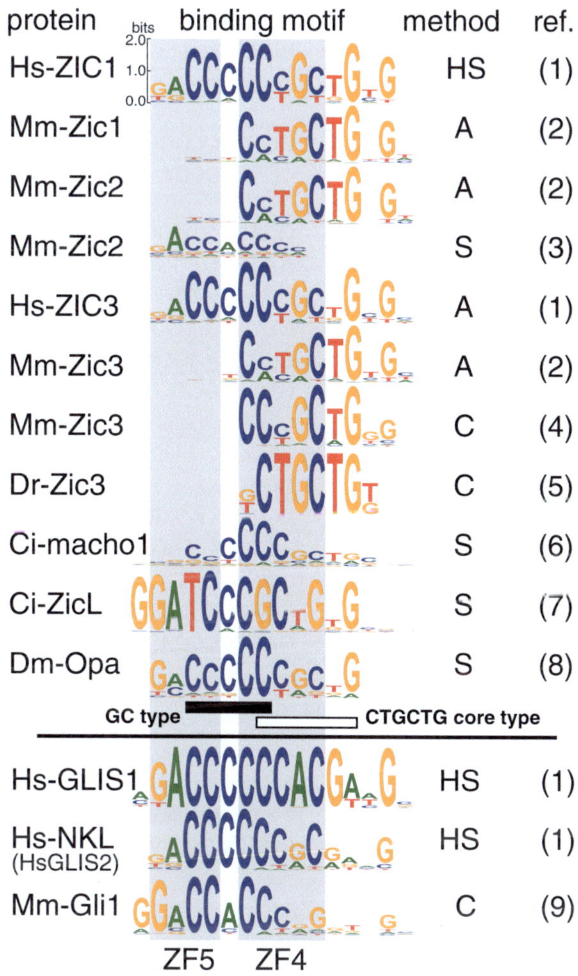

Fig. 18.4 Zic-binding sequence based on position weight matrix. The position weighted matrixes of the Zic, Gli, Glis, and Nkl proteins are indicated by WebLogo (Crooks et al. 2004). Experimental types are indicated as follows: *A*, protein-binding microarray; *C*, chromatin immunoprecipitation; *HS*, high-throughput SELEX; *S*, SELEX. References are as follows: (1) (Jolma et al. 2013); (2) (Badis et al. 2009); (3) (Mizugishi et al. 2001); (4) (Lim et al. 2010); (5) (Winata et al. 2015);(6) (Yagi et al. 2004a);(7) (Yagi et al. 2004b); (8) (Sen et al. 2010); (9) (Peterson et al. 2012). *Closed bar* and *open bar* indicate the regions containing GC-stretch-type and CTGCTG-core-type sequences, respectively

in vitro irrespective of the DNA sequences in cells. An oligonucleotide is typically used in which two defined sequences that can serve as PCR primers flank a degenerate core. After making this population of molecules double-stranded by priming DNA synthesis with a 3′ primer, the DNA is mixed with protein, and DNA–protein complexes are isolated in an experiment on binding such as an electrophoretic mobility shift assay. By this method, the binding sequences for mouse Zic1, Zic2, and Zic3, ascidian ZicL and macho-1, and *Drosophila* odd-paired have been analyzed (Fig. 18.4) (Mizugishi et al. 2001; Yagi et al. 2004a, b; Sawada et al. 2005). A recent study involving high-throughput SELEX analysis of 411 human or mouse transcription factors revealed the binding sequences for human ZIC1, ZIC3, and ZIC4 and mouse Zic3 (Fig. 18.4) (Jolma et al. 2013). Another synthetic DNA-based method involves a protein-binding microarray (PBM) in which DNA-binding domains of transcription factors are reacted with the PBM conjugated with 10-bp

Table 18.1 Zic-binding sequence identified in specific targets

Zic family	Target gene	Binding sequence	References
Mm-Zic1–5	Gli-binding seq.	TGGGTGGTC	Aruga et al. (1994) and Ishiguro et al. (2004)
Mm-Zic1	*Math1* (Y1H)	GCTCCCCGGGGAGCT	Ebert et al. (2003)
Mm-Zic1	*Lamin A/C* (Y1H)	CCACCCCCT	Okumura et al. (2004)
Mm-Zic1/2	*ApoE* (Y1H)	GGACTGTGGGGGGTGGTCAA	Salero et al. (2001)
		AAACTGTGGGGGGTGGTCAA	
		GGACTGTGGGGGGTGAAAAA	
		CTATCCCTGGGGGAGGGGGC	
Hs-ZIC2	*D1A* (Y1H)	CCCCCAGGGCA	Yang et al. (2000)
Mm-Zic2	*CaMK II* (Y1H)	GTGTGGGC	Sakurada et al. (2005)
Mm-Zic2	*Pax3*	CTGCTGGGG	Sanchez-Ferras et al. (2014)
Mm-Zic2	*Oct4*	−2550 ~ −2430 of Oct4 promoter	Zhu et al. (2015)
Hs-ZIC3	*α-actin*	GGAGGG	Zhu et al. (2007)
Mm-Zic3	*Nanog*	CC(C/T)GCTGGG	Lim et al. (2010)
Hs-ZIC5	*E-cadherin*	−283 ~ −71 of CDH1 promoter	Satow et al. (2017)
Ci-ZicL	*Brachyury*	CCAGCTGTG	Yagi et al. (2004b)
Ci-Macho1	*E(spl)/hairy-b*	CCCCCCGCT	Yagi et al. (2004a)
Hr-Macho1	Synthetic	GACCCCCCA	Sawada et al. (2005)

Ci Ciona intestinalis, *Dr Danio rerio*, *Hr Halocynthia roretzi*, *Hs Homo sapiens*, *Mm Mus musculus*. *Y1H* target sequence identified by yeast one-hybrid screening

sequences. This method enables researchers to identify the binding sequences of many transcription factors in a comprehensive manner. Badis et al. (2009) included Zic1, Zic2, and Zic3 in their PBM analysis of 104 murine DNA-binding proteins (Fig. 18.4).

On the other hand, the Zic-binding sites in genome-derived sequences have been analyzed either by chromatin immunoprecipitation (ChIP) or by the analysis of the *cis*-regulatory elements near a transcription start site (TSS) of individual genes (promoter analysis). The ChIP-based analyses have been applied to mouse embryonic stem (ES) cells (Zic3, Lim et al. 2010), mouse cerebellar granule cells (Zic1 and Zic2, Frank et al. 2015), and zebrafish embryos (zic3, Winata et al. 2013). In particular, the Zic3-binding sequence deduced from a Zic3 ChIP-enriched sequence (Lim et al. 2010) yielded the CC(C/T)GCTGGG consensus sequence (Fig. 18.4 and Table 18.1), which has high affinity (Kd=2.4 × 10^{-9} M) for mouse Zic3 ZFNC-ZFD with N-terminally epitope tagged with poly-histidine and maltose-binding protein. Similar ChIP-based consensus sequence CCTGCTGG was revealed by studying targeting of zebrafish zic3 at the embryonic stage (Fig. 18.4) (Winata et al. 2013). The analysis of *cis*-acting regulatory sequences has also contributed to the research into binding properties.

Additionally, Zic1- or Zic2-binding sites in the mammalian promoter or enhancer have been identified by yeast one-hybrid screening in five independent studies (Yang et al. 2000; Salero et al. 2001; Ebert et al. 2003; Okumura et al. 2004;

Sakurada et al. 2005). The list of Zic-binding sequences identified by the analysis of individual *cis*-regulatory elements is summarized in Table 18.1.

In Fig. 18.4 and Table 18.1, readers can notice that there are two types of core sequence motifs according to the synthetic-target-based methods. One motif features a stretch of G or C bases, and the other contains CTGCTG (CAGCAG) as a core sequence. The combination of the two types of motifs can be seen in the target sequence reported by Jolma et al. (2013) in which long (14–40 bp) random sequences are present. The core sequences based on the G/C stretch have been constructed for the analysis of mouse Zic1–Zic3; human ZIC1, ZIC3, and ZIC 4; *Drosophila* Opa; and ascidian macho-1 and were detected in the yeast one-hybrid screening-derived *cis*-acting regulatory sequences for DRD1, ApoE (Salero et al. 2001), Math1 (Ebert et al. 2003), lamin A/C (Okumura et al. 2004), and Camk2a (Sakurada et al. 2005). On the other hand, the CTGCTG core sequences, initially constructed by PBM analysis, have been found to be enriched in Zic-binding enhancers (Winata et al. 2013; Sankar et al. 2016; Sone et al. 2017) or in the promoter of *Nanog* (Lim et al. 2010), *FoxD3* (Simoes-Costa et al. 2012), and *Pax3* (Sanchez-Ferras et al. 2014).

An intriguing feature of the binding sequence can be the binding sequence similarity among the Gli, Glis, Nkl, and Zic family proteins (Chap. 1). A classical GLI-binding sequence (TGGGTGGT) was used as a target site in a preliminary study to show DNA-binding activity of a Zic protein (Aruga et al. 1994), and mouse Zic1, Zic2, and Zic3 bind the GLI target sequence with the affinity (K_d) of 5.2×10^{-8}, 4.8×10^{-8}, and 7.1×10^{-8} M, respectively, fivefold to eightfold weaker than that of GLI3 (8.5×10^{-9} M) (Mizugishi et al. 2001). The relation of binding properties is conserved in Opa and Ci: fly orthologs of Zic and Ci, respectively. Besides the optimal Opa-binding sequence (CGGGGGGTC), Opa can bind a Ci target sequence with low affinity (Sen et al. 2010). The high-throughput SELEX analysis by Jolma et al. (2013) provided more comprehensive evidence for the optimal binding sequences for Gli, Glis, Nkl, and Zic family proteins. The G/C stretch sequences are also present in GLIS3 and NKL (GLIS2), and the ZIC1-, ZIC3-, and ZIC4-binding sequences seem to be most similar to that of NKL (GLIS2) (Fig. 18.4). Nevertheless, the functional relation between Glis and Zic proteins remains to be explored in contrast to that of Gli and Zic proteins (Brewster et al. 1998; Aruga et al. 1999; Koyabu et al. 2001; Quinn et al. 2012).

Zic family protein-binding sequences are now deposited in databases of transcription factor-binding sequences (e.g., the JASPAR database, http://jaspar.genereg.net/, or UniPROBE database, http://the_brain.bwh.harvard.edu/uniprobe/index.php) and are often used for genome-wide analyses (Buecker et al. 2014; Frank et al. 2015; Matsuda et al. 2017). In terms of binding affinity, the ChIP-derived sequence CC(C/T)GCTGGG (Lim et al. 2010) is the best among the several known K_d values. Nonetheless, there have been no studies dealing with the similarities and differences in the sequence selectivity among the Zic family proteins or among the Gli, Glis, Nkl, and Zic superfamily proteins. There is still ambiguity concerning the differential usage of GC stretch-type and CTGCTG-type core sequences (e.g., promoter vs enhancer). We await further delineation of biomedically critical Zic targets including the results of genome-wide analyses.

18.2.2 Protein-Binding Properties of Zic Family Proteins

Studies on Zic-binding proteins have expanded our understanding of the behaviors and functions of Zic family proteins. We summarized the Zic-binding proteins along with the interacting domains and the known biological significance of binding (Table 18.2). Each biological context will be reviewed later.

Zic-binding proteins have been studied through purification of a Zic-containing complex by conventional protein purification and/or immunoprecipitation either with an anti-Zic antibody or an anti-epitope tag antibody combined with an epitope-tagged Zic protein (Ishiguro et al. 2007; Luo et al. 2015), by means of the yeast two-hybrid system (Mizugishi et al. 2004; Ogawa et al. 2008; Zhu et al. 2015) or by focusing on particular candidate proteins in a specific biological context (Koyabu et al. 2001; Zhu et al. 2007, 2008; Hatayama et al. 2008; Pourebrahim et al. 2011; Fujimi et al. 2012; Sanchez-Ferras et al. 2014; Houtmeyers et al. 2016). C2H2 ZFDs mediate various protein–protein interactions, whereas DNA binding is a more restricted function of the ZFD (Brayer et al. 2008). This is also the case for Zic proteins.

Table 18.2 Zic-binding proteins

Zics	Target	Domain	Function	References
Zic1	Pax3	ZOC	Transcription+	Himeda et al. (2013)
Zic1/2	Gli1/2	ZF	Nuclear import	Koyabu et al. (2001)
			Transcription+	Chan et al. (2011)
Zic2	I-mfa	ZOC⇔ZF	Nuclear import	Mizugishi et al. (2004)
			Transcription–	
Zic2	Tcf4	ZF	Wnt signal regulation	Pourebrahim et al. (2011)
Zic2	Ku70/Ku80/PARP/DNA-PKcs/RNA helicase	ZF	Transcription+	Ishiguro et al. (2007)
Zic2	Cdx	ZF	Transcription+	Sanchez-Ferras et al. (2014)
Zic2	NURF complex (RBBP4, BPTF, SNF2L)	Not ZF Remodeling factor complex	Recruiting chromatin	Zhu et al. (2015)
Zic2	Oct4	?	?	Pardo et al. (2010)
Zic2	Rines	NT	Degradation	Ogawa et al. (2008)
Zic2	NuRD complex	?	Chromatin remodeling	Luo et al. (2015)
Zic2	SMAD2/3	NT	Nodal signaling	Houtmeyers et al. (2016)
Zic3	Gli3	?	Transcription+	Zhu et al. (2008)
Zic3	Tcf1/XIC	?	Wnt signal regulation	Fujimi et al. (2012)
ZIC3	Importin-α5/7	ZF	Nuclear import	Hatayama et al. (2008)
ZIC3	SRF	ZF	Transcription+ (α-actin)	Zhu et al. (2007)

+ activation; – repression

18.3 Control of Zic Protein Behavior

18.3.1 Control of Subcellular Localization

Several studies have revealed nuclear localization of Zic proteins: mouse and human Zic1 proteins in cerebellar granule cells (Aruga et al. 1994; Yokota et al. 1996) (Chap. 13); epitope-tagged mouse Zic1, Zic2, and Zic3 in cultured cells (Koyabu et al. 2001); and epitope-tagged macho-1 in cell nuclei of ascidian embryos (Nishida and Sawada 2001). These findings may have contributed to the prevalent notion that Zic proteins are primarily located in the cell nucleus. In contrast, mouse Zic2 protein can be detected in both the nucleus and cytoplasm of preimplantation blastocysts (Brown and Brown 2009), and macho-1 was detected in a centrosome-attaching body (the centrosome is a microtubule-organizing center containing the centriole) in ascidian embryos (Kumano et al. 2010), thus shedding new light on the roles of extranuclear Zic proteins.

Concerning the nucleocytoplasmic shuttling mechanism, ZIC3 has been studied well. As we mentioned above, ZIC3 NLS is dispersed in the ZF2 and ZF3 regions and is folded into a bipartite-cluster-like structure (Fig. 18.2a) (Hatayama et al. 2008). In addition, ZF4 and ZF5 have a nuclear localization activity (Bedard et al. 2007; Hatayama et al. 2008). ZIC3 ZF2 and ZF3 interact with importin-α5 or importin-α7 (Hatayama et al. 2008). The importin system actively transports target proteins into the nucleus because macromolecules cannot pass through the nuclear pore complexes. Cargo proteins containing a classical NLS, e.g., ZIC3, bind to importin-α, which is the adapter protein between the cargo proteins and importin-β. The cargo protein–importin-α–importin-β heterotrimeric complex formed in the cytoplasm is translocated to the nuclear pore complexes and then into the nucleus (Fig. 18.5).

A nuclear export signal (NES) consensus sequence also exists in ZIC3 ZF2, and mutation of the NES-forming residues increases the nuclear localization of a ZIC3-derived protein (Bedard et al. 2007). Because CRM-1 overexpression reduces the ZIC3 nuclear localization and CRM-1 inhibitor leptomycin B increases the ZIC3 nuclear localization, CRM-1 is thought to mediate ZIC3 nuclear export. Collectively, the NLS and NES both of which are located in the multifunctional ZFD are involved in the nucleocytoplasmic shuttling machineries. The importin import system and CRM-1 export system may determine the basal nucleocytoplasmic distribution of the ZIC3 protein (Fig. 18.5).

As a regulatory system of the ZIC3 nucleocytoplasmic shuttling, SUMO-based modification of ZIC3 (i.e., SUMOylation; Sect. 18.1.3) has caught out attention. Chen et al. identified a lysine residue (K248) as the target of SUMOylation. The residue is crucial for the nuclear retention of ZIC3 because a substitution of this lysine with arginine (K248R) decreases nuclear localization. After additional experiments, they proposed that SUMOylation status of ZIC3 affects its nucleocytoplasmic distribution, and this mechanism may be associated with cytoplasmic diffusion of ZIC3 mutants derived from patients with heterotaxy (Chen et al. 2013). Because the target of SUMOylation is conserved in the conserved type of Zic proteins

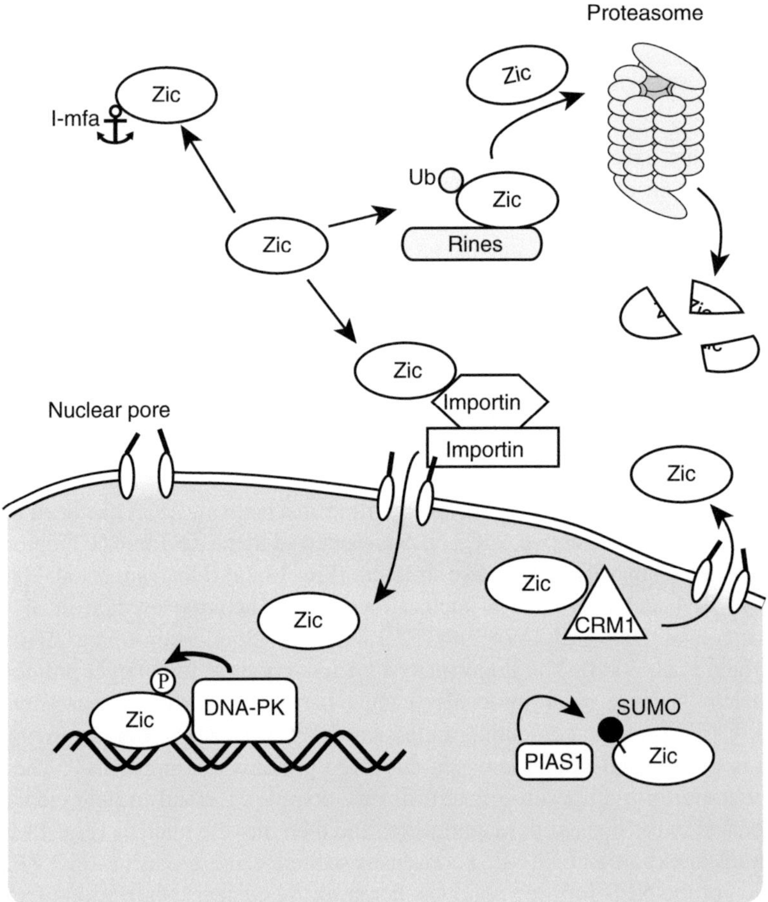

Fig. 18.5 Transport, modification, and degradation of Zic proteins. Nuclear import and export are mediated by the importin-α–importin-β complex and CRM-1-dependent machinery. Zic proteins can be trapped by I-mfa in a cytoplasmic region. In the nucleus, SUMO-based modification by PIAS1 may impair the export. Phosphorylation by DNA-PK can change the binding partners. The degradation pathway is mediated by the ubiquitin proteasome system, partly mediated by the ubiquitin ligase Rines

(Chap. 1) including human ZIC1, ZIC2, ZIC4, and ZIC5, it would be interesting to test whether the SUMOylation is a general mechanism underlying the functional control of Zic proteins.

On the other hand, another protein, I-mfa (inhibitor of MyoD family, also called Mdfi), may participate in the regulation of developmental-context-dependent nucleocytoplasmic shuttling of Zic proteins. I-mfa binds a Zic2 region between ZOC and ZFD and inhibits nuclear import and transcriptional activity of mouse Zic1–Zic3 by anchoring them in the cytoplasm (Fig. 18.5) (Mizugishi et al. 2004). I-mfa is strongly expressed in the sclerotome, a somite derivative in the mouse embryo on embryonic day 11.5, and Zic expression partially overlaps with that of I-mfa in the sclerotome and limb bud (Kraut et al. 1998; Mizugishi et al. 2004). It is also known that I-mfa-deficient mice show abnormal fusion of vertebral arches, fusion and bifurcation of ribs, and delayed neural tube closure (Kraut et al. 1998). Some of these phenotypes are similar to those observed in Zic mutants (Fig. 12.2). It is possible that the Zic–I-mfa antagonism is implicated in the compartment formation or boundary formation during skeletal patterning.

18.3.2 Control of Zic Protein Degradation

Degradation of Zic proteins has been addressed only in a few papers. Nonetheless, Rines (also known as RNF180) was identified as a Zic2-binding protein by yeast two-hybrid screening (Ogawa et al. 2008). Rines is a membrane-bound E3 ubiquitin ligase, most of which is exposed to the extraluminal side of the endoplasmic reticulum and nuclear envelope. Rines can ubiquitinate Zic2 and promote its proteasomal degradation (Fig. 18.5). Although the physiological significance of Rines-mediated Zic2 degradation has not been clarified, Rines is expressed in the ventricular layer of the cerebral cortex, and its expression partly overlaps with that of Zic2 (Ogawa et al. 2008). In another study, stability of the ZIC3 protein or ZIC3 mutants derived from heterotaxy patients was evaluated in the presence of cycloheximide (Chhin et al. 2007). The results indicated increased degradation of a mutant ZIC3 protein (W255G) in comparison with the intact ZIC3 protein, suggesting that the unification of ZF1 and ZF2 increases the stability of ZIC3. This is because W255 is an essential residue for the ZF1–ZF2 unification (Sect. 18.1.1).

18.3.3 An Unsolved Mechanism

Lastly, as an unsolved mechanism that involves the downregulation of Zic proteins, we can mention the mechanism mediated by teneurin-2. The latter is a single-membrane-span protein, and genetic interaction between its *Drosophila* homolog and *Drosophila* Zic homolog Opa has been described (Baumgartner et al. 1994). In vertebrates, both Zic2 and teneurin-2 play important roles in ipsilateral projection of binocular visual circuits. Nevertheless, a teneurin-2 knockout does not affect Zic1 and Zic2 expression in the retina (Siegenthaler et al. 2013). Regarding transcriptional regulation by Zic proteins, a cleaved intracellular domain of teneurin-2 can suppress Zic1-mediated transactivation of the *ApoE* promoter (Bagutti et al. 2003). The physical interaction between teneurin-2 and Zic1 is unclear at this point. Teneurin-2 may be a transcriptional regulator for a Zic protein during optic nerve development in vertebrates. In addition, teneurin-2 expression is observed in limbs and somites and possibly regulates the development of these tissues (Bagutti et al. 2003; Tucker et al. 2001).

18.4 The Molecular Function of Zic Proteins in Various Biomedical Areas

18.4.1 Regulation of Transcription via a Promoter

The involvement of Zic proteins in the transcriptional regulation has been investigated in some studies using cotransfection of Zic-expressing and reporter gene-containing plasmids in cultured cells. Mouse Zic1, Zic2, and Zic3 activate the transcription of reporter genes in plasmids under the control of the adenoviral major late promoter, herpes simplex virus thymidine kinase (TK) promoter, or SV40 early promoter, and the presence of Zic-binding sequences enhances their transcriptional activation (Mizugishi et al. 2001). In contrast, Zic4 and Zic5 do not have any significant transcription-activating ability according to the same assay (Ishiguro et al. 2004). Because the adenoviral major late promoter contains a minimal promoter element (a TATA box and a TSS), it was proposed that Zic family proteins may function as transcriptional coactivators rather than being the transcription factors that regulate gene expression by direct binding to DNA (Mizugishi et al. 2001). The SV40 promoter is highly activated by Zic2 and Zic3 (20- to 30-fold) and has been utilized to evaluate Zic proteins' transcription-activating ability in some studies. The "partial" dependence on the binding sequence is in contrast to that of GLI1, which does not exert the sequence-independent transcriptional activation (Mizugishi et al. 2001).

On the other hand, binding-sequence-dependent transcriptional regulation has been reported for several nonviral gene promoters. First, the promoter of human D1 dopamine receptor gene *DRD1* contains a ZIC2-binding sequence, and binding of ZIC2 to this sequence inhibits the promoter activation by displacing another transcription factor: Sp1 (Yang et al. 2000). The ZIC2-mediated repression of *DRD1* depends on the presence of a ZIC2-binding sequence. Second, the *APOE* gene promoter, which contains three Zic-binding sequences, is activated by proteins Zic1 and Zic2. Nonetheless, transcriptional activation is abrogated in a mutant *APOE* promoter lacking the three Zic-binding sites (Salero et al. 2001). Third, ascidian Brachyury gene promoter that contains two ZicL-binding sequences can direct gene expression in notochord cells, but a mutant promoter lacking the two ZicL-binding sequences does not direct the gene expression in the notochord or is not activated by ZicL overexpression (Yagi et al. 2004b). Regulation affecting the promoter of ZicL-binding-sequence-dependent Brachyury was also observed in another ascidian species (Matsumoto et al. 2007). In other studies, the presence of a Zic-binding sequence and the Zic-dependent transcriptional regulation have been demonstrated, but the studies lack analysis of the necessity of the binding sequence for Zic-dependent regulation.

Besides the above promoters, Zic-dependent transcriptional regulation has been described for many genes including lamin A/C (Okumura et al. 2004), *Camk2a* (Sakurada et al. 2005), cardiac α-actin (Zhu et al. 2007), and serotonin transporter *Slc6a4* (Garcia-Frigola and Herrera 2010). Zic proteins exert transcriptional activation except on *DRD1* and lamin A/C promoters, in both of which, binding site competition with Sp1 is evident (Yang et al. 2000; Okumura et al. 2004). In the case of Zic5, it acts as a transcriptional suppressor for E-cadherin. A Zic5-binding site was identified in the region −283 to −71 of the E-cadherin promoter, and a Zic5 knockdown increases E-cadherin expression in a human melanoma cell line (Satow et al. 2017). Zic5's binding sequence in the E-cadherin promoter and suppression mechanisms are unclear, but one biological implication is that Zic5 is involved in melanoma metastasis and proliferation.

Are there any common mechanisms for the Zic-mediated transcriptional regulation? To answer this question, mapping of the transcription activation domains was conducted. Mizugishi et al. (2004) analyzed the transcription-activating ability of a C-terminally and N-terminally deleted Zic2 series and of such proteins carrying an independent DNA-binding domain (GAL4 DBD) in C3H10T1/2 cells. They found that the region containing Zic ZF3–ZF5 and a part of the C terminus functions as a transcriptional activator domain, whereas the ZOC domain serves as either an activator domain or a repressor domain, depending on the assay system used. On the other hand, the transcriptional repression domain has been mapped to the carboxy terminus of Zic1 (Kuo et al. 1998) (see Chap. 13), ZIC2 (Brown et al. 2005), and Zic3 (Zhu et al. 2007).

To clarify the molecular machineries driving the transcriptional regulation by Zic proteins, Ishiguro et al. (2007) purified two types of Zic2-containing high-molecular-weight complexes. Complex I is composed of a DNA-dependent protein kinase catalytic subunit (DNA-PKcs or Prkdc), Ku70/80 (Xrcc6, Xrcc5), and poly (ADP-ribose) polymerase (Parp1). Complex II contains Ku70/80 and RNA helicase A (RHA or Dhx9). All the components interact directly with the Zic2 protein. The binding sites for DNA-PKcs are the ZF N-terminus adjacent region (141–255), and those for RHA, PARP, Ku70, and Ku80 are ZF3 (334–363). Inclusion of ZF3 in the binding sites may account for the transactivation domain-mapping results.

DNA-PKcs in complex I plays an essential part in the assembly of complex II. Stepwise switching from complex I to complex II depends on phosphorylation of Zic2 by DNA-PKcs and Parp1 (Fig. 18.6) (Ishiguro et al. 2007). Serine at aa position 200 of Zic2 was identified as the phosphorylation target of DNA-PKcs, and a phosphorylation-defective mutant (S200A) was found to be deficient in RHA binding, and its transcription-activating ability is diminished (Ishiguro and Aruga 2008). Because RHA was discovered as a bridging factor between the transcriptional coactivator CREB-binding protein (CBP)/p300 and RNA polymerase II (Nakajima et al. 1997), its presence in the Zic2 containing complex may be a mechanism that underlies Zic2-mediated transcriptional regulation. In agreement with this idea, not only RHA but also general transcription factors (TBP and TFIIB) coprecipitate with Zic2 in an S200-dependent manner (Ishiguro and Aruga 2008). This study is suggestive of a hypothetical model of Zic2-mediated promoter regulation. Because DNA-PKcs is well investigated in the context of DNA repair and V(D)J recombination, its involvement in the Zic2 regulation would be surprising but may need further clarification of the biological context that involves this mechanism.

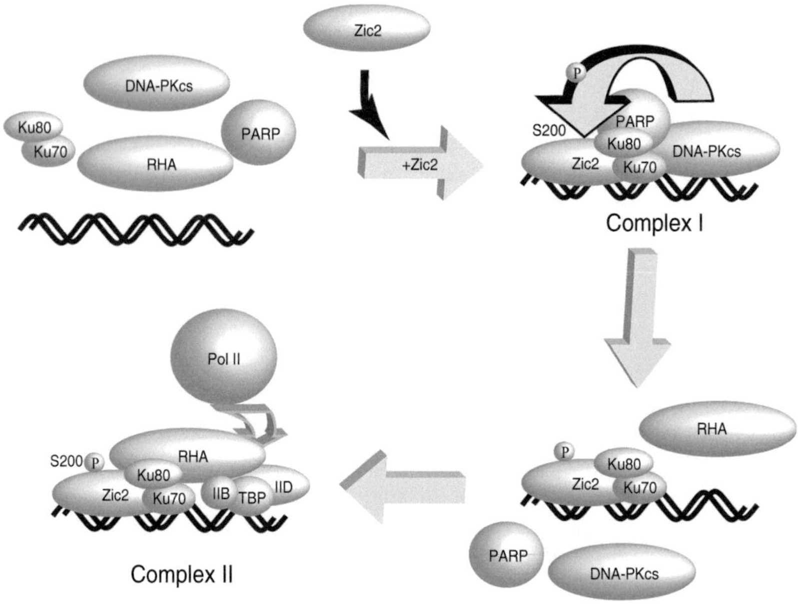

Fig. 18.6 The function of DNA-PK in Zic2-mediated transcriptional regulation. The ZF domain of the Zic2 protein binds to DNA upstream of a target gene promoter. At first, the Zic2 protein recruits the Ku70–Ku80 dimer, DNA-PKcs, and PARP. At this first step, complex I is constructed for phosphorylation of Zic2 protein by DNA-PK and PARP (top right). At the next step, RHA is recruited to the phosphorylated Zic2 protein, and then PARP and DNA-PKcs depart (bottom right). Finally, RHA forms complex II with the phosphorylated Zic2 (bottom left) and comes into contact with general transcription factors TBP and TFIIB. Complex II can recruit polymerase II and next activates the Zic2-dependent transcription (Illustration is based on the results in Ishiguro et al. 2007)

The mechanism underlying Zic-mediated transcription has been suggested by means of the analysis of several developmentally critical promoters as shown below. Nevertheless, this genome part excludes DNA regions involved in the epigenetic modification, as we will see later.

18.4.2 Promoters Targeted by Zic1

Synergistic transcriptional activation by Zic1 and Pax3 (a paired-type transcription factor) has been reported for *Myf5*, *Snail1*, and *Pax3* promoters (Himeda et al. 2013; Plouhinec et al. 2014). A physical interaction between Zic1 and Pax3 is proposed to mediate the synergistic regulation in case of the *Myf5* promoter (Himeda et al. 2013). In addition, the Zic1–Pax3 interaction has been demonstrated to be critical for the generation of the neural crest lineage (Sato et al. 2005; Milet et al. 2013; Bae et al. 2014; Plouhinec et al. 2014).

18.4.3 Promoters Targeted by Zic2

Houtmeyers et al. (2016) investigated the role of Zic2 in Nodal signaling downstream of transcription factor SMAD2-dependent promoter regulation and found that Zic2 can repress the transcription mediated by a SMAD2- and SMAD4-responsive promoter (FoxH1, activin-responsive element-driven reporter). SMAD2 or SMAD3 physically interacts with the Zic2 N-terminal region. Nevertheless, both the N-terminal SMAD3-binding region and ZFD are required for the transcriptional control. In this case, the role of Zic2 in the bridging of SMAD2 or SMAD3 and transcriptional coactivators or chromatin modifiers is proposed (Houtmeyers et al. 2016). On the other hand, Zic2 activates neural crest enhancer (NCE2) located within the 1.6-kb proximal region of the *Pax3* promoter (Sanchez-Ferras et al. 2014). NCE2 contains a Zic-binding site, Cdx (caudal-related homeobox)-binding sites, and Sox-binding sites. NCE2-driven reporter gene expression is synergistically activated by Zic2, Cdx1, and Sox2 in Neuro2a cells. Furthermore, Zic2 ZFD physically interacts with the Cdx1 homeodomain. Their interaction is preserved in the presence of their respective target sequences, and gel-shift experiments have suggested formation of a Zic2–Cdx1–DNA complex (Sanchez-Ferras et al. 2014).

18.4.4 Promoters Targeted by Zic3

Zic3 activates the *Nanog* promoter, which includes a high-affinity Zic3-binding sequence and Oct4- and/or Sox2-binding sites (Fig. 18.7), and the regulatory effects on the endogenous *Nanog* gene were confirmed by Zic3 overexpression and knockdown in ES cells (Lim et al. 2010, 2007). The possible relations among the pluripotency-associated transcription factors are intriguing, and detailed protein interaction analyses are expected. In an independent study (Zhu et al. 2007), Zic3 was shown to bind the cardiac α-actin promoter and to repress basal transcription. Nonetheless, Zic3 can activate the promoter when it physically interacts with SRF (serum responsive factor). A Zic3-dose-dependent shift of the binding target from DNA sequence to SRF has been proposed to explain this phenomenon.

18.4.5 A Summary of Promoter Regulation by Zic Family Proteins

The above review of the promoter regulation by Zic proteins points to the existence of multiple mechanisms behind the Zic-mediated transcriptional regulation. Therefore, it would be premature to define "a major mode of action." On the contrary, "versatility" may describe the features of Zic proteins as transcriptional regulators rather well at present. This property may partly reflect the multifunctional

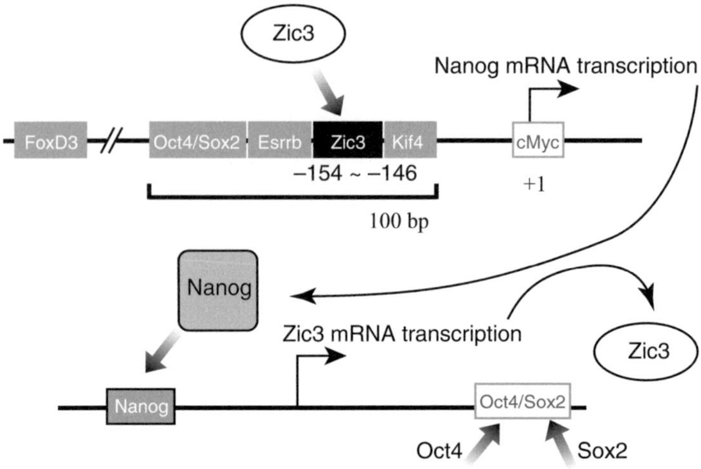

Fig. 18.7 Regulation of *Nanog* promoter activity by Zic3 and other pluripotency-associated transcription factors. The *Nanog* promoter contains multiple transcription factor-binding sites near its TSS. An *open box* indicates a binding site and binding factor. A Zic3-binding site is located in the region −154 to −146 relative to the TSS, and the other three binding sites are tandemly positioned within 100 bp. In stem cells, Zic3 promotes *Nanog* expression, and Nanog protein activates *Zic3* transcription. These factors are believed to form a positive feedback loop that maintains the pluripotency of embryonic stem cells (Illustration is based on the results in Lim et al. 2010.)

nature of ZFDs. There are still the unexplored partners as suggested by the complexity of the Oct4-binding protein interactome, where Zic2 is included (Pardo et al. 2010).

18.4.6 Enhancer Functioning and Epigenetic Regulation

Recent progress in genome-wide analysis of Zic-binding sequences highlighted the novel function of Zic proteins as chromatin-remodeling regulators. The initial genome-wide analysis of Zic target sites can be seen in Zic3 ChIP combined with a promoter microarray (ChIP on chip analysis) in which the investigators proposed 379 Zic3-occupied promoter candidates (Lim et al. 2010). On the other hand, a whole-genome targeted analysis was first reported by Winata et al. (2013), where they adopted a combined approach, ChIP-seq–microarray, to search for zebrafish Zic3 downstream targets (Chap. 9). In this study, Zic3-binding sites were found to be mostly located in distal regulatory elements. Only 8.6% and 4% of the bound sequences were located within 5 kb from TSSes at the gastrula stage (8 h postfertilization) and at the segmentation stage (24 h postfertilization), respectively. They next detected enhancer activities in 11 out of 15 evolutionarily conserved nonprotein-coding Zic3-binding sites. This result may have linked Zic3 function to developmentally critical distal regulatory regions (enhancers) for the first time.

Since 2013, evidence has been accumulating for the critical roles of Zic proteins in enhancers. Zic2- or Zic3-binding sequences were identified near the Oct4-binding sites, and Zic2 coprecipitates with Oct4 in epiblast-like cells (Buecker et al. 2014). Zic1 is recruited to H3K4me1-enriched sites created by a Lsh (SNF2 chromatin-remodeling protein) knockout in embryonic fibroblast cells (Yu et al. 2014). Zic2 ChIP-seq using mouse ES cells identified 10,273 Zic2-binding sites, 83% of which are located distally from TSSes (Luo et al. 2015). Zic2-binding sites were found in both an active and poised enhancer and are associated with genes involved, e.g., in the regulation of transcription from the RNA polymerase II promoter, embryonic morphogenesis or development, and regulation of cell proliferation (Luo et al. 2015). Those researchers next identified Zic2-binding proteins in mouse ES cells. The newly identified Zic2-binding proteins include the components of the MBD3-containing NuRD (nucleosome remodeling and histone deacetylase) activity complex, which is a chromatin-remodeling complex possessing both chromatin-opening and chromatin-closing activities. A Zic2 knockdown reduces H3K27me3-type histone modifications (a poised chromatin signature) and increases K3K27ac-type modifications (an open chromatin signature), and the chromatin state change was found to be similar to that in MBD3-null ES cells (Fig. 18.8a). According to additional results, Luo et al. (2015) suggested that Zic2 is broadly involved in marking the enhancers of genes in the ES cell state that are to be activated later during development. This study for the first time suggested a mechanism underlying Zic protein-mediated chromatin remodeling.

Mechanistic views of the Zic-mediated chromatin remodeling are also offered by the latest ChIP-based studies. Matsuda et al. (2017) highlighted the key function of Zic2 in the mouse epiblast stem cell regulatory network. Taking advantage of parallel ChIP-seq analysis with biotinylated transcription factors, they found that Zic2 ChIP peaks preferentially overlap with Otx2 among the four transcription factors (Otx2, Sox2, Pou5f1, and Pou3f1; see Chap. 17 for details). Sankar et al. (2016) clarified the association of Zic1 and geminin, which together with *Sox1–Sox3* genes cooperatively promote neural cell fate acquisition in *Xenopus* embryos. They carried out ChIP-seq analysis of mouse ES cells, where an epitope tag was introduced into either Zic1 or geminin-coding sequence by the Crispr/Cas9 technique. ChIP-seq analysis of ES cells or the cells in the neuroectodermal state revealed that the presence of both geminin- and Zic1-associated peaks is enriched in Zic1-binding sequences, and Zic1-bound peaks are also enriched in Sox3-binding sequences (Sankar et al. 2016). Because geminin itself does not contain a DNA-binding domain, Zic1 may be a partner that specifies targeting of geminin to particular sites in DNA. From the standpoint of Zic1, geminin may be an important partner for control of chromatin remodeling because geminin can interact with the chromatin-remodeling complex (SWI/SNF and Polycomb). The Zic1–geminin interaction may enhance the neural cell fate acquisition because Zic1 and geminin coexpression strongly upregulates neural cell fate-associated transcription factors (Ascl1, Irx3, and Pax7) (Sankar et al. 2016).

The involvement of Zic proteins in epigenetic regulation is not limited to the embryonic development and may have a major effect on the mature adult central

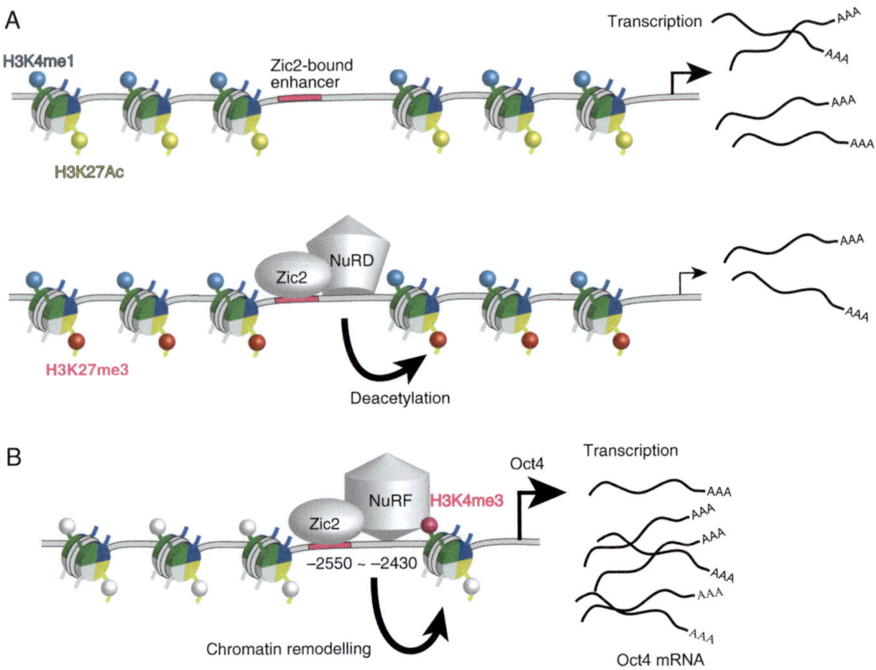

Fig. 18.8 Chromatin-remodeling activity of Zic2. (**a**) NuRD is a Zic2-binding protein and regulates the chromatin methylation state. Histone modification status near TSSes affects gene expression. Especially, the lysine residues at positions 4 and 27 (K4 and K27) of histone H3 are the key modification sites, and NuRD changes the acetylated K27 (H3K27Ac) to a trimethylated one (H3K27me3). This modification suppresses gene expression (Luo et al. 2015). (**b**) The NuRF chromatin-remodeling complex is involved in the *Oct4* promoter regulation in cancer stem cells (Zhu et al. 2015). The Zic2-binding site is located 2.5 kb upstream of the *Oct4* TSS. NuRF recognizes H3K4me3 and binds to the histone protein. The Zic3–NuRF complex conducts chromatin remodeling and upregulates *Oct4* expression

nervous system where Zic proteins are abundantly produced. This is because Zic1–Zic2 binding correlates with DNA accessibility and H3K27ac levels during postnatal differentiation of cerebellar granule neuron progenitors (Frank et al. 2015). They also reported that Zic1 and/or Zic2 controls the transcription of the genes near the Zic1- and Zic2-binding sites (see Chap. 13 for details).

Besides the above studies dealing with normal developmental processes, ChIP-seq and protein interaction analyses of cancer stem cells revealed a novel chromatin-remodeling mechanism of action of Zic2 (Zhu et al. 2015). Zic2 expression is high in hepatocellular carcinoma stem cells, and Zic2 increases H3K4me3-type histone modification in an *Oct4* promoter region (Zhu et al. 2015). H3K4me3 modification is a hallmark of open chromatin and upregulates transcriptional activity (Cedar and Bergman 2011; Chen et al. 2010). Searching for ZIC2-binding proteins by yeast two-hybrid screening resulted in the identification of RBBP4 as a Zic2 interactor. RBBP4 (retinoblastoma-binding protein 4) is a component of the chromatin-remodeling factor complex, NURF, which includes RBBP4, BPTF (bromodomain and PHD finger-containing transcription factor), and SNF2L (sucrose nonfermenting 2-like protein 1), and they are the key molecules for the H3K4me3 modification (Wysocka et al. 2006; Zhu et al. 2015). As a consequence, Zic2 promotes cancer cell self-renewal activity via Oct4 transcription (Zhu et al. 2015). This finding revealed a mechanism where Zic2 can induce chromatin remodeling by upregulating H3K4me3 locally by binding to NURF (Fig. 18.8b).

Looking back on the findings in the last 5 years, ChIP-seq and Crispr/Cas9 technologies have been rapidly uncovering the participation of Zic proteins in the priming of enhancer functions. To obtain a clearer picture, it would be necessary to elucidate the behavior of Zic proteins at sufficient spatiotemporal resolution and to uncover additional Zic-containing molecular complexes (if any), including the transitional or differential usage among them. An intriguing property of Zic proteins is their bipotential toward promoter and enhancer regulators. Although ChIP analysis revealed the presence of a large fraction of Zic peaks in a non-TSS region, they are still enriched near a TSS (Sankar et al. 2016).

18.5 Regulation of Signaling

The above sections dealt with the general molecular properties or functions of Zic proteins. Here, we briefly summarize the molecular functions in relation to major signaling cascades.

18.5.1 Hedgehog

Regulation of Hedgehog signaling-related gene expression (Figs. 14.4 and 15.2) is mediated by Gli proteins that contain five C2H2-type ZFDs similar to those of Zic proteins (see Chap. 1). Zic1–Zic3 proteins can bind a Gli target sequence with low affinity (Mizugishi et al. 2001) and can physically interact with Gli1–Gli3 proteins through their ZFDs (Koyabu et al. 2001; Zhu et al. 2007; Chan et al. 2011). Zic proteins enhance nuclear localization of Gli proteins and consequent transcription (Koyabu et al. 2001; Chan et al. 2011). Relevant pathophysiological phenomena have been demonstrated for pairs Zic1–Gli3, (Aruga et al. 1999), Zic3–Gli3 (Zhu et al. 2008), and Zic2–Gli1 (Chan et al. 2011).

18.5.2 Nodal

Nodal signaling is mediated by Smad2 and/or Smad3 proteins (Figs. 14.4 and 15.2); ZIC2 physically interacts with SMAD2 or SMAD3 in a Nodal-dependent manner (Houtmeyers et al. 2016). Such studies have shown that ZIC2 acts downstream of Nodal signaling via this mechanism (see Chap. 14 for details).

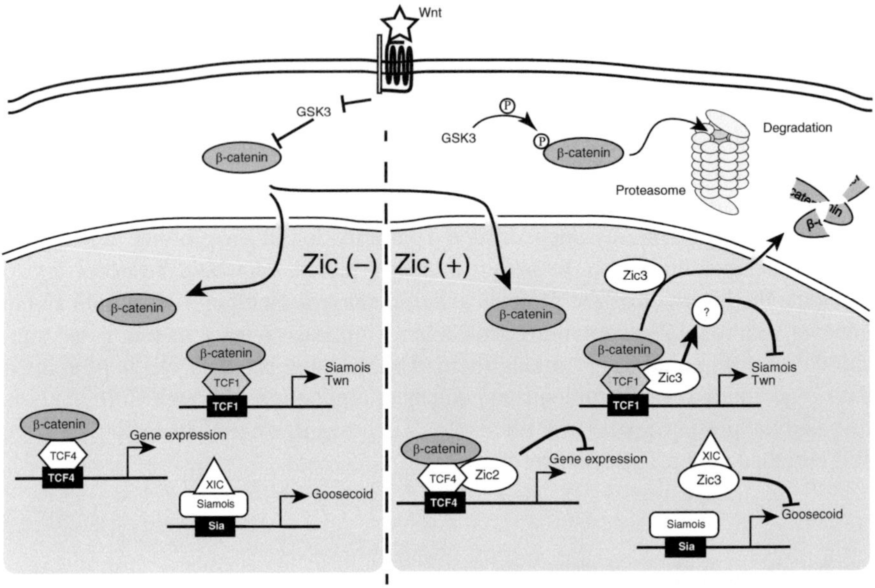

Fig. 18.9 Suppression of Wnt signaling by Zic2 and Zic3. Wnt signaling is mediated by a cell membrane receptor. Wnt ligand's binding to its receptors inhibits GSK3β, which phosphorylates β-catenin. Phosphorylated β-catenin is continuously degraded by the proteasome system. TCF1 and TCF4 are nuclear intermediaries for β-catenin and activate expression of multiple genes. Siamois (Sia) is an organizer marker gene and has important roles in the dorsal mesoderm formation together with Wnt signaling. A Zic protein can physically interact with these molecules and inhibit gene expression. Zic2's interaction with TCF4 does not inhibit DNA binding but suppresses a downstream reaction. Similarly, Zic3 also downregulates TCF1, but in this case, an unknown factor (*?*) is hypothesized to participate (Adapted from Fujimi et al. 2012 with permission, with the results in Pourebrahim et al. 2011)

18.5.3 *Wnt*

Wnt–β-catenin signaling is negatively regulated by Zic proteins interacting with TCF4 or the TCF1–XIC complex (Fujimi et al. 2012; Pourebrahim et al. 2011) (Fig. 18.9). All Zic proteins inhibit β-catenin signaling in an overexpression system based on frog embryos (Fujimi et al. 2012). This suppression is mediated by a physical interaction between TCF and a Zic protein. Pourebrahim et al. (2011) revealed that the ZF domain of Zic2 interacts with TCF4. DNA binding of the TCF protein is a crucial step for regulation of downstream genes. In the case of the TCF4–Zic2 interaction, ChIP using an anti-Zic2 antibody precipitates TCF4's target sequence. Nevertheless, Zic2 does not inhibit TCF4's binding to its target sequence, and Zic2 does not bind the target sequence. To be precise, the repressive effect of Zic2 proceeds without Zic2's DNA binding (Pourebrahim et al. 2011). In the case of Zic3 and of the Tcf1–XIC complex, Zic3 promotes degradation of β-catenin by competitively disrupting the Tcf1–β-catenin interaction. In addition, expression of Siamois target gene Goosecoid is inhibited by Zic3. The latter exerts this inhibition by removing the XIC protein from Siamois to form the XIC–Zic3 complex (Fujimi et al. 2012).

References

Aruga J, Yokota N, Hashimoto M, Furuichi T, Fukuda M, Mikoshiba K (1994) A novel zinc finger protein, zic, is involved in neurogenesis, especially in the cell lineage of cerebellar granule cells. J Neurochem 63(5):1880–1890

Aruga J, Mizugishi K, Koseki H, Imai K, Balling R, Noda T, Mikoshiba K (1999) Zic1 regulates the patterning of vertebral arches in cooperation with Gli3. Mech Dev 89(1–2):141–150

Badis G, Berger MF, Philippakis AA, Talukder S, Gehrke AR, Jaeger SA, Chan ET, Metzler G, Vedenko A, Chen X, Kuznetsov H, Wang CF, Coburn D, Newburger DE, Morris Q, Hughes TR, Bulyk ML (2009) Diversity and complexity in DNA recognition by transcription factors. Science 324(5935):1720–1723. https://doi.org/10.1126/science.1162327

Bae CJ, Park BY, Lee YH, Tobias JW, Hong CS, Saint-Jeannet JP (2014) Identification of Pax3 and Zic1 targets in the developing neural crest. Dev Biol 386(2):473–483. https://doi.org/10.1016/j.ydbio.2013.12.011

Bagutti C, Forro G, Ferralli J, Rubin B, Chiquet-Ehrismann R (2003) The intracellular domain of teneurin-2 has a nuclear function and represses zic-1-mediated transcription. J Cell Sci 116(Pt 14):2957–2966. https://doi.org/10.1242/jcs.00603

Baumgartner S, Martin D, Hagios C, Chiquet-Ehrismann R (1994) Tenm, a Drosophila gene related to tenascin, is a new pair-rule gene. EMBO J 13(16):3728–3740

Bedard JE, Purnell JD, Ware SM (2007) Nuclear import and export signals are essential for proper cellular trafficking and function of ZIC3. Hum Mol Genet 16(2):187–198. https://doi.org/10.1093/hmg/ddl461

Bedard JE, Haaning AM, Ware SM (2011) Identification of a novel ZIC3 isoform and mutation screening in patients with heterotaxy and congenital heart disease. PLoS One 6(8):e23755. https://doi.org/10.1371/journal.pone.0023755

Brayer KJ, Kulshreshtha S, Segal DJ (2008) The protein-binding potential of C2H2 zinc finger domains. Cell Biochem Biophys 51(1):9–19. https://doi.org/10.1007/s12013-008-9007-6

Brewster R, Lee J, Ruiz i Altaba A (1998) Gli/Zic factors pattern the neural plate by defining domains of cell differentiation. Nature 393(6685):579–583. https://doi.org/10.1038/31242

Brown L, Brown S (2009) Zic2 is expressed in pluripotent cells in the blastocyst and adult brain expression overlaps with makers of neurogenesis. Gene Expr Patterns 9(1):43–49. https://doi.org/10.1016/j.gep.2008.08.002

Brown L, Paraso M, Arkell R, Brown S (2005) In vitro analysis of partial loss-of-function ZIC2 mutations in holoprosencephaly: alanine tract expansion modulates DNA binding and transactivation. Hum Mol Genet 14(3):411–420. https://doi.org/10.1093/hmg/ddi037

Buecker C, Srinivasan R, Wu Z, Calo E, Acampora D, Faial T, Simeone A, Tan M, Swigut T, Wysocka J (2014) Reorganization of enhancer patterns in transition from naive to primed pluripotency. Cell Stem Cell 14(6):838–853. https://doi.org/10.1016/j.stem.2014.04.003

Cedar H, Bergman Y (2011) Epigenetics of haematopoietic cell development. Nat Rev Immunol 11(7):478–488. https://doi.org/10.1038/nri2991

Chan DW, Liu VW, Leung LY, Yao KM, Chan KK, Cheung AN, Ngan HY (2011) Zic2 synergistically enhances Hedgehog signalling through nuclear retention of Gli1 in cervical cancer cells. J Pathol 225(4):525–534. https://doi.org/10.1002/path.2901

Chen Z, Wang L, Wang Q, Li W (2010) Histone modifications and chromatin organization in prostate cancer. Epigenomics 2(4):551–560. https://doi.org/10.2217/epi.10.31

Chen L, Ma Y, Qian L, Wang J (2013) Sumoylation regulates nuclear localization and function of zinc finger transcription factor ZIC3. Biochim Biophys Acta 1833(12):2725–2733. https://doi.org/10.1016/j.bbamcr.2013.07.009

Chhin B, Hatayama M, Bozon D, Ogawa M, Schon P, Tohmonda T, Sassolas F, Aruga J, Valard AG, Chen SC, Bouvagnet P (2007) Elucidation of penetrance variability of a ZIC3 mutation in a family with complex heart defects and functional analysis of ZIC3 mutations in the first zinc finger domain. Hum Mutat 28(6):563–570. https://doi.org/10.1002/humu.20480

Christensen GL, Kelstrup CD, Lyngso C, Sarwar U, Bogebo R, Sheikh SP, Gammeltoft S, Olsen JV, Hansen JL (2010) Quantitative phosphoproteomics dissection of seven-transmembrane receptor signaling using full and biased agonists. Mol Cell Proteomics 9(7):1540–1553. https://doi.org/10.1074/mcp.M900550-MCP200

Crooks GE, Hon G, Chandonia JM, Brenner SE (2004) WebLogo: a sequence logo generator. Genome Res 14(6):1188–1190. https://doi.org/10.1101/gr.849004

Dephoure N, Zhou C, Villen J, Beausoleil SA, Bakalarski CE, Elledge SJ, Gygi SP (2008) A quantitative atlas of mitotic phosphorylation. Proc Natl Acad Sci U S A 105(31):10762–10767. https://doi.org/10.1073/pnas.0805139105

Dovat S, Ronni T, Russell D, Ferrini R, Cobb BS, Smale ST (2002) A common mechanism for mitotic inactivation of C2H2 zinc finger DNA-binding domains. Genes Dev 16(23):2985–2990. https://doi.org/10.1101/gad.1040502

Ebert PJ, Timmer JR, Nakada Y, Helms AW, Parab PB, Liu Y, Hunsaker TL, Johnson JE (2003) Zic1 represses Math1 expression via interactions with the Math1 enhancer and modulation of Math1 autoregulation. Development 130(9):1949–1959

Frank CL, Liu F, Wijayatunge R, Song L, Biegler MT, Yang MG, Vockley CM, Safi A, Gersbach CA, Crawford GE, West AE (2015) Regulation of chromatin accessibility and Zic binding at enhancers in the developing cerebellum. Nat Neurosci 18(5):647–656. https://doi.org/10.1038/nn.3995

Fujimi TJ, Hatayama M, Aruga J (2012) Xenopus Zic3 controls notochord and organizer development through suppression of the Wnt/beta-catenin signaling pathway. Dev Biol 361(2):220–231. https://doi.org/10.1016/j.ydbio.2011.10.026

Garcia-Frigola C, Herrera E (2010) Zic2 regulates the expression of Sert to modulate eye-specific refinement at the visual targets. EMBO J 29(18):3170–3183. https://doi.org/10.1038/emboj.2010.172

Grimsrud PA, Carson JJ, Hebert AS, Hubler SL, Niemi NM, Bailey DJ, Jochem A, Stapleton DS, Keller MP, Westphall MS, Yandell BS, Attie AD, Coon JJ, Pagliarini DJ (2012) A quantitative map of the liver mitochondrial phosphoproteome reveals posttranslational control of ketogenesis. Cell Metab 16(5):672–683. https://doi.org/10.1016/j.cmet.2012.10.004

Hatayama M, Aruga J (2010) Characterization of the tandem CWCH2 sequence motif: a hallmark of inter-zinc finger interactions. BMC Evol Biol 10:53. https://doi.org/10.1186/1471-2148-10-53

Hatayama M, Aruga J (2012) Gli protein nuclear localization signal. Vitam Horm 88:73–89. https://doi.org/10.1016/B978-0-12-394622-5.00004-3

Hatayama M, Tomizawa T, Sakai-Kato K, Bouvagnet P, Kose S, Imamoto N, Yokoyama S, Utsunomiya-Tate N, Mikoshiba K, Kigawa T, Aruga J (2008) Functional and structural basis of the nuclear localization signal in the ZIC3 zinc finger domain. Hum Mol Genet 17(22):3459–3473. https://doi.org/10.1093/hmg/ddn239

Hendriks IA, D'Souza RC, Yang B, Verlaan-de Vries M, Mann M, Vertegaal AC (2014) Uncovering global SUMOylation signaling networks in a site-specific manner. Nat Struct Mol Biol 21(10):927–936. https://doi.org/10.1038/nsmb.2890

Himeda CL, Barro MV, Emerson CP Jr (2013) Pax3 synergizes with Gli2 and Zic1 in transactivating the Myf5 epaxial somite enhancer. Dev Biol 383(1):7–14. https://doi.org/10.1016/j.ydbio.2013.09.006

Houtmeyers R, Tchouate Gainkam O, Glanville-Jones HA, Van den Bosch B, Chappell A, Barratt KS, Souopgui J, Tejpar S, Arkell RM (2016) Zic2 mutation causes holoprosencephaly via disruption of NODAL signalling. Hum Mol Genet 25(18):3946–3959. https://doi.org/10.1093/hmg/ddw235

Huttlin EL, Jedrychowski MP, Elias JE, Goswami T, Rad R, Beausoleil SA, Villen J, Haas W, Sowa ME, Gygi SP (2010) A tissue-specific atlas of mouse protein phosphorylation and expression. Cell 143(7):1174–1189. https://doi.org/10.1016/j.cell.2010.12.001

Ishiguro A, Aruga J (2008) Functional role of Zic2 phosphorylation in transcriptional regulation. FEBS Lett 582(2):154–158. https://doi.org/10.1016/j.febslet.2007.11.080

Ishiguro A, Inoue T, Mikoshiba K, Aruga J (2004) Molecular properties of Zic4 and Zic5 proteins: functional diversity within Zic family. Biochem Biophys Res Commun 324(1):302–307. https://doi.org/10.1016/j.bbrc.2004.09.052

Ishiguro A, Ideta M, Mikoshiba K, Chen DJ, Aruga J (2007) ZIC2-dependent transcriptional regulation is mediated by DNA-dependent protein kinase, poly(ADP-ribose) polymerase, and RNA helicase A. J Biol Chem 282(13):9983–9995. https://doi.org/10.1074/jbc.M610821200

Jantz D, Berg JM (2004) Reduction in DNA-binding affinity of Cys2His2 zinc finger proteins by linker phosphorylation. Proc Natl Acad Sci U S A 101(20):7589–7593. https://doi.org/10.1073/pnas.0402191101

Jolma A, Yan J, Whitington T, Toivonen J, Nitta KR, Rastas P, Morgunova E, Enge M, Taipale M, Wei G, Palin K, Vaquerizas JM, Vincentelli R, Luscombe NM, Hughes TR, Lemaire P, Ukkonen E, Kivioja T, Taipale J (2013) DNA-binding specificities of human transcription factors. Cell 152(1–2):327–339. https://doi.org/10.1016/j.cell.2012.12.009

Kettenbach AN, Schweppe DK, Faherty BK, Pechenick D, Pletnev AA, Gerber SA (2011) Quantitative phosphoproteomics identifies substrates and functional modules of Aurora and Polo-like kinase activities in mitotic cells. Sci Signal 4(179):rs5. https://doi.org/10.1126/scisignal.2001497

Koyabu Y, Nakata K, Mizugishi K, Aruga J, Mikoshiba K (2001) Physical and functional interactions between Zic and Gli proteins. J Biol Chem 276(10):6889–6892. https://doi.org/10.1074/jbc.C000773200

Kraut N, Snider L, Chen CM, Tapscott SJ, Groudine M (1998) Requirement of the mouse I-mfa gene for placental development and skeletal patterning. EMBO J 17(21):6276–6288. https://doi.org/10.1093/emboj/17.21.6276

Kumano G, Kawai N, Nishida H (2010) Macho-1 regulates unequal cell divisions independently of its function as a muscle determinant. Dev Biol 344(1):284–292. https://doi.org/10.1016/j.ydbio.2010.05.013

Kuo JS, Patel M, Gamse J, Merzdorf C, Liu X, Apekin V, Sive H (1998) Opl: a zinc finger protein that regulates neural determination and patterning in Xenopus. Development 125(15):2867–2882

Lim LS, Loh YH, Zhang W, Li Y, Chen X, Wang Y, Bakre M, Ng HH, Stanton LW (2007) Zic3 is required for maintenance of pluripotency in embryonic stem cells. Mol Biol Cell 18(4):1348–1358. https://doi.org/10.1091/mbc.E06-07-0624

Lim LS, Hong FH, Kunarso G, Stanton LW (2010) The pluripotency regulator Zic3 is a direct activator of the Nanog promoter in ESCs. Stem Cells 28(11):1961–1969. https://doi.org/10.1002/stem.527

Luo Z, Gao X, Lin C, Smith ER, Marshall SA, Swanson SK, Florens L, Washburn MP, Shilatifard A (2015) Zic2 is an enhancer-binding factor required for embryonic stem cell specification. Mol Cell 57(4):685–694. https://doi.org/10.1016/j.molcel.2015.01.007

Matsuda K, Mikami T, Oki S, Iida H, Andrabi M, Boss JM, Yamaguchi K, Shigenobu S, Kondoh H (2017) ChIP-seq analysis of genomic binding regions of five major transcription factors highlights a central role for ZIC2 in the mouse epiblast stem cell gene regulatory network. Development 144(11):1948–1958. https://doi.org/10.1242/dev.143479

Matsumoto J, Kumano G, Nishida H (2007) Direct activation by Ets and Zic is required for initial expression of the Brachyury gene in the ascidian notochord. Dev Biol 306(2):870–882. https://doi.org/10.1016/j.ydbio.2007.03.034

Milet C, Maczkowiak F, Roche DD, Monsoro Burq AH (2013) Pax3 and Zic1 drive induction and differentiation of multipotent, migratory, and functional neural crest in Xenopus embryos. Proc Natl Acad Sci U S A 110(14):5528–5533. https://doi.org/10.1073/pnas.1219124110

Mizugishi K, Aruga J, Nakata K, Mikoshiba K (2001) Molecular properties of Zic proteins as transcriptional regulators and their relationship to GLI proteins. J Biol Chem 276(3):2180–2188. https://doi.org/10.1074/jbc.M004430200

Mizugishi K, Hatayama M, Tohmonda T, Ogawa M, Inoue T, Mikoshiba K, Aruga J (2004) Myogenic repressor I-mfa interferes with the function of Zic family proteins. Biochem Biophys Res Commun 320(1):233–240. https://doi.org/10.1016/j.bbrc.2004.05.158

Nakajima T, Uchida C, Anderson SF, Lee CG, Hurwitz J, Parvin JD, Montminy M (1997) RNA helicase A mediates association of CBP with RNA polymerase II. Cell 90(6):1107–1112

Nishida H, Sawada K (2001) macho-1 encodes a localized mRNA in ascidian eggs that specifies muscle fate during embryogenesis. Nature 409(6821):724–729. https://doi.org/10.1038/35055568

Ogawa M, Mizugishi K, Ishiguro A, Koyabu Y, Imai Y, Takahashi R, Mikoshiba K, Aruga J (2008) Rines/RNF180, a novel RING finger gene-encoded product, is a membrane-bound ubiquitin ligase. Genes Cells 13(4):397–409. https://doi.org/10.1111/j.1365-2443.2008.01169.x

Okumura K, Hosoe Y, Nakajima N (2004) Zic1 is a transcriptional repressor trough the lamin A/C promoter and has an intrinsic repressor domain. J Health Sci 4:423–427

Olsen JV, Vermeulen M, Santamaria A, Kumar C, Miller ML, Jensen LJ, Gnad F, Cox J, Jensen TS, Nigg EA, Brunak S, Mann M (2010) Quantitative phosphoproteomics reveals widespread full phosphorylation site occupancy during mitosis. Sci Signal 3(104):ra3. https://doi.org/10.1126/scisignal.2000475

Pardo M, Lang B, Yu L, Prosser H, Bradley A, Babu MM, Choudhary J (2010) An expanded Oct4 interaction network: implications for stem cell biology, development, and disease. Cell Stem Cell 6(4):382–395. https://doi.org/10.1016/j.stem.2010.03.004

Pavletich NP, Pabo CO (1993) Crystal structure of a five-finger GLI-DNA complex: new perspectives on zinc fingers. Science 261(5129):1701–1707

Peterson KA, Nishi Y, Ma W, Vedenko A, Shokri L, Zhang X, McFarlane M, Baizabal JM, Junker JP, van Oudenaarden A, Mikkelsen T, Bernstein BE, Bailey TL, Bulyk ML, Wong WH, McMahon AP (2012) Neural-specific Sox2 input and differential Gli-binding affinity provide context and positional information in Shh-directed neural patterning. Genes Dev 26(24):2802–2816. https://doi.org/10.1101/gad.207142.112

Plouhinec JL, Roche DD, Pegoraro C, Figueiredo AL, Maczkowiak F, Brunet LJ, Milet C, Vert JP, Pollet N, Harland RM, Monsoro-Burq AH (2014) Pax3 and Zic1 trigger the early neural crest gene regulatory network by the direct activation of multiple key neural crest specifiers. Dev Biol 386(2):461–472. https://doi.org/10.1016/j.ydbio.2013.12.010

Pourebrahim R, Houtmeyers R, Ghogomu S, Janssens S, Thelie A, Tran HT, Langenberg T, Vleminckx K, Bellefroid E, Cassiman JJ, Tejpar S (2011) Transcription factor Zic2 inhibits Wnt/beta-catenin protein signaling. J Biol Chem 286(43):37732–37740. https://doi.org/10.1074/jbc.M111.242826

Quinn ME, Haaning A, Ware SM (2012) Preaxial polydactyly caused by Gli3 haploinsufficiency is rescued by Zic3 loss of function in mice. Hum Mol Genet 21(8):1888–1896. https://doi.org/10.1093/hmg/dds002

Rigbolt KT, Prokhorova TA, Akimov V, Henningsen J, Johansen PT, Kratchmarova I, Kassem M, Mann M, Olsen JV, Blagoev B (2011) System-wide temporal characterization of the proteome and phosphoproteome of human embryonic stem cell differentiation. Sci Signal 4(164):rs3. https://doi.org/10.1126/scisignal.2001570

Rizkallah R, Alexander KE, Hurt MM (2011) Global mitotic phosphorylation of C2H2 zinc finger protein linker peptides. Cell Cycle 10(19):3327–3336. https://doi.org/10.4161/cc.10.19.17619

Rizkallah R, Batsomboon P, Dudley GB, Hurt MM (2015) Identification of the oncogenic kinase TOPK/PBK as a master mitotic regulator of C2H2 zinc finger proteins. Oncotarget 6(3):1446–1461. 10.18632/oncotarget.2735

Sadoul K, Wang J, Diagouraga B, Khochbin S (2011) The tale of protein lysine acetylation in the cytoplasm. J Biomed Biotechnol 2011:970382. https://doi.org/10.1155/2011/970382

Sakai-Kato K, Ishiguro A, Mikoshiba K, Aruga J, Utsunomiya-Tate N (2008) CD spectra show the relational style between Zic-, Gli-, Glis-zinc finger protein and DNA. Biochim Biophys Acta 1784(7–8):1011–1019. https://doi.org/10.1016/j.bbapap.2008.01.013

Sakai-Kato K, Umezawa Y, Mikoshiba K, Aruga J, Utsunomiya-Tate N (2009) Stability of folding structure of Zic zinc finger proteins. Biochem Biophys Res Commun 384(3):362–365. https://doi.org/10.1016/j.bbrc.2009.04.151

Sakurada T, Mima K, Kurisaki A, Sugino H, Yamauchi T (2005) Neuronal cell type-specific promoter of the alpha CaM kinase II gene is activated by Zic2, a Zic family zinc finger protein. Neurosci Res 53(3):323–330. https://doi.org/10.1016/j.neures.2005.08.001

Salero E, Perez-Sen R, Aruga J, Gimenez C, Zafra F (2001) Transcription factors Zic1 and Zic2 bind and transactivate the apolipoprotein E gene promoter. J Biol Chem 276(3):1881–1888. https://doi.org/10.1074/jbc.M007008200

Sanchez-Ferras O, Bernas G, Laberge-Perrault E, Pilon N (2014) Induction and dorsal restriction of Paired-box 3 (Pax3) gene expression in the caudal neuroectoderm is mediated by integration of multiple pathways on a short neural crest enhancer. Biochim Biophys Acta 1839(7):546–558. https://doi.org/10.1016/j.bbagrm.2014.04.023

Sankar S, Yellajoshyula D, Zhang B, Teets B, Rockweiler N, Kroll KL (2016) Gene regulatory networks in neural cell fate acquisition from genome-wide chromatin association of Geminin and Zic1. Sci Rep 6:37412. https://doi.org/10.1038/srep37412

Sato T, Sasai N, Sasai Y (2005) Neural crest determination by co-activation of Pax3 and Zic1 genes in Xenopus ectoderm. Development 132(10):2355–2363. https://doi.org/10.1242/dev.01823

Satow R, Nakamura T, Kato C, Endo M, Tamura M, Batori R, Tomura S, Murayama Y, Fukami K (2017) ZIC5 drives melanoma aggressiveness by PDGFD-mediated activation of FAK and STAT3. Cancer Res 77(2):366–377. https://doi.org/10.1158/0008-5472.CAN-16-0991

Sawada K, Fukushima Y, Nishida H (2005) Macho-1 functions as transcriptional activator for muscle formation in embryos of the ascidian Halocynthia roretzi. Gene Expr Patterns 5(3):429–437. https://doi.org/10.1016/j.modgep.2004.09.003

Sen A, Stultz BG, Lee H, Hursh DA (2010) Odd paired transcriptional activation of decapentaplegic in the Drosophila eye/antennal disc is cell autonomous but indirect. Dev Biol 343(1–2):167–177. https://doi.org/10.1016/j.ydbio.2010.04.003

Siegenthaler JA, Choe Y, Patterson KP, Hsieh I, Li D, Jaminet SC, Daneman R, Kume T, Huang EJ, Pleasure SJ (2013) Foxc1 is required by pericytes during fetal brain angiogenesis. Biol Open 2(7):647–659. https://doi.org/10.1242/bio.20135009

Simoes-Costa MS, McKeown SJ, Tan-Cabugao J, Sauka-Spengler T, Bronner ME (2012) Dynamic and differential regulation of stem cell factor FoxD3 in the neural crest is Encrypted in the genome. PLoS Genet 8(12):e1003142. https://doi.org/10.1371/journal.pgen.1003142

Sone M, Morone N, Nakamura T, Tanaka A, Okita K, Woltjen K, Nakagawa M, Heuser JE, Yamada Y, Yamanaka S, Yamamoto T (2017) Hybrid cellular metabolism coordinated by Zic3 and Esrrb synergistically enhances induction of naive pluripotency. Cell Metab 25(5):1103–1117 e1106. https://doi.org/10.1016/j.cmet.2017.04.017

Song CZ, Keller K, Chen Y, Stamatoyannopoulos G (2003) Functional interplay between CBP and PCAF in acetylation and regulation of transcription factor KLF13 activity. J Mol Biol 329(2):207–215

Suzuki K, Sako K, Akiyama K, Isoda M, Senoo C, Nakajo N, Sagata N (2015) Identification of non-Ser/Thr-Pro consensus motifs for Cdk1 and their roles in mitotic regulation of C2H2 zinc finger proteins and Ect2. Sci Rep 5:7929. https://doi.org/10.1038/srep07929

Tucker RP, Chiquet-Ehrismann R, Chevron MP, Martin D, Hall RJ, Rubin BP (2001) Teneurin-2 is expressed in tissues that regulate limb and somite pattern formation and is induced in vitro and in situ by FGF8. Dev Dyn 220(1):27–39. https://doi.org/10.1002/1097-0177(2000)9999:9999<::AID-DVDY1084>3.0.CO;2-B

Winata CL, Kondrychyn I, Kumar V, Srinivasan KG, Orlov Y, Ravishankar A, Prabhakar S, Stanton LW, Korzh V, Mathavan S (2013) Genome wide analysis reveals Zic3 interaction with distal regulatory elements of stage specific developmental genes in zebrafish. PLoS Genet 9(10):e1003852. https://doi.org/10.1371/journal.pgen.1003852

Winata CL, Kondrychyn I, Korzh V (2015) Changing faces of transcriptional regulation reflected by Zic3. Curr Genomics 16(2):117–127. https://doi.org/10.2174/1389202916666150205124519

Wolfe SA, Nekludova L, Pabo CO (2000) DNA recognition by Cys2His2 zinc finger proteins. Annu Rev Biophys Biomol Struct 29:183–212. https://doi.org/10.1146/annurev.biophys.29.1.183

Wysocka J, Swigut T, Xiao H, Milne TA, Kwon SY, Landry J, Kauer M, Tackett AJ, Chait BT, Badenhorst P, Wu C, Allis CD (2006) A PHD finger of NURF couples histone H3 lysine 4 trimethylation with chromatin remodelling. Nature 442(7098):86–90. https://doi.org/10.1038/nature04815

Yagi K, Satoh N, Satou Y (2004a) Identification of downstream genes of the ascidian muscle determinant gene Ci-macho1. Dev Biol 274(2):478–489. https://doi.org/10.1016/j.ydbio.2004.07.013

Yagi K, Satou Y, Satoh N (2004b) A zinc finger transcription factor, ZicL, is a direct activator of Brachyury in the notochord specification of Ciona intestinalis. Development 131(6):1279–1288. https://doi.org/10.1242/dev.01011

Yang Y, Hwang CK, Junn E, Lee G, Mouradian MM (2000) ZIC2 and Sp3 repress Sp1-induced activation of the human D1A dopamine receptor gene. J Biol Chem 275(49):38863–38869. https://doi.org/10.1074/jbc.M007906200

Yokota N, Aruga J, Takai S, Yamada K, Hamazaki M, Iwase T, Sugimura H, Mikoshiba K (1996) Predominant expression of human zic in cerebellar granule cell lineage and medulloblastoma. Cancer Res 56(2):377–383

Yu W, Briones V, Lister R, McIntosh C, Han Y, Lee EY, Ren J, Terashima M, Leighty RM, Ecker JR, Muegge K (2014) CG hypomethylation in Lsh-/- mouse embryonic fibroblasts is associated with de novo H3K4me1 formation and altered cellular plasticity. Proc Natl Acad Sci U S A 111 (16):5890–5895. https://doi.org/10.1073/pnas.1320945111

Zhao S, Xu W, Jiang W, Yu W, Lin Y, Zhang T, Yao J, Zhou L, Zeng Y, Li H, Li Y, Shi J, An W, Hancock SM, He F, Qin L, Chin J, Yang P, Chen X, Lei Q, Xiong Y, Guan KL (2010) Regulation of cellular metabolism by protein lysine acetylation. Science 327(5968):1000–1004. https://doi.org/10.1126/science.1179689

Zhu L, Harutyunyan KG, Peng JL, Wang J, Schwartz RJ, Belmont JW (2007) Identification of a novel role of ZIC3 in regulating cardiac development. Hum Mol Genet 16(14):1649–1660. https://doi.org/10.1093/hmg/ddm106

Zhu L, Zhou G, Poole S, Belmont JW (2008) Characterization of the interactions of human ZIC3 mutants with GLI3. Hum Mutat 29(1):99–105. https://doi.org/10.1002/humu.20606

Zhu P, Wang Y, He L, Huang G, Du Y, Zhang G, Yan X, Xia P, Ye B, Wang S, Hao L, Wu J, Fan Z (2015) ZIC2-dependent OCT4 activation drives self-renewal of human liver cancer stem cells. J Clin Invest 125(10):3795–3808. https://doi.org/10.1172/JCI81979

MIX
Papier aus verantwortungsvollen Quellen
Paper from responsible sources
FSC® C105338

If you have any concerns about our products,
you can contact us on
ProductSafety@springernature.com

In case Publisher is established outside the EU,
the EU authorized representative is:
Springer Nature Customer Service Center GmbH
Europaplatz 3, 69115 Heidelberg, Germany

Printed by Libri Plureos GmbH
in Hamburg, Germany